Lecture Notes in Computer Science 15667

The series Lecture Notes in Computer Science (LNCS), including its subseries Lecture Notes in Artificial Intelligence (LNAI) and Lecture Notes in Bioinformatics (LNBI), has established itself as a medium for the publication of new developments in computer science and information technology research, teaching, and education.

LNCS enjoys close cooperation with the computer science R & D community, the series counts many renowned academics among its volume editors and paper authors, and collaborates with prestigious societies. Its mission is to serve this international community by providing an invaluable service, mainly focused on the publication of conference and workshop proceedings and postproceedings. LNCS commenced publication in 1973.

Tatiana A. Bubba · Romina Gaburro ·
Silvia Gazzola · Kostas Papafitsoros ·
Marcelo Pereyra · Carola-Bibiane Schönlieb
Editors

Scale Space and Variational Methods in Computer Vision

10th International Conference, SSVM 2025
Dartington, UK, May 18–22, 2025
Proceedings, Part I

Springer

Editors
Tatiana A. Bubba
University of Ferrara
Ferrara, Italy

Romina Gaburro
University of Limerick
Limerick, Ireland

Silvia Gazzola
University of Pisa
Pisa, Italy

Kostas Papafitsoros
Queen Mary University of London
London, UK

Marcelo Pereyra
Heriot-Watt University and Maxwell Institute
for Mathematical Sciences
Edinburgh, UK

Carola-Bibiane Schönlieb
University of Cambridge
Cambridge, UK

ISSN 0302-9743 ISSN 1611-3349 (electronic)
Lecture Notes in Computer Science
ISBN 978-3-031-92365-4 ISBN 978-3-031-92366-1 (eBook)
https://doi.org/10.1007/978-3-031-92366-1

This Springer imprint is published by the registered company Springer Nature Switzerland AG
The registered company address is: Gewerbestrasse 11, 6330 Cham, Switzerland

Preface

The International Conference on Scale Space and Variational Methods in Computer Vision (SSVM) series was born in 2007 in Ischia, from the fusion of the biannual Conference on Scale Space and the workshop on Variational, Geometric and Level Set Methods (VLSM). Since then, it has become a major and well-established event in the mathematical imaging community. Following the tradition of biannual events taking place at beautiful, remote places, the 10th edition of this conference (SSVM 2025, https://sites.google.com/view/ssvm-2025) took place from May 18–22, 2025 at the Dartington Estate, an idyllic retreat amidst 1,200 acres of South Devon parkland, just outside the ancient market town of Totnes, UK.

The conference provided a platform for state-of-the-art scientific exchange on topics related to computer vision and image analysis, including diverse themes such as 3D-vision, convex and non-convex variational modeling, image analysis, inverse problems in imaging, optimization methods in imaging, machine learning in imaging, PDEs in image processing, registration, restoration and reconstruction, scale-space methods, and segmentation. The 63 contributions (24 selected for oral and 39 for poster presentation) in these proceedings volumes demonstrate a strong and lively community. They underwent a rigorous double-blind peer review process similar to that of a high-ranking journal in the field, with each submission receiving three reviews.

Following the tradition of previous SSVM conferences, four outstanding researchers were invited to give keynote presentations: Panagiota Birmpa (Heriot-Watt University, UK), Thomas Pock (TU Graz, Austria), Jean Ponce (École Normale Supérieure-PSL, France and NYU, USA) and Tanja Tarvainen (University of Eastern Finland, Finland).

We would like to thank all those who contributed to the success of this conference. We are grateful to the authors for their contributions. We are indebted to the reviewers for their commitment, hard work, and enthusiasm that made this event possible at the highest scientific standards. We thank the University of Bath, the Institute for Mathematical Innovation at Bath, and the European Mathematical Society for financial support. We especially thank Myla Watts for her continuous and dedicated administrative support. Lastly, we thank the steering committee and the organizers of all the previous SSVM conferences.

March 2025

Tatiana A. Bubba
Romina Gaburro
Silvia Gazzola
Kostas Papafitsoros
Marcelo Pereyra
Carola-Bibiane Schönlieb

Organization

Conference Chairs

Tatiana A. Bubba	University of Ferrara, Italy
Romina Gaburro	University of Limerick, Ireland
Silvia Gazzola	University of Pisa, Italy
Kostas Papafitsoros	Queen Mary University of London, UK
Marcelo Pereyra	Heriot-Watt University, UK
Carola-Bibiane Schönlieb	University of Cambridge, UK

Program Committee

Simon Arridge	University College London, UK
Andrea Aspri	University of Milan, Italy
Jean-François Aujol	Université de Bordeaux, France
Danielle Bednarski	DESY, Germany
Robert Beinert	TU Berlin, Germany
Martin Benning	University College London, UK
Marcelo Bertalmío	Universitat Pompeu Fabra, Spain
Marta Betcke	University College London, UK
Davide Bianchi	Sun Yat-sen University, China
Laure Blanc-Féraud	CNRS, France
Johannes Bostelmann	University of Lübeck, Germany
Nicolas Boulle	Imperial College London, UK
Christina Brandt	University of Greifswald, Germany
Kristian Bredies	University of Graz, Austria
Michael Breuß	Brandenburg University of Technology Cottbus - Senftenberg, Germany
Leon Bungert	University of Würzburg, Germany
Martin Burger	DESY & Universität Hamburg, Germany
Luca Calatroni	University of Genoa, Italy
Marcello Carioni	University of Twente, Netherlands
Antonin Chambolle	CNRS CEREMADE, Université Paris-Dauphine, France
Raymond Chan	Lingnan University, China
Laetitia Chapel	IRISA, France
Laurent Cohen	Université Paris Dauphine-PSL, France

Agnes Desolneux	CNRS and ENS Paris-Saclay, France
Nicolas Dobigeon	University of Toulouse, France
Marco Donatelli	Università dell'Insubria, Italy
Yiqiu Dong	Technical University of Denmark, Denmark
Remco Duits	Eindhoven University of Technology, Netherlands
Vincent Duval	Inria & CEREMADE Université Paris Dauphine-PSL, France
Matthias Ehrhardt	University of Bath, UK
Jalal Fadili	National Graduate Engineering School of Caen, France
Michael Felsberg	Linköping University, Sweden
Guy Gilboa	Technion, Israel
Bart Goossens	Ghent University, Belgium
Yann Gousseau	Télécom Paris, France
Markus Grasmair	NTNU, Norway
Weihong Guo	Case Western Reserve University, USA
Tommi Heikkilä	LUT University, Finland
Johannes Hertrich	University College London, UK
Martin Holler	University of Graz, Austria
Samuel Hurault	ENS Paris, France
Martin Huska	University of Bologna, Italy
José A. Iglesias	University of Twente, Netherlands
Sung Ha Kang	Georgia Institute of Technology, USA
Hanne Kekkonen	Delft University of Technology, Netherlands
Zeljko Kereta	University College London, UK
Erich Kobler	University of Bonn, Germany
Yury Korolev	University of Bath, UK
Arjan Kuijper	Fraunhofer IGD & TU Darmstadt, Germany
Alessandro Lanza	University of Bologna, Italy
Jonas Latz	University of Manchester, UK
François Lauze	University of Copenhagen, Denmark
Damiana Lazzaro	University of Bologna, Italy
Carole Le Guyader	INSA Rouen Normandie, France
Jan Lellmann	University of Lübeck, Germany
Tony Lindeberg	KTH Royal Institute of Technology, Sweden
Elena Loli Piccolomini	University of Bologna, Italy
Felix Lucka	CWI, Netherlands
Maria Magnusson	Linköping University, Sweden
Jan Modersitzki	University of Lübeck, Germany
Michael Moeller	University of Siegen, Germany
Serena Morigi	University of Bologna, Italy
Subhadip Mukherjee	IIT-Kharagpur, India

Sebastian Neumayer	TU Chemnitz, Germany
Alasdair James Newson	ISIR, France
Thomas Oberlin	ISAE-SUPAERO, France
Nicolas Papadakis	CNRS, France
Barbara Pascal	CNRS, France
Jean-Christophe Pesquet	Université Paris-Saclay, France
Pascal Peter	Saarland University, Germany
Romain Petit	CNRS and ENS Paris, France
Stefania Petra	University of Augsburg, Germany
Tom Pock	University of Graz, Austria
Clarice Poon	University of Warwick, UK
Federica Porta	University of Modena and Reggio Emilia, Italy
Monica Pragliola	University of Naples Federico II, Italy
Marco Prato	University of Modena and Reggio Emilia, Italy
Yvain Quéau	CNRS, France
Luca Ratti	University of Bologna, Italy
Simone Rebegoldi	University of Modena and Reggio Emilia, Italy
Audrey Repetti	Heriot-Watt University, UK
Tim Roith	DESY, Germany
Matteo Santacesaria	University of Genoa, Italy
Otmar Scherzer	University of Vienna, Austria
Christoph Schnörr	Heidelberg University, Germany
Andrea Sebastiani	University of Modena and Reggio Emilia, Italy
Ferdia Sherry	University of Cambridge, UK
Samuli Siltanen	University of Helsinki, Finland
Stefan Sommer	University of Copenhagen, Denmark
Emmanuel Soubies	CNRS, France
Jean-Luc Starck	CEA Saclay, France
Gabriele Steidl	TU Berlin, Germany
Xue-Cheng Tai	Norwegian Research Center, Norway
Matthew Thorpe	University of Warwick, UK
Yann Traonmilin	CNRS, France
Tuomo Valkonen	MODEMAT, Ecuador
Silvia Villa	University of Genoa, Italy
Marco Viola	Dublin City University, Ireland
Joachim Weickert	Saarland University, Germany
Pierre Weiss	CNRS & Université de Toulouse, France
Martin Welk	UMIT Tirol, Austria

Invited Speakers

Panagiota Birmpa	Heriot-Watt University, UK
Thomas Pock	TU Graz, Austria
Jean Ponce	École Normale Supérieure-PSL, France and NYU, USA
Tanja Tarvainen	University of Eastern Finland, Finland

Administrative Support

Myla Watts	University of Bath, UK
Alison Humphries	University of Bath, UK
Michael Morgan	University of Bath, UK

Sponsoring Institutions

University of Bath, UK
Institute for Mathematical Innovation, Bath, UK
European Mathematical Society

Contents – Part I

Contents – Part II

Optimization for Imaging: Theory and Methods

Scale Space, PDES, Flow, Motion and Registration

Inverse Problems in Imaging

Contrast Highlighting of TV-Based Reconstructed Polarimetric Images

Carole Le Guyader[1(✉)] and Fabien Pierre[2]

[1] INSA Rouen Normandie, Normandie Univ, LMI UR 3226, F-76000 Rouen, France
carole.le-guyader@insa-rouen.fr
[2] LORIA UMR 7503, Université de Lorraine, CNRS, INRIA, Nancy, France
fabien.pierre@univ-lorraine.fr

Abstract. Using classical smoothers in restoration processes, such as total variation regularisation, while capturing well discontinuities, is known to induce an estimation bias in the final result, materialised by a loss of contrast. If the literature is prolific when dealing with standard modalities of images (grayscale or RGB images), it is more tenuous when the involved modality encodes some intrinsic geometrical properties, requiring the design of specific purpose-built algorithms. In this work, focused on such a specific modality, namely polarimetric imaging, we address the joint restoration and contrast re-enhancement (equivalently referred to as debiasing or refitting) question within an extension of the CLEAR framework (Covariant LEAst-square Refitting, [5]), emphasising the importance of preserving the Jacobian (with respect to the observed signal) of the original estimator.

Keywords: Inverse problems · Polarimetric imaging · Debiasing · Total variation

1 Introduction

1.1 Considered Imaging Modality and Motivations

One commonly agrees that polarisation-encoded images (also named Stokes images), obtained from raw radiance images, exhibit a wealth of information since providing access to parameters describing the vector nature of light (see, *e.g.*, [11]). In a linear configuration, a Stokes imaging polarimeter yields radiance images I related to Stokes vectors S, where one Stokes 3×1 vector is attached to each image pixel (i,j) (see Fig. 1). These data encode the outgoing radiance of a scene seized by L ($L = 4$ in our setting) independent probing states of a Polarisation State Analyser (PSA) positioned in front of a camera ([8]). More precisely, at each pixel of a regular Cartesian grid $\mathcal{G}$ of size $N \times N$ (to simplify) defined by $\mathcal{G} = \{(i,j) \mid 1 \leq i \leq N,\, 1 \leq j \leq N\}$, the L radiance measurements $I_{i,j}$ and the Stokes vector $S_{i,j} = ((S_0)_{i,j}, (S_1)_{i,j}, (S_2)_{i,j})^T$ are linked by the relation $I_{i,j} = AS_{i,j}$, $A \in M_{L,3}(\mathbb{R})$ designating the calibration matrix whose rows constitute the L probing states. Our dataset comes from

T. A. Bubba et al. (Eds.): SSVM 2025, LNCS 15667, pp. 3–15, 2025.
https://doi.org/10.1007/978-3-031-92366-1_1

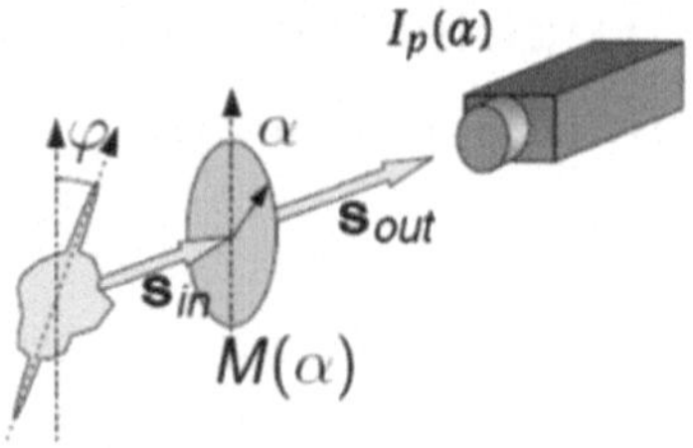

Fig. 1. The polarisation device principle: the Stokes vector represents the reflected light from the object that passes through the polariser filter before reaching the camera.

a Polarcam 4D Technology polarimetric camera, yielding four radiance images encoded in $I \in \mathbb{R}^{N \times N \times 4}$ obtained from four different polarisers oriented at $(\alpha_i)_{1 \leq i \leq 4} = (0°, 90°, 45°, 135°)$. The resulting calibration matrix A is thus such that its ith row is defined by $A_i = \frac{1}{2}\left(1 \ \cos(2\alpha_i) \ \sin(2\alpha_i)\right)$.

As always in forming, transmitting or recording processes, systematic calibration errors and Charge-Coupled Device noises occur (generally assumed additive), contaminating, in this particular case, the observed radiance measurements I and then propagating to Stokes vector evaluation. Uncertainty thus appends itself on the observation model that eventually reads $I_{i,j} = AS_{i,j} + n_{i,j}$, $n_{i,j}$ encompassing the noise term. In the end, the problem, falling within the class of inverse problems, amounts to estimating the Stokes vector S from (deteriorated) radiance measurements I while complying with the physical admissibility constraint (condition of belonging to the Lorentz cone) $\forall (i,j) \in \mathcal{G}$, $(S_0)_{i,j} \geq \sqrt{(S_1)_{i,j}^2 + (S_2)_{i,j}^2}$, condition ensuring the consistent physical interpretation of the scene. Suited Stokes vector reconstruction models must thus be customised to tackle those two requirements, namely explicit enforcement of admissibility constraints and smoothness/spatial consistency, not addressed by former classical methods like [17,18] or [16]. Recently, in [11], which constitutes the ground of the current work, the authors have proposed reconciling these two goals in a variational setting: (i) ensuring the smoothness of the vector constituents together with spatial consistency (homogeneity in materials manifesting similar physical properties) while preserving sharp edges; (ii) guaranteeing that the Stokes vectors belong to the Lorentz cone. The former task is achieved using a joint total variation (meaning here that regularisation on the channels S_1 and S_2 is jointly performed: a coupling is enforced on S_1 and S_2 to leverage discontinuity alignment, reflecting more faithfully the scene physical features), whilst the latter is completed by the introduction of explicit geometrical constraints in the objective function. This yields the following constrained minimisation problem (the notations are those of [11])

$$\inf_{S \in \mathcal{K}} \mathcal{J}(S) = \gamma \, TV_{\mathcal{C}}(S_1, S_2) + \gamma \, TV(S_0) + \mathcal{F}(S), \qquad (\mathcal{P}_b)$$

S_0 with our method | Zoom on details | Zoom on details

S_0 with [11] | Zoom on details | Zoom on details

Fig. 2. Comparison and zooms of our extended CLEAR [5] method with the state-of-the-art approach [11] for the reconstruction of the Stokes vector S_0. The use of CLEAR brings out more details. For instance, the car shapes in the foreground or the tree meshes in the background are enhanced. The contrast emphasis is achieved jointly with the reconstruction and does not require complementary post-processing steps.

with $X = \mathbb{R}^{N \times N}$, $\gamma > 0$, a tuning parameter, $\mathcal{K}$ the closed convex set defined by

$$\mathcal{K}=\left\{S=(S_0,S_1,S_2)\in X\times X\times X \mid \forall(i,j)\in\mathcal{G},\, (S_0)_{i,j}\geq\sqrt{(S_1)^2_{i,j}+(S_2)^2_{i,j}}\right\}, \quad (1)$$

$\mathcal{F}$, the data term expressed as $\mathcal{F}(S) = \frac{\mu}{2}\sum_{(i,j)\in\mathcal{G}} \|AS_{i,j} - I_{i,j}\|^2_{\mathbb{R}^4}$, and respectively, the joint total variation between channels S_1 and S_2, $TV_{\mathcal{C}}(S_1, S_2)$, and the total variation on channel S_0, $TV(S_0)$.

1.2 Debiasing to Counteract the Contrast Loss

Whereas experiments demonstrate significant improvements, both quantitatively and qualitatively over standard state-of-the-art methods, conformity with physical reality, and robustness regarding degraded real outdoor scenes, the model, as expected when utilising l_{12} regularisers, suffers from a systematic contrast loss (see Fig. 2). Debiasing approaches ([1–3,6,7,9,10,12,14,15], also known as refitting ones, intend to counteract this effect by enhancing the results provided by statistical estimators. A general refitting approach (CLEAR—which the present work is widely based on—) adapted and tractable for most variational image processing methods (total variation, BM3D, non-local means, etc.) has been proposed in [5], and successfully applied and extended to various problems [4].

The introduced estimator is motivated by the identification of meaningful covariant information that should be preserved after the re-enhancement process and by the importance of preserving the Jacobian of the original estimator with respect to the observed image. Here, the term covariant means that the estimator and the original data share the same properties and underlying statistical characteristics. For example, when using total variation, the data and the estimator have meaningful gradients (edges) at the same positions. The Jacobian of the estimator with respect to the original data is well adapted to measuring similar properties while discarding irrelevant changes. While suited to standard modalities, this method is not straightforwardly adaptable to our polarimetric setting involving intrinsic geometrical constraints (condition of belonging to the Lorentz cone). In this paper, we thus provide a theoretical framework based on the computation of an oblique projection onto the Lorentz cone, guaranteeing the recovery of a debiased estimator that complies with the physical constraints. The improvement which is raised in this paper can be observed in Fig. 2. This idea of oblique projection was first introduced by Pierre *et al.* in [13] in a heuristic way, unfortunately not substantiated by any theoretical arguments.

The remainder of the paper is structured as follows: in Sect. 2, we zoom in on the seminal CLEAR framework on which our work is based and emphasise the scientific barrier it unveils that motivates our contribution, namely the lack of constraint handling. Section 3 elaborates on our generic model involving constraints and its theoretical aspects, then focuses on our specific polarimetric problem, while Sect. 4 is devoted to numerical experiments and their analysis.

2 CLEAR Framework

The notations are similar to those in [5] and [13]: $\Phi : \mathbb{R}^p \to \mathbb{R}^n$ denotes a linear operator, whilst w is the realisation of a noisy random vector. Applied to the problem of estimating an unknown $x_0 \in \mathbb{R}^p$ from noisy and/or incomplete observations $\mathbb{R}^n \ni y = \Phi x_0 + w$ by a variational formulation phrased as

$$\widehat{x}(y) \in \operatorname*{arg\,min}_{x\in\mathbb{R}^p} \quad \underbrace{F(x,y)}_{\text{convex-in-}x\text{ data term}} \quad + \quad \underbrace{G(x)}_{\text{convex regulariser}} \quad , \tag{2}$$

the CLEAR method removes parts of the systematic errors affecting restoration processes. For instance, taking as regulariser G the total variation yields an estimator $\widehat{x}(y)$ recovering well the underlying signal jumps while blunting their amplitudes. To mitigate this amplitude under-estimation, the CLEAR method intends to produce a suitable refitting, *i.e.*, to define, from the original (biased) estimator $\widehat{x}(y)$ and the observation y, a new estimator satisfying several desirable properties and inheriting some relevant structural properties encoded in $\widehat{x}(y)$ (in fact, sharing some first-order properties with $\widehat{x}(y)$). Those properties are intrinsically captured in the directional derivative of the biased estimator $\widehat{x}$ at y in the direction d defined by $J_{\widehat{x}}(y)d = \lim_{\varepsilon\to 0} \frac{\widehat{x}(y+\varepsilon\, d)-\widehat{x}(y)}{\varepsilon}$. Precisely, the CLEAR method relies on two steps:

(**Step 1**) First, it builds, from a guess $z \in \mathbb{R}^n$ of Φx_0 and the biased estimator $\widehat{x}$, a new estimator $\mathcal{D}_{\widehat{x},z} : \mathbb{R}^n \to \mathbb{R}^p$ (called *guess-based covariant refitting of* $\widehat{x}$ *for* z) such that $\mathcal{D}_{\widehat{x},z} \in \arg\min_{h \in \mathcal{H}} \|\Phi h(z) - z\|_2^2$, where $\mathcal{H}$ is the set of maps $h : \mathbb{R}^n \to \mathbb{R}^p$ targeted to satisfy **(i)** an **affine structure** property ($\exists A \in \mathbb{R}^{p \times n}$, $\exists b \in \mathbb{R}^p, \forall s \in \mathbb{R}^n, h(s) = As + b$), **(ii)** a **covariant preserving** feature ($\exists \rho \in \mathbb{R},\ J_h(z) = \rho J_{\widehat{x}}(z)$), and **(iii)** a **coherence property** ($h(\Phi\widehat{x}(z)) = \widehat{x}(z)$), preserving in particular the structure of $\widehat{x}(z)$. Denoting by $J = J_{\widehat{x}}(z)$ and by $\delta = z - \Phi\widehat{x}(z)$, [5, Theorem 6] supplies a unique closed-form expression of $\mathcal{D}_{\widehat{x},z}$ as

$$\mathcal{D}_{\widehat{x},z}(y) = \widehat{x}(z) + \rho J(y - \Phi\widehat{x}(z)) \quad \text{with } \rho = \begin{cases} \frac{\langle \Phi J\delta, \delta\rangle}{\|\Phi J\delta\|_2^2} & \text{if } \Phi J\delta \neq 0 \\ 1 & \text{otherwise} \end{cases}. \tag{3}$$

(**Step 2**) Second, it sets $z = y$, yielding the CLEAR (*Covariant LEast-square Refitting*) associated to an *a.e.* differentiable estimator $y \mapsto \widehat{x}(y)$ defined, for almost all $y \in \mathbb{R}^n$, by

$$\mathcal{R}_{\widehat{x}}(y) = \mathcal{D}_{\widehat{x},y}(y) = \widehat{x}(y) + \rho J(y - \Phi\widehat{x}(y)), \tag{4}$$

with ρ defined similarly to (3) and with $\delta = y - \Phi\widehat{x}(y)$, $J = J_{\widehat{x}}(y)$.

The covariant refitting thus applies an additive correction to $\widehat{x}(y)$ with a fraction of the directional derivative $J\delta$ in the direction of the residual δ. Remark that $y \mapsto \mathcal{D}_{\widehat{x},z}(y)$ is affine while $y \mapsto \mathcal{R}_{\widehat{x}}(y)$ is not, and that ρ is constant for the map $\mathcal{D}_{\widehat{x},z}$ whilst varying with respect to y in $\mathcal{R}_{\widehat{x}}$. At last, $\mathcal{R}_{\widehat{x}}(y) \in \widehat{x}(y) + im\ (J_{\widehat{x}}(y))$, where im denotes the image.

Unlike [5], our problem requires incorporating a geometrical constraint: as the CLEAR estimator does not encompass this possibility, a novel model is derived in the following.

3 Purpose-Built Estimation Model to Include Constraints

3.1 Generic Model

For now, we comply with the previous notations and assume that $\mathcal{K}$ is an arbitrary closed convex set in $\mathbb{R}^p$ with non-empty interior. To include the constraints, compared to (2), the estimation of x is now provided by the following minimisation problem

$$\widehat{x}(y) \in \arg\min_{x \in \mathcal{K}} F(x, y) + G(x). \tag{5}$$

Classical estimation strategies rely on a structural prior of the data expressed in terms of model subsets.

Definition 1. *The model subset associated to an a.e. differentiable estimator $\widehat{x}$ is defined at almost all points $y \in \mathbb{R}^n$ by the intersection of an affine subspace and the constraint set $\mathcal{K}$:*

$$\mathcal{M}_{\widehat{x}}(y) = \left\{ \widehat{x}(y) + im\ (J_{\widehat{x}}(y)) \right\} \cap \mathcal{K}. \tag{6}$$

Let us notice that $\mathcal{M}_{\widehat{x}}(y)$ is non-empty since $\widehat{x}(y)$ exists and $\widehat{x}(y) \in \mathcal{K}$. Also, the definition of $J_{\widehat{x}}(y)d$ is still valid since both $\widehat{x}(y)$ and $\widehat{x}(y + \varepsilon d)$ belong to $\mathcal{K}$. Our goal is now to define, from the original estimator $\hat{x}$, a counterpart of $\mathcal{D}_{\widehat{x},z}$ (*resp.* $\mathcal{R}_{\widehat{x}}$) denoted by $\mathcal{D}^{\mathcal{K}}_{\widehat{x},z}$ (*resp.* $\mathcal{R}^{\mathcal{K}}_{\widehat{x}}$) in the constrained case.

Definition 2. *Let $\widehat{x} : \mathbb{R}^n \to \mathbb{R}^p$ be differentiable at almost every $y \in \mathbb{R}^n$ and such that for all $y \in \mathbb{R}^n$, $\widehat{x}(y) \in \mathcal{K}$. Let us assume that $\hat{x}$ is differentiable at $z \in \mathbb{R}^n$. An estimator $\mathcal{D}^{\mathcal{K}}_{\widehat{x},z}$ is a constrained guess-based covariant least-square refitting of $\widehat{x}$ for z if*

$$\mathcal{D}^{\mathcal{K}}_{\widehat{x},z} \in \underset{h \in \overline{\mathcal{H}}}{\arg\min}\ \|\Phi\, h(z) - z\|_2^2, \tag{7}$$

where $\overline{\mathcal{H}}$ is a set of maps $h : \mathbb{R}^n \to \mathbb{R}^p$ satisfying criteria **(i)**–**(iii)** *and* **(iv)** *stated as, $h(z) \in \mathcal{K}$, this latter point being the proposed estimator novelty.*

Theorem 1. *Let $\widehat{x}$ be an estimator from $\mathbb{R}^n$ to $\mathbb{R}^p$, differentiable at $z \in \mathbb{R}^n$, let $\delta = z - \Phi\widehat{x}(z)$ and $J = J_{\widehat{x}}(z)$. Let $\mathcal{D}_{\widehat{x},z}$ be the guess-based covariant least-square refitting estimator as defined in (3) (unconstrained case). Then the constrained counterpart $\mathcal{D}^{\mathcal{K}}_{\widehat{x},z}$ introduced in Definition 2 exists and is unique when $\Phi J\delta \neq 0$. It is given by $\mathcal{D}^{\mathcal{K}}_{\widehat{x},z}(y) = \widehat{x}(z) + \rho^*\, J(y - \Phi\widehat{x}(z))$, with ρ^* such that $\mathcal{D}^{\mathcal{K}}_{\widehat{x},z}(z)$ is the oblique projection of $\mathcal{D}_{\widehat{x},z}(z)$ on $\mathcal{K}$ in the direction of the line $(\widehat{x}(z), \mathcal{D}_{\widehat{x},z}(z))$.*

Proof. As previously stressed (**Step 1**), with criteria **(i)**–**(iii)**, the sought estimator reads $\mathcal{D}^{\mathcal{K}}_{\widehat{x},z}(y) = \widehat{x}(z) + \rho\, J(y - \Phi\widehat{x}(z))$. The value of ρ is obtained by recasting (7) as a one-dimensional optimisation problem in ρ, namely,

$$\min_{\rho \in \mathbb{R}} \|\Phi\widehat{x}(z) + \rho\Phi\, J(z - \Phi\widehat{x}(z)) - z\|_2^2 \text{ such that } \widehat{x}(z) + \rho\ J(z - \Phi\widehat{x}(z)) \in \mathcal{K}.$$

Let us denote by $\bar{\rho}$, the optimal ρ of the unconstrained CLEAR estimator as defined in (3). If $\widehat{x}(z) + \bar{\rho}\ J(z - \Phi\widehat{x}(z)) \in \mathcal{K}$, then the searched solution is $\mathcal{D}^{\mathcal{K}}_{\widehat{x},z}(y) = \widehat{x}(z) + \rho^*\ J(y - \Phi\widehat{x}(z))$ with $\rho^* = \bar{\rho}$, corresponding then to the unconstrained CLEAR estimator. Now, if $\widehat{x}(z) + \bar{\rho}\ J(z - \Phi\widehat{x}(z)) \notin \mathcal{K}$—keeping in mind that $\widehat{x}(z) \in \mathcal{K}$ and that the quantity to be minimised in ρ is a polynomial of degree 2 reaching its minimum for $\rho = \bar{\rho}$—, the optimal ρ^* is characterised by the fact that $\widehat{x}(z) + \rho^*\ J(z - \Phi\widehat{x}(z))$ is the oblique projection of $\mathcal{D}_{\widehat{x},z}(z) = \widehat{x}(z) + \bar{\rho}\ J(z - \Phi\widehat{x}(z))$ on $\mathcal{K}$ in the direction of the line $(\widehat{x}(z), \mathcal{D}_{\widehat{x},z}(z))$.

Like in [5], the constrained CLEAR refitting associated to an *a.e.* differentiable estimator $y \mapsto \widehat{x}(y)$ is then built, for almost all $y \in \mathbb{R}^n$, as

$$\mathcal{R}^{\mathcal{K}}_{\widehat{x}}(y) = \mathcal{D}^{\mathcal{K}}_{\widehat{x},y}(y). \tag{8}$$

3.2 Back to Our Specific Polarimetric Problem

The considered polarimetric problem ($\mathcal{P}_b$) falls within the framework (5) consisting in estimating an unknown S—the set of Stokes vectors, modelled through the variable x previously— from noisy observations I—radiance images, modelled through y previously —and stated as

$$\widehat{S}(I) \in \operatorname*{arg\,min}_{S\in X^3} F(S,I) + G(S),$$

with $\begin{cases} F(S,I) = & \frac{\mu}{2}\sum_{(i,j)\in\mathcal{G}} \|AS_{i,j} - I_{i,j}\|^2_{\mathbb{R}^4} \quad =: \frac{\mu}{2}\|I - \varPhi S\|^2 \\ G(S) & = \gamma TV_{\mathcal{C}}(S_1,S_2) + \gamma TV(S_0) + i_{\mathcal{K}}(S) \end{cases}$, and where $i_{\mathcal{K}}$ denotes the indicator function of the closed convex set $\mathcal{K}$. The next lemma yields a simplified expression of $J\varPhi\widehat{S}(I)$, and subsequently, of $\mathcal{R}^{\mathcal{K}}_{\widehat{S}}(I)$.

Lemma 1. *Let $\widehat{S}(I)$ be the (unique) solution of*

$$\widehat{S}(I) \in \operatorname*{arg\,min}_{S\in X^3} \frac{\mu}{2}\|I - \varPhi S\|^2 + G(S).$$

Then for all $\varepsilon \in [0,1]$, the following holds :

$$(1-\varepsilon)\,\widehat{S}(I) \in \operatorname*{arg\,min}_{S\in X^3} \frac{\mu}{2}\|I - \varepsilon\varPhi\,\widehat{S}(I) - \varPhi S\|^2 + G(S) \tag{9}$$

and $(1-\varepsilon)\,\widehat{S}(I) \in \mathcal{K}$ from the convex cone stability property.

Proof. We argue by contradiction and use the one-homogeneity feature of G and the convex cone stability property.

Corollary 1. *The uniqueness argument gives $(1-\varepsilon)\,\widehat{S}(I) = \widehat{S}(I - \varepsilon\varPhi\,\widehat{S}(I))$, so that*

$$J\varPhi\widehat{S}(I) = \lim_{\varepsilon\to 0} \frac{\widehat{S}(I) - \widehat{S}(I - \varepsilon\varPhi\widehat{S}(I))}{\varepsilon} = \widehat{S}(I),$$

with $J := J_{\widehat{S}}(I)$, and $\mathcal{R}^{\mathcal{K}}_{\widehat{S}}(I) = (1-\rho^*)\,\widehat{S}(I) + \rho^*\,JI$, ρ^* being as prescribed in Theorem 1. The resulting algorithm for practically computing ρ^* is thus derived as follows. Let us denote by $\bar{\rho}$ the optimal ρ of the unconstrained CLEAR estimator as defined in (3). The problem being separable in (i,j), for each $(i,j) \in \mathcal{G}$ such that the Lorentz cone constraint on $\left(\mathcal{R}_{\widehat{S}}(I)\right)_{i,j} =$ $(1-\bar{\rho})\left(\widehat{S}(I)\right)_{i,j} + \bar{\rho}\,(JI)_{i,j} = \begin{pmatrix} (r_0)_{i,j} = (1-\bar{\rho})\,(\widehat{s_0})_{i,j} + \bar{\rho}\,(j_0)_{i,j} \\ (r_1)_{i,j} = (1-\bar{\rho})\,(\widehat{s_1})_{i,j} + \bar{\rho}\,(j_1)_{i,j} \\ (r_2)_{i,j} = (1-\bar{\rho})\,(\widehat{s_2})_{i,j} + \bar{\rho}\,(j_2)_{i,j} \end{pmatrix}$ is violated, i.e., $(r_0)_{i,j} < \sqrt{(r_1)^2_{i,j} + (r_2)^2_{i,j}}$, compute $\rho^*_{i,j}$, which amounts to identifying the unique root between 0 and $\bar{\rho}$ of a degree 2 polynomial. Precisely, it consists in finding $\rho^*_{i,j}$ such that $\left((1-\rho^*_{i,j})(\widehat{s_0})_{i,j} + \rho^*_{i,j}(j_0)_{i,j}\right)^2 = \left((1-\rho^*_{i,j})(\widehat{s_1})_{i,j} + \rho^*_{i,j}(j_1)_{i,j}\right)^2 + \left((1-\rho^*_{i,j})(\widehat{s_2})_{i,j} + \rho^*_{i,j}(j_2)_{i,j}\right)^2$. Then set $\rho^* = \min_{i,j}\rho^*_{i,j}$ if $\bar{\rho} > 0$, $\max_{i,j}\rho^*_{i,j}$ otherwise and subsequently, $\mathcal{R}^{\mathcal{K}}_{\widehat{S}}(I) = (1-\rho^*)\,\widehat{S}(I) + \rho^*\,JI$.

The next subsection details the numerical algorithm based on algorithmic differentiation to estimate J. The notations are those of [5].

3.3 Practical Implementation

Like [5], our algorithm falls within the generic iterative scheme of the form

$$\begin{cases} x^k &= \gamma(a^k) \\ a^{k+1} &= \Psi(a^k, y) \end{cases}.$$

Here, $a^k \in \mathcal{A}$ is a sequence of auxiliary variables, $\Psi : \mathcal{A} \times \mathbb{R}^n \to \mathcal{A}$ is a fixed point operator in the sense that a^k converges to a^*, and $\gamma : \mathcal{A} \to \mathbb{R}^p$ is non-expansive implying that x^k converges to $x^* = \gamma(a^*)$. As a result, for almost all y and for any direction $d \in \mathbb{R}^n$, the directional derivatives $\mathcal{D}_x^k = J_{\hat{x}^k}(y)d$ and $\mathcal{D}_a^k = J_{a^k}(y)d$ can be jointly obtained with x^k and a^k as:

$$\begin{cases} x^k &= \gamma(a^k) \\ a^{k+1} &= \Psi(a^k, y) \\ \mathcal{D}_x^k &= \Gamma_a \mathcal{D}_a^k \\ \mathcal{D}_a^{k+1} &= \Psi_a \mathcal{D}_a^k + \Psi_y d, \end{cases} \qquad (\mathcal{S})$$

where $\Gamma_a = \left.\dfrac{\partial \gamma(a)}{\partial a}\right|_{a^k}$, $\Psi_a = \left.\dfrac{\partial \Psi(a,y)}{\partial a}\right|_{a^k}$, and $\Psi_y = \left.\dfrac{\partial \Psi(a^k,y)}{\partial y}\right|_{y}$. In our specific case, (i) $a^k = (p^k, S^k, v^k)^T$, with S^k, the estimated biased Stokes vectors, p^k, the dual variable of the involved Chambolle-Pock (C.-P.) algorithm and v^k, the variable resulting from the overrelaxation technique employed for the primal variable S^k; (ii) $x^k = S^k$. Also, due to the property inherited from Lemma 1, the identification of $\mathcal{R}_{\widehat{S}}^{\mathcal{K}}(I)$ requires computing $\widehat{S}(I)$ and $J_{\widehat{S}}(I)I$, this latter quantity being assessed by $\mathcal{D}_S^k = J_{S^k}(I)I$. In the case of algorithmic differentiation, the schematic algorithm $(\mathcal{S})$ can be run once to get $\mathcal{R}_{\widehat{S}}^{\mathcal{K}}(I)$ since setting $d = I$ yields $\widehat{S}(I)(= S^\infty(I))$, $J_{\widehat{S}}(I)I = \mathcal{D}_S^\infty$, and subsequently ρ^* by Corollary 1. In the algorithm, we denote by $\mathcal{D}_a^k = \left(\tilde{p}^k, \tilde{S}^k, \tilde{v}^k\right)^T$. The limited number of pages prevents us from elaborating on the numerical details and being completely self-contained. Thus, to flow the reading better, we have chosen (i) first to comply with the notations of [11] (whether it be for the parameters involved in the Chambolle-Pock algorithm (τ, σ), or for complementary auxiliary variables or parameters introduced to smooth the reading ($\alpha = \frac{1}{\tau} + \mu$, $\beta = \frac{1}{\tau} + \frac{\mu}{2}$, z_0, $\bar{z}$, etc.)), (ii) second, to deal with the update of S^k and $\tilde{S}^k$ separately, yielding the below schematic algorithm combined with Tables 1 and 2. In addition, we denote by

$$P_\gamma(\tilde{z})_{i,j} = \begin{cases} \tilde{z}_{i,j} & \text{if } \|\tilde{z}_{i,j}\|_{\mathbb{R}^2 \text{ or } \mathbb{R}^4} \leq \gamma \\ \gamma \dfrac{\tilde{z}_{i,j}}{\max(\gamma, \|\tilde{z}_{i,j}\|)} & \text{otherwise} \end{cases}$$

Table 1. Explicit expressions for the update of S^k. To lighten the notations, $b = (b_0, b_1, b_2)^T = A^T I$, z_0 (explicit dependence on k omitted) $= \frac{1}{\alpha\,\tau}\,(S_0^k + \tau \operatorname{div} p_0^{k+1}) + \frac{\mu}{\alpha}\, b_0$, $z_i = \frac{1}{\beta\,\tau}\,(S_i^k + \tau \operatorname{div} p_i^{k+1}) + \frac{\mu}{\beta}\, b_i$ and $\bar{z} = (z_1, z_2)$.

Sequential schematic algorithm	
C.-P.-biased estimator assessment	$\mathcal{D}_a^k$ computation (debiasing part)
$\begin{cases} p^{k+1} = \left(p_0^{k+1}, p_1^{k+1}, p_2^{k+1}\right) = \left(p_0^{k+1}, \bar{p}^{k+1}\right) \\ p_0^{k+1} = P_\gamma(p_0^k + \sigma\,\nabla v_0^k) \\ \bar{p}^{k+1} = P_\gamma(\bar{p}^k + \sigma \nabla \bar{v}^k) \end{cases}$	$\begin{cases} \tilde{p}^{k+1} = \left(\tilde{p_0}^{k+1}, \tilde{p_1}^{k+1}, \tilde{p_2}^{k+1}\right) = \left(\tilde{p_0}^{k+1}, \tilde{\bar{p}}^{k+1}\right) \\ \tilde{p_0}^{k+1} = \Pi_{p_0^k + \sigma\,\nabla v_0^k}(\tilde{p_0}^k + \sigma\,\nabla \tilde{v_0}^k) \\ \tilde{\bar{p}}^{k+1} = \Pi_{\bar{p}^k + \sigma\,\nabla \bar{v}^k}\left(\tilde{\bar{p}}^k + \sigma\,\nabla \tilde{\bar{v}}^k\right) \end{cases}$
$\begin{cases} S^{k+1} = \left(S_0^{k+1}, S_1^{k+1}, S_2^{k+1}\right) \\ \text{according to Table 1} \end{cases}$	$\begin{cases} \tilde{S}^{k+1} = \left(\tilde{S_0}^{k+1}, \tilde{S_1}^{k+1}, \tilde{S_2}^{k+1}\right) \\ \text{according to Table 2} \end{cases}$
$v^{k+1} = (v_0^{k+1}, \bar{v}^{k+1}) = 2S^{k+1} - S^k$	$\tilde{v}^{k+1} = (\tilde{v_0}^{k+1}, \tilde{\bar{v}}^{k+1}) = 2\tilde{S}^{k+1} - \tilde{S}^k$

Numerical details for S^k update		
Subcase $z_0 \geq 0$ and $\lVert\bar{z}\rVert \leq z_0$		$\left({S_0}^{k+1}, {S_1}^{k+1}, {S_2}^{k+1}\right) = (z_0, z_1, z_2)$
Subcase $z_0 < 0$ and $\beta\,\lVert\bar{z}\rVert \leq -\alpha z_0$		$\left({S_0}^{k+1}, {S_1}^{k+1}, {S_2}^{k+1}\right) = (0, 0, 0)$
Subcases	$z_0 \geq 0$ and $\lVert\bar{z}\rVert > z_0$ or $z_0 < 0$ and $\beta\lVert\bar{z}\rVert > -\alpha\, z_0$	$\begin{cases} {S_0}^{k+1} = \frac{\alpha}{\alpha+\beta}\, z_0 + \frac{\beta}{\alpha+\beta}\,\lVert\bar{z}\rVert \\ {S_1}^{k+1} = \left(\frac{\beta}{\alpha+\beta} + \frac{\alpha}{\alpha+\beta}\,\frac{z_0}{\lVert\bar{z}\rVert}\right) z_1 \\ {S_2}^{k+1} = \left(\frac{\beta}{\alpha+\beta} + \frac{\alpha}{\alpha+\beta}\,\frac{z_0}{\lVert\bar{z}\rVert}\right) z_2 \end{cases}$

Table 2. Explicit expressions for the update of $\tilde{S}^k$. To lighten the notations, $\tilde{z_0} = \frac{1}{\alpha\,\tau}\,(\tilde{S_0}^k + \tau \operatorname{div} \tilde{p_0}^{k+1}) + \frac{\mu}{\alpha}\, b_0$ and $\tilde{z}_i = \frac{1}{\beta\,\tau}\,(\tilde{S_i}^k + \tau \operatorname{div} \tilde{p_i}^{k+1}) + \frac{\mu}{\beta}\, b_i$, $i \in \{1, 2\}$.

Numerical details for $\tilde{S}^k$ update		
Subcase $z_0 \geq 0$ and $\lVert\bar{z}\rVert \leq z_0$		$\left(\tilde{S_0}^{k+1}, \tilde{S_1}^{k+1}, \tilde{S_2}^{k+1}\right) = (\tilde{z_0}, \tilde{z_1}, \tilde{z_2})$
Subcase $z_0 < 0$ and $\beta\,\lVert\bar{z}\rVert \leq -\alpha z_0$		$\left(\tilde{S_0}^{k+1}, \tilde{S_1}^{k+1}, \tilde{S_2}^{k+1}\right) = (0, 0, 0)$
Subcases	$z_0 \geq 0$ and $\lVert\bar{z}\rVert > z_0$ or $z_0 < 0$ and $\beta\lVert\bar{z}\rVert > -\alpha\, z_0$	$\begin{cases} \tilde{S_0}^{k+1} = \frac{\alpha}{\alpha+\beta}\,\tilde{z_0} + \frac{\beta}{\alpha+\beta}\,\frac{z_1}{\lVert\bar{z}\rVert}\,\tilde{z_1} + \frac{\beta}{\alpha+\beta}\,\frac{z_2}{\lVert\bar{z}\rVert}\,\tilde{z_2} \\ \tilde{S_1}^{k+1} = \frac{\alpha}{\alpha+\beta}\,\frac{z_1}{\lVert\bar{z}\rVert}\,\tilde{z_0} + \left(\frac{\beta}{\alpha+\beta} + \frac{\alpha}{\alpha+\beta}\,\frac{z_0}{\lVert\bar{z}\rVert} - \frac{\alpha}{\alpha+\beta}\,\frac{z_0 z_1^2}{\lVert\bar{z}\rVert^3}\right)\tilde{z_1} \\ \qquad - \frac{\alpha}{\alpha+\beta}\,\frac{z_0 z_1 z_2}{\lVert\bar{z}\rVert^3}\,\tilde{z_2} \\ \tilde{S_2}^{k+1} = \frac{\alpha}{\alpha+\beta}\,\frac{z_2}{\lVert\bar{z}\rVert}\,\tilde{z_0} - \frac{\alpha}{\alpha+\beta}\,\frac{z_0 z_1 z_2}{\lVert\bar{z}\rVert^3}\,\tilde{z_1} \\ \qquad + \left(\frac{\beta}{\alpha+\beta} + \frac{\alpha}{\alpha+\beta}\,\frac{z_0}{\lVert\bar{z}\rVert} - \frac{\alpha}{\alpha+\beta}\,\frac{z_0 z_2^2}{\lVert\bar{z}\rVert^3}\right)\tilde{z_2} \end{cases}$

and

$$\Pi_z(\tilde{z})_{i,j} = \begin{cases} \tilde{z}_{i,j} & \text{if } \|z_{i,j}\|_{\mathbb{R}^2 \text{ or } \mathbb{R}^4} \leq \gamma \\ \dfrac{\gamma}{\|z_{i,j}\|}\left(\tilde{z}_{i,j} - \langle \tilde{z}_{i,j}, \dfrac{z_{i,j}}{\|z_{i,j}\|}\rangle\, \dfrac{z_{i,j}}{\|z_{i,j}\|}\right) & \text{otherwise.} \end{cases}$$

4 Experiments and Results

As a validation process of our approach, we compare the debiased algorithm with the original version available in [11], this approach achieving competitive results in the current literature. The experiments have been carried out on both artificial and real data, and quantitative/qualitative comparisons exemplify the improvement supplied by the refitting procedure. The dataset stems from the work of [11] in which the authors have used a trioptics Polar4D camera providing the four images I_0, I_{45}, I_{90}, and I_{135}. Both indoor and outdoor images are available. In the case of outdoor data, one of the targeted applications is car detection enhancement within the autonomous vehicle setting. The sequential algorithm of the previous section has been implemented in Matlab without hardware optimisation. Processing a 620×960 image takes approximately 2 min on a standard laptop.

For quantitative comparisons, one needs the ground truth, requiring thus synthetic data. To generate artificial images, we use the same algorithm as in [11]. As a reminder, these data have been produced with a code provided by Dr. S.

S_0 with our method S_1 with our method S_2 with our method

S_0 with [11] S_1 with [11] S_2 with [11]

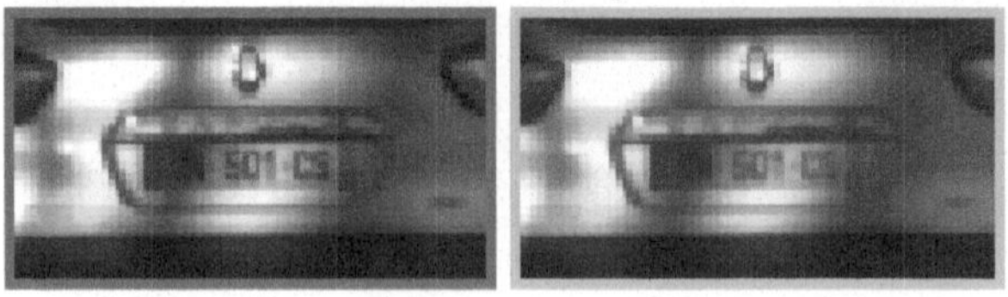

Zoom on our method Zoom on [11]

Fig. 3. Comparison between our method and [11] on outdoor images.

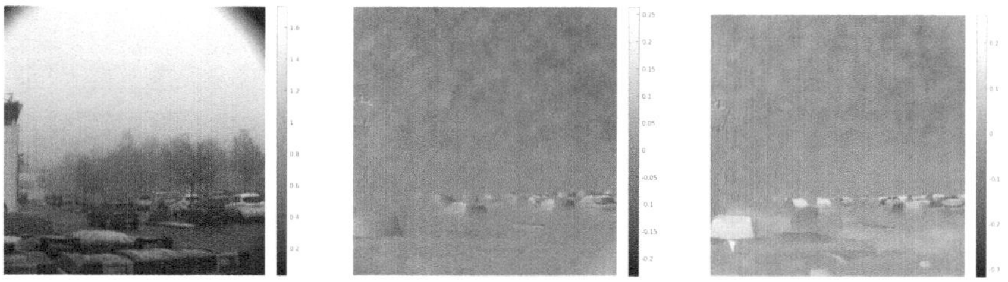

S_0 with our method S_1 with our method S_2 with our method

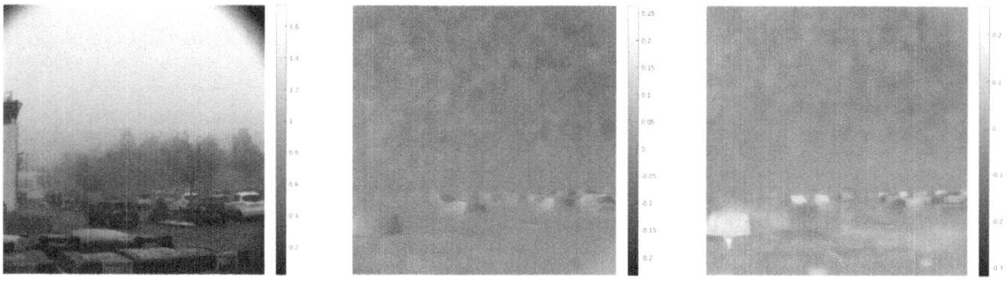

S_0 with [11] S_1 with [11] S_2 with [11]

Fig. 4. Comparison on outdoor images.

Faisan (ICube - MIV, Université de Strasbourg) devoted to full Stokes vector reconstruction ([8]). This algorithm creates a $256 \times 256 \times 4$ -pixel image composed of polarisation signatures. A background and a foreground with smooth variation and known polarisation signatures compose the image. Since the polarisation depends on the object surface material in the scene, the total variation proves to be adapted for urban scenarios, and the generated artificial data are relevant to the targeted application. In the data generation algorithm, the Stokes vectors are computed with smoothly varying components, yielding the radiance images $[I_0, I_{45}, I_{90}, I_{135}]$ derived from the formula $I = AS$ which are then corrupted by an additive white Gaussian noise with variance $\sigma^2 = 0.02$. A specific PSNR based on a weighted average of the standard PSNR on each channel is introduced for polarimetric images in [8]. This measure has been used on artificial data, see, *e.g.*, [11, Fig. 3]. Given this input condition, the computations, both with the original algorithm and its debiased counterpart, are performed with the prescribed parameters $\sigma = 0.3$, $\gamma = 1$, $\tau = 1/(8\sigma)$, $\theta = 1$, $\mu = 1.5$. After 1000 iterations, the result is obtained jointly with the PSNR values. While the original biased version [11] produces a PSNR of 37.3601, the debiased algorithm provides a higher PSNR equal to 41.5768.

Whilst the PSNR values encode the extended-CLEAR debiased version improvement over the original method [11], the main advantage of the proposed method is contrast highlighting, yielding increased visibility of the objects on the scene. For real-data-images, adapting the parameters is required, achieved through a grid search technique, yielding stable values: $\sigma = 0.1$, $\gamma = 1$, $\tau = 1/(8\sigma)$, $\theta = 1$, $\mu = 25$. Figure 2 shows some cars on a hazy road. The car

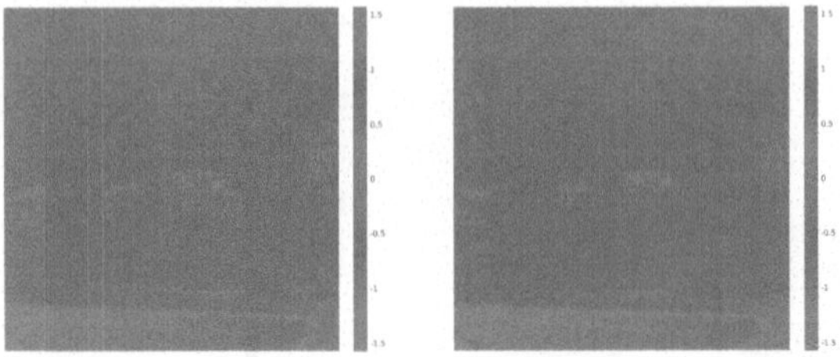

AOLP with our method *AOLP* with [11]

Fig. 5. Comparison of AOLP maps between our method and [11].

shapes are captured more faithfully in the extended-CLEAR debiased version compared to [11]. Figure 3 also displays such results. Each channel unveils contrast highlight with the refitting method compared to the standard method [11]. In particular, a significant improvement can be observed on the licence plate in the S_2 Stokes vector component, where the refitting approach highly increases the contrast of the visible letters. Figure 4 displays complementary results on outdoor images. Again, the contrast enhancement applies to the whole of the Stokes vectors. In this paper, we do not display the radiance images I_0, I_{45}, I_{90}, and I_{135}. If the contrast is slightly enhanced, it is not significant. This effect is more of a spillover since the refitting model aims to improve the contrast of the Stokes vectors but not directly the radiance images. This outcome is thus secondary in our model. The DOLP (Degree Of Linear Polarisation defined as $\sqrt{S_1^2 + S_2^2}/S_0$) and the AOLP (Angle Of Linear Polarisation characterised as $(1/2)\,\text{atan2}\,(S_2, S_1)$) are relevant metrics to check whether the method is efficient at recovering Stokes vectors. The latter indicator provides information about the object plane (surface normal) that reflects the incoming light. In the absence of any form of noise, the AOLP should be constant in homogeneous areas. It is illustrated in Fig. 5 for both our method and [11]. Nevertheless, Fig. 5 demonstrates that the refitting procedure has also the advantage to enhance the details while avoiding to introduce artefacts and maintaining the homogeneity of the calculated map. The DOLP is not displayed, but the results are similar for these maps with the guarantee that it belongs $[0, 1]$ (condition of belonging to the Lorentz cone).

5 Conclusion

This paper analyses a refitting approach to model [11]. Since the direct application of CLEAR [5] does not fulfil the admissibility constraints, a novel framework has been developed and applied to polarimetric images. The proposed algorithm has been quantitatively tested on synthetic data while qualitatively validated on real data. The proposed results outperform those obtained from previous approaches. In future works, heuristically-driven constraint enforcement (*e.g.*, design of an adaptive ρ like in [13]) will be compared with our approach.

References

1. Belloni, A., Chernozhukov, V.: Least squares after model selection in high-dimensional sparse models. Bernoulli **19**, 521–547 (2013)
2. Brinkmann, E.M., Burger, M., Rasch, J., Sutour, C.: Bias reduction in variational regularization. J. Math. Imag. Vis. **59**(3), 534–566 (2017)
3. Bühlmann, P., Yu, B.: Boosting with the l^2 loss: regression and classification. J. Am. Stat. Assoc. **98**(462), 324–339 (2003)
4. Burger, M., Korolev, Y., Rasch, J.: Convergence rates and structure of solutions of inverse problems with imperfect forward models. Inverse Probl. **35**(2), 024006 (2019)
5. Deledalle, C.A., Papadakis, N., Salmon, J., Vaiter, S.: CLEAR: covariant LEAst-square refitting with applications to image restoration. SIAM J. Imaging Sci. **10**(1), 243–284 (2017)
6. Deledalle, C.-A., Papadakis, N., Salmon, J., Vaiter, S.: Refitting solutions promoted by ℓ_{12} sparse analysis regularizations with block penalties. In: Lellmann, J., Burger, M., Modersitzki, J. (eds.) SSVM 2019. LNCS, vol. 11603, pp. 131–143. Springer, Cham (2019). https://doi.org/10.1007/978-3-030-22368-7_11
7. Efron, B., Hastie, T., Johnstone, I., Tibshirani, R.: Least angle regression. Ann. Stat. **32**(2), 407–499 (2004)
8. Faisan, S., Heinrich, C., Rousseau, F., Lallement, A., Zallat, J.: Joint filtering estimation of stokes vector images based on a nonlocal means approach. J. Opt. Soc. Am. A **29**(9), 2028–2037 (2012)
9. Friedman, J., Hastie, T., Tibshirani, R.: Additive logistic regression: a statistical view of boosting (with discussion and a rejoinder by the authors). Ann. Stat. **28**(2), 337–407 (2000)
10. Hastie, T., Tibshirani, R.: Generalized additive models. Stat. Sci. **1**(3), 297–310 (1986)
11. Le Guyader, C., Ainouz, S., Canu, S.: A physically admissible stokes vector reconstruction in linear polarimetric imaging. J. Math. Imaging Vis. **65**(4), 592–617 (2023)
12. Pascal, B., Pustelnik, N., Abry, P., Serres, M., Vidal, V.: Joint estimation of local variance and local regularity for texture segmentation. Application to multiphase flow characterization. In: 2018 25th IEEE International Conference on Image Processing (ICIP), pp. 2092–2096 (2018)
13. Pierre, F., Aujol, J.-F., Deledalle, C.-A., Papadakis, N.: Luminance-guided chrominance denoising with debiased coupled total variation. In: Pelillo, M., Hancock, E. (eds.) EMMCVPR 2017. LNCS, vol. 10746, pp. 235–248. Springer, Cham (2018). https://doi.org/10.1007/978-3-319-78199-0_16
14. Romano, Y., Elad, M.: Boosting of image denoising algorithms. SIAM J. Imaging Sci. **8**(2), 1187–1219 (2015)
15. Stein, C.M.: Estimation of the mean of a multivariate normal distribution. Ann. Stat. **9**(6), 1135–1151 (1981)
16. Valenzuela, J.R., Fessler, J.A.: Joint reconstruction of stokes images from polarimetric measurements. J. Opt. Soc. Am. A **26**(4), 962–968 (2009)
17. Zallat, J., Heinrich, C.: Polarimetric data reduction: a Bayesian approach. Opt. Express **15**(1), 83–96 (2007)
18. Zallat, J., Heinrich, C., Petremand, M.: A Bayesian approach for polarimetric data reduction: the Mueller imaging case. Opt. Express **16**(10), 7119–7133 (2008)

Denoising Hyperbolic-Valued Data by Relaxed Regularizations

Robert Beinert and Jonas Bresch(✉)

Institut für Mathematik, Technische Universität Berlin, Straße des 17. Juni 136, 10623 Berlin, Germany
{beinert,bresch}@math.tu-berlin.de

Abstract. We introduce a novel relaxation strategy for denoising hyperbolic-valued data. The main challenge is here the non-convexity of the hyperbolic sheet. Instead of considering the denoising problem directly on the hyperbolic space, we exploit the Euclidean embedding and encode the hyperbolic sheet using a novel matrix representation. For denoising, we employ the Euclidean Tikhonov and total variation (TV) model, where we incorporate our matrix representation. The major contribution is then a convex relaxation of the variational ansätze allowing the utilization of well-established convex optimization procedures like the alternating directions method of multipliers (ADMM). The resulting denoisers are applied to a real-world Gaussian image processing task, where we simultaneously restore the pixelwise mean and standard deviation of a retina scan series.

Keywords: Hyperbolic-valued data · Signal and image processing · TV denoising · Tikhonov denoising · Convex relaxation methods

1 Introduction

The geometry of hyperbolic spaces finds applications across diverse research fields spanning from theoretical studies in mathematics and physics to practical implementations in machine learning, network analysis, and computer vision. One of the most famous example is the specific relativity theory, where hyperbolic spaces and their geometry are often used to model the spacetime [24]. Moreover, hyperbolic spaces are particularly well-suited for capturing hierarchical structures, such as those found in social and biological networks as well as in the topology of the internet [7,18]. In order to analyse hierarchical or tree-like data, the hyperbolic geometry has been incorporated into machine learning architectures leading to so-called hyperbolic neural networks [13,22], which, for instance, play a major role in hierarchical clustering [21,23]. From the view point of image analysis, hyperbolic spaces become interesting due to their utilization in image segmentation [2], image denoising [6], and morphological image processing [1]. Hyperbolic-valued images, for instance, occur as structure tensor in edge and texture processing [9] and in probabilistic image analysis [1,6]. The main idea behind the latter is that instead of a single image, a series of heavily noisy instances is acquired, where each pixel follows a Gaussian distribution with

T. A. Bubba et al. (Eds.): SSVM 2025, LNCS 15667, pp. 16–29, 2025.
https://doi.org/10.1007/978-3-031-92366-1_2

certain mean and standard deviation. The set of Gaussians equipped with the so-called Fisher metric is isometric to the two-dimensional hyperbolic space [6,20], which builds the bridge to the main concern of this paper—the denoising of hyperbolic-valued signals, images, and data. During the numerical experiments, we return to the probabilistic image denoising task for real-world retina scans.

Mathematically, the *hyperbolic space* $\mathbb{H}_d$ is the unique, simply connected, d-dimensional Riemannian manifold with constant curvature of -1, see [15,17]. The hyperbolic spaces possess several different embeddings [20]. Throughout this paper, we rely on the Euclidean one:

$$\mathbb{H}_d := \{\boldsymbol{x} \in \mathbb{R}^{d+1} \colon \eta(\boldsymbol{x}, \boldsymbol{x}) = -1 \quad \text{and} \quad x_{d+1} > 0\}$$

where $\eta \colon \mathbb{H}_d \times \mathbb{H}_d \to \mathbb{R}$, $(\boldsymbol{x}, \boldsymbol{y}) \mapsto \sum_{i=1}^{d} x_i y_i - x_{d+1} y_{d+1}$ is the *Minkowski bilinearfrom* defining the *hyperbolic distance* $\operatorname{dist}_{\mathbb{H}_d}(\boldsymbol{x}, \boldsymbol{y}) := \operatorname{arcosh}(-\eta(\boldsymbol{x}, \boldsymbol{y}))$. Note that the Euclidean embedding of the hyperbolic space $\mathbb{H}_d$, which is sometimes called the *hyperbolic sheet*, is contained in the $(d+1)$-dimensional, affine halfspace $\mathbb{A}_{d+1} := \mathbb{R}^d \times \mathbb{R}_{\geq 1}$, i.e. $\mathbb{H}_d \subset \mathbb{A}_{d+1}$.

In this paper, we approach the problem of hyperbolic image denoising. Here, the main issue comes form the non-convexity of the hyperbolic sheet. This is why hyperbolic image denoising becomes highly demanding from a numerical point of view. For instance, hyperbolic images can be denoised using total variation (TV) models based on the hyperbolic distance and applying a manifold version of the Douglas–Rachford algorithm [6]. The major contribution of this paper is that, instead of minimizing a variational denoising model directly on the hyperbolic sheet $\mathbb{H}_d$, we exploit the Euclidean embedding and propose a novel convex relaxation that allows the application of well-established convex optimization methods on $\mathbb{R}^{d+1}$. The paper is organized as follows: In § 2, we introduce (Euclidean) TV- and Tikhonov-like regularization models and convex relaxations that encode the hyperbolic sheet using positive semi-definite matrices in the spirit of [4,5,12]. In § 3, we use the alternating directions method of multipliers (ADMM) [14] to derive an actual denoising algorithm. Finally, in § 4, we apply the proposed method to denoising problems from Gaussian image processing.

2 Denoising of Hyperbolic-Valued Data

The aim of this paper is to denoise hyperbolic-valued data on a connected, undirected graph $G = (V, E)$, where $V := \{1, \dots, N\}$ denotes the set of N vertices and $E \subseteq \{(n, m) \in V \times V : n < m\}$ the set of $M := |E|$ edges. We are now interested in the restoration of hyperbolic-valued data $\boldsymbol{x} := (\boldsymbol{x}_n)_{n \in V} \in \mathbb{H}_d^N$ form noisy measurements $\boldsymbol{y} := (\boldsymbol{y}_n)_{n \in V}$ in $\mathbb{H}_d^N$ or $(\mathbb{R}^{d+1})^N$. For applications in signal and image processing, G corresponds to the line or grid graph.

2.1 Tikhonov-Like Regularization

For denoising relatively smooth hyperbolic data, we consider the variational Tikhonov-like denoising model:

$$\operatorname*{arg\,min}_{\boldsymbol{x}\in\mathbb{H}_d^N} \quad \frac{1}{2}\sum_{n\in V}\|\boldsymbol{x}_n-\boldsymbol{y}_n\|_2^2+\frac{\lambda}{2}\sum_{(n,m)\in E}\|\boldsymbol{x}_n-\boldsymbol{x}_m\|_2^2, \tag{1}$$

where $\|\cdot\|_2$ denotes the Euclidean norm, and $\lambda > 0$ the regularization parameter. Note that the model (1) is non-convex due to the minimization of the hyperbolic sheet. In order to convexify (1), we adapt the strategies in [5,12], where the Tikhonov denoising for the hypersphere and SO(3) is considered. For this, we introduce the auxiliary variables $\boldsymbol{f}\in\mathbb{R}^M$ and $\boldsymbol{v}\in\mathbb{R}^N$ and the linear objective

$$\mathcal{J}(\boldsymbol{x},\boldsymbol{f},\boldsymbol{v}) := \frac{1}{2}\sum_{n\in V}(\boldsymbol{v}_n-2\langle\boldsymbol{x}_n,\boldsymbol{y}_n\rangle)+\frac{\lambda}{2}\sum_{(n,m)\in E}(\boldsymbol{v}_n+\boldsymbol{v}_m-2\boldsymbol{f}_{(n,m)}) \tag{2}$$

with the Euclidean inner product $\langle\cdot,\cdot\rangle$. Therewith, (1) becomes equivalent to

$$\operatorname*{arg\,min}_{\boldsymbol{x}\in\mathbb{H}_d^N} \quad \mathcal{J}(\boldsymbol{x},\boldsymbol{f},\boldsymbol{v}) \quad \text{s.t.} \quad \left\{\begin{matrix}\boldsymbol{v}_n=\|\boldsymbol{x}_n\|_2^2 & \forall n\in V\\ \boldsymbol{f}_{(n,m)}=\langle\boldsymbol{x}_n,\boldsymbol{x}_m\rangle & \forall(n,m)\in E\end{matrix}\right\}. \tag{3}$$

For any $\boldsymbol{\xi}\in\mathbb{R}^{d+1}$, we define $\tilde{\boldsymbol{\xi}} := (\xi_1,\dots,\xi_d,-\xi_{d+1})^*$. Further, we denote the $(d+1)\times(d+1)$ identity by $\boldsymbol{I}_{d+1}$. To encode the minimization over the hyperbolic sheet, we introduce—based on the additional variables $\boldsymbol{\ell}\in\mathbb{R}^M$—the series of matrices

$$\boldsymbol{Q}_{(n,m)} := \left[\begin{array}{c|cc|cc}\boldsymbol{I}_{d+1} & \boldsymbol{x}_n & \tilde{\boldsymbol{x}}_n & \boldsymbol{x}_m & \tilde{\boldsymbol{x}}_m\\ \hline \boldsymbol{x}_n^* & \boldsymbol{v}_n & -1 & \boldsymbol{f}_{(n,m)} & \boldsymbol{\ell}_{(n,m)}\\ \tilde{\boldsymbol{x}}_n^* & -1 & \boldsymbol{v}_n & \boldsymbol{\ell}_{(n,m)} & \boldsymbol{f}_{(n,m)}\\ \hline \boldsymbol{x}_m^* & \boldsymbol{f}_{(n,m)} & \boldsymbol{\ell}_{(n,m)} & \boldsymbol{v}_m & -1\\ \tilde{\boldsymbol{x}}_m^* & \boldsymbol{\ell}_{(n,m)} & \boldsymbol{f}_{(n,m)} & -1 & \boldsymbol{v}_m\end{array}\right].$$

Proposition 1. *Let $\boldsymbol{x}_n,\boldsymbol{x}_m\in\mathbb{A}_{d+1}$. Then $\boldsymbol{x}_n,\boldsymbol{x}_m\in\mathbb{H}_d$, $\boldsymbol{v}_n=\|\boldsymbol{x}_n\|_2^2$, $\boldsymbol{v}_m=\|\boldsymbol{x}_m\|_2^2$, $\boldsymbol{\ell}_{(n,m)}=\eta(\boldsymbol{x}_n,\boldsymbol{x}_m)$, and $\boldsymbol{f}_{(n,m)}=\langle\boldsymbol{x}_n,\boldsymbol{x}_m\rangle$ if and only if $\boldsymbol{Q}_{(n,m)}\succeq\boldsymbol{0}$ and* $\operatorname{rank}(\boldsymbol{Q}_{(n,m)})=d+1$.

Proof. Because of the identity in the upper left, $\boldsymbol{Q}_{(n,m)}$ has at least rank $d+1$. Now, the last four rows can be represented by the first $d+1$ rows if and only if $\boldsymbol{v}_n=\|\boldsymbol{x}_n\|_2^2$, $\boldsymbol{v}_m=\|\boldsymbol{x}_m\|_2^2$, $\boldsymbol{\ell}_{(n,m)}=\eta(\boldsymbol{x}_n,\boldsymbol{x}_m)$, $\boldsymbol{f}_{(n,m)}=\langle\boldsymbol{x}_n,\boldsymbol{x}_m\rangle$ as well as $\eta(\boldsymbol{x}_n,\boldsymbol{x}_n)=-1$ and $\eta(\boldsymbol{x}_m,\boldsymbol{x}_m)=-1$, i.e. $\boldsymbol{x}_n,\boldsymbol{x}_m\in\mathbb{H}_d$. Applying Schur's theorem [16], we have $\boldsymbol{Q}_{(n,m)}\succeq\boldsymbol{0}$ if and only if the Schur complement

$$\boldsymbol{Q}_{(n,m)}/\boldsymbol{I}_{d+1} = \left[\begin{array}{cc|cc}\boldsymbol{v}_n-\|\boldsymbol{x}_n\|_2^2 & -1-\eta(\boldsymbol{x}_n,\boldsymbol{x}_n) & \boldsymbol{f}_{(n,m)}-\langle\boldsymbol{x}_n,\boldsymbol{x}_m\rangle & \boldsymbol{\ell}_{(n,m)}-\eta(\boldsymbol{x}_n,\boldsymbol{x}_m)\\ -1-\eta(\boldsymbol{x}_n,\boldsymbol{x}_n) & \boldsymbol{v}_n-\|\boldsymbol{x}_n\|_2^2 & \boldsymbol{\ell}_{(n,m)}-\eta(\boldsymbol{x}_n,\boldsymbol{x}_m) & \boldsymbol{f}_{(n,m)}-\langle\boldsymbol{x}_n,\boldsymbol{x}_m\rangle\\ \hline \boldsymbol{f}_{(n,m)}-\langle\boldsymbol{x}_n,\boldsymbol{x}_m\rangle & \boldsymbol{\ell}_{(n,m)}-\eta(\boldsymbol{x}_n,\boldsymbol{x}_m) & \boldsymbol{v}_m-\|\boldsymbol{x}_m\|_2^2 & -1-\eta(\boldsymbol{x}_m,\boldsymbol{x}_m)\\ \boldsymbol{\ell}_{(n,m)}-\eta(\boldsymbol{x}_n,\boldsymbol{x}_m) & \boldsymbol{f}_{(n,m)}-\langle\boldsymbol{x}_n,\boldsymbol{x}_m\rangle & -1-\eta(\boldsymbol{x}_m,\boldsymbol{x}_m) & \boldsymbol{v}_m-\|\boldsymbol{x}_m\|_2^2\end{array}\right]$$

is positive semi-definite. Under the given assumptions, $\boldsymbol{Q}_{(n,m)}/\boldsymbol{I}_{d+1}$ becomes zero and hence positive semi-definite. □

In light of Proposition 1, the Tikhonov-like denoising model (3) is equivalent to

$$\mathop{\arg\min}_{\substack{\boldsymbol{x}\in\mathbb{A}_{d+1}^N,\, \boldsymbol{\ell}\in\mathbb{R}^M,\\ \boldsymbol{f}\in\mathbb{R}^M,\, \boldsymbol{v}\in\mathbb{R}^N}} \mathcal{J}(\boldsymbol{x},\boldsymbol{f},\boldsymbol{v}) \quad \text{s.t.} \quad \left\{\begin{matrix} \boldsymbol{Q}_{(n,m)} \succeq \boldsymbol{0} \\ \operatorname{rank}(\boldsymbol{Q}_{(n,m)}) = d+1 \end{matrix}\right\} \ \forall (n,m) \in E. \tag{4}$$

In the manner of [5], we convexify (4) by neglecting the non-convex rank-constraints. Thus, we propose to solve the convex *relaxed Tikhonov denoising model*:

$$\mathop{\arg\min}_{\substack{\boldsymbol{x}\in\mathbb{A}_{d+1}^N,\, \boldsymbol{\ell}\in\mathbb{R}^M,\\ \boldsymbol{f}\in\mathbb{R}^M,\, \boldsymbol{v}\in\mathbb{R}^N}} \mathcal{J}(\boldsymbol{x},\boldsymbol{f},\boldsymbol{v}) \quad \text{s.t.} \quad \boldsymbol{Q}_{(n,m)} \succeq \boldsymbol{0} \quad \forall (n,m) \in E. \tag{5}$$

Instead of the chosen, rather strong relaxation, an alternative could be to employ the nuclear norm heuristic, i.e. to penalize the objective $\mathcal{J}$ additionally with the trace $\operatorname{tr}(\boldsymbol{Q}_{(n,m)}) = (d+1) + 2(\boldsymbol{v}_n + \boldsymbol{v}_m)$ for each $(n,m) \in E$. Notice that the terms $\boldsymbol{v}_n + \boldsymbol{v}_m$ are included on the right-hand side of (2), which, to some extend, already incorporates the effect of the nuclear norm relaxation into (5).

2.2 TV Regularization

For denoising cartoon-like data, the Tikhonov denoising model in § 2.1 is not suitable. Instead, we consider the variational TV denoising model:

$$\mathop{\arg\min}_{\boldsymbol{x}\in\mathbb{H}_d^N} \frac{1}{2}\sum_{n\in V} \|\boldsymbol{x}_n - \boldsymbol{y}_n\|_2^2 + \mu \operatorname{TV}(\boldsymbol{x}), \text{ were } \operatorname{TV}(\boldsymbol{x}) := \sum_{(n,m)\in E} \|\boldsymbol{x}_n - \boldsymbol{x}_m\|_1 \tag{6}$$

and $\|\cdot\|_1$ denotes the Manhattan norm, and $\mu > 0$ is the regularization parameter. In the manner of [4], where the TV denoising on the hypersphere and SO(3) is considered, we introduce the auxiliary parameter $\boldsymbol{v} \in \mathbb{R}^N$ and the (non-linear) objective

$$\mathcal{K}(\boldsymbol{x},\boldsymbol{v}) := \frac{1}{2}\sum_{n\in V} (\boldsymbol{v}_n - 2\langle \boldsymbol{x}_n, \boldsymbol{y}_n\rangle) + \mu \operatorname{TV}(\boldsymbol{x}). \tag{7}$$

Hence, (6) becomes equivalent to

$$\mathop{\arg\min}_{\boldsymbol{x}\in\mathbb{H}_d^N} \mathcal{K}(\boldsymbol{x},\boldsymbol{v}) \quad \text{s.t.} \quad \boldsymbol{v}_n = \|\boldsymbol{x}_n\|_2^2 \quad \forall n \in V. \tag{8}$$

Applying Proposition 1, we may rewrite (8) as

$$\mathop{\arg\min}_{\substack{\boldsymbol{x}\in\mathbb{A}_{d+1}^N,\boldsymbol{\ell}\in\mathbb{R}^M\\ \boldsymbol{f}\in\mathbb{R}^M,\boldsymbol{v}\in\mathbb{R}^N}} \mathcal{K}(\boldsymbol{x},\boldsymbol{v}) \quad \text{s.t.} \quad \left\{\begin{matrix} \boldsymbol{Q}_{(n,m)} \succeq \boldsymbol{0} \\ \operatorname{rank}(\boldsymbol{Q}_{(n,m)}) = d+1 \end{matrix}\right\} \quad \forall (n,m) \in E. \tag{9}$$

Different from § 2.1, where $\boldsymbol{f}$ is essential to linearize the objective, and where $\boldsymbol{Q}_{(n,m)}$ encodes $\boldsymbol{f}_{(n,m)} = \langle \boldsymbol{x}_n, \boldsymbol{x}_m\rangle$, the auxiliary variables $\boldsymbol{f}$ and $\boldsymbol{\ell}$ here seem to be superfluous. We derive an alternative encoding of the hyperbolic sheet.

Proposition 2. *Let $\boldsymbol{x}_n \in \mathbb{A}_{d+1}$. Then $\boldsymbol{x}_n \in \mathbb{H}_d$ and $\boldsymbol{v}_n = \|\boldsymbol{x}_n\|_2^2$ if and only if*

$$\boldsymbol{V}_n := \left[\begin{array}{c|cc} \boldsymbol{I}_{d+1} & \boldsymbol{x}_n & \tilde{\boldsymbol{x}}_n \\ \hline \boldsymbol{x}_n^* & \boldsymbol{v}_n & -1 \\ \tilde{\boldsymbol{x}}_n^* & -1 & \boldsymbol{v}_n \end{array}\right] \succeq \boldsymbol{0} \quad \textit{and} \quad \operatorname{rank}(\boldsymbol{V}_n) = d+1.$$

The statement can be established in analogy to Proposition 1 using Schur's theorem. On the basis of Proposition 2, the TV denoising model (8) is also equivalent to

$$\operatorname*{arg\,min}_{\boldsymbol{x} \in \mathbb{A}_{d+1}^N, \boldsymbol{v} \in \mathbb{R}^N} \quad \mathcal{K}(\boldsymbol{x}, \boldsymbol{v}) \quad \text{s.t.} \quad \left\{ \begin{array}{c} \boldsymbol{V}_{(n)} \succeq \boldsymbol{0} \\ \operatorname{rank}(\boldsymbol{V}_{(n)}) = d+1 \end{array} \right\} \quad \text{for all} \quad n \in V.$$

Neglecting the rank constraint, we propose to solve the convex *relaxed TV denoising model*:

$$\operatorname*{arg\,min}_{\boldsymbol{x} \in \mathbb{A}_{d+1}^N, \boldsymbol{v} \in \mathbb{R}^N} \quad \mathcal{K}(\boldsymbol{x}, \boldsymbol{v}) \quad \text{s.t.} \quad \boldsymbol{V}_n \succeq \boldsymbol{0} \quad \forall n \in V. \tag{10}$$

Notice that the relaxed version of (9)—without the rank constraint—and the proposed model (10) are equivalent: if $(\hat{\boldsymbol{x}}, \hat{\boldsymbol{\ell}}, \hat{\boldsymbol{f}}, \hat{\boldsymbol{v}})$ is a solution of the relaxed version of (9), then $(\hat{\boldsymbol{x}}, \hat{\boldsymbol{v}})$ is a solution of (10); if $(\hat{\boldsymbol{x}}, \hat{\boldsymbol{v}})$ is a solution of (10), then $(\hat{\boldsymbol{x}}, (\eta(\hat{\boldsymbol{x}}_n, \hat{\boldsymbol{x}}_m))_{(n,m) \in E}, (\langle \hat{\boldsymbol{x}}_n, \hat{\boldsymbol{x}}_m \rangle)_{(n,m) \in E}, \hat{\boldsymbol{v}})$ is a solution of the relaxed version of (9). The positive semi-definiteness can be transferred between $\boldsymbol{Q}_{(n,m)}$, $\boldsymbol{V}_n$, and $\boldsymbol{V}_m$ utilizing Schur's theorem and Sylvester's criterion [16]. Similarly to above, the trace $\operatorname{tr}(\boldsymbol{V}_n) = (d+1) + 2\boldsymbol{v}_n$ occurs on the left-hand side of (7); so that the nuclear norm heuristic is again incorporated into the relaxed formulation (10).

3 Denoising Models and Algorithms

In this section we solve both convex relaxed problems, defined in (5) and (10), employing the alternating directions methods of multipliers (ADMM) [14]. In general, ADMM allows to solve convex minimization problems of the form

$$\operatorname{arg\,min}_{\boldsymbol{Y}, \boldsymbol{U}} \quad \mathcal{F}(\boldsymbol{Y}) + \mathcal{G}(\boldsymbol{U}) \quad \text{s.t.} \quad \mathcal{L}(\boldsymbol{Y}) = \boldsymbol{U}, \tag{11}$$

where $\mathcal{F}$ and $\mathcal{G}$ are convex, and $\mathcal{L}$ is linear, using the iteration:

$$\boldsymbol{Y}^{(k+1)} := \operatorname{arg\,min}_{\boldsymbol{Y}} \quad \mathcal{F}(\boldsymbol{Y}) + \tfrac{\rho}{2} \|\mathcal{L}(\boldsymbol{Y}) - \boldsymbol{U}^{(k)} + \boldsymbol{Z}^{(k)}\|^2, \tag{12a}$$

$$\boldsymbol{U}^{(k+1)} := \operatorname{arg\,min}_{\boldsymbol{U}} \quad \mathcal{G}(\boldsymbol{U}) + \tfrac{\rho}{2} \|\mathcal{L}(\boldsymbol{Y}^{(k+1)}) - \boldsymbol{U} + \boldsymbol{Z}^{(k)}\|^2, \tag{12b}$$

$$\boldsymbol{Z}^{(k+1)} := \boldsymbol{Z}^{(k)} + \mathcal{L}(\boldsymbol{Y}^{(k+1)}) - \boldsymbol{U}^{(k+1)}. \tag{12c}$$

3.1 Solving the Relaxed Tikhonov-Like Denoising Model

To solve (5), we rely on the following splitting:

$$\operatorname*{arg\,min}_{\boldsymbol{x},\boldsymbol{\ell},\boldsymbol{f},\boldsymbol{v},\boldsymbol{U}} \mathcal{J}(\boldsymbol{x},\boldsymbol{f},\boldsymbol{v}) + \mathcal{I}(\boldsymbol{x}) + \mathcal{G}(\boldsymbol{U}) \quad \text{s.t.} \quad \mathcal{Q}(\boldsymbol{x},\boldsymbol{\ell},\boldsymbol{f},\boldsymbol{v}) = \boldsymbol{U} \in \mathcal{S}_{d+5}^{M}, \tag{13}$$

where $\mathcal{S}_{d+5}$ denotes the symmetric $(d+5)\times(d+5)$ matrices. Using the indicator function $\iota_\bullet$, which is zero on the indicated set and $+\infty$ otherwise, we define

$$\mathcal{I}(\boldsymbol{x}) := \sum_{n\in V} \iota_{\mathbb{A}_{d+1}}(\boldsymbol{x}_n) \quad \text{and} \quad \mathcal{G}(\boldsymbol{U}) := \sum_{(n,m)\in E} \iota_{\mathcal{C}}(\boldsymbol{U}_{(n,m)}) \tag{14}$$

with $\boldsymbol{E} := \operatorname{diag}(\boldsymbol{I}_{d+1}, \left[\begin{smallmatrix} 0 & -1 \\ -1 & 0 \end{smallmatrix}\right], \left[\begin{smallmatrix} 0 & -1 \\ -1 & 0 \end{smallmatrix}\right])$ and $\mathcal{C} := \{\boldsymbol{A} \in \mathcal{S}_{d+5} : \boldsymbol{A} \succeq -\boldsymbol{E}\}$. Finally, the linear operator $\mathcal{Q} := (\mathcal{Q}_{(n,m)})_{(n,m)\in E}$ is defined as

$$\mathcal{Q}_{(n,m)}(\boldsymbol{x},\boldsymbol{\ell},\boldsymbol{f},\boldsymbol{v}) := \boldsymbol{Q}_{(n,m)} - \boldsymbol{E}.$$

To apply (12), we define $\|\boldsymbol{U}\|^2 := \sum_{(n,m)\in E}\|\boldsymbol{U}_{(n,m)}\|_{\mathrm{F}}^2$, where $\|\cdot\|_{\mathrm{F}}$ denotes the Frobenius norm. Further, we require the adjoint of $\mathcal{Q}$. Based on the Kronecker delta $\delta_{i,j}$, the adjoint of $\mathcal{Q}$ with respect to the argument $\boldsymbol{x}$ is given by

$$\begin{aligned}\big((\mathcal{Q}_{\boldsymbol{x}}^*(\boldsymbol{U}))_n\big)_i = 2\Big[&\sum_{(n,m)\in E} (\boldsymbol{U}_{(n,m)})_{i,d+2} + (-1)^{\delta_{i,d+1}}(\boldsymbol{U}_{(n,m)})_{i,d+3}\Big] \\ + 2\Big[&\sum_{(m,n)\in E} (\boldsymbol{U}_{(m,n)})_{i,d+4} + (-1)^{\delta_{i,d+1}}(\boldsymbol{U}_{(m,n)})_{i,d+5}\Big]\end{aligned}$$

for $n \in V$ and $1 \le i \le d+1$; the adjoint with respect to $\boldsymbol{\ell}$ and $\boldsymbol{f}$ are given by

$$\begin{aligned}(\mathcal{Q}_{\boldsymbol{\ell}}^*(\boldsymbol{U}))_{(n,m)} &= 2\left[(\boldsymbol{U}_{(n,m)})_{d+5,d+2} + (\boldsymbol{U}_{(n,m)})_{d+4,d+3}\right] \\ (\mathcal{Q}_{\boldsymbol{f}}^*(\boldsymbol{U}))_{(n,m)} &= 2\left[(\boldsymbol{U}_{(n,m)})_{d+4,d+2} + (\boldsymbol{U}_{(n,m)})_{d+5,d+3}\right]\end{aligned}$$

for $(n,m) \in E$; and the adjoint with respect to $\boldsymbol{v}$ is given by

$$\begin{aligned}(\mathcal{Q}_{\boldsymbol{v}}^*(\boldsymbol{U}))_n = \Big[&\sum_{(n,m)\in E} (\boldsymbol{U}_{(n,m)})_{d+2,d+2} + (\boldsymbol{U}_{(n,m)})_{d+3,d+3}\Big] \\ &+ \Big[\sum_{(m,n)\in E} (\boldsymbol{U}_{(m,n)})_{d+4,d+4} + (\boldsymbol{U}_{(m,n)})_{d+5,d+5}\Big].\end{aligned}$$

for $n \in V$. ADMM for (13) has the following explicit form.

Theorem 1. *Let* $\nu_n := |\{m \in V : (n,m) \in E| + |\{m \in V : (m,n) \in E\}|$ *for* $n \in V$. *For* $\rho > 0$, ADMM (12) *applied to* (13) *reads as*

$$\begin{aligned}
\boldsymbol{x}_n^{(k+1)} &:= \operatorname{proj}_{\mathbb{A}_{d+1}}\big(\tfrac{1}{4\nu_n}\,(\mathcal{Q}_{\boldsymbol{x}}^*(\boldsymbol{U}^{(k)} - \boldsymbol{Z}^{(k)})_n + \tfrac{1}{\rho}\,\boldsymbol{y}_n)\big) \quad \forall n \in V,\\
\boldsymbol{\ell}^{(k+1)} &:= \tfrac{1}{4\rho}\,\mathcal{Q}_{\boldsymbol{\ell}}^*(\boldsymbol{U}^{(k)} - \boldsymbol{Z}^{(k)}),\\
\boldsymbol{f}^{(k+1)} &:= \tfrac{1}{4}\,(\mathcal{Q}_{\boldsymbol{f}}^*(\boldsymbol{U}^{(k)} - \boldsymbol{Z}^{(k)}) + \tfrac{\lambda}{\rho}\,\mathbf{1}_M)\\
\boldsymbol{v}_n^{(k+1)} &:= \tfrac{1}{2\nu_n}\,(\mathcal{Q}_{\boldsymbol{v}}^*(\boldsymbol{U}^{(k)} - \boldsymbol{Z}^{(k)})_n - \tfrac{1+\nu_n\lambda}{2\rho}) \quad \forall n \in V,\\
\boldsymbol{U}_{(n,m)}^{(k+1)} &:= \operatorname{proj}_{\succeq \mathbf{0}}\big(\mathcal{Q}_{(n,m)}([\boldsymbol{x},\boldsymbol{\ell},\boldsymbol{f},\boldsymbol{v}]^{(k+1)}) + \boldsymbol{E} + \boldsymbol{Z}_{(n,m)}^{(k)}\big) - \boldsymbol{E} \quad \forall (n,m) \in E,\\
\boldsymbol{Z}^{(k+1)} &:= \boldsymbol{Z}^{(k)} + \mathcal{Q}(\boldsymbol{U}^{(k+1)}) - \boldsymbol{U}^{(k+1)},
\end{aligned}$$

where $\operatorname{proj}_{\mathbb{A}_{d+1}}$ *is the projection onto* $\mathbb{A}_{d+1}$, $\operatorname{proj}_{\succeq \mathbf{0}}$ *onto the positive semi-definite matrices, and* $\mathbf{1}_\bullet$ *is the all ones vector.*

The projections can be computed explicitly, cf. [5, Thm. 5.2] and [8, p. 399]. For $\boldsymbol{\xi} \in \mathbb{R}^{d+1}$ and the eigenvalue decomposition $\boldsymbol{A} = \boldsymbol{V}\boldsymbol{\Sigma}\boldsymbol{V}^* \in \mathcal{S}_{d+5}$, we have

$$\operatorname{proj}_{\mathbb{A}_{d+1}}(\boldsymbol{\xi}) = (\xi_1, \dots, \xi_d, \max\{\xi_{d+1}, 1\})^*, \tag{15a}$$

$$\operatorname{proj}_{\succeq \mathbf{0}}(\boldsymbol{A}) = \boldsymbol{V} \max(\boldsymbol{\Sigma}, 0)\boldsymbol{V}^*. \tag{15b}$$

Proof. The minimizer of (12a) can be easily computed as in [5, Thm. 5.1]. The objective with $\mathcal{F}$ chosen as $\mathcal{J} + \mathcal{I}$ is quadratic and decouples all variables. In order to compute the derivative, we can exploit

$$\mathcal{Q}_{\boldsymbol{x}}^*(\mathcal{Q}(\boldsymbol{x},\boldsymbol{\ell},\boldsymbol{f},\boldsymbol{v}))_n = 4\boldsymbol{x}_n, \qquad \mathcal{Q}_{\boldsymbol{\ell}}^*(\mathcal{Q}(\boldsymbol{x},\boldsymbol{\ell},\boldsymbol{f},\boldsymbol{v}))_{(n,m)} = 4\boldsymbol{\ell}_{(n,m)},$$
$$\mathcal{Q}_{\boldsymbol{f}}^*(\mathcal{Q}(\boldsymbol{x},\boldsymbol{\ell},\boldsymbol{f},\boldsymbol{v}))_{(n,m)} = 4\boldsymbol{f}_{(n,m)}, \qquad \mathcal{Q}_{\boldsymbol{v}}^*(\mathcal{Q}(\boldsymbol{x},\boldsymbol{\ell},\boldsymbol{f},\boldsymbol{v}))_n = 2\boldsymbol{v}_n.$$

For $(\boldsymbol{x}_n)_{d+1}$, we have the constraint $(\boldsymbol{x}_n)_{d+1} \geq 1$, which results in the projection onto $\mathbb{A}_{d+1}$. In 15, $\boldsymbol{U}_{(n,m)}$ are decoupled, and we have

$$\begin{aligned}
\boldsymbol{U}_{(n,m)}^{(k+1)} &= \operatorname*{arg\,min}_{U \succeq -E} \|\mathcal{Q}_{(n,m)}([\boldsymbol{x},\boldsymbol{\ell},\boldsymbol{f},\boldsymbol{v}]^{(k+1)}) - \boldsymbol{U} + \boldsymbol{Z}_{(n,m)}^{(k)}\|_{\mathrm{F}}^2\\
&= \Big[\operatorname*{arg\,min}_{\tilde{U} := U + E \succeq \mathbf{0}} \|(\mathcal{Q}_{(n,m)}([\boldsymbol{x},\boldsymbol{\ell},\boldsymbol{f},\boldsymbol{v}]^{(k+1)}) + \boldsymbol{E} + \boldsymbol{Z}_{(n,m)}^{(k)}) - \tilde{\boldsymbol{U}}\|_{\mathrm{F}}^2\Big] - \boldsymbol{E}\\
&= \operatorname{proj}_{\succeq \mathbf{0}}(\boldsymbol{Q}_{(n,m)}^{(k+1)} + \boldsymbol{Z}_{(n,m)}^{(k)}) - \boldsymbol{E}.
\end{aligned}$$

□

3.2 Solving the Relaxed TV Denoising Model

Similar to § 3.1, we relay on the following splitting, to solve (10):

$$\operatorname*{arg\,min}_{\boldsymbol{x},\boldsymbol{v},\boldsymbol{u},\boldsymbol{U}} \mathcal{K}(\boldsymbol{x},\boldsymbol{v}) + \mathcal{I}(\boldsymbol{u}) + \mathcal{H}(\boldsymbol{U}) \quad \text{s.t.} \quad \boldsymbol{x} = \boldsymbol{u} \text{ and } \mathcal{V}(\boldsymbol{x},\boldsymbol{v}) = \boldsymbol{U} \in \mathcal{S}_{d+3}^N. \tag{16}$$

Here, $\mathcal{I}$ is defined as in (14) and

$$\mathcal{H}(\boldsymbol{U}) := \sum_{n\in V} \iota_{\hat{\mathcal{C}}}(\boldsymbol{U}_n),$$

where $\hat{\boldsymbol{E}} := \operatorname{diag}(\boldsymbol{I}_{d+1}, \left[\begin{smallmatrix} 0 & -1 \\ -1 & 0 \end{smallmatrix}\right])$ and $\hat{\mathcal{C}} := \{\boldsymbol{A} \in \mathcal{S}_{d+3} : \boldsymbol{A} + \hat{\boldsymbol{E}} \succeq \boldsymbol{0}\}$. The linear operator $\mathcal{V} := (\mathcal{V}_n)_{n\in V}$ is defined as

$$\mathcal{V}_n(\boldsymbol{x}, \boldsymbol{v}) := \boldsymbol{V}_n - \hat{\boldsymbol{E}}.$$

To apply (12), we define $\|\boldsymbol{U}\|^2 := \sum_{n\in V} \|\boldsymbol{U}_n\|_{\mathrm{F}}^2$ and derive the adjoint of $\mathcal{V}$ that—with respect to the arguments $\boldsymbol{x}$ and $\boldsymbol{v}$—is given by

$$\big((\mathcal{V}_{\boldsymbol{x}}^*(\boldsymbol{U}))_n\big)_i = 2\left[(\boldsymbol{U}_n)_{i,d+2} + (-1)^{\delta_{i,d+1}}(\boldsymbol{U}_n)_{i,d+3}\right], \tag{17a}$$

$$(\mathcal{V}_{\boldsymbol{v}}^*(\boldsymbol{U}))_n = (\boldsymbol{U}_n)_{d+2,d+2} + (\boldsymbol{U}_n)_{d+3,d+3} \tag{17b}$$

for $n \in V$ and $1 \le i \le d+1$. ADMM for (16) has the following form.

Theorem 2. *For* $\rho > 0$, ADMM (12) *applied to* (16) *reads as*

$$\begin{aligned}
\boldsymbol{x}^{(k+1)} &:= \operatorname{prox}_{\mu/5\rho\,\mathrm{TV}}\big(\tfrac{1}{5}[\mathcal{V}_{\boldsymbol{x}}^*(\boldsymbol{U}^{(k)} - \boldsymbol{Z}^{(k)}) + \tfrac{1}{\rho}\boldsymbol{y} + (\boldsymbol{u}^{(k)} - \boldsymbol{z}^{(k)})]\big),\\
\boldsymbol{v}^{(k+1)} &:= \tfrac{1}{2\rho}\big(\rho\mathcal{V}_{\boldsymbol{v}}^*(\boldsymbol{U}^{(k)} - \boldsymbol{Z}^{(k)}) - \tfrac{1}{2}\boldsymbol{1}_N\big),\\
\boldsymbol{U}_n^{(k+1)} &:= \operatorname{proj}_{\succeq \boldsymbol{0}}(\boldsymbol{V}_n^{(k+1)} + \boldsymbol{Z}_n^{(k)}) - \hat{\boldsymbol{E}} \quad \forall n \in V,\\
\boldsymbol{u}_n^{(k+1)} &:= \operatorname{proj}_{\mathbb{A}_{d+1}}(\boldsymbol{x}_n^{((k+1)} + \boldsymbol{z}_n^{(k)}) \quad \forall n \in V,\\
\boldsymbol{Z}^{(k+1)} &:= \boldsymbol{Z}^{(k)} + \mathcal{V}([\boldsymbol{x}, \boldsymbol{v}]^{(k+1)}) - \boldsymbol{U}^{(k+1)},\\
\boldsymbol{z}^{(k+1)} &:= \boldsymbol{z}^{(k)} + \boldsymbol{x}^{(k+1)} - \boldsymbol{u}^{(k+1)},
\end{aligned}$$

where $\operatorname{prox}_f := \arg\min_{\boldsymbol{y}}\{f(\boldsymbol{y}) + \frac{1}{2}\|\cdot - \boldsymbol{y}\|^2\}$ *for convex, lower semi-continuous, proper* f.

The projection onto $\mathbb{A}_{d+1}$ and the positive semi-definite matrices is known from (15). Further, we rely on an anisotropic TV regularization. Therefore, we can apply the fast TV programs [10,11] coordinatewise to efficiently compute $\operatorname{prox}_{\mu/4\rho\,\mathrm{TV}}$ for line and grid graphs, i.e. for signals and images, cf. [4, Rem. 4.1].

Proof. In the proof, we denote by $[\![\boldsymbol{A}]\!]_j$ the first $d+1$ entries of the j-th column of $\boldsymbol{A} \in \mathcal{S}_{d+3}$. Furthermore, let $\boldsymbol{e} := (1, \cdots, 1, -1)^* \in \mathbb{R}^{d+1}$. Then, the adjoint in (17a) may be written as

$$(\mathcal{V}_{\boldsymbol{x}}^*(\boldsymbol{U}))_n = 2\left([\![\boldsymbol{U}_n]\!]_{d+2} + \boldsymbol{e} \odot [\![\boldsymbol{U}_n]\!]_{d+3}\right) \quad \forall n \in V,$$

where $\odot$ denotes the Hadamard product. In the first ADMM step (12a) for our splitting (16), the minimization with respect to $\boldsymbol{x}$ and $\boldsymbol{v}$ decouples. Using the symmetry of $\boldsymbol{U}^{(k)}$ and $\boldsymbol{Z}^{(k)}$, exploiting $\tilde{\boldsymbol{x}}_n = \boldsymbol{e} \odot \boldsymbol{x}_n$, and combining the squared norms and inner products, we can compute the minimizer with respect to $\boldsymbol{x}$ by

$$\begin{aligned}
&\operatorname*{arg\,min}_{\boldsymbol{x} \in (\mathbb{R}^{d+1})^N} \Big\{ \tfrac{1}{2} \sum_{n \in V} (v_n - 2\langle \boldsymbol{x}_n, \boldsymbol{y}_n \rangle) + \mu \operatorname{TV}(\boldsymbol{x}) + \tfrac{\rho}{2} \sum_{n \in V} \|\boldsymbol{x}_n - \boldsymbol{u}_n^{(k)} + \boldsymbol{z}_n^{(k)}\|_2^2 \\
&\qquad\qquad + \tfrac{\rho}{2} \sum_{n \in V} \|\mathcal{V}_n(\boldsymbol{x}, \boldsymbol{v}) - \boldsymbol{U}_n^{(k)} + \boldsymbol{Z}_n^{(k)}\|_{\mathrm{F}}^2 \Big\} \\
&= \operatorname*{arg\,min}_{\boldsymbol{x} \in (\mathbb{R}^{d+1})^N} \Big\{ \sum_{n \in V} \langle \boldsymbol{x}_n, \boldsymbol{y}_n \rangle + \mu \operatorname{TV}(\boldsymbol{x}) + \rho \sum_{n \in V} \Big[\tfrac{1}{2} \|\boldsymbol{x}_n - \boldsymbol{u}_n^{(k)} + \boldsymbol{z}_n^{(k)}\|_2^2 \\
&\quad + \|\boldsymbol{x}_n - [\![\boldsymbol{U}_n^{(k)}]\!]_{d+2} + [\![\boldsymbol{Z}_n^{(k)}]\!]_{d+2}\|_2^2 + \|\tilde{\boldsymbol{x}}_n - [\![\boldsymbol{U}_n^{(k)}]\!]_{d+3} + [\![\boldsymbol{Z}_n^{(k)}]\!]_{d+3}\|_2^2 \Big] \Big\} \\
&= \operatorname*{arg\,min}_{\boldsymbol{x} \in (\mathbb{R}^{d+1})^N} \Big\{ - \sum_{n \in V} \langle \boldsymbol{x}_n, \boldsymbol{y}_n \rangle + \mu \operatorname{TV}(\boldsymbol{x}) \\
&\quad + \tfrac{5\rho}{2} \sum_{n \in V} \|\boldsymbol{x}_n - \tfrac{1}{5} [\mathcal{V}_{\boldsymbol{x}}^*(\boldsymbol{U} - \boldsymbol{Z})_n + (\boldsymbol{u}^{(k)} - \boldsymbol{z}^{(k)})_n]\|_2^2 \Big\} \\
&= \operatorname*{arg\,min}_{\boldsymbol{x} \in (\mathbb{R}^{d+1})^N} \Big\{ \tfrac{5\rho}{2} \sum_{n \in V} \|\boldsymbol{x}_n - \tfrac{1}{5} [\mathcal{V}_{\boldsymbol{x}}^*(\boldsymbol{U} - \boldsymbol{Z})_n + \tfrac{\boldsymbol{y}_n}{\rho} + (\boldsymbol{u}_n - \boldsymbol{z}_n)]\|_2^2 + \mu \operatorname{TV}(\boldsymbol{x}) \Big\},
\end{aligned}$$

Because of the differentiability of the objective with respect to $\boldsymbol{v}$, and since $\mathcal{V}_{\boldsymbol{v}}^*(\mathcal{V}(\boldsymbol{x}, \boldsymbol{v})) = 2\boldsymbol{v}$, the minimizer is given by

$$\begin{aligned}
\mathbf{0}_N &= \tfrac{1}{2} \mathbf{1}_N + \rho \mathcal{V}_{\boldsymbol{v}}^*(\mathcal{V}_{\boldsymbol{v}}(\boldsymbol{x}, \boldsymbol{v}) - \boldsymbol{U}^{(k)} + \boldsymbol{Z}^{(k)}), \quad \text{and thus} \\
\boldsymbol{v}^{(k+1)} &= \tfrac{1}{2} [\mathcal{V}_{\boldsymbol{v}}^*(\boldsymbol{U}^{(k)} - \boldsymbol{Z}^{(k)}) - \tfrac{1}{2\rho} \mathbf{1}_N].
\end{aligned}$$

The two subproblems with respect to $\boldsymbol{U}$ and $\boldsymbol{u}$ of the second ADMM step (12b) regarding (16) can be treated similarly to the proof of Theorem 1 yielding the projection onto the positive semi-definite matrices and $\mathbb{A}_{d+1}$. □

4 Numerical Experiments

The derived algorithms in Theorem 1 and 2 are implemented[1] in Python 3.11.4 using Numpy 1.25.0 and Scipy 1.11.1. The experiments are performed on an off-the-shelf iMac 2020 with Apple M1 Chip (8Core CPU, 3.2 GHz) and 8 GB RAM. The convergence of both algorithms is ensured by [3, Cor. 28.3]. For all experiments, the methods are initialized by zeros for all variables. The resulting implementations are very similar to the ones proposed for sphere- and SO(3)-valued data in [4,5].

Synthetic Image Processing. In this subsection, we give a proof of the concept for both models—the Tikhonov-like and TV model—based on synthetic data.

[1] The code is available at GitHub: https://github.com/JJEWBresch/relaxed_tikhonov_regularization/.

For the ground truth $\boldsymbol{x}_{\text{true}}$ of the two experiments, see Fig. 1, we interpolate points in $\mathbb{R}$ or $\mathbb{R}^2$ and apply the smooth mappings

$$\boldsymbol{x}_{\mathbb{H}_1} : \mathbb{R} \to \mathbb{H}_1, \qquad r \mapsto (\sinh(r), \cosh(r))^*, \tag{18a}$$

$$\boldsymbol{x}_{\mathbb{H}_2} : \mathbb{R}^2 \to \mathbb{H}_2, \qquad (r, s) \mapsto (\sinh(r)\cos(s), \sinh(r)\sin(s), \cosh(r))^*. \tag{18b}$$

In the first experiment, where an one-dimensional signal of length 400, i.e. a signal on a line graph, is considered, the noisy measurements $\boldsymbol{y} \in \mathbb{H}_1^{400}$ are generated employing the tangential normal distribution [19] with standard deviation $\sigma := 0.6$, see Fig. 1 (top). The $\mathbb{H}_1$-valued signals are here illustrated via the parameter r in (18a). Applying ADMM from Theorem 1 and 2 for the Tikhonov-like and TV model respectively, we observe convergence to the hyperbolic sheet. In both cases, the mean absolute error (MAE) with respect to the Minkowski bilinear form $\eta \equiv -1$ is 10^{-4} at the most. The restored signals after 12 sec. and the used parameters are recorded in Fig. 1 (top).

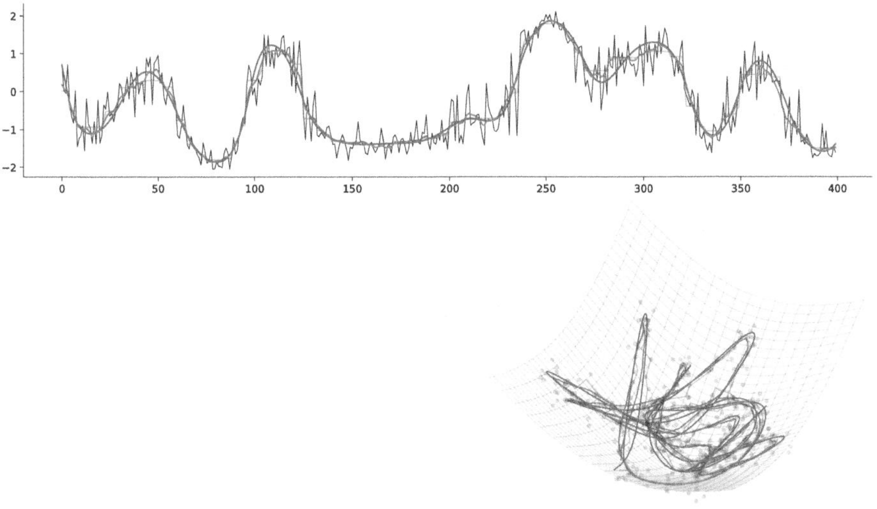

Fig. 1. Restoration of smooth synthetic line signals (blue) from noisy measurements (black/gray). For the Tikhonov model (red), we choose $\rho = 10^{-1}$, $\lambda = 6$ (top) and $\lambda = 5$ (right). For the TV model (green), we choose $\rho = 1$, $\mu = 0.75$ (top), and $\mu = 0.1$ (right). Note that the $\mathbb{H}_1$ signal (top) is visualized using the parametrization in (18a). The $\mathbb{H}_2$ signal (right) is directly visualized on the hyperbolic sheet in $\mathbb{A}_3 \subset \mathbb{R}^3$. (Color figure online)

In the second experiment, we consider a line signal of length 400 with values in $\mathbb{H}_2$. The noisy measurements are generated via additive Gaussian noise, i.e. $\boldsymbol{y} \sim \mathcal{N}(\boldsymbol{x}_{\text{true}}, \sigma^2 \boldsymbol{I})$ with $\sigma := 0.3$. Note that the measurements lie in $\mathbb{R}^3$ and not necessarily on $\mathbb{H}_2$. The denoised signals using ADMM for both models are shown in Fig. 1 (right). After 80 sec., we recorded an MAE with respect to $\eta \equiv -1$ of

10^{-5} at the most. Because of convergence back to the hyperbolic sheet, the numerical solutions are actual solutions of the non-convex model (1) and (6) respectively.

Fig. 2. Empirical mean (top) and standard deviation (bottom) for the Brandenburg Gate image. Here, $K = 20$ shots are randomly sampled by adding white noise with $\sigma = 0.15$. The empirical quantities are shown on the left, the Tikhonov denoised images (Theorem 1, $\rho = 10$, $\lambda = 4$) in the middle, and the TV denoised images (Theorem 2, $\rho = 1$, $\mu = 0.6$) on the right.

Gaussian Image Processing. The idea behind Gaussian image processing is that, instead of a single image, we measure a series of K noisy instances, where we assume that each pixel $x_{i,j}^{(k)}$ follows a Gaussian distribution. The empirical mean and variance may be estimated by the maximum likelihood estimators

$$\hat{\mu}_{ij} := \frac{1}{K}\sum_{k=1}^{K} x_{i,j}^{(k)}, \qquad \hat{\sigma}_{ij}^2 := \frac{1}{K}\sum_{k=1}^{K}(x_{i,j}^{(k)} - \hat{\mu}_{ij})^2. \tag{19}$$

The set of Gaussians $\mathcal{N}(\mu,\sigma)$ parameterized by $(\mu,\sigma) \in \mathbb{R} \times \mathbb{R}_{>0}$ and equipped with the so-called Fisher metric is isometric to the hyperbolic space $\mathbb{H}_2$. More precisely, the isometry is given by $\pi_3 \circ \pi_2 \circ \pi_1 : \mathcal{N} \to \mathbb{H}_2$ with

$$\pi_1 : \mathcal{N} \to \mathbb{P}_2, \qquad (\mu,\sigma) \mapsto \tfrac{1}{\sqrt{2}}(\mu, \sqrt{2}\sigma)^*, \tag{20a}$$

$$\pi_2 : \mathbb{P}_2 \to \mathbb{D}_2, \qquad \boldsymbol{y} \mapsto \tfrac{1}{y_1^2 + (y_2+1)^2}(2y_1, \|\boldsymbol{y}\|_2^2 - 1)^*, \tag{20b}$$

$$\pi_3 : \mathbb{D}_2 \to \mathbb{H}_2, \qquad \boldsymbol{y} \mapsto \tfrac{1}{1-\|\boldsymbol{y}\|_2^2}(2y_1, 2y_2, 1 + \|\boldsymbol{y}\|_2^2)^*, \tag{20c}$$

where $\mathbb{D}_2 := \{\boldsymbol{x} \in \mathbb{R}^2 : \|\boldsymbol{x}\|_2 < 1\}$ is the Pioncaré disc, and the Pioncaré half-space is given by $\mathbb{P}_2 := \{\boldsymbol{x} \in \mathbb{R}^2 : x_2 > 0\}$, see [6,20].

In the first experiment, as proof of concept, we consider a gray-valued image of the Brandenburg Gate and generate a series of 20 images by adding Gaussian white noise with constant standard deviation. Calculating the empirical mean

Table 1. Comparison of the TV denoising (Theorem 2) and PDRA (Douglas–Rachford) [6] regarding the SNR for the restored mean and standard deviation. The marked instance ($\mu = 0.15$) is visualized in Fig. 3.

ADMM from Theorem 2 ($\rho = 1$)						PDRA ($\eta = 0.5, \lambda = 0.9$)			
SNR($\hat{\mu}$)	SNR($\hat{\sigma}$)	μ	$\mathcal{K}$	MAE $\eta \equiv -1$	time (sec.)	SNR($\hat{\mu}$)	SNR($\hat{\sigma}$)	α	time (sec.)
6.121	12.211	0.05	$2.61 \cdot 10^5$	$4.73 \cdot 10^{-4}$	160	5.963	12.400	0.1	120
6.001	12.369	0.10	$2.54 \cdot 10^5$	$4.78 \cdot 10^{-4}$	180	5.911	12.618	0.3	140
5.986	**12.301**	0.15	$1.81 \cdot 10^5$	$4.47 \cdot 10^{-4}$	190	5.874	12.748	0.5	280
5.966	12.455	0.20	$1.19 \cdot 10^5$	$4.67 \cdot 10^{-4}$	220	5.843	12.849	0.7	430

and variance pixelwise, and using the isometry in (20), we obtain an $\mathbb{H}_2$-valued image. Afterwards we denoise mean and variance simultaneously using the algorithms from Theorem 1 and 2. The results are recorded in Fig. 2. Analyzing the qualitative outcomes, we conclude that both models are adequate ansätze. The Tikhonov-like model, however, overestimates the standard deviation.

In the second experiment, we consider a real-world application of Gaussian image processing, where 20 highly noisy images of a retina are given. The experiment originates from [1], and the goal is again to denoise mean and variance simultaneously. The qualitative results of the Tikhonov-like and TV denoising are shown in Fig. 3. The Tikhonov-like model again overestimates the standard deviation. During the denoising, we observe numerical convergence to the hyperbolic sheet. More precisely, we reach an MAE of at most 10^{-4} with respect to $\eta \equiv -1$ in 15 min. for the Tikhonov-like model and in 3 min. for the TV model.

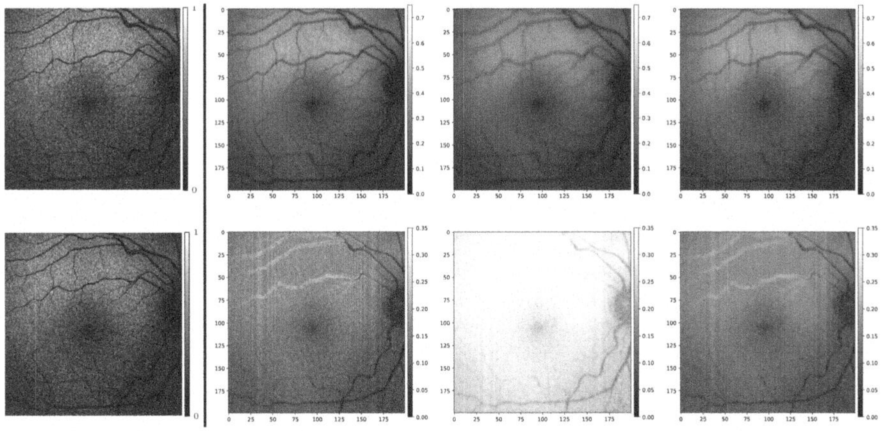

Fig. 3. Clearing the estimated mean and standard deviation of $K = 20$ retina scans. *Leftmost:* the 1st (top) and 20th (bottom) image of the retina scans. *From left to right:* the empirical mean (top) and standard deviation (bottom); Tikhonov denoised images (Theorem 1, $\lambda = 1.5$, $\rho = 10$); TV denoised images (Theorem 2, $\mu = 0.15$, $\rho = 1$).

The numerical solutions are thus solutions of the non-convex models (1) and (6), too. For the TV proximity mapping in Theorem 2, we use the TV program (ADRA) [10] with $\gamma = 1$ as subprocedure. A quantitative comparison with the Proximal Douglas–Rachford Algorithm (PDRA) [6] for more generally manifold-valued data is given in Table 1. Our relaxed Euclidean models here yield comparable results with respect to the SNR (signal-to-noise ratio) and the computation time.

5 Conclusion

In this paper we aim to denoise hyperbolic-valued data. For this, we propose a variational Tikhonov and TV model, which employs the distance of the surrounding Euclidean space instead of the geodesic distance. The non-convexity of the hyperbolic space is encoded using specific positive semi-definite matrices with fixed rank. After a relaxation the problem is solved by ADMM. One of the main applications is the denoising task in Gaussian image processing, which, for instance, occurs in the study of highly noisy data series. In the numerical experiments, we observe the convergence of the iterates to the hyperbolic sheet; so the computed solution of the relaxed problem is actually a solution of the original, non-convex formulation. A rigorous study of this behavior is left for future work, as well as the adaption of this relaxation to other non-convex manifolds.

Acknowledgements. We would like to thank R. Bergmann, J. Persch and G. Steidl for providing the data of the retina experiment, and G. Steidl for drawing the attention to this experiments. Originally, the data was provided by J. Angulo. Furthermore, we thank the anonymous reviewers for their valuable comments.

References

1. Angulo, J., Velasco-Forero, S.: Morphological processing of univariate gaussian distribution-valued images based on Poincaré upper-half plane representation. In: Nielsen, F. (ed.) Geometric Theory of Information. SCT, pp. 331–366. Springer, Cham (2014). https://doi.org/10.1007/978-3-319-05317-2_12
2. Atigh, M.G., Schoep, J., Acar, E., Van Noord, N., Mettes, P.: Hyperbolic image segmentation. In: Proceedings of the CVPR '22, pp. 4443–4452 (2022). https://doi.org/10.1109/CVPR52688.2022.00441
3. Bauschke, H.H., Combettes, P.L.: Convex Analysis and Monotone Operator Theory in Hilbert Spaces. Springer, New York (2011). https://doi.org/10.1007/978-1-4419-9467-7
4. Beinert, R., Bresch, J.: Denoising sphere-valued data by relaxed total variation regularization. Proc. Appl. Math. Mech. (2024). https://doi.org/10.1002/pamm.202400016
5. Beinert, R., Bresch, J., Steidl, G.: Denoising of sphere- and SO(3)-valued data by relaxed Tikhonov regularization. Inverse Probl. Imaging (2024). https://doi.org/10.3934/ipi.2024026

6. Bergmann, R., Persch, J., Steidl, G.: A parallel Douglas-Rachford algorithm for minimizing ROF-like functionals on images with values in symmetric Hadamard manifolds. SIAM J. Imaging Sci. **9**(3), 901–937 (2016). https://doi.org/10.1137/15M1052858
7. Boguñá, M., Papadopoulos, F., Krioukov, D.: Sustaining the internet with hyperbolic mapping. Nat. Commun. **1**(1), 62 (2010). https://doi.org/10.1038/ncomms1063
8. Boyd, S., Vandenberghe, L.: Convex Optimization. Cambridge University Press, Cambridge (2004). https://doi.org/10.1017/CBO9780511804441
9. Chossat, P., Faugeras, O.: Hyperbolic planforms in relation to visual edges and textures perception. PLoS Comput. Biol. **5**(12), 1–16 (2009). https://doi.org/10.1371/journal.pcbi.1000625
10. Condat, L.: A direct algorithm for 1D total variation denoising (2012). https://hal.science/hal-00675043
11. Condat, L.: A direct algorithm for 1D total variation denoising. IEEE Signal Process. Lett. **20**(11), 1054–1057 (2013). https://doi.org/10.1109/LSP.2013.2278339
12. Condat, L.: Tikhonov regularization of circle-valued signals. IEEE Trans. Signal Process. **70**, 2775–2782 (2022). https://doi.org/10.1109/TSP.2022.3179816
13. Ganea, O.E., Bécigneul, G., Hofmann, T.: Hyperbolic neural networks. In: Advances in Neural Information Processing Systems, pp. 5350—5360. Curran Associates, New York (2018). https://proceedings.neurips.cc/paper_files/paper/2018/file/dbab2adc8f9d078009ee3fa810bea142-Paper.pdf
14. Glowinski, R., Marroco, A.: Sur l'approximation, par éléments finis d'ordre un, et la résolution, par pénalisation-dualité d'une classe de problèmes de dirichlet non linéaires. R.A.I.R.O. Analyse Numérique **9**(R2), 41–76 (1975). https://doi.org/10.1051/m2an/197509R200411
15. Hopf, H.: Zum Clifford-Kleinschen Raumproblem. Math. Ann. **95**, 313–339 (1926)
16. Horn, R.A., Johnson, C.R.: Matrix Analysis. Cambridge University Press, Cambridge, 2nd edn. (2012). https://doi.org/10.1017/CBO9780511810817
17. Killing, W.: Ueber die Clifford-Klein'schen Raumformen. Math. Ann. **39**, 257–278 (1891)
18. Krioukov, D., Papadopoulos, F., Kitsak, M., Vahdat, A., Boguñá, M.: Hyperbolic geometry of complex networks. Phys. Rev. E **82**, 036106 (2010). https://doi.org/10.1103/PhysRevE.82.036106
19. Laus, F.J.: Statistical Analysis and Optimal Transport for Euclidean and Manifold-Valued Data. Dr. Hut, München (2020)
20. Lee, J.: Riemannian Manifolds: An Introduction to Curvature. Springer, New York (1997). https://doi.org/10.1007/b98852
21. Lin, F., Bai, B., Guo, Y., Chen, H., Ren, Y., Xu, Z.: MHCN: A hyperbolic neural network model for multi-view hierarchical clustering. In: Proceedings of the ICCV '23, pp. 16479–16489 (2023). https://doi.org/10.1109/ICCV51070.2023.01515
22. Liu, Q., Nickel, M., Kiela, D.: Hyperbolic graph neural networks. In: Advances in Neural Information Processing Systems. vol. 32. Curran Associates (2019). https://proceedings.neurips.cc/paper_files/paper/2019/file/103303dd56a731e377d01f6a37badae3-Paper.pdf
23. Nickel, M., Kiela, D.: Poincaré embeddings for learning hierarchical representations. In: Advances in Neural Information Processing Systems, vol. 30. Curran Associates (2017). https://proceedings.neurips.cc/paper_files/paper/2017/file/59dfa2df42d9e3d41f5b02bfc32229dd-Paper.pdf
24. Wald, R.M.: General Relativity. Chicago University Press, Chicago (1984). https://doi.org/10.7208/chicago/9780226870373.001.0001

Fast Inexact Bilevel Optimization for Analytical Deep Image Priors

Mohammad Sadegh Salehi[1(✉)], Tatiana A. Bubba[2], and Yury Korolev[1]

[1] Department of Mathematical Sciences, University of Bath, Bath BA27AY, UK
{mss266,ymk30}@bath.ac.uk

[2] Department of Mathematics and Computer Science, University of Ferrara, Ferrara, Italy
tatiana.bubba@unife.it

Abstract. The analytical deep image prior (ADP) introduced by Dittmer et al. (2020) establishes a link between deep image priors and classical regularization theory via bilevel optimization. While this is an elegant construction, it involves expensive computations if the lower-level problem is to be solved accurately. To overcome this issue, we propose to use adaptive inexact bilevel optimization to solve ADP problems. We discuss an extension of a recent inexact bilevel method called the method of adaptive inexact descent of Salehi et al. (2024) to an infinite-dimensional setting required by the ADP framework. In our numerical experiments we demonstrate that the computational speed-up achieved by adaptive inexact bilevel optimization allows one to use ADP on larger-scale problems than in the previous literature, e.g. in deblurring of 2D color images.

Keywords: Data-driven regularization · regularization by architecture · inexact bilevel optimization · semi-blind deblurring

1 Introduction

The Deep Image Prior (DIP) was introduced in [15] as an unsupervized learning technique for image processing tasks such as denoising, super-resolution, and inpainting. The regularity of the reconstruction is achieved by a combination of a network architecture and early stopping during training. This method was later called *regularization by architecture* in [9]. The analysis of the DIP approach has received a fair deal of interest in the literature: [3] interpreted it as the optimization of an analytical prior such as Total Variation; [5] studied convergence of the method and derived error bounds; a Bayesian viewpoint was taken in [8]; and reconstruction capabilities were analyzed in [11].

Using the similarity to classical regularization methods such as Landweber iteration, [9] proposed a bilevel optimization formulation of the DIP approach for an unrolled proximal gradient architecture (we postpone the precise formulation

Supplementary Information The online version contains supplementary material available at https://doi.org/10.1007/978-3-031-92366-1_3.

T. A. Bubba et al. (Eds.): SSVM 2025, LNCS 15667, pp. 30–42, 2025.
https://doi.org/10.1007/978-3-031-92366-1_3

to Sect. 2). This reformulation—albeit restricted to a particular architecture—allowed carrying out a theoretical analysis in the spirit of regularization theory. This formulation was called the *analytic deep prior* (ADP) [9]. The ADP was further analyzed in [2], where equivalence to classical Ivanov regularization was established. Furthermore, it was emphasized that early stopping plays a crucial role in the original DIP approach, which, if applied to the ADP, would break the equivalence with Ivanov regularization. To mimic this effect, [2] proposed a Sobolev-regularized version of the ADP, cf. (2.4).

Although the bilivel formulation of the ADP has theoretical advantages, it is hard to solve numerically. Two main approaches have been used, one based on the implicit function theorem (IFT) and another on algorithm unrolling [1, 10]. While the IFT approach has convergence guarantees, in practice it requires a large number of iterations of the lower-level problem to achieve the desired accuracy, which have to be repeated at every iteration of the upper-level problem. Unrolling is faster, but lacks convergence guarantees. These alternatives were tested within the ADP framework in both [2] and [9], albeit only for simple 1D signals.

Recent advances in bilevel optimization, however, have produced methods that do not require solving the lower-level problem to high accuracy and can tolerate a controlled error in the lower-level solution while retaining convergence guarantees. These methods present a computationally efficient alternative to the IFT approach for the ADP problem. In this work, we propose to use the Method of Adaptive Inexact Descent (MAID), recently introduced in [13], for the Sobolev-regularized ADP problem (2.4). On the one hand, MAID does not require solving the lower-level problem exactly and thus avoids prohibitively expensive numbers of iterations in the lower-level problem. On the other hand, it comes with convergence guarantees. The considerable computational speed-up offered by inexact bilevel optimization opens up the possibility to test ADP on more challenging problems, such as deblurring of 2D color images.

The remainder of this paper is organized as follows. In Sect. 2, we formulate the problem and provide a summary of existing results. In Sect. 3, we recap the inexact bilevel algorithm from [13] and extend the convergence proof from the Euclidean setting to that of separable Hilbert spaces. We discuss the application of MAID to ADP and its limitations. Next, in Sect. 4 we compare our results with those in [2] on deblurring of 1D signals (demonstrating a significant computational speed-up) and extend our numerical experiments to various deblurring scenarios of 2D color images.

2 Problem Statement and Summary of Existing Results

We consider linear inverse problems given by

$$Ax = y, \tag{2.1}$$

where $A\colon X \to Y$ is an injective forward operator acting between separable Hilbert spaces X and Y. The exact right-hand side y is assumed to be unavailable; as it is customary in deterministic regularization methods, we assume that we have access to $y^\delta \in Y$ such that $\|y - y^\delta\| \leq \delta$, where $\delta > 0$ is the error magnitude. We denote by $x^\dagger$ the solution of the noise-free problem (2.1).

The analytic deep prior approach (ADP) [9] seeks to approximate the solution of (2.1) by a solution of the following bilevel problem

$$\min_{B \in \mathcal{L}(X,Y)} \left\|A\hat{x}(B) - y^\delta\right\|^2 \tag{2.2a}$$

$$\hat{x}(B) := \arg\min_x \left\|Bx - y^\delta\right\|^2 + \alpha \mathcal{J}(x), \tag{2.2b}$$

where $\mathcal{J} \colon X \to \mathbb{R} \cup \{+\infty\}$ is a regularization functional chosen so that the minimizer in the lower-level problem exists and is unique, and $\alpha > 0$ a regularization parameter. The lack of coercivity of the upper-level problem makes the question of the existence of minimizers non-obvious; however, it was shown in [2] that under certain conditions (2.2) can be reformulated as a classical Ivanov regularization problem (with a modified regularization term), for which existence of solutions can be shown. Under the additional assumption that $\mathcal{J}(x) = \|x\|^2$, the authors were also able to prove stability with respect to the data y^δ.

In practice, however, solutions of (2.2) are never computed because early stopping is used. To mimic this effect, [2] proposed to add a regularization term to the upper-level problem. The most obvious candidate, the operator norm $\|B - A\|_{\mathcal{L}(X,Y)}$, fails to guarantee the existence of minimisers. Roughly speaking, the reason is that regularization with $\|B - A\|_{\mathcal{L}(X,Y)}$ ensures convergence of the minimizing sequence only in the weak operator topology, with respect to which the term $\left\|Bx - y^\delta\right\|^2$ isn't continuous. This prevents the term $\left\|A\hat{x}(B) - y^\delta\right\|^2$ in the upper-level problem from being lower-semicontinuous and hinders the application of the direct method of the calculus of variations.

To ensure the existence of minimizers in the upper-level problem, the following modification was proposed in [2]. Let $\Omega \subset \mathbb{R}^n$ be an open bounded Lipschitz domain, $X = Y = L^2(\Omega)$, and A an integral operator with a kernel $a \in L^2(\Omega \times \Omega)$

$$Ax(t) = \int_\Omega a(t,\tau)x(\tau)\,\mathrm{d}\tau =: [T(a,x)](t), \tag{2.3}$$

where T is a bilinear operator. By restricting oneself to operators with Sobolev kernels $b \in H^1(\Omega \times \Omega)$, one can make use of the the compact embedding $H^1 \subset\subset L^2$ (valid for any n). This leads to the following bilevel problem

$$\min_{b \in H^1(\Omega \times \Omega)} \left\|T(a, \hat{x}(b)) - y^\delta\right\|_{L^2}^2 + \beta \|b - a\|_{H^1}^2 \tag{2.4a}$$

$$\hat{x}(b) := \arg\min_x \left\|T(b,x) - y^\delta\right\|_{L^2}^2 + \alpha \mathcal{J}(x), \tag{2.4b}$$

where $a \in H^1(\Omega \times \Omega)$ is the kernel of the forward operator A. This problem was called ADP-β in [2]; its efficient solution is the subject of the current paper. Existence, stability, and convergence rates for solutions of this modified problem can be shown with standard arguments, see [2, Sec. 3.4].

3 Inexact Bilevel Optimization

In its standard form (i.e., not reformulated as an Ivanov or Tikhonov problem, as suggested in [2]), the ADP is a bilevel optimization problem. Research has primarily focused on two main approaches: the use of the implicit function theorem (IFT) and algorithm unrolling [9]. While the IFT approach offers convergence guarantees, in practice it requires a large number of iterations of the lower-level problem to compute the exact solution $\hat{x}(B)$, as well as its gradient with respect to B. These computations must be repeated at every iteration of the upper-level problem. Achieving such high accuracy in solving the lower-level problem can be computationally prohibitive. Moreover, when approximations are used, errors in solving the lower-level problem can propagate to the gradient obtained via IFT, potentially compromising the convergence of the bilevel scheme. On the other hand, algorithm unrolling uses a predefined number of iterations in the lower-level problem to obtain an approximation of $\hat{x}(B)$, with gradients computed via backpropagation. Although this approach is more practical than IFT with an exact lower-level solution, the accuracy of the approximation cannot be controlled, and convergence guarantees are lost.

Inexact bilevel optimization methods (e.g., [12,13]), however, can tolerate a controlled amount of error in the solution of the lower-level problem (and its gradients) while retaining convergence guarantees, hence presenting an attractive alternative to the IFT approach in ADP. In the next section we summarize the inexact bilevel method with adaptive step sizes from [13], which we propose to use for solving the ADP-β problem (2.4).

3.1 Method of Adaptive Inexact Descent

Consider the following bilevel problem

$$\min_{b\in H}\{g(\hat{x}(b)) + r(b)\} \qquad \text{s.t.} \quad \hat{x}(b) := \arg\min_{x\in X}\ h(x, b), \tag{3.1}$$

where H and X are separable Hilbert spaces and the functions g, r and h satisfy the following assumptions:

(A1) g is convex and both g and r are bounded below;
(A2) $\nabla_x h$ and $\nabla_x^2 h$ are continuous in b and there exist constants $\mu(b)$ and $L(b)$, $0 < \mu^* \leq \mu(b) \leq L(b)$, such that $\mu(b)I \preceq \nabla_x^2 h(x,b) \preceq L(b)I\ \forall x$ (this implies, in particular, that the lower-level objective function is strongly convex);
(A3) g is $L_{\nabla g}$-smooth and r is $L_{\nabla r}$-smooth, which means g and r are continuously differentiable with $L_{\nabla g}$ and $L_{\nabla r}$ Lipschitz gradients, respectively.

Under (A1)-(A3), the gradient of the upper-level objective with respect to b takes the following form, for any $x \in X$,

$$\nabla(g\circ\hat{x})(b) + \nabla r(b) = -\nabla_{xb}^2 h(x,b)^T \nabla_x^2 h(x,b)^{-1} \nabla g(\hat{x}(b)) + \nabla r(b), \tag{3.2}$$

which can be obtained using the IFT. Inexact methods avoid solving the lower-level problem exactly (which may be prohibitively expensive) and instead work

Algorithm 1. Calculating an inexact hypergradient

Input: $b \in H$, accuracies $\epsilon, \delta > 0$.
function InexactGradient(b, ϵ, δ)
  Solve lower-level problem to find $\tilde{x}(b)$ such that $\|\nabla_x h(\tilde{x}(b), b)\| \leq \epsilon\mu(b)$.
  Solve $\nabla_x^2 h(\tilde{x}(b), b)q = \nabla g(\tilde{x}(b))$ with residual δ.
  Calculate $z = -(\nabla_{xb}^2 h(\tilde{x}(b), b))^T q + \nabla r(b)$.

Algorithm 2.[13, Algorithm 3.1] Method of Adaptive Inexact Descent (MAID). Hyperparameters: $\underline{\rho}, \underline{\nu} \in (0,1)$, $\overline{\rho}, \overline{\nu} > 1$, and $\max_{\text{BT}} \in \mathbb{N}$.

1: Input $b_0 \in H$, accuracies ϵ_0, δ_0, step size $\alpha_0 > 0$.
2: **for** $k = 0, 1, \ldots$ **do**
3:   **for** $j = \max_{\text{BT}}, \max_{\text{BT}} + 1, \ldots$ **do**
4:     $z_k, \epsilon_k, \delta_k \leftarrow$ InexactGradient$(b_k, \epsilon_k, \delta_k)$
5:     **for** $i = 0, 1, \ldots, j - 1$ **do**
6:       **if** inexact sufficient decrease $\psi(\alpha_k) \leq 0$ holds **then** ▷ Cond. (3.4)
7:         Go to line 10 ▷ Backtracking successful
8:       $\alpha_k \leftarrow \underline{\rho}\alpha_k$ ▷ Adjust the starting step size
9:     $\epsilon_k, \delta_k \leftarrow \underline{\nu}\epsilon_k, \underline{\nu}\delta_k$ ▷ Backtracking failed, needs higher accuracy
10:   $b_{k+1} \leftarrow b_k - \alpha_k z_k$ ▷ Gradient descent update
11:   $\epsilon_{k+1}, \delta_{k+1}, \alpha_{k+1} \leftarrow \overline{\nu}\epsilon_k, \overline{\nu}\delta_k, \overline{\rho}\alpha_k$ ▷ Increasing $\epsilon_k, \delta_k, \alpha_k$

with an approximation $\tilde{x}(b)$ such that $\|\tilde{x}(b) - \hat{x}(b)\| \leq \epsilon$, whose error can be controlled by the gradient of the lower-level objective:

$$\|\tilde{x}(b) - \hat{x}(b)\| \leq \frac{1}{\mu(b)} \|\nabla_x h(\tilde{x}(b), b)\|,$$

where μ is the strong convexity constant of the lower-level objective. This has been proved in [13], but can also be seen as a direct consequence of Lemma 2. In addition, the inversion of the Hessian can be avoided by solving the linear system $\nabla_x^2 h(\tilde{x}(b), b)q = \nabla g(\tilde{x}(b))$ up to a prescribed accuracy δ, e.g., using the Conjugate Gradient (CG) method. Denoting the inexact hypergradient (i.e. gradient of the upper-level problem) by $z := -\nabla_{xb} h(\tilde{x}(b), b)^T q + \nabla r(b)$, and the corresponding error by $e := z - (\nabla(g \circ \hat{x})(b) + \nabla r(b))$, we get the procedure outlined in Algorithm 1.

Now consider the inexact descent update

$$b_{k+1} = b_k - \alpha_k z_k, \tag{3.3}$$

where α_k denotes the step size. To find a suitable sequence $\{\alpha_k\}$ and the accuracies of computing the hypergradient ϵ_k and δ_k adaptively, [13] uses the following condition. Choose $\lambda > 0$ and, for each upper-level iteration $k = 0, 1, \ldots$, set $u_k := (\tilde{x}(b_k), b_k)$ and $\bar{\epsilon}_{k+1} := \max\{\epsilon_k, \epsilon_{k+1}\}$. Let $\ell(x(b), b) := g(x(b)) + r(b)$ and

$$\psi(\alpha_k) := \ell(u_{k+1}) + \|\nabla_x \ell(u_{k+1})\|\bar{\epsilon}_{k+1} + \frac{L_{\nabla_x \ell}}{2}\,\bar{\epsilon}_{k+1}^2 - \ell(u_k) + \|\nabla_x \ell(u_k)\|\bar{\epsilon}_{k+1} + \lambda\alpha_k\|z_k\|^2. \tag{3.4}$$

Lemma 1. (Sufficient Decrease Condition, *[13]*). *Suppose that the condition $\psi(\alpha_k) \leq 0$ is satisfied at α_k. Then the sufficient decrease in the exact upper-level function $g(\hat{x}(b_{k+1})) + r(b_{k+1}) - g(\hat{x}(b_k)) - r(b_k) \leq -\lambda\alpha_k\|z_k\|^2$ holds.*

In practice, since the components in (3.4) depend on the accuracy, both the accuracy $\bar{\epsilon}_{k+1}$ and the step size α_k are decreased until the inequality holds. If the inequality is already satisfied, they are increased to reduce computational cost and take potentially larger steps. The resulting algorithm, called the Method of Adaptive Inexact Descent (MAID), is summarized in Algorithm 2.

Theorem 1. (Convergence to a Stationary Point, *[13]*). *Suppose that the sub-level sets of $r(\cdot)$ from (3.1) are strongly compact in H. Then, under (A1)-(A3) and with adaptive parameters chosen as in Lemma 1, the iterates b_k of Algorithm 2 converge to a stationary point b^* strongly in H. Hence, we have*

$$\lim_{k\to\infty} \|\nabla(g \circ \hat{x})(b_k) + \nabla r(b_k)\| = 0.$$

Remark 1. The proofs of [13] are in the finite-dimensional setting where H and X are Euclidean spaces, but they are easily adapted to the case of separable Hilbert spaces. The only caveat is guaranteeing strong convergence of the iterates $\tilde{x}(b_k)$ to the solution $\hat{x}(b^*)$ of the limiting lower-level problem, as the next lemma shows.

Lemma 2. *Let $\Phi : X \to \mathbb{R}$, where X is a separable Hilbert space, be μ-strongly convex and differentiable. Denote the minimizer of Φ by $x^* \in X$. Then, we have*

$$\|x^* - x\| \leq \frac{1}{\mu}\,\|\nabla\Phi(x)\|, \qquad \forall x \in X.$$

Proof. A direct application of [16, Lemma 3 (ii)], noting that $\nabla\Phi(x^*) = 0$. □

Remark 2. Of course, Lemma 2 remains true if Φ is only *sub*-differentiable.

Corollary 1. *Under the assumptions of Theorem 1, the iterates $\tilde{x}(b_k)$ converge strongly to to the solution $\hat{x}(b^*)$ of the limiting lower-level problem.*

Proof. It is easy to check that the minimizers of the lower-level problem converge strongly, $\hat{x}(b_k) \to \hat{x}(b^*)$. Furthermore, by Lemma 2, we have that

$$\|\tilde{x}(b_k) - \hat{x}(b_k)\| \leq \frac{1}{\mu}\|\nabla_x h(\tilde{x}(b_k), b_k)\| \leq \epsilon_k \to 0.$$

An application of the triangle inequality completes the proof. □

3.2 Inexact Bilevel Optimization Applied to ADP-β

We now discuss briefly how the ADP-β problem (2.4) fits into the general setting of (3.1). Here we have $H = X = L^2(\Omega)$, $g(\cdot) = \left\|T(a,\cdot) - y^\delta\right\|_{L^2}^2$, $r(\cdot) = \beta \left\|\cdot - a\right\|_{H^1}^2$ and $h(\cdot) = \left\|T(b,\cdot) - y^\delta\right\|_{L^2}^2 + \alpha\mathcal{J}(\cdot)$. By the Sobolev embedding theorem, the sub-level sets of $r(\cdot)$ are compact in L^2. The validity of (A1) is obvious. For (A2) we have the following proposition.

Proposition 1. *Suppose that the regularizer $\mathcal{J}$ is $\mu_\mathcal{J}$-strongly convex, twice continuously differentiable and $\sup_x \left\|\nabla_x^2 \mathcal{J}(x)\right\|_{\mathcal{L}(L^2)} \leq C_\mathcal{J}$ for some constants $\mu_\mathcal{J}, C_\mathcal{J} > 0$. Then $\mu(b)$ and $L(b)$ in (A2) can be taken as*

$$\mu(b) = \alpha\mu_\mathcal{J}, \quad L(b) = \|b\|_{L^2(\Omega\times\Omega)}^2 + \alpha C_\mathcal{J}.$$

Proof. The choice of $\mu(b)$ is obvious. For $L(b)$, we start by computing the Hessian of the lower-level problem:

$$\nabla_x \left(\left\|T(b,x) - y^\delta\right\|_{L^2}^2 + \alpha\mathcal{J}(x)\right) = B^*Bx - B^*By^\delta + \alpha\nabla_x\mathcal{J}(x), \qquad \text{and}$$
$$\nabla_x^2 \left(\left\|T(b,x) - y^\delta\right\|_{L^2}^2 + \alpha\mathcal{J}(x)\right) = B^*B + \alpha\nabla_x^2\mathcal{J}(x),$$

where B is defined analoguously to (2.3). Taking the operator norm and a supremum in x we get, with the triangle inequality, that

$$\sup_x \left\|\nabla_x^2 \left(\left\|T(b,x) - y^\delta\right\|_{L^2}^2 + \alpha\mathcal{J}(x)\right)\right\|_{\mathcal{L}(L^2)} \leq \|B\|_{\mathcal{L}(L^2)}^2 + \alpha C_\mathcal{J}.$$

It remains to note that $\|B\|_{\mathcal{L}(L^2)} = \|b\|_{L^2(\Omega\times\Omega)}$. □

Remark 3. Because of the regularization term $\|b - a\|_{H^1}^2$ in the upper-level problem in (2.4), the values $L(b_k)$ are bounded uniformly in k.

Assumption (A3) is more problematic. The term $r(b) = \|b - b_0\|_{H^1}^2$ is not differentiable on L^2 (it is only subdifferentiable) and Theorem 1 does not apply. It is possible to interpret Algorithm 2 as a subgadient method for the upper-level objective (with a linearization of the non-convex but smooth term $\left\|T(a,\hat{x}(b)) - y^\delta\right\|_{L^2}^2$ at each step) but ensuring convergence of the subgradient method typically requires either diminishing step sizes (which in not satisfied by MAID) or a uniform bound on the subgradients [14], which fails in ADP-β (2.4).

Proximal methods such as PALM [4] would be applicable, but, to the best of our knowledge, an adaptive inexact version of this algorithm à la [13] is not available. This is subject of our future work. In our numerical experiments, which we view as a proof of concept, we stay in the finite-dimensional setting where the non-differentiability issue does not arise.

4 Numerical Experiments

In this section, we present the numerical results of applying our method, which we call ADP-β-MAID, to the solution of the ADP-β bilevel problem (2.4). First, we compare the efficiency of ADP-β-MAID with Algorithms 1 and 2 in [2] on the same 1D deconvolution problem as in [2] and using the same settings. We use the implementation from the supplementary material[1] of [2].

Then we consider deblurring and semi-blind deblurring problems for 2D color images with smoothed total variation (TV) regularization [6], in place of the elastic-net regularizer considered for the 1D deconvolution problem. All experiments are conducted on an Apple Silicon M1 Pro chip using PyTorch[2].

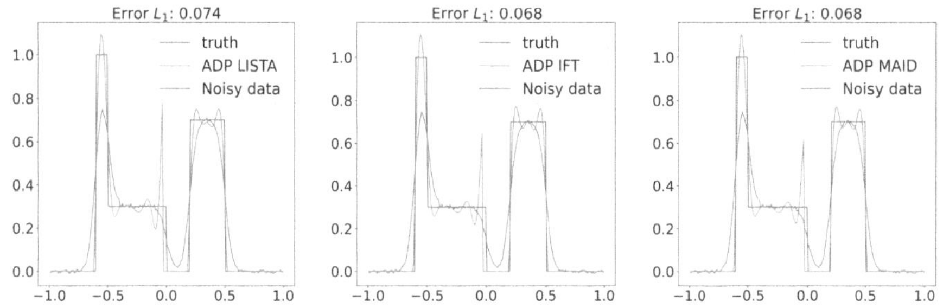

Fig. 1. Deblurring of a 1D signal using ADP with elastic-net regularizer, solved via ADP LISTA (left), ADP IFT (center), and ADP-β-MAID (right).

4.1 1D Deconvolution

We consider a 1D signal deconvolution problem where A is the forward operator corresponding to the 1D convolution with a Gaussian kernel with standard deviation $\widetilde{\sigma} = 5$. To run the experiments, we work with the full matrix A (see Figure S1(a) in the supplementary material) rather than its kernel. The corresponding blurry data are additionally corrupted with Gaussian noise with zero mean and standard deviation $\sigma = 5 \times 10^{-3}$ (cf. green line in Fig. 1).

For a fair comparison with the experiments in [2], we set $\beta = 0$ in (2.4) and use the elastic-net regularizer $\mathcal{J}(x) = \alpha_1 \|x\|_1 + \alpha_2 \|x\|_2^2$, with $\alpha_1 = 1.2 \times 10^{-3}$ and $\alpha_2 = 4 \times 10^{-3}$ as in [2], for the lower-level problem. Algorithms 1 and 2 in [2], to which we compare ADP-β-MAID, are hereafter denoted by ADP IFT and ADP LISTA[3], respectively. Following [2], we solve the lower-level problem using the proximal-gradient descent method with step size $\lambda_x = 0.1$, and compute an approximate hypergradient. The first approach, based on IFT as in [2],

1 https://gitlab.informatik.uni-bremen.de/carndt/analytic_deep_prior.
2 Code: https://github.com/MohammadSadeghSalehi/Analytical-Deep-Priors.
3 In [2], the notation "LISTA $L = \infty$" was used to denote unrolling with a high number of iterations. We drop "$L = \infty$" for ease of notation.

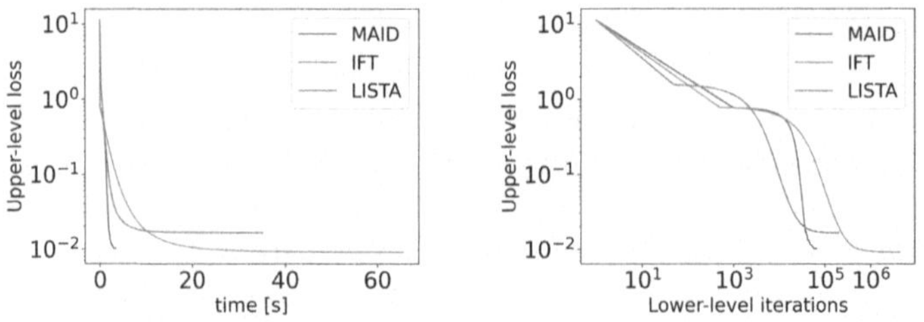

Fig. 2. Comparison of ADP IFT, ADP LISTA and ADP-β-MAID in terms of upper-level loss as a function of wall-clock time and lower-level iterations.

solves the lower-level problem with 500 iterations to achieve high accuracy; the resulting approximate hypergradient is then used to perform inexact gradient descent. The second approach, LISTA [10], solves the lower-level problem with 50 iterations, retaining the computational graph and using automatic differentiation. Additionally, in both IFT and LISTA, the initial solution of the lower-level problem for each successive upper-level iteration is set to the solution from the previous iteration to enable warm-starting. In ADP-β-MAID, we initialize the accuracy for inexactness to $\epsilon_0 = 10^{-2}$ and set a maximum of 300 lower-level iterations; however, this iteration limit is never reached in our experiments. The hyperparameters for ADP-β-MAID are set as $\overline{\rho} = 1.1$, $\underline{\rho} = 0.5$, $\overline{\tau} = 1.25$, and $\underline{\tau} = 0.5$. Finally, to ensure a fair comparison, the initial upper-level step size $\alpha_0 = \lambda_B = 0.1$ is the same across all three methods, and the same initialization was used for both the lower-level and upper-level problems.

As shown in Fig. 1, ADP-β-MAID and ADP IFT achieve a similar reconstruction quality with comparable learned parameters B^* (cf. also Figure S1 in the supplementary material), while ADP LISTA results in a slightly higher error. However, the computational cost and CPU time, shown in Fig. 2, differ significantly. In terms of CPU time, ADP-β-MAID requires approximately half the time compared to ADP LISTA and only a quarter compared to ADP IFT with high fixed accuracy. Although ADP LISTA converged to a suboptimal stationary point, possibly due to insufficient lower-level accuracy in early upper-level iterations, ADP-β-MAID reached a better stationary point with significantly lower computational cost than ADP IFT.

4.2 2D Deconvolution and Semi-blind Deconvolution

In this section, we implement ADP-β with Sobolev regularization for the upper-level (i.e., (2.4) with $\beta \neq 0$), and 2D smoothed TV for the lower-level regularizer, i.e. $\mathcal{J}(x) = \|\nabla x\|_\nu$, where $\|x\|\nu = \sum_{i=1}^{N} \sqrt{x_i^2 + \nu^2} - \nu$ with $\nu = 10^{-4}$. To ensure μ-strong convexity of the lower-level objective, we also add a quadratic term $\frac{\mu}{2}\| \cdot \|^2$ with $\mu = 10^{-8}$. Given the higher dimensionality of the problem, in the implementation, we take a kernel b_0 instead of the entire convolution operator B_0 as the initial kernel. We consider color images of size $300 \times 400 \times 3$. The regularization parameters for the lower-level and upper-level problems are set as

Fig. 3. Reconstructions in deblurring with a motion blur

$\alpha = 0.6$ and $\beta = 0.3$, respectively. The value of α and β are determined using grid search. To solve ADP-β, we set the same parameters for ADP-β-MAID as in Sect. 4.1. We use the Accelerated Gradient Descent Algorithm (AGD) for smooth functions [7, Algorithm 3], to solve the lower-level problem. In step 4 of Algorithm 1, we use CG with tolerance $\delta = \epsilon$, where ϵ denotes the accuracy of the lower-level solution.

Motion Blur. In this case, A is the 2D convolution operator associated with a motion kernel of size 5×5, with non-zero diagonal elements (cf. b_0 in Figure S2 in the supplementary material). The corresponding blurry image is corrupted with additional Gaussian noise with zero mean and standard deviation $\sigma = 0.02$.

(a) Ground truth (b) Blurry and noisy (c) TV (d) ADP-β-MAID

Fig. 4. Reconstructions in semi-blind deblurring (initial kernel: motion blur).

The Sobolev norm $\|B\|_{H^1}^2 = C(\|b\|_{L_2}^2 + \|\nabla_x b\|_{L_2}^2 + \|\nabla_y b\|_{L_2}^2)$ in (2.4a), where ∇_x and ∇_y denote the forward difference operators in the horizontal and vertical directions, respectively, and $C > 0$ is some constant absorbed into the regularization parameter, is computed using the kernel b associated with B, that is, $\|b\|_{H^1}^2 = \|\nabla b\|_2^2$, where ∇ is the forward difference operator. The upper-level iterations are stopped when the accuracy in ADP-β-MAID reaches $\epsilon = 10^{-6}$.

As illustrated in Fig. 3, the deblurred images obtained by solving the ADP-β bilevel problem using ADP-β-MAID are superior, both visually and in terms of quality metrics, to the results obtained by solving only the lower-level variational problem with the initial kernel b_0. PSNR and SSIM values for all reconstructions are reported on top of each image in Fig. 3. By way of example, we report in Figure S2 in the supplementary material the first channel of the kernel b^* obtained by solving the ADP-β problem, along with the absolute difference $|b^* - b_0|$, corresponding to the reconstructions in the first row of Fig. 3. These results indicate off-diagonal changes in the initial kernel.

Semi-blind Deblurring. In this case, we consider two different blur operators, A_1 and A_2, with corresponding kernels $\widehat{b}_1$ and $\widehat{b}_2$, representing motion blur and Gaussian blur, respectively. The diagonal motion kernel is defined as in the previous section, while for the Gaussian blur, we use a 5×5 kernel with $\widetilde{\sigma} = 0.5$. Since only the motion blur is assumed to be known, we set b_0 as the motion blur kernel $\widehat{b}_1$, as shown in Figure S3 in the supplementary material. To obtain the blurry image, we first apply A_1 and then A_2. Moreover, additive Gaussian noise with $\sigma = 0.02$ is added to obtain the blurry and noisy image.

The deblurred image for this mixed blur is shown in Fig. 4, where the last column demonstrates that ADP-β-MAID achieves higher reconstruction quality, both visually and in terms of quality metrics, compared to smoothed TV. Examining the changes in the (first channel of the) kernel (reported in Figure S3 of the supplementary material and corresponding to the images in Fig. 4), reveals a Gaussian-like difference that ADP-β-MAID was able to recover successfully.

5 Conclusions

In this paper, we discussed the applicability of an adaptive inexact bilevel optimization method (MAID) to the Sobolev-regularized analytical deep image prior problem. We have seen that MAID admits a straightforward generalization to an infinite-dimensional Hilbert-space setting, yet the non-differentiability of the Sobolev regularization term in the L^2 setting requires further development of the MAID algorithm. In particular, an adaptive inexact version of the PALM algorithm seems desirable, which we plan to investigate in our future work. In the finite-dimensional setting, however, we demonstrate that adaptive inexact bilevel optimization can achieve significant computational speed-ups and allows us to use ADP in larger-scale problems than in previous literature.

Acknowledgement. We used Hex, the GPU Cloud in the Department of Computer Science at the University of Bath. The work of MSS is supported by a scholarship from the EPSRC Centre for Doctoral Training in Statistical Applied Mathematics at Bath (SAMBa), under the project EP/S022945/1. TAB acknowledges partial support by the European Union - NextGeneration EU through the Italian Ministry of University and Research as part of the PNRR - M4C2, Investment 1.3, FAIR "Future Partnership Artificial Intelligence Research", Proposal Code PE00000013 - CUP J33C22002830006), and INdAM-GNCS project code CUP_E53C24001950001. YK acknowledges the support of the EPSRC (Fellowship EP/V003615/2 and Programme Grant EP/V026259/1).

References

1. Adler, J., Öktem, O.: Solving ill-posed inverse problems using iterative deep neural networks. Inverse Prob. **33**(12), 124007 (2017)
2. Arndt, C.: Regularization theory of the analytic deep prior approach. Inverse Prob. **38**(11), 115005 (2022). https://doi.org/10.1088/1361-6420/ac9011
3. Baguer, D.O., Leuschner, J., Schmidt, M.: Computed tomography reconstruction using deep image prior and learned reconstruction methods. Inverse Prob. **36**(9), 094004 (2020). https://doi.org/10.1088/1361-6420/aba415
4. Bolte, J., Sabach, S., Teboulle, M.: Proximal alternating linearized minimization for nonconvex and nonsmooth problems. Math. Program. **146**(1), 459–494 (2014)
5. Buskulic, N., Fadili, J., Quéau, Y.: Convergence and recovery guarantees of unsupervised neural networks for inverse problems. J. Math. Imag. Vis. **66**(4), 584–605 (2024)
6. Chambolle, A., Caselles, V., Cremers, D., Novaga, M., Pock, T.: An introduction to total variation for image analysis, pp. 263–340. De Gruyter, Berlin (2010)
7. Chambolle, A., Pock, T.: An introduction to continuous optimization for imaging. Acta Numer. **25**, 161–319 (2016)
8. Cheng, Z., Gadelha, M., Maji, S., Sheldon, D.: A Bayesian perspective on the deep image prior. In: Conference on Computer Vision and Pattern Recognition (2019)
9. Dittmer, S., Kluth, T., Maass, P., Otero Baguer, D.: Regularization by architecture: a deep prior approach for inverse problems. J. Math. Imag. Vis. **62**(3), 456–470 (2020)
10. Gregor, K., LeCun, Y.: Learning fast approximations of sparse coding, pp. 399–406. Omnipress, Madison, WI, USA (2010)
11. Jagatap, G., Hegde, C.: Algorithmic guarantees for inverse imaging with untrained network priors. In: Advances in Neural Information Processing Systems. Curran Associates Inc., NY, USA (2019)
12. Pedregosa, F.: Hyperparameter optimization with approximate gradient. In: Balcan, M.F., Weinberger, K.Q. (eds.) Proceedings of The 33rd International Conference on Machine Learning, vol. 48, pp. 737–746. New York, USA (2016)
13. Salehi, M.S., Mukherjee, S., Roberts, L., Ehrhardt, M.J.: An adaptively inexact first-order method for bilevel optimization with application to hyperparameter learning (2024). https://arxiv.org/abs/2308.10098
14. Shor, N.Z.: Minimization methods for non-differentiable functions, vol. 3. Springer Science & Business Media (2012). https://doi.org/10.1007/978-3-642-82118-9
15. Ulyanov, D., Vedaldi, A., Lempitsky, V.: Deep image prior. In: Conference on Computer Vision and Pattern Recognition (2018)
16. Zhou, X.: On the Fenchel duality between strong convexity and Lipschitz continuous gradient (2018). https://arxiv.org/abs/1803.06573

Why do we Regularise in Every Iteration for Imaging Inverse Problems?

Evangelos Papoutsellis[1(✉)], Zeljko Kereta[2], and Kostas Papafitsoros[3]

[1] Finden Ltd., Oxford, UK
epapoutsellis@gmail.com
[2] University College London, London, UK
z.kereta@ucl.ac.uk
[3] School of Mathematical Sciences, Queen Mary University of London, London, UK
k.papafitsoros@qmul.ac.uk

Abstract. Regularisation is a common method in iterative solutions for imaging inverse problems. The majority of algorithms evaluate the proximal operator of the regulariser in every iteration, leading to a significant computational overhead, as such evaluations can be costly. In this context, we investigate skipping the regulariser to reduce the frequency of proximal operator computations. This approach shows a reduction in computational time without compromising convergence or image quality. Here we study for the first time the efficacy of regularisation skipping on a variety of imaging inverse problems. We build upon the ProxSkip algorithm and we also propose a novel skip-version of the PDHG algorithm. Extensive numerical results highlight the potential of these methods to accelerate computations while maintaining high-quality reconstructions.

Keywords: Inverse Problems · Iterative Regularisation · Proximal Operator · Stochastic Optimisation

1 Introduction

Inverse problems involve the process of estimating an unknown quantity $\boldsymbol{u}^\dagger \in \mathbb{X}$ from indirect and often noisy measurements $\boldsymbol{b} \in \mathbb{Y}$ obeying $\boldsymbol{b} = \boldsymbol{A}\boldsymbol{u}^\dagger + \boldsymbol{\eta}$. Here $\mathbb{X}, \mathbb{Y}$ denote finite Euclidean dimensional spaces, $\boldsymbol{A} : \mathbb{X} \to \mathbb{Y}$ is a linear forward operator, $\boldsymbol{u}^\dagger$ is the ground truth and $\boldsymbol{\eta}$ is a random noise component. Given $\boldsymbol{b}$ and $\boldsymbol{A}$, the goal is to approximate $\boldsymbol{u}^\dagger$. Since inverse problems are typically ill-conditioned, prior information about $\boldsymbol{u}$ is often incorporated in the form of regularisation. The inverse problem solution is then approximated by solving

$$\operatorname*{arg\,min}_{\boldsymbol{u}\in\mathbb{X}} \mathcal{D}(\boldsymbol{A}\boldsymbol{u}, \boldsymbol{b}) + \alpha\mathcal{R}(\boldsymbol{u}). \tag{1}$$

Here $\mathcal{D}$ denotes the fidelity term, measuring the distance between $\boldsymbol{b}$ and the solution $\boldsymbol{u}$ under $\boldsymbol{A}$. Regularisation term $\mathcal{R}$ promotes properties such as smoothness, sparsity, edge preservation, and low-rankness of the solution, and is weighted by

T. A. Bubba et al. (Eds.): SSVM 2025, LNCS 15667, pp. 43–55, 2025.
https://doi.org/10.1007/978-3-031-92366-1_4

a parameter $\alpha > 0$. Established examples for $\mathcal{R}$ in imaging include Total Variation (TV), the Total Generalised Variation [8], the Total Nuclear Variation (TNV) [17], and more general tensor-based structure regularisation [20].

In order to obtain a solution for (1), one employs iterative algorithms such as Gradient Descent (GD) for smooth objectives or Forward-Backward Splitting (FBS) [13] for non-smooth ones. Moreover, under the general framework

$$\min_{\boldsymbol{x}\in\mathbb{X}} f(\boldsymbol{x}) + g(\boldsymbol{x}), \tag{2}$$

the Proximal Gradient Descent (PGD) algorithm, also known as Iterative Shrinkage Thresholding Algorithm (ISTA), and its accelerated version FISTA [6,25] are commonly used when f is convex and L-smooth, and g is proper and convex. Saddle-point methods such as the Primal Dual Hybrid Gradient (PDHG) [11] are commonly used for non-smooth f.

A common property of most of these methods, see Algorithms 2–3, is the evaluation of proximal operators for the regulariser in every iteration, which for $\tau > 0$ is defined as

$$\operatorname{prox}_{\tau} g(\boldsymbol{x}) := \arg\min_{\boldsymbol{z}\in\mathbb{X}} \left\{ \frac{1}{2}\|\boldsymbol{z}-\boldsymbol{x}\|_2^2 + \tau g(\boldsymbol{z}) \right\}. \tag{3}$$

The proximal operator either has a closed form solution, e.g., when $\mathcal{R}(\boldsymbol{u}) = \|\boldsymbol{u}\|_1$, or requires an iterative solver, e.g., when $\mathcal{R}(\boldsymbol{u}) = \mathrm{TV}(\boldsymbol{u})$. In the latter case, solving (3) can incur a significant computational cost for large data, particularly in medical imaging and material science applications.

Algorithm 1. GD

Parameters: $\gamma > 0$
Initialise: $\boldsymbol{x}_0 \in \mathbb{X}$
for $k = 0, \dots, K-1$ **do**
  $\boldsymbol{x}_{k+1} = \boldsymbol{x}_k - \gamma \nabla f(\boldsymbol{x}_k)$
end for

Algorithm 2. ISTA/PGD/FBS

Parameters: $\gamma > 0$
Initialise: $\boldsymbol{x}_0 \in \mathbb{X}$
for $k = 0, \dots, K-1$ **do**
  $\boldsymbol{x}_{k+1} = \operatorname{prox}_{\gamma g}(\boldsymbol{x}_k - \gamma \nabla f(\boldsymbol{x}_k))$
end for

Algorithm 3. FISTA

Parameters: $\gamma > 0$, $t_0 = 1$
Initialise: $\boldsymbol{x}_0 \in \mathbb{X}$
for $k = 0, \dots, K-1$ **do**
  $\boldsymbol{x}_{k+1} = \operatorname{prox}_{\gamma g}(\boldsymbol{y}_{k+1} - \gamma \nabla f(\boldsymbol{y}_{k+1}))$
  $t_{k+1} = \frac{1+\sqrt{1+4t_k^2}}{2}$, $a_k = \frac{t_k - 1}{t_{k+1}}$
  $\boldsymbol{y}_{k+1} = \boldsymbol{x}_{k+1} + a_k(\boldsymbol{x}_{k+1} - \boldsymbol{x}_k)$
end for

Algorithm 4. ProxSkip

Parameters: $\gamma > 0$, probability $p > 0$
Initialise: $\boldsymbol{x}_0, \boldsymbol{h}_0 \in \mathbb{X}$
for $k = 0, 1, \dots, K-1$ **do**
  $\hat{\boldsymbol{x}}_{k+1} = \boldsymbol{x}_k - \gamma(\nabla f(\boldsymbol{x}_k) - \boldsymbol{h}_k)$
  Sample $\theta_k \sim \mathrm{Bernoulli}(p)$
  if $\theta_k = 1$ **then**
    $\boldsymbol{x}_{k+1} = \operatorname{prox}_{\frac{\gamma}{p} g}\left(\hat{\boldsymbol{x}}_{k+1} - \frac{\gamma}{p}\boldsymbol{h}_k\right)$
  else $\boldsymbol{x}_{k+1} = \hat{\boldsymbol{x}}_{k+1}$
  end if
  $\boldsymbol{h}_{k+1} = \boldsymbol{h}_k + \frac{p}{\gamma}(\boldsymbol{x}_{k+1} - \hat{\boldsymbol{x}}_{k+1})$
end for

The ProxSkip algorithm

Skipping the computation of the proximal operator in certain iterations can accelerate an algorithm without impacting its convergence, as noted in [24]. There, the ProxSkip algorithm was developed to handle federated learning tasks involving costly proximal operator computations. ProxSkip introduces a control variate $\boldsymbol{h}_k$ and a probability parameter p, see Algorithm 4. When the proximal step is not applied, the control variate is unchanged. For successive iterations where the proximal operator is not applied, the computational cost is not increased. However, this leads to error accumulation. Applying the proximal operator and updating the control variate in certain iterations reduces the error.

In [24] it was shown that ProxSkip converges provided that f in (2) is L-smooth and μ-strongly convex, and the probability p satisfies

$$p \geq \sqrt{\mu/L}. \tag{4}$$

In the case of equality in (4), the algorithm converges (in expectation) at a linear rate with $\gamma = \frac{1}{L}$ and the iteration complexity is $\mathcal{O}(\frac{L}{\mu}\log(\frac{1}{\varepsilon}))$. In addition, the total number of proximal evaluations (in expectation) is only $\mathcal{O}(\frac{1}{\sqrt{p}}\log(\frac{1}{\varepsilon}))$.

Our Contribution

Our aim is conduct the first numerical study, via extensive experiments, of the computational benefits of ProxSkip for various imaging inverse problems, including challenging real-world tomographic applications. In particular, we show that ProxSkip can outperform even FISTA. Moreover, we introduce PDHGSkip, a novel skip version of PDHG, see Algorithm 6, which we motivate via numerical experiments. We anticipate that this will spark further research around developing skip-versions of a variety of proximal-based algorithms used nowadays.

For all our imaging experiments we consider a quadratic distance term as data fidelity and as a regulariser we use the (isotropic) total variation

$$\mathrm{TV}(\boldsymbol{u}) = \|\boldsymbol{D}\boldsymbol{u}\|_{2,1} = \sum \sqrt{(\boldsymbol{D}_y\boldsymbol{u})^2 + (\boldsymbol{D}_x\boldsymbol{u})^2},$$

where $\boldsymbol{D} = (\boldsymbol{D}_y, \boldsymbol{D}_x)$ is the discrete forward finite differences operator. Problem (1) is thus written as

$$\operatorname*{arg\,min}_{u\in\mathbb{X}} \frac{1}{2}\|\boldsymbol{A}\boldsymbol{u} - \boldsymbol{b}\|_2^2 + \alpha\mathrm{TV}(\boldsymbol{u}). \tag{5}$$

2 Skipping Light Proximals in Imaging Problems

2.1 Dual TV Denoising

To showcase the algorithmic properties we consider a toy example, with the dual formulation of the classical TV denoising (ROF) which reads

$$\min_{\|q\|_{2,\infty}\leq\alpha} \left\{\mathcal{F}(\boldsymbol{q}) := \frac{1}{2}\|\mathbf{div}\boldsymbol{q} + \boldsymbol{b}\|_2^2 + \frac{1}{2}\|\boldsymbol{b}\|_2^2\right\} \qquad \text{(Dual-ROF)}. \tag{6}$$

Fig. 1. Left to right: Ground truth $\boldsymbol{u}^\dagger \in \mathbb{R}^{200\times 300}$. Noisy image $\boldsymbol{b}$, $\sigma = 0.05$. Dual-ROF $\boldsymbol{u}^*$ with $\alpha = 0.5$. Dual-Huber-ROF $\boldsymbol{u}^*$ (see Sect. 2.2) with $\alpha = 0.55$, $\hat{\varepsilon} = 0.1$. The parameters α are optimised with respect to SSIM.

Here $\mathbf{div}$ is the discrete divergence operator such that $\mathbf{div} = -\boldsymbol{D}^T$. The solutions $\boldsymbol{u}^*$ and $\boldsymbol{q}^*$ of the primal and dual ROF problems respectively are linked via $\boldsymbol{u}^* = \boldsymbol{b} + \mathbf{div}\boldsymbol{q}^*$. A simple algorithm to solve (6) was introduced in [12], based on a Projected Gradient Descent (ProjGD) iteration which $\boldsymbol{q}_{k+1} = \mathcal{P}_C(\boldsymbol{q}_k - \gamma \boldsymbol{D}(\mathbf{div}\boldsymbol{q}_k + \boldsymbol{b}))$, which is globally convergent under a fixed step size $\gamma \leq \frac{2}{\|\boldsymbol{D}\|^2}$ [3], with

$$\mathcal{P}_C(\boldsymbol{x}) = \frac{\boldsymbol{x}}{\max\{\alpha, \|\boldsymbol{x}\|_2\}}, \quad C = \left\{\|\boldsymbol{q}\|_{2,\infty} \leq \alpha : \boldsymbol{q} \in \mathbb{R}^{2\times d}\right\}, \tag{7}$$

and d being the image dimension. The projection $\mathcal{P}_C$ can be identified as the proximal operator of the indicator function of the feasibility set C. Thus, ProjGD is a special case of ISTA/PGD and ProxSkip can be applied. Note that due to the divergence operator this problem is not strongly convex. In fact, this is the case for the majority of the problems of the type (1), typically due to the non-injectivity of $\boldsymbol{A}$. Thus, this example also shows that the strong convexity assumption could potentially be relaxed for imaging inverse problems.

To ensure that any biases from algorithms under evaluation are avoided, the "exact" solution $\boldsymbol{u}^*$ is calculated using an independent high-precision solver, in particular, the MOSEK solver from the CVXpy library [1,15], see Fig. 1.

Both ISTA and ProxSkip use the step size $\gamma = \frac{1}{L} = \frac{1}{8}$, where L is the Lipschitz constant of $\mathcal{F}'(\cdot)$. For every iteration we monitor the ℓ_2 error $\|\boldsymbol{u}_k - \boldsymbol{u}^*\|_2$ between the iterate $\boldsymbol{u}_k = \boldsymbol{b} + \mathbf{div}\boldsymbol{q}_k$ and estimated exact solution $\boldsymbol{u}^*$. We use $p = [0.01, 0.1, 0.3, 0.5]$ and 50000 iterations as a stopping criterion.

In Fig. 2 (top-left), we observe that the two algorithms are almost identical in terms of the ℓ_2 error. Note that ProxSkip and ISTA coincide only when $p = 1$. Indeed, one can detect some discrepancies during the first 100 iterations, which quickly dissipate with only a few applications of the projection $\mathcal{P}_C$, see bottom row of Fig. 2 where we compare ISTA and ProxSkip with $p = 0.1$ for the first and last 100 iterations. In Fig. 2 (top-right), we plot the ℓ_2 error with respect to CPU time (computed as the average over 30 independent runs). We observe a clearly superior performance of ProxSkip, in terms of the CPU time, for all values of p. This serves as a first demonstration of the advantages of ProxSkip in terms of computational time, without compromising image quality. In Sect. 3, we present a more emphatic computational impact using computationally more expensive proximal steps.

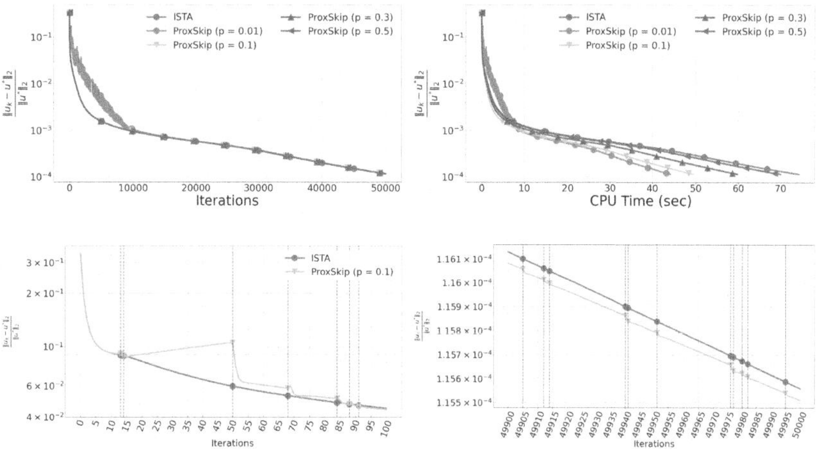

Fig. 2. Top: Comparison of ISTA and ProxSkip for multiple values of p for (6) with respect to iterations (left) and CPU time (right). Bottom: Detailed versions for the first 100 (left) and the last 100 iterations (right) when $p = 0.1$. The vertical dotted lines indicate the iterations where $\mathcal{P}_C(\cdot)$ is applied.

2.2 Dual TV Denoising with Strong Convexity

In order to be consistent with the convergence theory of ProxSkip where strong convexity is a requirement, one can add a small quadratic term to the objective function. This is commonly used in imaging applications and allows the use of accelerated versions of first-order methods. For problem (6) this results in

$$\min_{\|\boldsymbol{q}\|_{2,\infty} \leq \alpha} \left\{ \mathcal{F}(\boldsymbol{q}) := \frac{1}{2}\|\mathbf{div}\boldsymbol{q} + \boldsymbol{b}\|_2^2 + \frac{1}{2}\|\boldsymbol{b}\|_2^2 + \frac{\hat{\varepsilon}}{2\alpha}\|\boldsymbol{q}\|_2^2 \right\}. \tag{8}$$

It is known that the corresponding primal problem of (8) is the standard Huber-TV denoising [11] which involves a quadratic smoothing of the $\|\cdot\|_{2,1}$-norm around an $\hat{\varepsilon}$-neighbourhood of the origin. Among other effects, this reduces the staircasing artifacts of TV, see the rightmost image in Fig. 1.

We repeat the experiments of Sect. 2.1 with 30 independent runs of the algorithm to solve (8) with 5000 iterations and using the probability p given by the lower bound in (4). Using $(\alpha, \hat{\varepsilon}) = (0.55, 0.01)$, we have $p = \sqrt{\frac{\mu}{L}} = \sqrt{\frac{\hat{\varepsilon}}{\alpha L}} = 0.04767$. This results in an average of only 215 projection steps being used over 5000 iterations. In addition to ISTA, we also look at FISTA. For the dual problems (6) and (8) this algorithm is commonly found in the literature under alternative names such as Accelerated ProjGD or Fast Gradient Projection, [5]. In Fig. 3 (left), we observe that FISTA performs better than both ISTA and ProxSkip with respect to iteration number. While a similar behaviour is observed with respect to CPU time, see Fig. 3 (right), the average time for FISTA and

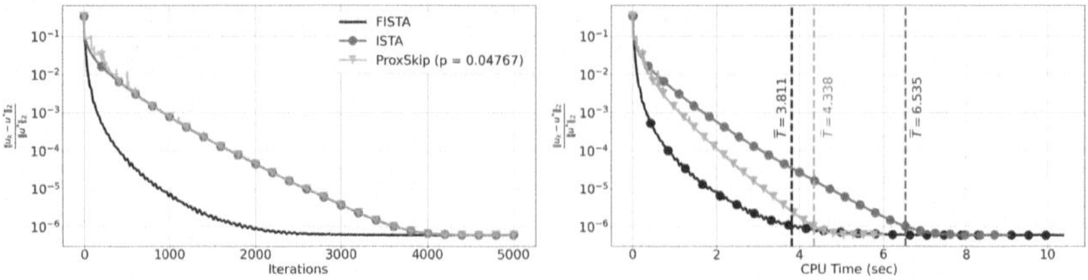

Fig. 3. ISTA, FISTA and ProxSkip with optimal p value for the problem (8) with respect to iterations (left) and CPU time (right). $\bar{T}$ indicates the average time required to reach a relative error less that 10^{-6}

ProxSkip to reach a relative error less than 10^{-6} is nearly identical. A similar speed-up would be expected when FISTA is combined with skipping techniques but we leave this for future work.

3 Skipping Heavy Proximals in Imaging Problems

3.1 TV Deblurring

We now consider more challenging imaging tasks, involving proximal operators that do not admit closed form solutions, and thus require computationally intensive iterative solvers, making them perfect candidates for skipping techniques.

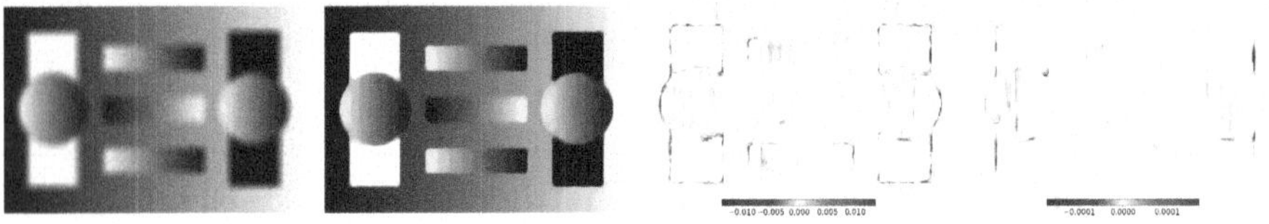

Fig. 4. Left to right: Noisy and blurry image. TV deblurred image $\boldsymbol{u}^*$ with $\alpha = 0.025$. Difference error for FISTA when it is less than 10^{-3} and 10^{-5}.

We first consider the deblurring task (5), where $\boldsymbol{A}$ is a convolution operator and $\boldsymbol{b}$ a noisy and blurred image, see Fig. 4. We solve (5) and compare ISTA, FISTA and ProxSkip with different values of p. Here the proximal operator of g corresponds to a TV denoising problem applied on the dual formulation (6). We employ the FISTA algorithm with a fixed number of iterations as an inner iterative solver, see next section for alternatives.

In the framework of inexact regularisation, another option is to terminate the inner solver based on some metric and predefined threshold [29]. As noted therein, the number of required inner iterations typically increases up to 10^3, as the outer algorithm progresses, leading to higher computational costs over time. To explore both the computationally easy and hard cases, we run the inner solver with 10 and 100 iterations, and use a warm-start strategy. Warm-start is vital for inexact regularisation as it avoids semi-convergence, where the error stagnates and fails to reach high precision solutions. The "exact" solution $\boldsymbol{u}^*$ is computed with 200000 iterations of PDHG with diagonal preconditioning [28]. Outer iterations are terminated if either 3000 iterations are reached or if the relative error is less than 10^{-5}. The CPU time is averaged over 10 runs.

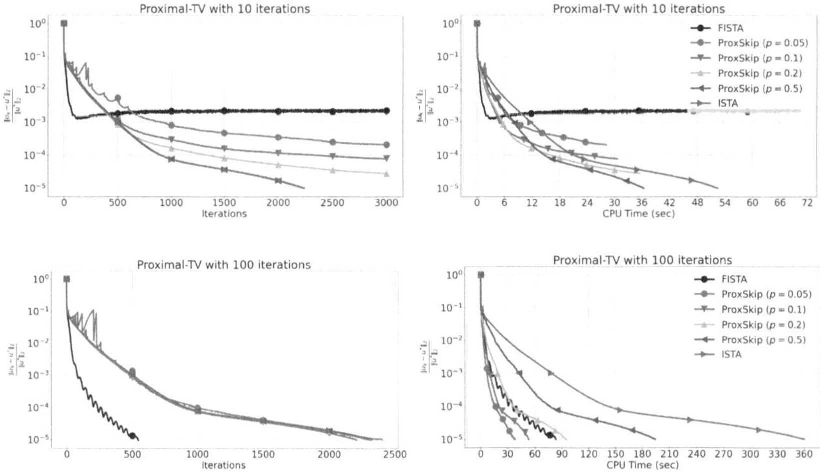

Fig. 5. Comparing ISTA, FISTA and ProxSkip for multiple values of p for TV deblurring. The proximal of TV is solved using FISTA with 10 and 100 iterations applied on (6). ProxSkip outperforms FISTA when $p = 0.05$ and 0.1.

In Fig. 5, we first observe that solving the inner TV problem with 10 iterations significantly affects the convergence of FISTA. Notably, ProxSkip versions are less affected, even though besides the error introduced by the inexact solver, there is also an error from skipping the proximal. By raising the number of inner iterations to 100, and thus increasing the accuracy of the inner solver, we observe that FISTA exhibits a fast initial convergence, albeit with some oscillations. It terminates after around 500 iterations, reaching an accuracy of 10^{-5}, see Fig. 5 (bottom-left). On the other hand, ISTA and ProxSkip require many more iterations to reach the same level of accuracy. However, remarkably, in this regime ProxSkip is significantly faster, in terms of CPU time, than ISTA and even outperforms FISTA when $p = 0.05$ and 0.1, see Fig. 5 (bottom-right).

We note that there are versions of FISTA [4,10,21,26] with improved performance, also avoiding oscillations. However, our goal here is not an exhaustive comparison but rather to show that a simple version of ProxSkip outperforms a simple version of FISTA. We anticipate the future development of more sophisticated skip-based algorithms including ones based on accelerated methods.

Algorithm 5. PDHGSkip-1 [14]

Parameters: $\sigma, \tau, \omega \geq 0$, probability p
Initialise: $\boldsymbol{x}_0 \in \mathbb{X}$, $\boldsymbol{y}_0 \in \mathbb{Y}$
for $k = 0, \dots, K-1$ **do**
  Sample $\theta_k \sim \text{Bernoulli}(p)$
  if $\theta_k = 1$ **then**
    $\hat{\boldsymbol{x}}_k = \frac{1}{p}(\text{prox}_{\sigma g}(\boldsymbol{x}_k - \sigma \mathcal{K}^T \boldsymbol{y}_k) - \boldsymbol{x}_k)$
  else
    $\hat{\boldsymbol{x}}_k = 0$
  end if
  $\boldsymbol{x}_{k+1} = \boldsymbol{x}_k + \frac{1}{1+\omega}\hat{\boldsymbol{x}}_k$
  $\boldsymbol{y}_{k+1} = \text{prox}_{\tau f^*}(\boldsymbol{y}_k + \tau \mathcal{K}(\boldsymbol{x}_{k+1} + \hat{\boldsymbol{x}}_k))$
end for

Algorithm 6. PDHGSkip-2 (ours)

Parameters: $\sigma, \tau > 0$, probability p
Initialise: $\boldsymbol{x}_0, \boldsymbol{h}_0 \in \mathbb{X}$, $\boldsymbol{y}_0 \in \mathbb{Y}$
for $k = 0, \dots, K-1$ **do**
  $\hat{\boldsymbol{x}}_k = \boldsymbol{x}_k - \sigma(\mathcal{K}^T \boldsymbol{y}_k - \boldsymbol{h}_k)$
  Sample $\theta_k \sim \text{Bernoulli}(p)$
  if $\theta_k = 1$ **then**
    $\boldsymbol{x}_{k+1} = \text{prox}_{\frac{\sigma}{p} g}\left(\hat{\boldsymbol{x}}_k - \frac{\sigma}{p}\boldsymbol{h}_k\right)$
  else $\boldsymbol{x}_{k+1} = \hat{\boldsymbol{x}}_k$
  end if
  $\bar{\boldsymbol{x}}_{k+1} = 2\boldsymbol{x}_{k+1} - \boldsymbol{x}_k$
  $\boldsymbol{y}_{k+1} = \text{prox}_{\tau f^*}(\boldsymbol{y}_k + \tau \mathcal{K}\bar{\boldsymbol{x}}_k)$
  $\boldsymbol{h}_{k+1} = \boldsymbol{h}_k + \frac{p}{\sigma}(\boldsymbol{x}_{k+1} - \hat{\boldsymbol{x}}_{k+1})$
end for

3.2 PDHGSkip

Skipping the proximal can also be applied to primal-dual type algorithms. Here, for convex f, g and a bounded linear operator $\mathcal{K}$, the optimisation framework is

$$\min_{\boldsymbol{x} \in \mathbb{X}} f(\mathcal{K}\boldsymbol{x}) + g(\boldsymbol{x}). \tag{9}$$

In [14], a skip-version of PDHG [11] was proposed, which we denote here by PDHGSkip-1. It allows not only to skip one of the proximal steps but also the forward or backward operations of $\mathcal{K}$, depending on the order of the proximal steps, see Algorithm 5. Again, strong convexity of both f and g is required for convergence. However, our imaging experiments – in the absence of strong convexity – reveal a relatively slow performance, even with optimised step sizes σ, τ and $\omega + 1 = 1/p$ according to [14, Theorem 7].

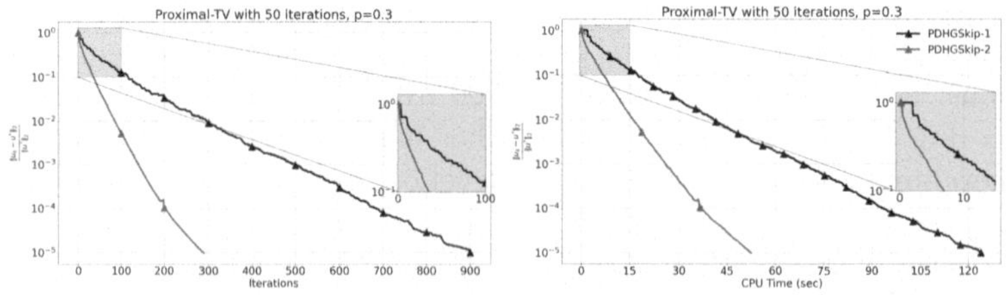

Fig. 6. Comparing PDHGSkip-1 by [14] and our proposed PDHGSkip-2 for $p = 0.3$ for the tomography problem of Sect. 3.3, using 50 iterations for the inner solver of the proximal TV denoising problem.

To address the slow convergence, we propose a modification: PDHGSkip-2, see Algorithm 6. The difference to PDHGSkip-1 is that the adjoint $\mathcal{K}^T$ and the

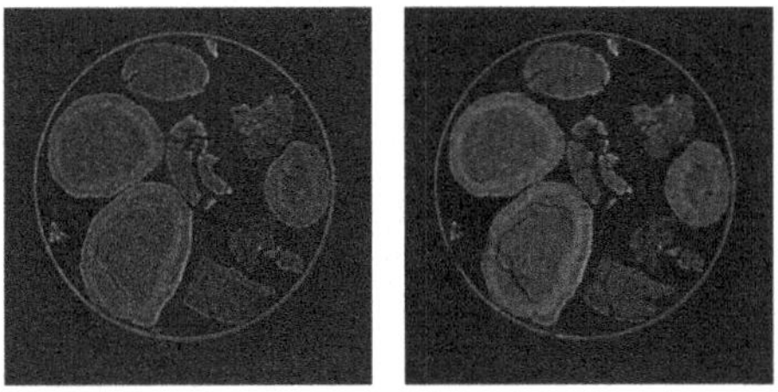

Fig. 7. FBP (left) and TV (right) reconstruction using diagonal preconditioned PDHG for 200000 iterations. Regularisation parameter is manually set to balance noise reduction with feature preservation.

proximal operator of g are now separated, see steps 4–7. Note that the control variate $\boldsymbol{h}$ vanishes when $p = 1$ and PDHG is recovered. To illustrate their practical differences, in Fig. 6 we compare the two versions on tomography, see Sect. 3.3 for details. Apart from the clear acceleration of PDHGSkip-2 over PDHGSkip-1, we also observe a staircasing pattern for the relative error of PDGHSkip-1, see detailed zoom of the first iterations. This is expected since in most iterations, where the proximal step is skipped, one variable vanishes without contributing to the next iterate. Hence, the update remains unchanged.

In general, ℓ^2-TV problems can be solved using implicit or explicit formulations of PDHG. In the implicit case, we have $f(\cdot) = \|\cdot\|_2^2$, $\mathcal{K} = \boldsymbol{A}$ and $g(\cdot) = \mathrm{TV}(\cdot)$. In the explicit case, f is a separable sum of $\|\cdot\|_2^2$ and $\|\cdot\|_{2,1}$ composed with the block operator $\mathcal{K} = [\boldsymbol{A}; \boldsymbol{D}]$, and g can be the zero function or a non-negativity constraint. Here every proximal step has an analytical solution. This significantly reduces the per-iteration cost, but also requires more iterations to reach a desired accuracy [9]. Inexact regularisation is usually preferred and helps reduce the number of iterations. Hence, the number of calls of the forward and backward operations of $\boldsymbol{A}$ is reduced, which in certain applications gives a considerable speed-up.

3.3 TV Tomography Reconstruction with Real-World Data

For our final case study, we solve (5) under a non-negativity constraint on $\boldsymbol{u}$ for a real-world tomographic reconstruction. Here $\boldsymbol{A}$ is the discrete Radon transform and $\boldsymbol{b}$ is a noisy sinogram of a real chemical imaging tomography dataset, representing post partial oxidation of methane reaction Ni-Pd/CeO2-ZrO2/Al2O3 catalyst [22], see Fig. 7. The initial dataset was acquired for 800 projection angles with 695×695 detector size and 700 vertical slices. For demonstration purposes and to be able to perform multiple runs for computing more representative CPU times, we consider one vertical slice with half the projections and 2× rebinned detector size. Same conditions for the inexact solver and stopping rules are used as in the previous section. For these optimisation problems, one can use algorithms that fit within both general frameworks (2) and (9), see [2].

The two algorithms use the same values for step sizes σ and τ, satisfying $\sigma\tau\|\mathcal{K}\|^2 \leq 1$. Comparisons in Fig. 8 show a similar trend for proximal-gradient-based algorithms as in Sect. 3.1. When we use 10 iterations for the inner solver, FISTA fails to reach the required accuracy and stagnates, which is not the case for the ProxSkip algorithms even for the smallest p. This is accompanied by a significant CPU time speed-up, which further increases as more inner iterations are used. In fact, by increasing the number of inner iterations, the computational gain is evident with around 90% speed-up compared to FISTA.

The best performance with respect to CPU time is achieved by PDHGSkip-2. There, distance errors are identical to the case $p = 1$, in terms of iterations, except for the extreme case $p = 0.1$ which oscillates in the early iterations. For 10 inner iterations and $p \in \{0.3, 0.5\}$ we observe a slight delay in terms of the convergence rate and the desired accuracy, due to error accumulation caused by skipping the proximal operator and limited accuracy of the inner solver. This is corrected by increasing the number of inner iterations. Moreover, we see no difference with respect to CPU time for 10 inner iterations and $p \in \{0.7, 1.0\}$. This is expected since the computational cost to run 10 iterations of FISTA is relatively low. However, it demonstrates that we can have the same reconstruction using the proximal operator 70% of the time. Finally, the computational gain is more apparent when we increase the number of inner iterations, thereby increasing the computational cost of the inner solver.

We note that such expensive inner steps are used by open source imaging libraries and are solved with different algorithms and stopping rules. In CIL [18, 27], FISTA is used to solve (6) with 100 iterations. In PyHST2 [23], FISTA is used with 200 iterations and a duality gap is evaluated every 3 iterations as a stopping rule. In TIGRE [7], the PDHG algorithm with adaptive step sizes, [30], is applied to (6), with 50 iterations. In DeepInv[1], PDHG is applied to (5) (with $\boldsymbol{A} = \mathbf{Id}$) with 1000 iterations and error distance between two consecutive iterates. Finally, in Tomopy [16], PDHG is applied to (5) (with $\boldsymbol{A} = \mathbf{Id}$) and the number of iterations is specified by the user. All these default options are optimised and tested for particular real-world data cases like the ones encountered in synchroton facilities for instance. Alternatively, one can avoid inexact inner solvers for TV denoising. In [19] the proximal is replaced by a combination of wavelet and scaling transforms and is computed using a componentwise shrinkage operator. Even in this case, we expect a computational gain by skipping this operator.

4 Code Reproducibility

The code and datasets needed to reproduce the results are available at https://github.com/epapoutsellis/Why-do-we-regularise-in-every-iteration. For all the experiments we use an Apple M2 Pro, 16Gb without GPU usage to avoid measuring data transferring time between the host and the device which can be misleading.

[1] https://github.com/deepinv/deepinv.

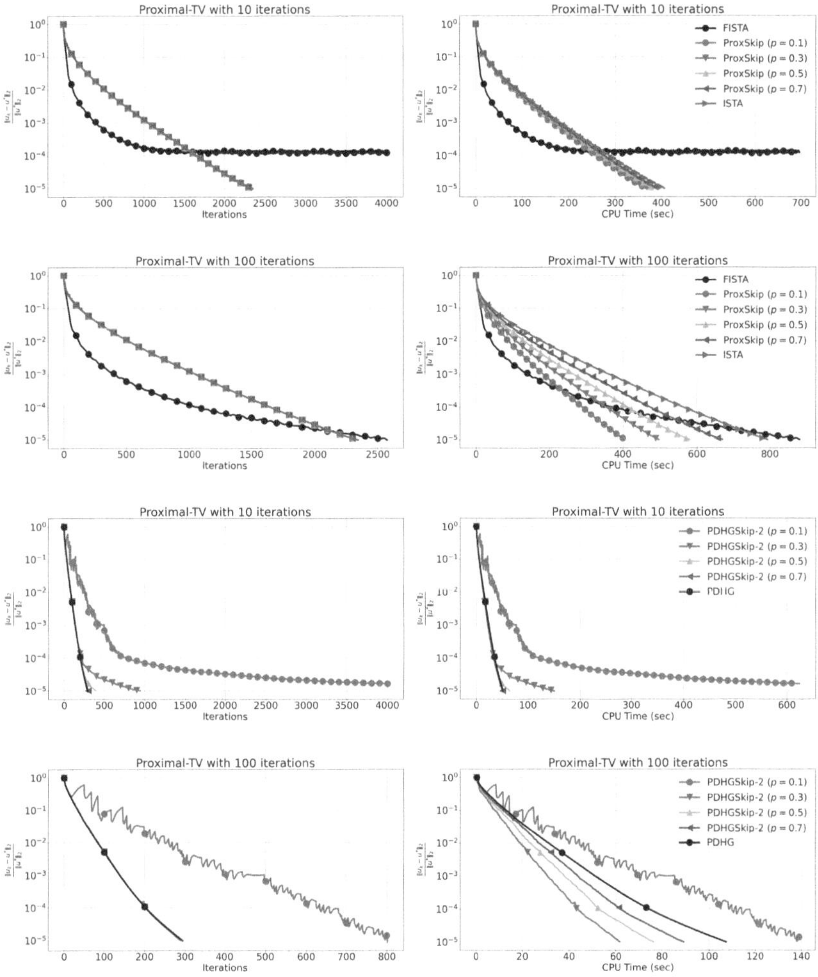

Fig. 8. Comparing ISTA, FISTA, PDHG, ProxSkip and PDHGSkip-2 for multiple values of p for the TV tomography problem. The proximal of TV is solved using FISTA applied on (6) for 10 and 100 iterations.

5 Discussion and Future Work

In this paper, we explored the use of the ProxSkip algorithm for imaging inverse problems. ProxSkip allows skipping costly proximal operators, usually related to the regulariser, without impacting the convergence and the solution. In addition, we presented a new skipped version of PDHG. This can be useful when the L-smoothness assumption is not satisfied, e.g., when using the Kullback-Leibler divergence. Here we focused on showcasing its advantages in practice rather than proving its convergence which we leave it for future work. Although we demonstrated that avoiding computing the proximal leads to better computational times, this speed-up can be further increased when dealing with larger datasets and more costly regularisers, e.g., TGV and TNV. In addition to skipping, one can combine stochastic optimisation methods and variance reduction estimators that use only a subset of the data per iteration. For example, in tomography applications, where the per-iteration cost is dominated by the forward and backward operations, one can randomly select a smaller subset of projection angles in addition to a random evaluation of the proximal operator. In this scenario, the per-iteration cost is significantly decreased and preliminary experiments have shown that it outperforms deterministic algorithms in terms of CPU time. Finally, we note that further computational gains could be achieved by applying skipping techniques to the inner solver as well. Notice that Algorithm (6) by default skips the proximal of g. For example, in the case of TV denoising, one can use the approach of Sect. 2.1, employing ProxSkip on its dual.

In summary, proximal-based algorithms presented herein, along with potential future extensions, open the door to revisiting a range of optimisation solutions for a plethora of imaging modalities developed over the last decades, with the potential to greatly reduce actual computation times.

Acknowledgments. E. Pap acknowledges funding through the Innovate UK Analysis for Innovators (A4i) program: *Denoising of chemical imaging and tomography data* under which the experiments were initially conducted. E. Pap acknowledges also CCPi (EPSRC grant EP/T026677/1). Ž. K. is supported by the UK EPSRC grant EP/X010740/1.

References

1. The MOSEK optimization toolbox for MATLAB manual. Version 10.1. (2024)
2. Anthoine, S., Aujol, J.F., Boursier, Y., Melot, C.: Some proximal methods for Poisson intensity CBCT and PET. Inverse Probl. Imaging **6**(4), 565–598 (2012)
3. Aujol, J.F.: Some first-order algorithms for total variation based image restoration. J. Math. Imaging Vis. **34**(3), 307–327 (2009)
4. Aujol, J.F., Calatroni, L., et al.: Parameter-free FISTA by adaptive restart and backtracking. SIAM J. Optim. **34**(4), 3259–3285 (2024)
5. Beck, A., Teboulle, M.: Fast gradient-based algorithms for constrained total variation image denoising and deblurring problems. IEEE Trans. Image Process. **18**(11), 2419–2434 (2009)
6. Beck, A., Teboulle, M.: A fast iterative shrinkage-thresholding algorithm for linear inverse problems. SIAM J. Imaging Sci. **2**(1), 183–202 (2009)
7. Biguri, A., Dosanjh, M., et al.: TIGRE: a MATLAB-GPU toolbox for CBCT image reconstruction. Biomed. Phys. Eng. Express **2**(5), 055010 (2016)
8. Bredies, K., Kunisch, K., Pock, T.: Total generalized variation. SIAM J. Imaging Sci. **3**(3), 492–526 (2010)
9. Burger, M., Sawatzky, A., Steidl, G.: First order algorithms in variational image processing. In: Glowinski, R., Osher, S.J., Yin, W. (eds.) Splitting Methods in Communication, Imaging, Science, and Engineering. SC, pp. 345–407. Springer, Cham (2016). https://doi.org/10.1007/978-3-319-41589-5_10

10. Chambolle, A., Dossal, C.: On the convergence of the iterates of the "fast iterative shrinkage/thresholding algorithm". J. Optim. Theory Appl. **166**(3), 968–982 (2015)
11. Chambolle, A., Pock, T.: A first-order primal-dual algorithm for convex problems with applications to imaging. J. Math. Imaging Vis. **40**(1), 120–145 (2011)
12. Chambolle, A.: An algorithm for total variation minimization and applications. J. Math. Imaging Vis. **20**(1), 89–97 (2004)
13. Combettes, P.L., Wajs, V.R.: Signal recovery by proximal forward-backward splitting. Multiscale Model. Sim. **4**(4), 1168–1200 (2005)
14. Condat, L., Richtárik, P.: Randprox: primal-dual optimization algorithms with randomized proximal updates. In: 11th ICLR (2023)
15. Diamond, S., Boyd, S.: CVXPY: a python-embedded modeling language for convex optimization. J. Mach. Learn. Res. **17**(83), 1–5 (2016)
16. Gürsoy, D., De Carlo, F., et al.: TomoPy: a framework for the analysis of synchrotron tomographic data. J. Synchrotron Radiat. **21**(5), 1188–1193 (2014)
17. Holt, K.M.: Total nuclear variation and Jacobian extensions of total variation for vector fields. IEEE Trans. Image Process. **23**(9), 3975–3989 (2014)
18. Jørgensen, J.S., Ametova, E., et al.: Core Imaging Library - part I: a versatile Python framework for tomographic imaging. Philos. Trans. A Math. Phys. Eng. Sci. **379**(2204), 20200192 (2021)
19. Kamilov, U.S.: Minimizing isotropic total variation without subiterations. Mitsubishi Electric Research Laboratories (MERL) (2) (2016)
20. Lefkimmiatis, S., Roussos, A., Unser, M., Maragos, P.: Convex generalizations of total variation based on the structure tensor with applications to inverse problems. In: Scale Space and Variational Methods in Computer Vision, pp. 48–60 (2013)
21. Liang, J., Luo, T., et al.: Improving fast iterative shrinkage-thresholding algorithm: faster, smarter, and greedier. SIAM J. Sci. Comput. **44**(3), A1069–A1091 (2022)
22. Matras, D., Vamvakeros, A., et al.: Multi-length scale 5D diffraction imaging of Ni-Pd/CeO2-ZrO2/Al2O3 catalyst during partial oxidation of methane. J. Mater. Chem. A **9**(18), 11331–11346 (2021)
23. Mirone, A., Brun, E., et al.: The PyHST2 hybrid distributed code for high speed tomographic reconstruction with iterative reconstruction and a priori knowledge capabilities. Nucl. Instrum. Methods Phys. Res. B **324**, 41–48 (2014)
24. Mishchenko, K., Malinovsky, G., et al.: ProxSkip: yes! local gradient steps provably lead to communication acceleration! Finally! In: PMLR 39, vol. 162, pp. 15750–15769 (2022)
25. Nesterov, Y.E.: A method of solving a convex programming problem with convergence rate $o\left(\frac{1}{k^2}\right)$. Dokl. Akad. Nauk SSSR **269**(3), 543–547 (1983)
26. O'Donoghue, B., Candès, E.: Adaptive restart for accelerated gradient schemes. Found. Comput. Math. **15**(3), 715–732 (2013)
27. Papoutsellis, E., Ametova, E., et al.: Core Imaging Library - part II: multichannel reconstruction for dynamic and spectral tomography. Philos. Trans. A Math. Phys. Eng. Sci. **379**(2204), 20200193 (2021)
28. Pock, T., Chambolle, A.: Diagonal preconditioning for first order primal-dual algorithms in convex optimization. In: 2011 International Conference on Computer Vision, pp. 1762–1769 (2011)
29. Rasch, J., Chambolle, A.: Inexact first-order primal-dual algorithms. Comput. Optim. Appl. **76**(2), 381–430 (2020)
30. Zhu, M., Chan, T.: An efficient primal-dual hybrid gradient algorithm for total variation image restoration. CAM Reports 08-34, UCLA, Center for Applied Math., 2008. (2) (2008)

On an Analytical Inversion Formula for the Modulo Radon Transform

Matthias Beckmann[1,2](✉) and Carla Dittert[1]

[1] Center for Industrial Mathematics, University of Bremen Bibliothekstraße 5, 28359 Bremen, Germany
research@mbeckmann.de, cadi@uni-bremen.de

[2] Department of Electrical and Electronic Engineering, Imperial College London, London SW72AZ, U.K.

Abstract. This paper proves a novel analytical inversion formula for the so-called modulo Radon transform (MRT), which models a recently proposed approach to one-shot high dynamic range tomography. It is based on the solution of a Poisson problem linking the Laplacian of the Radon transform (RT) of a function to its MRT in combination with the classical filtered back projection formula for inverting the RT. Discretizing the inversion formula using Fourier techniques leads to our novel Laplacian Modulo Unfolding - Filtered Back Projection algorithm, in short LMU-FBP, to recover a function from fully discrete MRT data. Our theoretical findings are finally supported by numerical experiments.

Keywords: X-ray computerized tomography · high dynamic range · Radon transform · modulo non-linearity · analytical inversion formula

1 Introduction

Computerized tomography (CT) deals with the recovery of the interior of scanned objects from given X-ray measurements taken at different views. By design, the detector's dynamic range is limited leading to saturation artifacts in the reconstruction of high-contrast objects, cf. Fig. 1. The goal of increasing the dynamic range points towards the recent topic of high dynamic range (HDR) tomography. Inspired by consumer HDR photography, typical multi-exposure approaches combine several low dynamic range measurements into a single image with increased dynamic range [6]. To avoid the drawback of manual calibration of the detector for each exposure, an automated calibration approach is proposed in [7,9]. Unfortunately, multi-exposure approaches suffer from the same difficulties as HDR photography: ghosting artifacts due to movements of the investigated object; increased acquisition time; unknown detector's sensor response necessary for satisfactory image fusion. Moreover, the mentioned approaches are solely based on empirical experiments lacking mathematical recovery guarantees.

As opposed to this, inspired by the Unlimited Sampling (US) framework [4,5], a recently introduced single-exposure approach based on a co-design of hardware and algorithms allows for mathematically backed recovery strategies for HDR CT [2,3]. Here, instead of truncating measurements that exceed the dynamic range,

T. A. Bubba et al. (Eds.): SSVM 2025, LNCS 15667, pp. 56–69, 2025.
https://doi.org/10.1007/978-3-031-92366-1_5

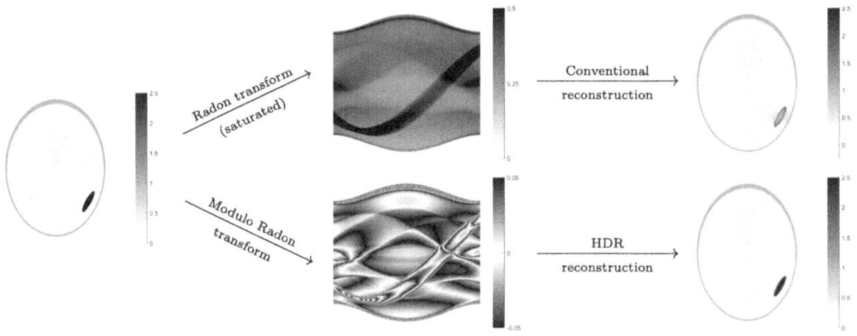

Fig. 1. Demonstration of reconstruction artifacts caused by saturation in Radon data. Left: modified Shepp-Logan phantom with additional ellipsoidal block of density 2.5. Middle: saturated Radon data with reduced range $[0, 0.5]$ (top) and modulo Radon data with dynamic range $[-0.05, 0.05]$. Right: reconstruction from saturated data showing severe artifacts (top) and HDR reconstruction from modulo data (bottom).

these are folded into the limited range of the detector via a modulo operation implemented in hardware. Thereon, these folded measurements are unfolded algorithmically. Mathematically, this can be modeled by the so-called modulo Radon transform (MRT) [3], which concatenates the modulo operation with the well-known Radon transform (RT), yielding a nonlinear reconstruction problem. For illustration, Fig. 1 shows a MRT with dynamic range reduction of 10 times along with the corresponding HDR reconstruction, yielding near perfect recovery. Albeit algorithmic recovery strategies exist [2,3], an explicit inversion formula for the MRT is still missing. In this work, we address this point by proposing an analytical inversion formula for the MRT, that combines the solution of a Poisson equation to unfold the modulo operation with the well-known filtered back projection (FBP) formula to invert the RT. The formulation of the Poisson equation is inspired by an approach to the classical phase unwrapping problem in [13,15]. By discretizing our analytical inversion formula, we deduce the novel LMU-FBP algorithm for recovering a target from MRT data, which operates simultaneously on the angle and radial variable in contrast to the US-FBP algorithm [3]. Moreover, our inversion formula does not require bandlimited Radon data so that LMU-FBP does not rely on bandlimitedness as opposed to [2].

The paper starts with an overview of the MRT in §2. This is followed by our main theoretical result: the deduction of the Poisson equation with suitable boundary conditions in §3 and the proof of the analytical inversion in §4. Finally, we present our LMU-FBP algorithm in §5 and showcase numerical experiments with non-bandlimited test cases in §6 including a comparison with US-FBP [3].

2 Modulo Radon Transform

For $f \in \mathrm{L}^1(\mathbb{R}^2)$, we define its *Radon transform* (RT) $\mathcal{R}f \in \mathrm{L}^1((-\pi,\pi) \times \mathbb{R})$ via

$$\mathcal{R}f(\vartheta,t) := \int_{\mathbb{R}} f(t\cos(\vartheta) - s\sin(\vartheta), t\sin(\vartheta) + s\cos(\vartheta))\,\mathrm{d}s.$$

Moreover, for $g \in \mathrm{L}^\infty((-\pi,\pi)\times\mathbb{R})$, we define its *back projection* $\mathcal{B}g \in \mathrm{L}^\infty(\mathbb{R}^2)$ via

$$\mathcal{B}g(x) := \int_{-\pi}^{\pi} g(\vartheta, x_1\cos(\vartheta) + x_2\sin(\vartheta))\,\mathrm{d}\vartheta.$$

Note that $\mathcal{R}$ is injective on $\mathrm{L}^1(\mathbb{R}^2)$ and, for $f \in \mathrm{L}^1(\mathbb{R}^2) \cap \mathrm{C}(\mathbb{R}^2)$ with integrable Fourier transform $\mathcal{F}f \in \mathrm{L}^1(\mathbb{R}^2)$, defined as

$$\mathcal{F}f(\omega) = \int_{\mathbb{R}^2} f(x)\,\mathrm{e}^{-\mathrm{i}x^T\omega}\,\mathrm{d}x,$$

the inversion of $\mathcal{R}$ is given by the classical *filtered back projection formula*

$$f(x) = \frac{1}{4\pi}\mathcal{B}\left(\mathcal{F}^{-1}\left[|S|\,\mathcal{F}\left(\mathcal{R}f\right)(\vartheta,S)\right]\right)(x), \tag{1}$$

which holds pointwise for all $x \in \mathbb{R}^2$ and where $\mathcal{F}$ acts on the second variable.

The FBP formula (1), however, requires exact knowledge of $\mathcal{R}f$ and, in particular, saturation effects due to range limitations lead to severe artefacts in the reconstruction. To circumvent this, in [3], the following *modulo Radon transform* is introduced, which folds $\mathcal{R}f$ into a given range interval $[-\lambda,\lambda]$ with $\lambda > 0$.

Definition 1 (Modulo Radon transform). For $\lambda > 0$ and $f \in \mathrm{L}^1(\mathbb{R}^2)$ the *modulo Radon transform (MRT)* $\mathcal{R}^\lambda f : (-\pi,\pi) \times \mathbb{R} \to [-\lambda,\lambda]$ is defined as

$$\mathcal{R}^\lambda f(\vartheta,t) := \mathcal{M}^\lambda(\mathcal{R}f(\vartheta,t)),$$

where $\mathcal{M}^\lambda : \mathbb{R} \to [-\lambda,\lambda]$ is the 2λ-*modulo operator* with $\mathcal{M}^\lambda(t) := t - 2\lambda\left\lfloor \frac{t+\lambda}{2\lambda} \right\rfloor$.

In [3], it is shown that $\mathcal{R}^\lambda$ is injective on the space $\mathscr{B}_L^1(\mathbb{R}^2)$ of bandlimited integrable functions with bandwidth L for any $L > 0$ and on the space $\mathrm{C}_c(\mathbb{R}^2)$ of continuous functions with compact support, which implies the invertibility of $\mathcal{R}^\lambda$ on suitable spaces. An analytical inversion formula, however, is not known so far. In the following sections, we take steps towards closing this gap by first relating the Laplacian of $\mathcal{R}f$ to $\mathcal{R}^\lambda f$ and, afterwards, showing unique solvability of the corresponding Poisson problem. This in combination with the FBP formula (1) then gives a new analytical inversion formula for $\mathcal{R}^\lambda$ under suitable assumptions.

3 Poisson Problem for the Modulo Radon Transform

Relating the Laplacian of $\mathcal{R}f$ to $\mathcal{R}^\lambda f$ via a Poisson problem is inspired by [15] from the context of phase unwrapping. Let $f \in \mathrm{L}^1(\mathbb{R}^2)$ such that $\mathcal{R}f$ is twice continuously differentiable on $(-\pi, \pi) \times \mathbb{R}$. Furthermore, let $\lambda > 0$ be the modulo threshold. Then, the Laplacian of $\exp(\mathrm{i}\frac{\pi}{\lambda}\mathcal{R}f)$ can be computed as

$$\Delta \exp\Big(\mathrm{i}\frac{\pi}{\lambda}\mathcal{R}f\Big) = \left[\mathrm{i}\frac{\pi}{\lambda}\Delta\mathcal{R}f - \Big(\frac{\pi}{\lambda}\frac{\partial}{\partial\vartheta}\mathcal{R}f\Big)^2 - \Big(\frac{\pi}{\lambda}\frac{\partial}{\partial t}\mathcal{R}f\Big)^2\right] \exp\Big(\mathrm{i}\frac{\pi}{\lambda}\mathcal{R}f\Big),$$

and, thereon, for the Laplacian of $\mathcal{R}f$ follows that

$$\Delta\mathcal{R}f = \frac{\lambda}{\pi}\left[\cos\Big(\frac{\pi}{\lambda}\mathcal{R}f\Big)\Delta\sin\Big(\frac{\pi}{\lambda}\mathcal{R}f\Big) - \sin\Big(\frac{\pi}{\lambda}\mathcal{R}f\Big)\Delta\cos\Big(\frac{\pi}{\lambda}\mathcal{R}f\Big)\right]. \tag{2}$$

Observe that $\mathcal{R}f$ can be decomposed pointwise into $\mathcal{R}^\lambda f$ and a piecewise constant residual function with values in $2\lambda\mathbb{Z}$, i.e.,

$$\mathcal{R}f = \mathcal{R}^\lambda f + 2\lambda\varepsilon_{\mathcal{R}f} \quad \text{with} \quad \varepsilon_{\mathcal{R}f} = \sum\nolimits_{i\in\mathcal{I}} c_i 1_{D_i}. \tag{3}$$

Here, $\mathcal{I}$ denotes an arbitrary index set such that $\mathbb{R}^2$ is covered by the pairwise disjoint sets $\{D_i\}_{i\in\mathcal{I}}$, and the coefficients $c_i \in \mathbb{Z}$ are integers. Exploiting the 2π-periodicity of the sine and cosine function in (2) results in the *Poisson equation*

$$\Delta\mathcal{R}f = \frac{\lambda}{\pi}\left[\cos\Big(\frac{\pi}{\lambda}\mathcal{R}^\lambda f\Big)\Delta\sin\Big(\frac{\pi}{\lambda}\mathcal{R}^\lambda f\Big) - \sin\Big(\frac{\pi}{\lambda}\mathcal{R}^\lambda f\Big)\Delta\cos\Big(\frac{\pi}{\lambda}\mathcal{R}^\lambda f\Big)\right]. \tag{4}$$

To compute $\mathcal{R}f$ from given $\mathcal{R}^\lambda f$, we aim to solve the Poisson equation (4). For this, we restrict the domain of $\mathcal{R}f$ to a bounded rectangle and impose boundary conditions. Note that, for applications like CT, it is reasonable to assume that the function f is compactly supported. Hence, in the following, let $f \in \mathrm{L}^1(\mathbb{R}^2)$ have compact support in the open unit ball $\mathbb{B}_1(0) \subseteq \mathbb{R}^2$. As the compact support of f transfers to $\mathcal{R}f$, we obtain $\operatorname{supp}\mathcal{R}f(\vartheta,\cdot) \subseteq (-1,1)$ for all $\vartheta \in (-\pi,\pi)$. Consequently, we assume $f \in \mathrm{L}^1(\mathbb{B}_1(0))$ as in [10,12] and extend f by zero if needed. On the restricted Lipschitz domain $\Omega = (-\pi,\pi) \times (-1,1)$, the RT $\mathcal{R}f : \Omega \to \mathbb{R}$ can be represented as

$$\mathcal{R}f(\vartheta, t) = \int_{\ell_{\theta(\vartheta),t}\cap\mathbb{B}_1(0)} f(x)\,\mathrm{d}x$$

with $\theta(\vartheta) = (\cos(\vartheta), \sin(\vartheta))^\perp$. With this, $\mathcal{R}f$ satisfies homogeneous Dirichlet boundary conditions on $\Gamma_D := (-\pi,\pi) \times \{-1\} \cup (-\pi,\pi) \times \{1\} \subseteq \partial\Omega$, where $\partial\Omega$ denotes the boundary of Ω. Furthermore, $\mathcal{R}f$ is 2π-periodic with respect to ϑ. Let $\mathcal{R}f$ also be continuously differentiable with respect to the first argument on $[-\pi,\pi] \times \mathbb{R}$. Then, $\mathcal{R}f$ and its first partial derivative with respect to ϑ have to satisfy 2π-periodic boundary conditions on $\Gamma_P := \{-\pi\}\times(-1,1)\cup\{\pi\}\times(-1,1) \subseteq \partial\Omega$. Altogether, $\mathcal{R}f$ has to fulfill the *mixed Dirichlet-periodic boundary conditions*

$$\begin{cases} \mathcal{R}f(\vartheta,-1) = \mathcal{R}f(\vartheta,1) = 0 & \text{for } \vartheta \in (-\pi,\pi) \\ \mathcal{R}f(-\pi,t) = \mathcal{R}f(\pi,t) & \text{for } t \in (-1,1) \\ \frac{\partial}{\partial\vartheta}\mathcal{R}f(-\pi,t) = \frac{\partial}{\partial\vartheta}\mathcal{R}f(\pi,t) & \text{for } t \in (-1,1). \end{cases} \tag{5}$$

In total, by abbreviating the right-hand side in (4) as

$$\mathcal{R}^{\lambda}_{\Delta}(f) := \left(\tfrac{\lambda}{\pi}\left[\cos(\tfrac{\pi}{\lambda}\mathcal{R}^{\lambda}f)\Delta\sin(\tfrac{\pi}{\lambda}\mathcal{R}^{\lambda}f) - \sin(\tfrac{\pi}{\lambda}\mathcal{R}^{\lambda}f)\Delta\cos(\tfrac{\pi}{\lambda}\mathcal{R}^{\lambda}f)\right]\right)$$

and setting $\Omega_\vartheta = (-\pi,\pi)$, $\Omega_t = (-1,1)$, we obtain the *Poisson problem*

$$\begin{cases} \Delta\mathcal{R}f = \mathcal{R}^{\lambda}_{\Delta}(f) & \text{on } \Omega \\ \mathcal{R}f(\cdot,-1) = 0,\ \mathcal{R}f(\cdot,1) = 0 & \text{on } \Omega_\vartheta \\ \mathcal{R}f(-\pi,\cdot) = \mathcal{R}f(\pi,\cdot);\ \frac{\partial}{\partial\vartheta}\mathcal{R}f(-\pi,\cdot) = \frac{\partial}{\partial\vartheta}\mathcal{R}f(\pi,\cdot) & \text{on } \Omega_t. \end{cases} \tag{6}$$

4 Analytical Inversion of the Modulo Radon Transform

The solution of the boundary value problem (6) enables the inversion of the modulo operation. To guarantee the existence of a unique solution, we employ certain periodic function spaces. A periodic function is defined on the torus $\mathrm{T}^2 = [-\pi,\pi]\times[-1,1]$ such that opposite points are identified with each other, analogously to [17, § 9.1.1]. With this, the *Lebesgue space* $\mathrm{L}^2_{\mathrm{per}}(\mathrm{T}^2)$ of periodic and square-integrable functions is defined as

$$\mathrm{L}^2_{\mathrm{per}}(\mathrm{T}^2) := \left\{ f : \mathrm{T}^2 \to \mathbb{R} \text{ periodic, measurable} \,\middle|\, \|f\|_{\mathrm{L}^2_{\mathrm{per}}} < \infty \right\}$$

with norm

$$\|f\|_{\mathrm{L}^2_{\mathrm{per}}} := \left(\int_{\mathrm{T}^2} |f(x)|^2 \,\mathrm{d}x\right)^{\frac{1}{2}}$$

while identifying functions which agree almost everywhere on T^2. Let ∂_ϑ and ∂_t denote the first weak partial derivative with respect to ϑ and t, respectively. Then, the *periodic Sobolev space* $\mathrm{H}^1_{\mathrm{per}}(\mathrm{T}^2)$ on T^2 is defined as

$$\mathrm{H}^1_{\mathrm{per}}(\mathrm{T}^2) := \left\{ f \in \mathrm{L}^2_{\mathrm{per}}(\mathrm{T}^2) \,\middle|\, \partial_\vartheta f,\ \partial_t f \in \mathrm{L}^2_{\mathrm{per}}(\mathrm{T}^2) \right\}$$

with

$$\|f\|_{\mathrm{H}^1_{\mathrm{per}}} := \left(\|f\|^2_{\mathrm{L}^2_{\mathrm{per}}} + \|\partial_\vartheta f\|^2_{\mathrm{L}^2_{\mathrm{per}}} + \|\partial_t f\|^2_{\mathrm{L}^2_{\mathrm{per}}}\right)^{\frac{1}{2}}.$$

Furthermore, we define the subspace $\mathrm{H}^1_{\mathrm{per},D}(\mathrm{T}^2)$ that encodes homogeneous Dirichlet boundary conditions on a part of the boundary as the closure

$$\mathrm{H}^1_{\mathrm{per},D}(\mathrm{T}^2) := \overline{\mathrm{C}^\infty_{\mathrm{per},D}(\mathrm{T}^2)}^{\mathrm{H}^1_{\mathrm{per}}}$$

of the space $\mathrm{C}^\infty_{\mathrm{per},D}(\mathrm{T}^2)$ of periodic smooth functions vanishing in a neighborhood of Γ_D

$$\mathrm{C}^\infty_{\mathrm{per},D}(\mathrm{T}^2) := \left\{ \varphi \in \mathrm{C}^\infty_{\mathrm{per}}(\mathrm{T}^2) \,\middle|\, \begin{array}{c} \exists\, 0 < r < 1\ \forall \vartheta \in (-\pi,\pi): \\ \operatorname{supp}\varphi(\vartheta,\cdot) \subseteq (-r,r) \end{array} \right\}.$$

Employing standard arguments, the well-known Poincaré inequality can be adapted to the setting of $\mathrm{H}^1_{\mathrm{per},D}(\mathrm{T}^2)$.

Proposition 1 (Poincaré inequality) . *There exists a constant $C_p > 0$ such that*

$$\|u\|_{\mathrm{L}^2_{\mathrm{per}}} \le C_p \|\nabla u\|_{\mathrm{L}^2_{\mathrm{per}}} \quad \forall\, u \in \mathrm{H}^1_{\mathrm{per},D}(\mathrm{T}^2).$$

Due to space limitations, we omit the proof. Instead, we now formulate our main theoretical result: an analytical inversion formula for the MRT.

Theorem 1 (Inversion formula for $\mathcal{R}^\lambda$). *Let $f \in \mathrm{L}^1(\mathbb{B}_1(0)) \cap \mathrm{C}(\overline{\mathbb{B}_1(0)})$ be compactly supported in $\mathbb{B}_1(0)$ with $\mathcal{F}f \in \mathrm{L}^1(\mathbb{R}^2)$. Moreover, let $\mathcal{R}f \in \mathrm{C}^2(\overline{\Omega})$ with $\Omega = (-\pi,\pi) \times (-1,1)$ and set*

$$\mathcal{R}^\lambda_\Delta(f) := \left(\tfrac{\lambda}{\pi}\left[\cos(\tfrac{\pi}{\lambda}\mathcal{R}^\lambda f)\Delta \sin(\tfrac{\pi}{\lambda}\mathcal{R}^\lambda f) - \sin(\tfrac{\pi}{\lambda}\mathcal{R}^\lambda f)\Delta \cos(\tfrac{\pi}{\lambda}\mathcal{R}^\lambda f)\right]\right).$$

Then,

$$f(x) = \frac{1}{4\pi}\mathcal{B}\left(\mathcal{F}_1^{-1}\left[|S|\,\mathcal{F}_1\left(\Delta^{-1}\mathcal{R}^\lambda_\Delta(f)\right)(\vartheta,S)\right]\right)(x) \tag{7}$$

holds for all $x \in \mathbb{R}^2$, where $\Delta^{-1}\mathcal{R}^\lambda_\Delta(f)$ is the weak solution to (6) *for given $\mathcal{R}^\lambda f$.*

For the proof of the Theorem 1, we first show that the Poisson problem (6) admits a unique weak solution. To this end, we deduce a weak formulation of the boundary value problem, where our calculations are inspired by [1, §5.2, 6.5, 7.4] and [14, §6.8.2, 8.3]. Assume that $f \in \mathrm{L}^1(\mathbb{B}_1(0))$, $\mathcal{R}^\lambda_\Delta(f) \in \mathrm{C}(\overline{\Omega})$ and $\mathcal{R}f \in \mathrm{C}^2(\overline{\Omega})$. Then, $\mathcal{R}f$ is a classical solution to the boundary value problem in (6). Analogously to [12], we periodize $\mathcal{R}f$ with respect to t, since the values on the opposing boundary parts $(-\pi,\pi) \times \{-1\}$ and $(-\pi,\pi) \times \{1\}$ agree due to the Dirichlet conditions in (5). Consequently, multiplying the Poisson equation (4) with a test function $\varphi \in \mathrm{C}^\infty_{\mathrm{per},D}(\mathrm{T}^2)$ and applying Green's first identity results in

$$-\int_{\mathrm{T}^2} \varphi\,\mathcal{R}^\lambda_\Delta(f)\,\mathrm{d}(\vartheta,t) = \int_\Omega \nabla\varphi \cdot \nabla\mathcal{R}f\,\mathrm{d}(\vartheta,t) - \int_{\partial\Omega} \varphi\,\nabla\mathcal{R}f \cdot \nu\,\mathrm{d}\sigma,$$

where the integral along the boundary vanishes. To see this, we use that the test function φ is periodic and vanishes in a neighborhood of Γ_D, as well as the condition that the partial derivative $\partial_\vartheta \mathcal{R}f$ is 2π-periodic with respect to ϑ. Finally, we enlarge the set of test functions to $\mathrm{H}^1_{\mathrm{per},D}(\mathrm{T}^2)$ to obtain the desired weak formulation: For $\mathcal{R}^\lambda_\Delta(f) \in \mathrm{L}^2_{\mathrm{per}}(\mathrm{T}^2)$, a function $u \in \mathrm{H}^1_{\mathrm{per},D}(\mathrm{T}^2)$ is called a *weak solution* to the Poisson problem (6) if

$$\int_{\mathrm{T}^2} \nabla u(\vartheta,t) \cdot \nabla v(\vartheta,t)\,\mathrm{d}(\vartheta,t) = -\int_{\mathrm{T}^2} v(\vartheta,t)\,\mathcal{R}^\lambda_\Delta(f)(\vartheta,t)\,\mathrm{d}(\vartheta,t) \tag{8}$$

holds for all $v \in \mathrm{H}^1_{\mathrm{per},D}(\mathrm{T}^2)$.

Lemma 1 (Weak solution) . *Let $f \in \mathrm{L}^1(\mathbb{B}_1(0))$ with $\mathcal{R}^\lambda_\Delta(f) \in \mathrm{L}^2_{\mathrm{per}}(\mathrm{T}^2)$. Then, the Poisson problem* (6) *has a unique weak solution $u_0 \in \mathrm{H}^1_{\mathrm{per},D}(\mathrm{T}^2)$ satisfying* (8)*. Moreover, there exists a constant $C > 0$ such that the stability estimate*

$$\|u_0\|_{\mathrm{H}^1_{\mathrm{per}}} \le C \left\|\mathcal{R}^\lambda_\Delta(f)\right\|_{\mathrm{L}^2_{\mathrm{per}}} \tag{9}$$

holds.

Proof We follow a standard approach based on the classical Lax-Milgram theorem, see e.g. [1, Chapter 5.2, 6.5 7], [14, Chapter 8.3]. To this end, we first define the bilinear form $\alpha : \mathrm{H}^1_{\mathrm{per},D}(\mathrm{T}^2) \times \mathrm{H}^1_{\mathrm{per},D}(\mathrm{T}^2) \to \mathbb{R}$ via

$$\alpha(u,v) := \int_{\mathrm{T}^2} \nabla u(\vartheta,t) \cdot \nabla v(\vartheta,t) \,\mathrm{d}(\vartheta,t).$$

Due to the Poincaré inequality, Proposition 1, the mapping α is an inner product on $\mathrm{H}^1_{\mathrm{per},D}(\mathrm{T}^2)$ and the Cauchy-Schwarz inequality implies the continuity of α by

$$|a(u,v)| \leq \|\nabla u\|_{\mathrm{L}^2_{\mathrm{per}}} \|\nabla v\|_{\mathrm{L}^2_{\mathrm{per}}} \leq \|u\|_{\mathrm{H}^1_{\mathrm{per}}} \|v\|_{\mathrm{H}^1_{\mathrm{per}}}.$$

Furthermore, the Poincaré inequality guarantees the existence of $C_p > 0$ such that

$$\alpha(u,u) \geq \frac{1}{2}\|\nabla u\|^2_{\mathrm{L}^2_{\mathrm{per}}} + \frac{1}{2C_p^2}\|u\|^2_{\mathrm{L}^2_{\mathrm{per}}} \geq \min\left\{\frac{1}{2}, \frac{1}{2C_p^2}\right\}\|u\|^2_{\mathrm{H}^1_{\mathrm{per}}}.$$

This shows the coercivity of α on $\mathrm{H}^1_{\mathrm{per},D}(\mathrm{T}^2)$. Secondly, we define the linear form $F : \mathrm{H}^1_{\mathrm{per},D}(\mathrm{T}^2) \to \mathbb{R}$ by

$$F(v) := -\int_{\mathrm{T}^2} v(\vartheta,t)\, \mathcal{R}^\lambda_\Delta(f)(\vartheta,t) \,\mathrm{d}(\vartheta,t).$$

Again, the Cauchy-Schwarz inequality implies the continuity of F by

$$|F(v)| \leq \left\|\mathcal{R}^\lambda_\Delta(f)\right\|_{\mathrm{L}^2_{\mathrm{per}}} \|v\|_{\mathrm{L}^2_{\mathrm{per}}} \leq \left\|\mathcal{R}^\lambda_\Delta(f)\right\|_{\mathrm{L}^2_{\mathrm{per}}} \|v\|_{\mathrm{H}^1_{\mathrm{per}}}.$$

Consequently, all requirements of the Lax-Milgram theorem are satisfied, and there exists a unique $u_0 \in \mathrm{H}^1_{\mathrm{per},D}(\mathrm{T}^2)$ such that

$$\int_{\mathrm{T}^2} \nabla u_0 \cdot \nabla v \,\mathrm{d}(\vartheta,t) = \alpha(u_0,v) = F(v) = -\int_{\mathrm{T}^2} v\, \mathcal{R}^\lambda_\Delta(f) \,\mathrm{d}(\vartheta,t)$$

is satisfied for all $v \in \mathrm{H}^1_{\mathrm{per},D}(\mathrm{T}^2)$, i.e., u_0 is the unique weak solution to the Poisson problem (6). Furthermore, the coercivity of α and the continuity of F imply that

$$\min\left\{\frac{1}{2}, \frac{1}{2C_p^2}\right\}\|u_0\|^2_{\mathrm{H}^1_{\mathrm{per}}} \leq |\alpha(u_0,u_0)| = |F(u_0)| \leq \left\|\mathcal{R}^\lambda_\Delta(f)\right\|_{\mathrm{L}^2_{\mathrm{per}}} \|u_0\|_{\mathrm{H}^1_{\mathrm{per}}}$$

and, thus, dividing by $\|u_0\|_{\mathrm{H}^1_{\mathrm{per}}}$ gives the stability estimate (9). □

Using the unique weak solution in Lemma 1, we can now prove our main theorem.

Proof (Theorem 1). In the first step, we invert the modulo operator by showing that the equation

$$\mathcal{R}f = \Delta^{-1}\left(\mathcal{R}^\lambda_\Delta(f)\right) \tag{10}$$

holds pointwise on Ω. Since $\mathcal{R}f \in C^2(\overline{\Omega})$ by assumption, the function $\mathcal{R}^\lambda_\Delta(f)$ is square-integrable. Moreover, since $\mathcal{R}^\lambda f$ is periodic with respect to ϑ and, after periodization, also periodic with respect to t, $\mathcal{R}^\lambda_\Delta(f)$ is periodic on Ω and, thus, $\mathcal{R}^\lambda_\Delta(f) \in L^2_{\mathrm{per}}(\mathrm{T}^2)$. Hence, Lemma 1 implies the existence of a unique weak solution $u_0 := \Delta^{-1}(\mathcal{R}^\lambda_\Delta(f)) \in H^1_{\mathrm{per},D}(\mathrm{T}^2)$ to the Poisson problem (6). It remains to argue that $\mathcal{R}f$ and u_0 coincide. By assumption, we have $\mathcal{R}f \in C^2_{\mathrm{per},D}(\mathrm{T}^2)$ and, hence, $\mathcal{R}f \in H^1_{\mathrm{per},D}(\mathrm{T}^2)$. Furthermore, $\mathcal{R}f$ satisfies the Poisson equation (4) and the boundary conditions (5). Following the deduction of (8), $\mathcal{R}f$ is also a weak solution of (6) and, due to the uniqueness of the weak solution, it follows that $u_0 = \mathcal{R}f \in H^1_{\mathrm{per},D}(\mathrm{T}^2)$. Consequently, the continuity of $\mathcal{R}f$ implies that (10) holds pointwise choosing a continuous representative.

In the second step, we need to invert the Radon operator. Since $f \in C(\overline{\mathbb{B}_1(0)})$ and $\mathcal{F}f \in L^1(\mathbb{R}^2)$ by assumption, the filtered back projection formula (1) holds pointwise on $\mathbb{R}^2$ and, consequently, the explicit inversion formula (7) for the MRT follows by combining (10) with (1). □

5 Numerical Inversion of the Modulo Radon Transform

We approximate the analytical inversion formula (7) numerically by solving the Poisson problem (6) with Fourier techniques to invert the modulo operation and applying the well-known discrete FBP algorithm to invert the RT. This combines into our novel *Laplacian Modulo Unfolding - Filtered Back Projection* (LMU-FBP) Algorithm 1, which we now explain in more detail.

The MRT is discretized using parallel beam geometry [10], where, due to the evenness of $\mathcal{R}f$, it suffices to consider the domain $[0, \pi) \times [-1, 1]$. For this, let $K, M \in \mathbb{N}$ and set $N = 2K + 1$. Moreover, let $T > 0$ be the radial sampling rate. Then, $\mathcal{R}^\lambda f$ is discretized by evaluating at the grid points $\vartheta_m = m\frac{\pi}{M}$, $m \in \{0, \ldots, M-1\}$, and $t_n = (n - K)T$, $n \in \{0, \ldots, N-1\}$, resulting in the discrete MRT data

$$\left\{p^\lambda[m,n] := \mathcal{R}^\lambda f(\vartheta_m, t_n) \,\middle|\, m \in \{0, \ldots, M-1\},\ n \in \{0, \ldots, N-1\}\right\}.$$

Inspired by [15], in the first *Laplacian Modulo Unfolding* (LMU) stage of Algorithm 1 we invert the modulo operator numerically by solving the Poisson equation (4) using discrete Fourier transforms. This is based on the observation that, under suitable assumptions, the Laplacian of a function p can be computed via

$$\Delta p = \mathcal{F}^{-1}\left(-|\cdot|^2\, \mathcal{F}p\right).$$

To incorporate the boundary conditions (5), the MRT data is extended. Then, LMU consists of five steps: In Step 1, the data is extended in ϑ to the interval $[0, 2\pi)$ such that it becomes 2π-periodic by employing the evenness property of $\mathcal{R}$. To ensure the homogeneous Dirichlet boundary conditions, in Step 2, the data is extended in t such that it becomes odd around $N + 1$. Using the discrete

Algorithm 1. LMU-FBP

Input: MRT data $p^\lambda[m,n]$ for $m \in \{0,\dots,M-1\}$, $n \in \{0,\dots,N-1\}$; low-pass filter A_L with bandwidth $L > 0$; interpolation method $\mathfrak{I}$

Laplacian Modulo Unfolding (LMU):

1: **for** $m \in \{0,\dots,2M-1\}$, $n \in \{0,\dots,N-1\}$ **do**

$$\tilde{p}^\lambda[m,n] \leftarrow \begin{cases} p^\lambda[m,n] & \text{if } m \in \{0,\dots,M-1\} \\ p^\lambda[m-M, N-1-n] & \text{if } m \in \{M,\dots,2M-1\} \end{cases}$$

2: **for** $m \in \{0,\dots,2M-1\}$, $n \in \{0,\dots,2N+1\}$ **do**

$$\hat{p}^\lambda[m,n] \leftarrow \begin{cases} 0 & \text{if } n \in \{0, N+1\} \\ \tilde{p}^\lambda[m,n-1] & \text{if } n \in \{1,\dots,N\} \\ -\tilde{p}^\lambda[m,2N+1-n] & \text{if } n \in \{N+2,\dots,2N+1\} \end{cases}$$

3: $\hat{p}_s^\lambda \leftarrow \mathrm{iDFT}[-(2\pi\,|\cdot|)^2\,\mathrm{DFT}[\sin(\frac{\pi}{\lambda}\hat{p}^\lambda)]]$, $\hat{p}_c^\lambda \leftarrow \mathrm{iDFT}[-(2\pi\,|\cdot|)^2\,\mathrm{DFT}[\cos(\frac{\pi}{\lambda}\hat{p}^\lambda)]]$,
$\hat{p}_\Delta^\lambda \leftarrow \frac{\lambda}{\pi}[\cos(\frac{\pi}{\lambda}\hat{p}^\lambda)\hat{p}_s^\lambda - \sin(\frac{\pi}{\lambda}\hat{p}^\lambda)\hat{p}_c^\lambda]$

4: $\hat{p} \leftarrow \mathrm{iDFT}[-\,(2\pi\,|\cdot|)^{-2}\,\mathrm{DFT}[\hat{p}_\Delta^\lambda]]$

5: **for** $m \in \{0,\dots,M-1\}$, $n \in \{0,\dots,N-1\}$ **do**

$$p_{\mathrm{LMU}}[m,n] \leftarrow \hat{p}[m,n+1]$$

Filtered Back Projection (FBP):

6: **for** $m \in \{0,\dots,M-1\}$, $i \in \mathcal{I}$ **do**

$$h(\vartheta_m,t_i) \leftarrow T\sum_{n=0}^{N-1}\mathcal{F}^{-1}A_L(t_i-t_n)\,p_{\mathrm{LMU}}[m,n]$$

7: **for** $(i,j) \in \mathcal{I}_x \times \mathcal{I}_y$ **do**

$$f_{\mathrm{LMU}}[i,j] \leftarrow \frac{1}{2M}\sum_{m=0}^{M-1}\mathfrak{I}h(\vartheta_m, x_i\cos(\vartheta_m)+y_j\sin(\vartheta_m))$$

Output: LMU-FBP reconstruction $f_{\mathrm{LMU}}(x_i,y_j)$ for $(i,j) \in \mathcal{I}_x \times \mathcal{I}_y$

Fourier transform (DFT) and its inverse (iDFT), the discrete version of the right-hand side of the Poisson equation (4) is computed in Step 3. Finally, in Step 4, the Poisson equation is solved numerically and the LMU solution p_{LMU} is found in Step 5 by restricting to the original index set $\{0,\dots,M-1\}\times\{0,\dots,N-1\}$.

In the second *Filtered Back Projection* (FBP) stage of Algorithm 1 we apply the discrete FBP algorithm to numerically invert the RT. This is based on the approximate FBP reconstruction formula

$$f_{\mathrm{FBP}} = \frac{1}{4\pi}\mathcal{B}(A_L * \mathcal{R}f), \tag{11}$$

where a low-pass filter A_L satisfying $\mathcal{F}A_L(S) = |S|\,W(\frac{S}{L})$ with an even window $W \in \mathrm{L}^\infty(\mathbb{R})$ supported in $[-1,1]$ and bandwidth $L > 0$ is incorporated to deal with the ill-posedness of the Radon inversion. The approximate FBP formula (11) is discretized using a standard approach, cf. [10], involving the discrete convolution in Step 6 followed by the discrete back projection in Step 7, where an interpolation method $\mathfrak{I}$ is applied to reduce the computational costs. To this end, the discrete convolution is computed at t_i, $i \in \mathcal{I}$, for a sufficiently large index set $\mathcal{I} \subset \mathbb{Z}$. The result is the LMU-FBP reconstruction f_{LMU} in grid points $(x_i,y_j) \in \mathbb{R}^2$ for index sets $\mathcal{I}_x, \mathcal{I}_y \subset \mathbb{N}$.

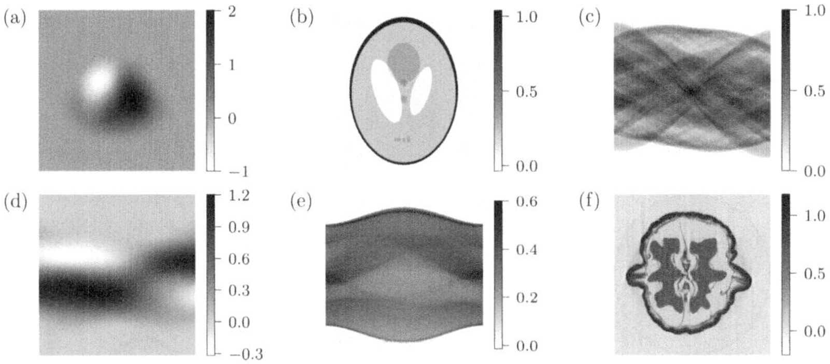

Fig. 2. Utilized test data for numerical experiments. (a) Smooth phantom from [11]. (b) Shepp-Logan phantom from [16]. (c) Walnut Radon data from [8]. (d) Radon data of (a). (e) Radon data of (b). FBP reconstruction of (c) serving as ground truth.

Improvement Step. In [15], it is proposed to apply an enhancement rounding step, which we adapt to our setting and include after the LMU stage in Algorithm 1. More precisely, our *improvement step* is defined as

$$p_{\mathrm{LMU}_+}[m,n] = p^{\lambda}[m,n] + 2\lambda \,\mathrm{round}\Big(\frac{p_{\mathrm{LMU}}[m,n] - p^{\lambda}[m,n]}{2\lambda}\Big) \tag{12}$$

for $m \in \{0, \dots, M-1\}$ and $n \in \{0, \dots, N-1\}$. This yields exact recovery of the Radon data $p[m,n] = \mathcal{R}f(\vartheta_m, t_n)$ if the absolute LMU reconstruction error satisfies $|p_{\mathrm{LMU}}[m,n] - p[m,n]| < \lambda$. Indeed, using the modulo decomposition property (3) with piecewise constant residual $\varepsilon_p[m,n] = \varepsilon_{\mathcal{R}f}(\vartheta_m, t_n) \in \mathbb{Z}$, it is

$$\begin{aligned} p_{\mathrm{LMU}_+}[m,n] &= p^{\lambda}[m,n] + 2\lambda \,\mathrm{round}\Big(\frac{p_{\mathrm{LMU}}[m,n] - p[m,n] + 2\lambda\varepsilon_p[m,n]}{2\lambda}\Big) \\ &= p^{\lambda}[m,n] + 2\lambda\varepsilon_p[m,n] = p[m,n]. \end{aligned}$$

However, if the absolute LMU reconstruction error is large, then the improvement step in (12) yields undesirable jumps in the recovered Radon data.

6 Numerical Experiments

We now present numerical experiments to demonstrate our inversion approach. To this end, we use the smooth phantom [11], depicted in Fig. 2(a) along with its Radon data in Fig. 2(d), and the classical Shepp-Logan phantom [16] in Fig. 2(b), whose Radon data is shown in Fig. 2(e). We also consider the open source walnut dataset [8], that includes realistic uncertainties arising from the tomography hardware. In all cases, we present reconstruction results on a grid of 512×512 pixels from noisy modulo Radon projections

$$\big\{p_\delta^{\lambda}[m,n] \,\big|\, m \in \{0, \dots, M-1\},\ n \in \{0, \dots, N-1\}\big\}$$

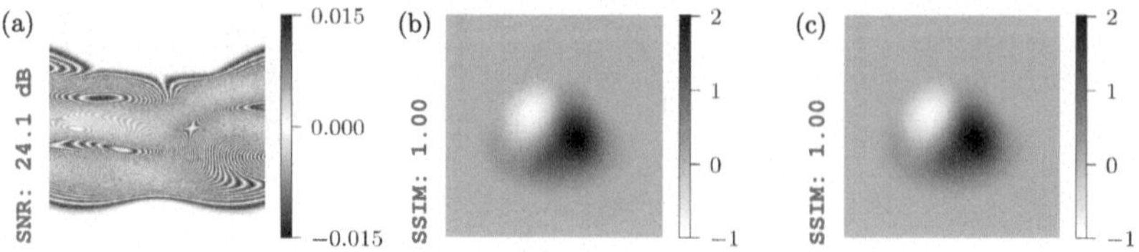

Fig. 3. Numerical experiments with smooth phantom. (a) Noisy modulo Radon data with $\lambda = 0.015$ and $\delta = 0.05 \cdot \lambda$. (b) US-FBP on (a). (c) LMU-FBP on (a).

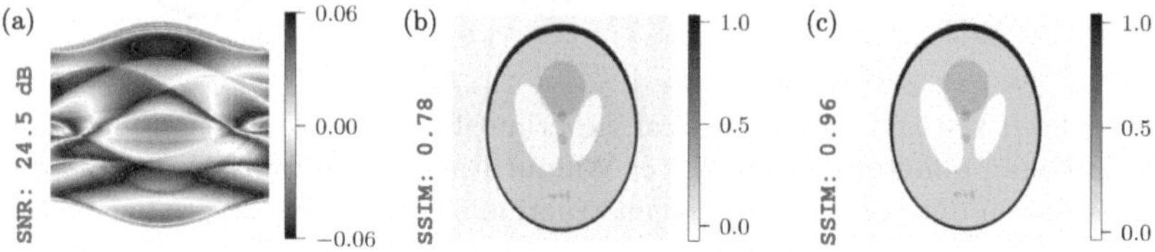

Fig. 4. Numerical experiments with Shepp-Logan phantom. (a) Noisy modulo Radon data with $\lambda = 0.06$ and $\nu = 0.05 \cdot \lambda$. (b) US-FBP on (a). (c) LMU_+-FBP on (a).

of noise level $\delta > 0$ in the sense that $\|p^\lambda - p^\lambda_\delta\|_\infty \leq \delta$ and use the cosine filter with window function $W(S) = \cos(\frac{\pi S}{2})\, \mathbb{1}_{[-1,1]}(S)$ and optimal bandwidth $L = M$. We compare our novel LMU-FBP algorithm with the US-FBP method from [3], which is based on Unlimited Sampling (US) [5]. Note that US-FBP is designed for recovering bandlimited functions but it can be adapted to non-bandlimited data by manually setting the order of forward differences. Here, we always choose order 1 as higher orders are observed to fail in our examples.

Smooth Phantom. In our first set of proof-of-concept simulations, we consider the smooth phantom from [11] with smoothness parameter $\nu = 2.5$ so that our assumptions of Theorem 1 are satisfied. Hence, we expect nearly perfect reconstruction via LMU-FBP. The simulated modulo Radon data with $\lambda = 0.015$ is shown in Fig. 3(a), compressing the dynamic range by 50 times and corrupted by uniform noise with noise level $\delta = 0.05 \cdot \lambda$ yielding a signal-to-noise ratio (SNR) of 24.1 dB. We use the parameter choices $M = 360$ and $K = 1958$ leading to $N = 3917$ so that $T = \frac{1}{K} \leq \frac{1}{2Le}$, which guarantees that US-FBP stably recovers an L-bandlimited function from MRT samples. Although the smooth phantom is not bandlimited, we see in Fig. 3(b) that US-FBP nearly perfectly recovers with a structural similarity index measure (SSIM) of 1.00. The same is true for our newly proposed LMU-FBP reconstruction scheme, see Fig. 3(c).

Shepp-Logan Phantom. To also deal with a non-smooth test case, we now consider the classical Shepp-Logan phantom, which is piecewise constant and has jump discontinuities so that our assumptions in Theorem 1 are *not* satisfied. The simulated MRT data with $\lambda = 0.06$ is shown in Fig. 4(a), compressing the

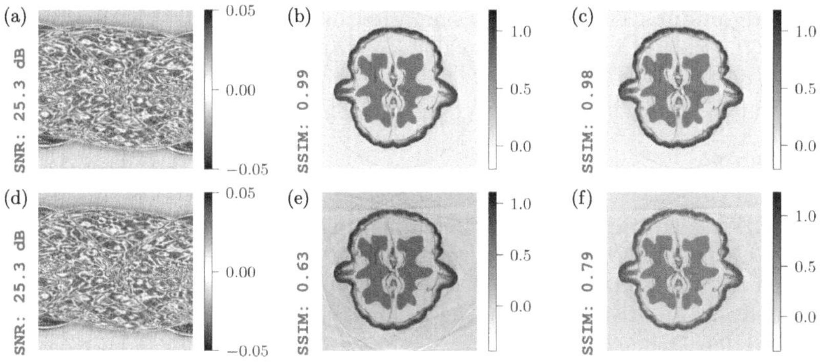

Fig. 5. Numerical experiments with walnut dataset. (a) Noisy modulo Radon data with $\lambda = 0.05$ and $\nu = 0.05 \cdot \lambda$. (b) US-FBP on (a). (c) LMU-FBP on (a). (d) 2-times downsampled noisy modulo Radon data. (e) US-FBP on (d). (f) LMU-FBP on (d).

dynamic range by about 5 times and corrupted by uniform noise with $\delta = 0.05 \cdot \lambda$ leading to an SNR of 24.5 dB. In this case, we see that US-FBP introduces artefacts in the reconstruction, cf. Fig. 4(b), while our improved LMU$_+$-FBP method yields a nearly perfect reconstruction with SSIM of 0.96, cf. Fig. 4(c).

Walnut Data. We finally present reconstruction results for the walnut dataset from [8], which is transformed to parallel beam geometry with $M = 600$ and $K = 1128$. Moreover, the Radon data is normalized to the dynamical range $[0, 1]$, see Fig. 2(c). The corresponding FBP reconstruction is shown in Fig. 2(f) and serves as ground truth for comparing our reconstruction results. Simulated modulo Radon projections with $\lambda = 0.05$ are displayed in Fig. 5(a), where we added uniform noise with $\delta = 0.05 \cdot \lambda$ to account for quantization errors leading to an SNR of 25.3 dB. The reconstruction with US-FBP is shown in Fig. 5(b) and with LMU-FBP in Fig. 5(c). Both algorithm yield a reconstruction of the walnut that is indistinguishable from the FBP reconstruction with an SSIM of 0.99 and 0.98, respectively, while compressing the dynamic range by 10 times. The results for twice radially downsampled noisy MRT data are shown in Fig. 5(d)–(f). While US-FBP produces severe artefacts, LMU-FBP still gives a decent reconstruction.

7 Conclusion

In this work, we proved a novel analytical inversion formula for the MRT closing a gap in the existing literature. Discretization with Fourier techniques lead to the new LMU-FBP algorithm, which can handle non-bandlimited Radon data and performs on par or even better than US-FBP in this case. Future work includes weakening the assumptions and analyzing recovery guarantees for discrete data.

Acknowledgement. This work was supported by the Deutsche Forschungsgemeinschaft (DFG) - Project numbers 530863002 and 281474342/GRK2224/2.

References

1. Arendt, W., Urban, K.: Partial Differential Equations: An Introduction to Analytical and Numerical Methods. Springer, Cham (2023)
2. Beckmann, M., Bhandari, A., Iske, M.: Fourier-domain inversion for the modulo Radon transform. IEEE Trans. Comput. Imaging **10**, 653–665 (2024). https://doi.org/10.1109/TCI.2024.3388871
3. Beckmann, M., Bhandari, A., Krahmer, F.: The modulo Radon transform: theory, algorithms and applications. SIAM J. Imag. Sci. **15**(2), 455–490 (2022). https://doi.org/10.1137/21M1424615
4. Bhandari, A., Krahmer, F., Poskitt, T.: Unlimited sampling from theory to practice: Fourier-Prony recovery and prototype ADC. IEEE Trans. Signal Process. **70**, 1131–1141 (2021). https://doi.org/10.1109/TSP.2021.3113497
5. Bhandari, A., Krahmer, F., Raskar, R.: On unlimited sampling and reconstruction. IEEE Trans. Signal Process. **69**, 3827–3839 (2020). https://doi.org/10.1109/tsp.2020.3041955
6. Chen, P., Han, Y., Pan, J.: High-dynamic-range CT reconstruction based on varying tube-voltage imaging. PLoS ONE **10**(11), e0141789 (2015). https://doi.org/10.1371/journal.pone.0141789
7. Chen, P., Yang, S., Han, Y., Pan, J., Li, Y.: High-dynamic-range x-ray CT imaging method based on energy self-adaptation between scanning angles. OSA Continuum **3**(2), 253 (2020). https://doi.org/10.1364/osac.380634
8. Hämäläinen, K., Harhanen, L., Kallonen, A., Kujanpää, A., Niemi, E., Siltanen, S.: Tomographic X-ray data of a walnut (2015). https://doi.org/10.5281/zenodo.1254206
9. Li, Y., Zhao, X., Han, Y., Chen, P.: Improving CT quality for complex objects with the novel autoexposure imaging of stepped voltage scanning. IEEE Trans. Instrum. Measur. **72** (2023). https://doi.org/10.1109/TIM.2022.3225022
10. Natterer, F.: The Mathematics of Computerized Tomography. SIAM, Philadelphia (2001)
11. Rieder, A., Faridani, A.: The semidiscrete filtered backprojection algorithm is optimal for tomographic inversion. SIAM J. Numer. Anal. **41**(3), 869–892 (2003). https://doi.org/10.1137/S0036142902405643
12. Rieder, A., Schuster, T.: The approximate inverse in action with an application to computerized tomography. SIAM J. Numer. Anal. **37**(6), 1909–1929 (2000). https://doi.org/10.1137/S0036142998347619
13. Robinson, S., Bredies, K., Khabipova, D., Dymerska, B., Marques, J., Schweser, F.: An illustrated comparison of processing methods for MR phase imaging and QSM: combining array coil signals and phase unwrapping. NMR Biomed. **30**(4), e3601 (2016). https://doi.org/10.1002/nbm.3601
14. Salsa, S., Verzini, G.: Partial Differential Equations in Action: From Modelling to Theory. Springer, Cham (2022)
15. Schofield, M., Zhu, Y.: Fast phase unwrapping algorithm for interferometric applications. Opt. Lett. **28**(14), 1194–1196 (2003). https://doi.org/10.1364/OL.28.001194

16. Shepp, L., Logan, B.: The Fourier reconstruction of a head section. IEEE Trans. Nucl. Sci. **21**(3), 21–43 (1974). https://doi.org/10.1109/TNS.1974.6499235
17. Triebel, H.: Theory of Function Spaces. Birkhäuser (1983)

MultiResolution Low-Rank Regularization of Dynamic Imaging Problems

Tommi Heikkilä(✉)

School of Engineering Science, LUT University, Lappeenranta, Finland
tommi.heikkila@lut.fi

Abstract. MultiResolution Low-Rank decomposition is formulated for regularization of dynamic image sequences. The decomposition applies a local low-rank decomposition on a sequence of discrete wavelet transforms. Its effective formulation as a regularization functional is discussed and numerically tested for dynamic X-ray tomography in comparison to other low-rank methods. The results suggest it is similar to traditional locally low-rank decomposition but produces less severe block artifacts.

1 Introduction

In dynamic imaging problems such as magnetic resonance imaging (MRI) and computed tomography (CT) of moving targets it is often better to drastically shorten the measurement time since noisy and undersampled data is easier to correct than the underlying unknown movement. Over the years many *low-rank methods* have been proposed and shown to be effective in resolving the static components of sequential data [7,8,21,29]. If the imaging problem at hand is linear, the slowly changing and static parts of the image sequence can be reconstructed independently and more robustly. This also lowers the complexity of reconstructing the dynamic components and adds robustness since a rough outline of the object of interest is already provided, assuming the changes are not too drastic.

Local low-rank (LLR) methods consider only small patches of the images at once and try to find low-rank approximations for the smaller regions. Originally introduced for dynamic MRI [22], the method has also been used in magnetic resonance fingerprinting (MRF) [4]. Many of these low-rank methods also include sparsity promoting terms, either at the same time [4,21] or in separate components [7,16,18]. See also [11] for different patch based temporal regularization method.

Recently many low-rank representations (LRRs) have also been developed in the context of tensors [6,12,13], in order to extend the familiar tools and ideas from matrices (i.e. 2-dimensional arrays) to higher dimensional objects. This way vital underlying structure of the data is still preserved in dynamic (location and time), hyperspectral (location and frequency) and phase-space (location and momentum) applications, just to name a few. Furthermore in data processing and machine learning, extracting some lower dimensional subspace or

T. A. Bubba et al. (Eds.): SSVM 2025, LNCS 15667, pp. 70–82, 2025.
https://doi.org/10.1007/978-3-031-92366-1_6

manifold from hugely large and complicated data sets is often helpful in keeping the computations and storage manageable. In this work the perspective is limited only to the traditional low-rank matrix factorizations (namely the singular value decomposition (SVD)), but there is no inherent reason why the proposed method could not be extended to tensors as in [13,14].

In this work a locally low-rank decomposition is formulated by applying the patched low-rank decomposition on the wavelet domain rather than the usual spatial domain. This idea has already been used in (hyperspectral) image denoising and deblurring [17,28], where it evolved from local wavelet coefficient thresholding methods. Alternate approach of wavelet-denoising the principal components has also been proposed even earlier [2].

Thanks to the multiscale nature of the wavelet decomposition, the low-rank decomposition will also cover multiple scales. Moreover, due to the smooth and overlapping wavelet elements, the patches should have less distinct boundary artifacts. We show that this multiresolution low-rank (MRLR) decomposition is unitary and briefly showcase its use as a regularization term with numerical examples in dynamic X-ray tomography application.

2 Low-Rank Methods

We consider a sequence of time-dependent inverse problems of the form

$$\mathcal{A}_t f_t = m_t + \varepsilon_t, \quad \text{for } t = 1, 2, \ldots, T. \tag{1}$$

Here each $f_t \in \mathbb{R}^N$ is a $N_r \times N_c$ image flattened into vector with $N = N_r N_c$ values in total. Extension to 3D volumes would follow in similar fashion. For simplicity, we denote the value at i'th row and j'th column of the image by $f_t[i, j]$, even though formally f_t is a one dimensional vector.

We can stack each problem in Eq. (1) by denoting $\boldsymbol{f} = \left(f_t\right)_{t=1}^{T}$, $\boldsymbol{m} + \boldsymbol{\varepsilon} = \left(m_t + \varepsilon_t\right)_{t=1}^{T}$ and $\boldsymbol{\mathcal{A}} = \operatorname{diag}\left(\mathcal{A}_1, \ldots, \mathcal{A}_T\right)$ the block diagonal operator. Then we can write (1) as $\boldsymbol{\mathcal{A}f} = \boldsymbol{m} + \boldsymbol{\varepsilon}$ and the least squares term is then

$$\left\|\boldsymbol{\mathcal{A}f} - \boldsymbol{m}\right\|_2^2 = \sum_{t=1}^{T} \left\|\mathcal{A}_t f_t - m_t\right\|_2^2.$$

Since each time step t is currently independent, we regularize the problem by requiring that the evolution of $\boldsymbol{f}$ over time must have a low-rank structure.

We form the *Casorati matrix* [8] using the associated operator $\mathcal{M}$ given as

$$\mathcal{M} : \boldsymbol{f} \longmapsto F = \left[f_1\, f_2 \ldots f_T\right] = \begin{bmatrix} f_1[1,1] & f_2[1,1] & \ldots & f_T[1,1] \\ f_1[2,1] & f_2[2,1] & \ldots & f_T[2,1] \\ \vdots & \vdots & \ddots & \vdots \\ f_1[N_r, N_c] & f_2[N_r, N_c] & \ldots & f_T[N_r, N_c] \end{bmatrix}, \tag{2}$$

where $F \in \mathbb{R}^{N \times T}$. Since this is simply a reordering of the terms, it can be reverted with

$$\mathcal{M}^{-1} : F \longmapsto \boldsymbol{f}. \tag{3}$$

We note also that for any $\boldsymbol{f}, \boldsymbol{g} \in \mathbb{R}^{NT}$ we have

$$\langle \boldsymbol{f}, \boldsymbol{g} \rangle = \langle \mathcal{M}\boldsymbol{f}, \mathcal{M}\boldsymbol{g} \rangle_F, \tag{4}$$

where $\langle \cdot, \cdot \rangle_F$ is the Frobenius inner product.

We could attempt penalizing the rank of the matrix $F = \mathcal{M}\boldsymbol{f}$, but this corresponds to minimizing the ℓ^0-"norm" of the singular values. The usual remedy from compressed sensing and other sparsity promoting regularization methods is to use the convex ℓ^1-norm instead. This leads to penalizing the *nuclear norm* of the matrix F:

$$\|F\|_* = \sum_{i=1}^{r} |\sigma(F)_i| = \|\sigma(F)\|_1, \tag{5}$$

where $r = \text{rank}(F) \leq \min\{N, T\}$ and $\sigma(F) = \text{diag}(\Sigma) \in \mathbb{R}^r$ denotes the singular values of F given by the singular value decomposition (SVD):

$$F = U \Sigma V^T = \begin{bmatrix} | & | & & | \\ u_1 & u_2 & \dots & u_r \\ | & | & & | \end{bmatrix} \begin{bmatrix} \sigma(F)_1 & & \\ & \ddots & \\ & & \sigma(F)_r \end{bmatrix} \begin{bmatrix} \text{—} \; v_1^T \; \text{—} \\ \text{—} \; v_2^T \; \text{—} \\ \vdots \\ \text{—} \; v_r^T \; \text{—} \end{bmatrix}. \tag{6}$$

Hence the global low-rank (GLR) method is given by minimizing the functional

$$GLR(\boldsymbol{f}) = \|\boldsymbol{A}\boldsymbol{f} - \boldsymbol{m}\|_2^2 + \lambda \|\mathcal{M}\boldsymbol{f}\|_*, \tag{7}$$

where $\lambda > 0$ is the regularization parameter.

In order to minimize Eq. (7) effectively, we need to analyze the low-rank operator properly. Let diag denote the linear operator which takes the diagonal values from any matrix. For rectangular $m \times n$ matrices, if $m > n$, the output is zero-padded to be in $\mathbb{R}^m$. Formally we can write $\text{diag}(M) = (M \odot I)\mathbb{1}$, where $\odot$ is the Hadamard product (elementwise product of matrix elements) and $\mathbb{1} \in \mathbb{R}^r$ a column vector of all ones. Finally for any $y \in \mathbb{R}^r$ we can also define $\text{diag}^*(y) = (y\mathbb{1}^T) \odot I$, which gives a diagonal matrix with elements of y on the diagonal.

Now consider

$$\mathcal{L} : \boldsymbol{f} \longmapsto \text{diag}\big(U^T \mathcal{M}(\boldsymbol{f}) V\big). \tag{8}$$

By definition $\mathcal{L}$ extracts the diagonal matrix Σ from the SVD of $\mathcal{M}\boldsymbol{f}$ and we obtain a vector of r singular values with the diag operator. It is easy to verify that $\mathcal{L} : \mathbb{R}^{NT} \to \mathbb{R}^r$ is a linear operator, but note that it only yields the singular values of the specific vector $\boldsymbol{f}$ since the unitary matrices U and V depend on the SVD of $\boldsymbol{f}$. However, we have the following result independent of the particular $\boldsymbol{f} \in \mathbb{R}^{NT}$.

Proposition 1. *The operator $\mathcal{L}$ as defined in* (8) *is unitary and for any $y \in \mathbb{R}^r$*

$$\mathcal{L}^* y = \mathcal{M}^{-1}\left(U \operatorname{diag}^*(y) V^T\right). \tag{9}$$

Proof. The result follows easily by considering the Frobenius inner product, for which $\langle AB, C\rangle_F = \langle B, A^T C\rangle_F$ for any compatible sized matrices A, B and C.

Let $\boldsymbol{x} = \left(x_t\right)_{t=1}^T \in \mathbb{R}^{NT}$ and $y \in \mathbb{R}^r$ be arbitrary and consider

$$\begin{aligned}\langle \mathcal{L}\boldsymbol{x}, y\rangle &= \langle \operatorname{diag}\big(U^T \mathcal{M}(\boldsymbol{x}) V\big), y\rangle = \langle U^T \mathcal{M}(\boldsymbol{x}) V, \operatorname{diag}^*(y)\rangle_F \\ &= \langle \mathcal{M}(\boldsymbol{x}), U \operatorname{diag}^*(y) V^T\rangle_F = \langle \boldsymbol{x}, \mathcal{M}^{-1}\big(U \operatorname{diag}^*(y) V^T\big)\rangle\end{aligned}$$

Therefore Eq. (9) holds. Then we can check that

$$\begin{aligned}\mathcal{L}\mathcal{L}^* y &= \operatorname{diag}\Big(U^T \mathcal{M}(\mathcal{M}^{-1}(U \operatorname{diag}^*(y) V^T)) V\Big) \\ &= \operatorname{diag}\big(U^T U \operatorname{diag}^*(y) V^T V\big) = \big((y \mathbb{1}^T) \odot I \odot I\big)\mathbb{1} = y,\end{aligned}$$

and

$$\begin{aligned}\mathcal{L}^* \mathcal{L}\boldsymbol{x} &= \mathcal{M}^{-1}\big(U \operatorname{diag}^*\big(\operatorname{diag}(U^T \mathcal{M}(\boldsymbol{x}) V)\big) V^T\big) \\ &= \mathcal{M}^{-1}\big(U U^T \mathcal{M}(\boldsymbol{x}) V V^T\big) = \mathcal{M}^{-1}\big(\mathcal{M}(\boldsymbol{x})\big) = \boldsymbol{x}.\end{aligned}$$

This concludes the proof.

Proposition 1 gives an alternative definition to Eq. (7) as

$$GLR(\boldsymbol{f}) = \left\|\boldsymbol{A}\boldsymbol{f} - \boldsymbol{m}\right\|_2^2 + \lambda \left\|\mathcal{L}\boldsymbol{f}\right\|_1, \tag{10}$$

which can be minimized with relatively standard methods.

Remark 1. An easier way of defining the diag-operator would be to simply use a vector of ones $\mathbb{1} \in \mathbb{R}^r$. For diagonal matrices the operations $\operatorname{diag}(\Sigma)$ and $\Sigma\mathbb{1}$ would give identical results, but the adjoints would be different since $\operatorname{diag}^*(\operatorname{diag}(\Sigma)) = \Sigma$, but $\Sigma\mathbb{1}\mathbb{1}^T \neq \Sigma$ and $\mathcal{L}$ would not be unitary but r-times redundant.

2.1 Locally Low-Rank Methods

By construction the *global low-rank* method in Eq. (7) can only provide a very coarse approximation of the image sequence $\boldsymbol{f}$ since each principal component is of same size as f_t. And since the SVD is limited by the dimensions N and T (whichever is smaller), the number of components may be very limited, even when $\boldsymbol{f}$ has a very high spatial resolution or is a 3D volume, for example.

One way to overcome these limitations is to divide each f_t into K patches of size $p \times p$ (or more generally size $p_1 \times \cdots \times p_d$ in d dimensions) [22]. We assume the patches are *non-overlapping*. We can then consider a sequence of K patched Casorati matrices, independently for each patch, as

$$\mathcal{M}_{p\times p} : \boldsymbol{f} \longmapsto (F_k)_{k=1}^{K}\,, \tag{11}$$

where for example, the first patch is

$$F_1 = \begin{bmatrix} f_1[1,1] & f_2[1,1] & \dots & f_T[1,1] \\ \vdots & \vdots & \ddots & \vdots \\ f_1[p,1] & f_2[p,1] & \dots & f_T[p,1] \\ f_1[1,2] & f_2[1,2] & \dots & f_T[1,2] \\ \vdots & \vdots & \ddots & \vdots \\ f_1[p,p] & f_2[p,p] & \dots & f_T[p,p] \end{bmatrix} \tag{12}$$

and so on. This process is also illustrated in Fig. 1.

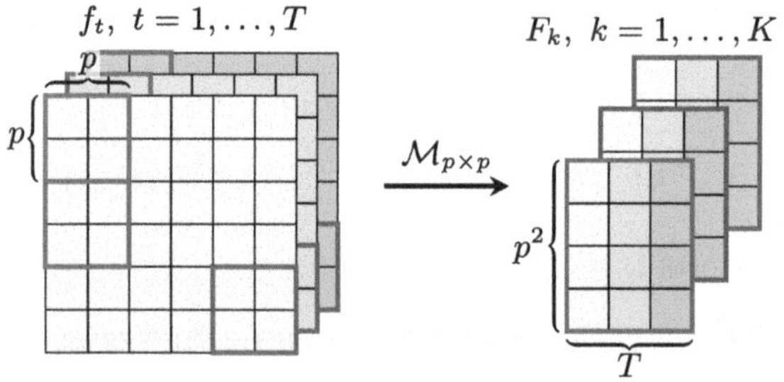

Fig. 1. Illustration how the patched Casorati matrices are formed.

Hence the local low-rank (LLR) method is given by

$$LLR(\boldsymbol{f}) = \left\|\boldsymbol{\mathcal{A}}\boldsymbol{f} - \boldsymbol{m}\right\|_2^2 + \lambda \sum_{k=1}^{K} \left\|(\mathcal{M}_{p\times p}\boldsymbol{f})_k\right\|_*\,, \tag{13}$$

where $\lambda > 0$ is the regularization parameter. Defining a patched variant $\mathcal{L}_p$ of the unitary operator $\mathcal{L}$ would be straight forward (with either square or more general rectangular patches) but the construction is omitted here.

Changing the patch sizes gives a lot of control over the regions where each low-rank approximation is carried out. Notice also that the patches can have different sizes as long as each patch is present in every time step f_t since the patches are separated in to the columns of matrices F_k in the sequence, and F_k and F_l need not have the same size for $k \neq l$. Let $\boldsymbol{p}(k)$ denote the size of the k'th patch from now on.

However, this simple separation suffers from distinct artifacts at the boundaries of the patches since each is handled independently of the neighboring patches. This can be partly resolved with *overlapping* patches, but this makes the associated operator $\mathcal{M}$ more complicated and then the minimization of Eq. (13) can become more difficult. Some overlapping patching strategies have been proposed, in particular for dynamic MRI [1,15].

3 Multiresolution Low-Rank Method

In the proposed *multiresolution low-rank* method the (patched) Casorati matrix is not formed from the image sequence $\boldsymbol{f} = (f_t)_{t=1}^T$ but from the corresponding 2D *wavelet decompositions* $(\mathcal{W}f_t)_{t=1}^T$. The coefficients from the wavelet decomposition retain some of the local spatial information from $\boldsymbol{f}$, but the smoothness of $\boldsymbol{f}$ is no longer tied to the values of neighboring matrix elements but the underlying wavelet system. In other words, even though the patched Casorati matrices split the input into independent, non-overlapping blocks, once the wavelet coefficients are reconstructed (via the inverse wavelet transform $\mathcal{W}^{-1} = \mathcal{W}^*$), the end result should not show sharp artifacts at the edges of the patches since the wavelet elements $\psi_{j,k}$ can overlap with nearby wavelets.

Moreover, since the wavelet transform covers multiple spatial scales, the local low-rank approximation of said transform will also cover multiple scales. Since we are free to change the patch size for the different decomposition levels (and directional subbands, but we omit this), this gives more control than the direct low-rank approximations considered earlier.

Formally we consider the 2D discrete wavelet transform $\mathcal{W}$ and write

$$\mathcal{W} : f \longmapsto (c_j)_{j=0}^J = \boldsymbol{c}, \tag{14}$$

where the scale $j = 0$ corresponds to the scaling coefficients and the scales $j \geqslant 1$ the detail coefficients. The different translates $\ell = (l_1, l_2) \in \mathbb{Z}^2$ are given as

$$c_j[\ell, d] = \langle f, \psi_{j,\ell}^{(d)} \rangle = \langle f, 2^j \psi^{(d)}(2^j \cdot -\ell) \rangle, \tag{15}$$

where the different directional wavelets are constructed from 1D wavelet and scaling functions in the following manner:

$$\begin{aligned} \text{horizontal: } \psi^{HL}(x_1, x_2) &= \psi(x_1)\phi(x_2), \\ \text{vertical: } \psi^{LH}(x_1, x_2) &= \phi(x_1)\psi(x_2), \\ \text{diagonal: } \psi^{HL}(x_1, x_2) &= \psi(x_1)\psi(x_2). \end{aligned}$$

The 2D scaling function is denoted $\psi^{LL}(x_1, x_2) = \phi(x_1)\phi(x_2)$.

The multiresolution low-rank (MRLR) method is given by

$$MRLR(\boldsymbol{f}) = \left\|\boldsymbol{A}\boldsymbol{f} - \boldsymbol{m}\right\|_2^2 + \lambda \sum_{j=0}^{J} \sum_{k=1}^{K_j} \left\| \left(\mathcal{M}_{\boldsymbol{p}(j,k)} \mathcal{W}\boldsymbol{f}\right)_k \right\|_*, \tag{16}$$

where at every scale j, the wavelet subbands are broken in to K_j local patches of size $\boldsymbol{p}_j(k) = p_1^{j,k} \times p_2^{j,k}$. Again the patch sizes could vary, but for simplicity let us consider square $p_j \times p_j$ patches chosen solely based on the scale j. Unlike [14], the different directional subbands are kept in separate Casorati matrix patches (accounted by the index k).

This regularization term can be written as

$$MRLR(\boldsymbol{f}) = \left\|\boldsymbol{\mathcal{A}}\boldsymbol{f} - \boldsymbol{m}\right\|_2^2 + \lambda \left\|\mathcal{L}_{\mathcal{W},\boldsymbol{p}}\boldsymbol{f}\right\|_1. \tag{17}$$

Here we define

$$\mathcal{L}_{\mathcal{W},\boldsymbol{p}} : (f_t)_{t=1}^T \longmapsto \begin{bmatrix} \mathcal{L}_{\boldsymbol{p}(0)}(\mathcal{W}\boldsymbol{f})_0 \\ \mathcal{L}_{\boldsymbol{p}(1)}(\mathcal{W}\boldsymbol{f})_1 \\ \vdots \\ \mathcal{L}_{\boldsymbol{p}(J)}(\mathcal{W}\boldsymbol{f})_J \end{bmatrix},$$

where each $\mathcal{L}_{\boldsymbol{p}(j)}(\mathcal{W}\boldsymbol{f})_j$ itself is a local low-rank decomposition in to K_j vectors of singular values.

Since the wavelet transform corresponding to orthogonal wavelets is unitary, the operator $\mathcal{L}_{\mathcal{W},\boldsymbol{p}}$ is also unitary. The process is illustrated in Fig. 2.

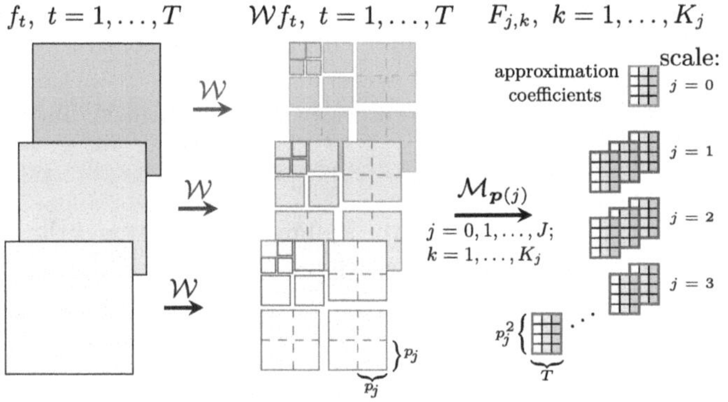

Fig. 2. Illustration of the multiresolution low-rank scheme using $J = 3$ scales. Note how there is a lot of flexibility in choosing the patch sizes for each scale and directional subband.

4 Numerical Examples

To assess the quality and viability of the different low-rank methods the following numerical tests are performed on dynamic X-ray tomography problems.

We consider an ill-posed inverse problem of reconstructing moving object of interest from a single measurement organized into a *sinogram*. Since the underlying movement is unknown, we approximate the problem as a sequence of static *limited angle* problems which are notoriously difficult to reconstruct and often suffer from streaking artifacts.

In addition to the proposed MRLR-method given in (17), all tested data is also reconstructed using three existing low-rank methods. First the Local Low-Rank (LLR) method [22] as formulated in (13). Then the Low-rank + Sparse (L+S) method [7,16], which is a different kind regularization scheme which decomposes the unknown dynamics into *separate* low-rank and sparse components (using some multiresolution representation system such as wavelets). The objective functional for the L+S method is given as

$$LplusS(L,S) = \|\boldsymbol{\mathcal{A}}(L+S) - \boldsymbol{m}\|_2^2 + \lambda_L \|\mathcal{M}L\|_* + \lambda_S \|\mathcal{W}S\|_1, \tag{18}$$

where $L, S \in \mathbb{R}^{NT}$, $\mathcal{M}$ is the operator defined earlier in Eq. (2) and $\mathcal{W}$ is again a 2D wavelet transform applied separately to each time step S_t as in (16). Once again, with a suitable unitary operator, the nuclear norm can be replaced with the ℓ^1-norm. The two regularization terms have separate regularization parameters $\lambda = (\lambda_L, \lambda_S)$. And finally the Fast algorithm for Total Variation and Nuclear Norm Regularization (FTVNNR) [26,27]. Its objective functional can be given as

$$FTVNNR(\boldsymbol{f}) = \|\boldsymbol{\mathcal{A}}\boldsymbol{f} - \boldsymbol{m}\|_2^2 + \lambda_1 \sum_{t=1}^{T} TV(f_t) + \lambda_2 \|\mathcal{M}\boldsymbol{f}\|_*. \tag{19}$$

In short, it incorporates a spatial anisotropic total variation penalty on each time step in addition to the global low-rank penalty term. The implementation is based on the codes available on http://github.com/uta-smile/FTVNNR_Dynamic_MRI_MEDIA, and only slightly modified for CT.

In [4] the authors propose using both locally low-rank and (wavelet) sparsity regularization terms on the unknown image sequence $\boldsymbol{f}$ for MRF. This is more closely related to the earlier LLR method in (13) than the L+S method, since the locally low-rank and sparse components are not separated.

The chosen patch sizes and different regularization and wavelet parameters are all listed in Table 1. The LLR and MRLR functionals are minimized using the primal-dual fixed point (PDFP) algorithm [3] with non-negativity constraints on the attenuation values. All algorithms are ran until the relative change between iterates drops below $5 \cdot 10^{-4}$. Daubechies-3 wavelets [5] are used with every wavelet-based method and data, with periodic boundary conditions to ease the choice of subband patch sizes. The Matlab implementation of the L+S algorithm is based on the example code from [10]. The repository with all the Matlab codes is available on github.com/tommheik/WaveletLowRank.

4.1 Data

Tests are performed both using simulated and real data:

Simulated data consists of multiple objects of different shapes and attenuation. One object is translated from left to right, the rest undergo periodic and affine deformation, and one object is slightly rotated counterclockwise at the same time. The reconstruction resolution is 256×256, but the data was initially

Table 1. Table of parameters used

Data	Method	Wavelet transform	Regularization parameters λ	No. of patches	Patch sizes
Simulated	MRLR	db3 ($J = 2$)	1.0	$[4\times4, 2\times2, 2\times2]$	$[64\times64, 64\times64, 64\times64]$
	LLR	-	0.1	32×32	8×8
	L+S	db3 ($J = 3$)	$(0.2, 0.08)$	-	-
	FTVNNR	-	$(0.001, 5)$	-	-
STEMPO	MRLR	db3 ($J = 3$)	1.0	$[4\times4, 2\times2, 1\times1, 1\times1]$	$[35\times35, 35\times35, 35\times35, 35\times35]$
	LLR	-	0.1	40×40	7×7
	L+S	db3 ($J = 3$)	$(0.35, 0.02)$	-	-
	FTVNNR	-	$(0.01, 2)$	-	-

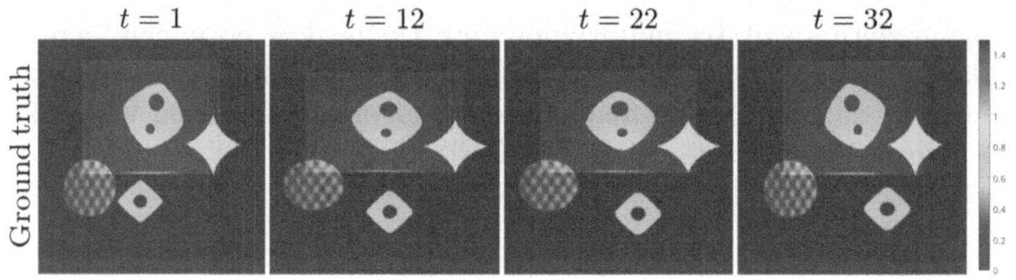

Fig. 3. Selected time steps of the synthetic ground truth object.

simulated at twice the spatial resolution and white Gaussian noise of 3% variance was added to avoid inverse crime. A total of 360 projections with 1° steps were computed using 180 unique time steps (i.e. 2 projections per time step).

For the reconstructions, the data is organized into just 32 batches, each with 19 consecutive projections and overlap of 11 projections. The forward operator follows a parallel beam geometry and is implemented using the ASTRA Toolbox [23,24].

32 reference time steps are chosen (out of the original 180) to match the average state of the batches and some of these are shown in Fig. 3.

Real data is `stempo_cont360_2d_b8.mat` from the STEMPO dataset [9,10] which consists of 360 unique fan-beam projections measured at 1° steps of a motorized dynamic tomography phantom. Detailed documentation is available in [9]. With the chosen binning level (8), the reconstruction resolution is set to 280×280.

For the reconstructions, the sinogram is organized into just 32 batches, each with 19 consecutive projections and overlap of 11 projections, just like the simulated data. The forward operator is implemented using the ASTRA Toolbox.

4.2 Results

Reconstructions of the simulated data are shown in Fig. 4 and reconstructions of the STEMPO data are shown in Fig. 5. In Fig. 4 the relative L^2-error, SSIM [25] and HaarPSI [19] values for each individual time step are also shown.

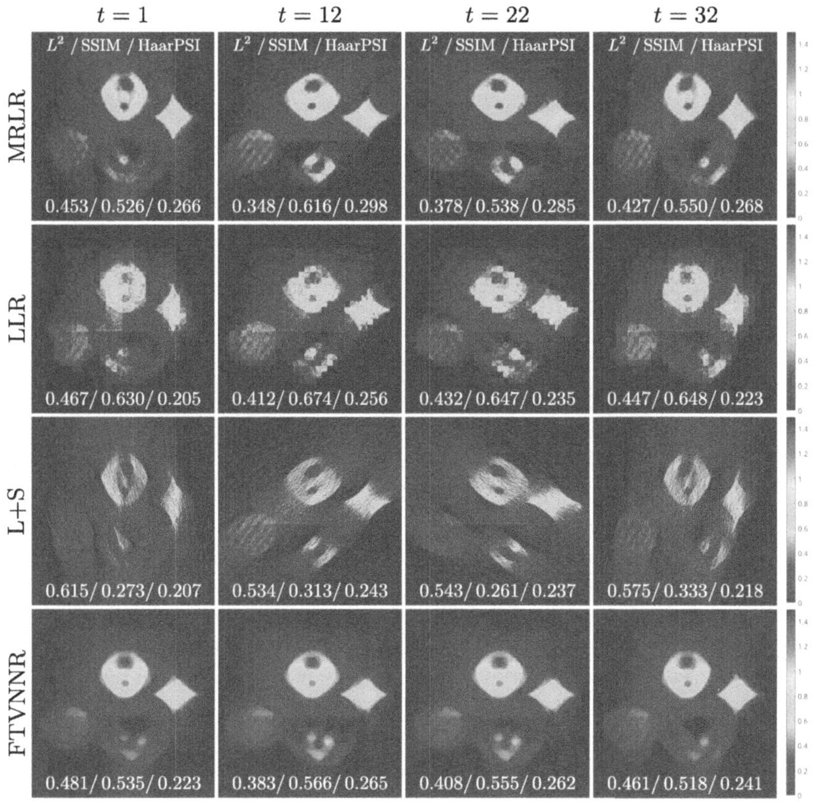

Fig. 4. Simulated data reconstructions using different low rank methods: MultiResolution Low-Rank, Local Low-Rank, Low-rank + Sparse decomposition and Total Variation and Nuclear Norm regularization. The L^2-error, SSIM and HaarPSI values of each time step is also shown.

With both data, all low-rank methods show similar movement artifacts and have particular difficulty reconstructing the translated object (e.g. object at the bottom center in Fig. 4). However, with the chosen wavelet decomposition and block sizes, the MRLR regularization does not produce visible block-artifacts unlike the traditional LLR method. The relative L^2-error and HaarPSI metrics also favor MRLR whereas the SSIM values are consistently better for LLR. FTVNNR performs almost as well and the images have less noise thanks to the TV. However the images are also more blurry. This is especially noticeable with the STEMPO reconstructions. None of the methods reconstruct the intermediate time steps well and there is a distinct jump in the translations seen between time steps 12 and 22.

The L+S method worked surprisingly poorly with the tested data both visually and based on the numerical error metrics. Before the non-negativity projection was included in the implementation (without any mathematical rigor), its

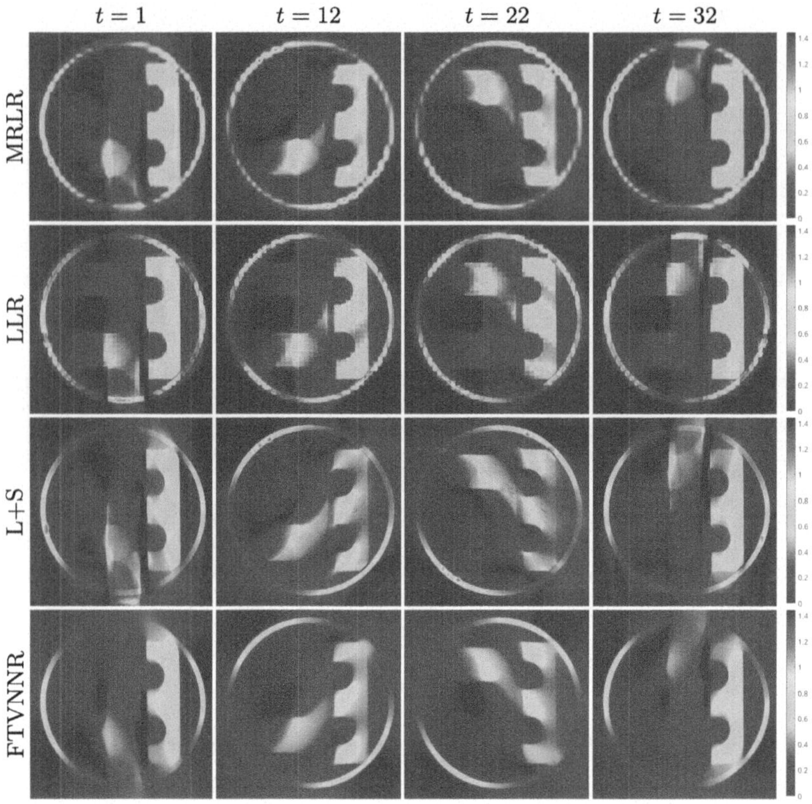

Fig. 5. STEMPO data reconstructions using different low rank methods: MultiResolution Low-Rank, Local Low-Rank, Low-rank + Sparse decomposition and Total Variation and Nuclear Norm regularization.

results were even poorer. Due to the two distinct regularization parameters (see (18)), it is somewhat harder to fine tune than the other methods which might also affect the results. The MRLR also requires choosing the wavelet decomposition and different patch sizes but the changes have clearer effect on the behavior.

Although the MRLR should be slower to compute than the LLR, in practice the difference in computational times seems negligible.

5 Conclusions

In this paper the multiresolution low-rank decomposition is formulated as a possible regularization strategy for reconstructing dynamic image sequences. As a combination of discrete wavelet transform and a local low-rank decomposition, it forms a unitary operator and should be easily applicable to many optimization strategies. The numerical examples indicate that heavy regularization produces less severe artifacts compared to traditional low-rank regularization methods.

However, in practice it still has very similar regularizing effect on the reconstructions and should be applied in place of traditional low-rank methods or in tandem with some different prior. In the future it would be interesting to pair it with stronger spatial regularization such as total variation [11,20,27].

Acknowledgments. The author is grateful of the funding from the Vilho, Yrjö and Kalle Väisälä Foundation; the Finnish Foundation for Technology Promotion; Academy of Finland (the Centre of Excellence in Inverse Modelling and Imaging, decision 312339); and Research Council of Finland (Flagship of Advanced Mathematics for Sensing Imaging and Modelling grant 359183).

References

1. Candes, E.J., Sing-Long, C.A., Trzasko, J.D.: Unbiased risk estimates for singular value thresholding and spectral estimators. IEEE Trans. Signal Process. **61**(19), 4643–4657 (2013)
2. Chen, G., Qian, S.E.: Denoising of hyperspectral imagery using principal component analysis and wavelet shrinkage. IEEE Trans. Geosci. Remote Sens. **49**(3), 973–980 (2010)
3. Chen, P., Huang, J., Zhang, X.: A primal-dual fixed point algorithm for convex separable minimization with applications to image restoration. Inverse Prob. **29**(2), 025011 (2013)
4. Lima da Cruz, G., Bustin, A., Jaubert, O., Schneider, T., Botnar, R.M., Prieto, C.: Sparsity and locally low rank regularization for MR fingerprinting. Magn. Reson. Med. **81**(6), 3530–3543 (2019)
5. Daubechies, I.: Ten lectures on wavelets. Society for industrial and applied mathematics (1992)
6. Du, S., Liu, B., Shan, G., Shi, Y., Wang, W.: Enhanced tensor low-rank representation for clustering and denoising. Knowl.-Based Syst. **243**, 108468 (2022)
7. Gao, H., Cai, J.F., Shen, Z., Zhao, H.: Robust principal component analysis-based four-dimensional computed tomography. Phys. Med. Biol. **56**(11), 3181 (2011)
8. Haldar, J.P., Liang, Z.P.: Spatiotemporal imaging with partially separable functions: a matrix recovery approach. In: 2010 IEEE International Symposium on Biomedical Imaging: From Nano to Macro, pp. 716–719. IEEE (2010)
9. Heikkilä, T.: Stempo-dynamic x-ray tomography phantom. In: INdAM Workshop: Advanced Techniques in Optimization for Machine Learning and Imaging, pp. 1–14. Springer (2022)
10. Heikkilä, T.: STEMPO - dynamic X-ray tomography phantom (2023). https://doi.org/10.5281/zenodo.8239013, v1.2.0 [data set]
11. Kazantsev, D., et al.: 4D-CT reconstruction with unified spatial-temporal patch-based regularization. Inverse Probl. Imaging **9**(2), 447–467 (2015)
12. Kilmer, M.E., Martin, C.D.: Factorization strategies for third-order tensors. Linear Algebra Appl. **435**(3), 641–658 (2011)
13. Liu, S., Lei, M., Cao, J., Yang, T.: Dynamic MRI reconstruction via multi-directional low-rank tensor regularization. Biomed. Signal Process. Control **99**, 106848 (2025)
14. Liu, S., Li, W., Cao, J., Zhang, K., Hu, S.: Image restoration via wavelet-based low-rank tensor regularization. Optik **273**, 170415 (2023)

15. Meyer, N.K., et al.: Locally low-rank denoising of multi-echo functional MRI data with application in resting-state analysis. Top. Magn. Reson. Imaging **32**(5), 37–49 (2023)
16. Otazo, R., Candes, E., Sodickson, D.K.: Low-rank plus sparse matrix decomposition for accelerated dynamic MRI with separation of background and dynamic components. Magn. Reson. Med. **73**(3), 1125–1136 (2015)
17. Palsson, F., Ulfarsson, M.O., Sveinsson, J.R.: Hyperspectral image denoising using a sparse low rank model and dual-tree complex wavelet transform. In: 2014 IEEE Geoscience and Remote Sensing Symposium, pp. 3670–3673. IEEE (2014)
18. Ravishankar, S., Moore, B.E., Nadakuditi, R.R., Fessler, J.A.: Low-rank and adaptive sparse signal (LASSI) models for highly accelerated dynamic imaging. IEEE Trans. Med. Imaging **36**(5), 1116–1128 (2017)
19. Reisenhofer, R., Bosse, S., Kutyniok, G., Wiegand, T.: A HAAR wavelet-based perceptual similarity index for image quality assessment. Signal Process. Image Commun. **61**, 33–43 (2018)
20. Rudin, L.I., Osher, S., Fatemi, E.: Nonlinear total variation based noise removal algorithms. Physica D **60**(1–4), 259–268 (1992)
21. Trémoulhéac, B., Dikaios, N., Atkinson, D., Arridge, S.R.: Dynamic MR image reconstruction-separation from undersampled (k, t)-space via low-rank plus sparse prior. IEEE Trans. Med. Imaging **33**(8), 1689–1701 (2014)
22. Trzasko, J., Manduca, A., Borisch, E.: Local versus global low-rank promotion in dynamic MRI series reconstruction. In: Proceedings of International Symposium on Magnetic Resonance Medicine, vol. 19, p. 4371 (2011)
23. Van Aarle, W., et al.: Fast and flexible X-ray tomography using the ASTRA toolbox. Opt. Express **24**(22), 25129–25147 (2016)
24. Van Aarle, W., et al.: The ASTRA toolbox: a platform for advanced algorithm development in electron tomography. Ultramicroscopy **157**, 35–47 (2015)
25. Wang, Z., Bovik, A.C., Sheikh, H.R., Simoncelli, E.P.: Image quality assessment: from error visibility to structural similarity. IEEE Trans. Image Process. **13**(4), 600–612 (2004)
26. Yao, J., Xu, Z., Huang, X., Huang, J.: Accelerated dynamic MRI reconstruction with total variation and nuclear norm regularization. In: International Conference on Medical Image Computing and Computer-Assisted Intervention, pp. 635–642. Springer (2015)
27. Yao, J., Xu, Z., Huang, X., Huang, J.: An efficient algorithm for dynamic MRI using low-rank and total variation regularizations. Med. Image Anal. **44**, 14–27 (2018)
28. Zhao, B., Sveinsson, J.R., Ulfarsson, M.O., Chanussot, J.: Wavelet-based block low-rank representations for hyperspectral denoising. In: 2021 IEEE International Geoscience and Remote Sensing Symposium IGARSS, pp. 2484–2487. IEEE (2021)
29. Zhao, B., Haldar, J.P., Brinegar, C., Liang, Z.P.: Low rank matrix recovery for real-time cardiac MRI. In: 2010 IEEE International Symposium on Biomedical Imaging: From Nano to Macro, pp. 996–999. IEEE (2010)

A Fractional Graph La+Ψ Approach to Image Reconstruction

Stefano Aleotti[1], Marco Donatelli[1], Rolf Krause[2,3,4], and Pietro Maurino[1(✉)]

[1] Department of Science and High Technology, University of Insubria, 22100 Como, Italy
{stefano.aleotti,marco.donatelli,pmaurino}@uninsubria.it

[2] AMCS, CEMSE, King Abdullah University of Science and Technology, Thuwal 23955-6900, Kingdom of Saudi Arabia
rolf.krause@kaust.edu.sa

[3] Faculty of Informatics, Euler Institute, Università della Svizzera Italiana, 6900 Lugano, Switzerland

[4] Faculty of Mathematics and Informatics, UniDistance, 3900 Brig, Switzerland

Abstract. We investigate a variational method for ill-posed problems, which embeds the fractional power of the standard graph Laplacian operator in the regularization term. We explore the dependence of the regularizer on a preliminary approximation of the solution, which is obtained using various existing reconstruction methods Ψ from the literature. As a result, the regularization term is both dependent on and adaptive to the observed data, noise, and the choice of the fractional exponent.

We present a selected numerical example problem on 2D computerized tomography, for which we consider various reconstruction techniques Ψ, including Filtered Back Projection, Total Variation, and a trained deep neural network. Incorporating the fractional power of the graph Laplacian operator into the regularization term significantly enhances the quality of the approximated solutions for each method Ψ. Additionally, we show that our proposal behaves as a regularization method and is also stable with respect to variations in the noise level.

Keywords: ill-posed problems · fractional graph Laplacian · image reconstruction

1 Introduction

In this work, we consider problems of the form

$$\arg\min_{\mathbf{x}\in\mathbb{R}^n} \|A\mathbf{x} - \mathbf{b}^\delta\|_2^2, \tag{1}$$

where $A \in \mathbb{R}^{m\times n}$ is a severely ill-conditioned operator, $\mathbf{b}^\delta \in \mathbb{R}^m$ represents the observed data corrupted by noise, and $\mathbf{x} \in \mathbb{R}^n$ is an unknown two-dimensional image with n pixels. We assume that the observed data $\mathbf{b}^\delta$ can be expressed

T. A. Bubba et al. (Eds.): SSVM 2025, LNCS 15667, pp. 83–95, 2025.
https://doi.org/10.1007/978-3-031-92366-1_7

as $\mathbf{b}^\delta = \mathbf{b} + \boldsymbol{\xi}$, where $\boldsymbol{\xi} \in \mathbb{R}^m$ is a white Gaussian perturbation such that $0 < \|\boldsymbol{\xi}\|_2 < \delta$, and $\mathbf{b}$ is the unavailable noise-free data. Due to the presence of noise in the observed data $\mathbf{b}^\delta$ and the severe ill-conditioning of A, the naive solution of (1) is a poor approximation of the desired solution $\mathbf{x}^\dagger = A^\dagger \mathbf{b}$.

To address this issue, we replace the original problem (1) with the $\ell_2 - \ell_1$ regularized problem

$$\arg\min_{\mathbf{x}\in\mathbb{R}^n} \frac{1}{2}\|A\mathbf{x} - \mathbf{b}^\delta\|_2^2 + \mu\|B\mathbf{x}\|_1, \tag{2}$$

where $B \in \mathbb{R}^{m' \times n}$ is referred to as the regularization operator, and $\mu > 0$ is called the regularization parameter. The use of the ℓ_2–norm in the data fidelity term is motivated by the presence of white Gaussian noise in the observed data, while the ℓ_1–norm in the regularization term promotes sparsity in the coefficients of the computed solution with respect to B.

In recent years, the use of graph-based regularization operators has gained significant attention [10,11]. In particular, in [3], the authors selected B as the graph Laplacian of a properly constructed graph obtained from a given approximation of $\mathbf{x}^\dagger$ and applied it to image deblurring. This dependence on a reasonably accurate initial approximation of $\mathbf{x}^\dagger$ was further investigated in [4] in the context of Computed Tomography (CT) problems.

Let Ψ_Θ be a reconstruction map, where $\Theta = \Theta(\delta, \mathbf{b}^\delta)$ is a family of parameters that can depend on δ and $\mathbf{b}^\delta$. Then, denoting by $L_{\Psi_\Theta^\delta}$ the graph Laplacian operator constructed on the initial approximation $\Psi_\Theta^\delta := \Psi_\Theta(\mathbf{b}^\delta)$, and replacing the regularization operator B in (2) with $L_{\Psi_\Theta^\delta}$, we obtain the variational problem

$$\arg\min_{\mathbf{x}\in\mathbb{R}^n} \frac{1}{2}\|A\mathbf{x} - \mathbf{b}^\delta\|_2^2 + \mu\|L_{\Psi_\Theta^\delta}\mathbf{x}\|_1. \tag{3}$$

The `graphLa+Ψ` method introduced in [4] consists in computing the solution of problem (3), and it has been proven to be a stable and convergent algorithm.

In this work, we explore the application of the fractional exponent to the graph Laplacian regularization operator $L_{\Psi_\Theta^\delta}$. The use of fractional operators in imaging has garnered interest due to their ability to enhance diffusion [2,14], while the fractional graph Laplacian has gained attention in the study of complex networks for its capacity to model non-local dynamics. For instance, in [1], the authors extended the approach presented in [3] to the fractional case and applied it to image restoration problems.

In what follows, we extend the `graphLa+Ψ` method to the fractional case, thus considering the variational problem

$$\arg\min_{\mathbf{x}\in\mathbb{R}^n} \frac{1}{2}\|A\mathbf{x} - \mathbf{b}^\delta\|_2^2 + \mu\|L^\alpha_{\Psi_\Theta^\delta}\mathbf{x}\|_1, \tag{4}$$

where $\alpha > 0$ is the fractional exponent. Let us remark that the `graphLa+Ψ` method already introduces a layer of complexity into the analysis, as the regularization term in (3) intrinsically depends on both the noise intensity δ and the observed data $\mathbf{b}^\delta$. In (4), we must also account for the fractional exponent α. For this reason, we are not yet able to provide theoretical results regarding convergence or stability; however, we experimentally verify that our proposed method is stable.

This paper is structured as follows. Section 2 is devoted to the `graphLa+`Ψ method. In particular, Subsect. 2.1 recalls the definition of the graph Laplacian operator and describes how graphs can be connected to images. Subsection 2.2 focuses on the MM-GKS algorithm for solving the variational problems (3) and (4). For completeness, in Subsect. 2.3, we summarize the theoretical results of the `graphLa+`Ψ method. Finally, Sect. 3 extends the graph Laplacian to the fractional case, and Sect. 4 presents selected numerical experiments on a 2D CT application.

2 GraphLa+Ψ Method

This section briefly describes the main properties of the `graphLa+`Ψ method.

2.1 Graph Laplacian

We first recall the standard definition of a weighted graph and outline the construction of the graph Laplacian operator, following the notation used in [9].

Definition 1 (Graph). *A graph over a set V is a pair $G = (w, \mu)$ given by:*

- *A non negative edge-weight function $w : V \times V \to [0, \infty)$ that satisfies:*
  - *Symmetry: $w(p, q) = w(q, p)$ for every $p, q \in V$;*
  - *No self-loops: $w(p, p) = 0$ for every $p \in V$.*
- *A positive node measure $\mu : V \to (0, \infty)$.*

We denote with $C(V)$ the set of real-valued functions over V and with $\mathbf{x}$ an element of $C(V)$.

Definition 2 (Graph Laplacian). *The graph Laplacian $L : C(V) \to C(V)$ associated to the graph $G(w, \mu)$ is defined by the action*

$$L\mathbf{x}(p) = \frac{1}{\mu(p)} \sum_{q \in V} w(p, q)(\mathbf{x}(p) - \mathbf{x}(q)). \tag{5}$$

Any grayscale image $I \in \mathbb{R}^{s \times d}$ consists of a collection of pixels $P \in V$ arranged on a grid of dimensions $s \times d$. It is natural to identify these pixels as ordered pairs $P = (i, j) \in \mathbb{Z}^2$ with $i = 1, \ldots, s$ and $j = 1, \ldots, d$. Moreover, a grayscale image is defined by the light intensities of its pixels. Specifically, it can be represented by a function $\mathbf{x}\colon V \to [0, 1]$, where 0 corresponds to black and 1 to white.

To model the correlation between pixels, we define the weight function $w\colon V \times V \to \mathbb{R}$, which connects pixels based on their "physical" distance and weights the connection by their light intensities. The weight function is given by

$$w(P, Q) = \chi_{[0,R)}(\|P - Q\|_\infty) \exp\left(\frac{-(\mathbf{x}(P) - \mathbf{x}(Q))^2}{\sigma^2}\right), \tag{6}$$

where $\chi_{[0,R)}$ is the characteristic function of the interval $[0, R)$.

In general, a weighted graph can be represented by its adjacency matrix $\Omega \in \mathbb{R}^{l \times l}$, where l is the cardinality of the vertex set V. By imposing a lexicographic ordering on the pixels of the image, we have $\Omega \in \mathbb{R}^{n \times n}$, where $n = sd$, allowing us to uniquely identify each pixel P in the image with an index $i \in \mathbb{N}$. Consequently, the elements of the adjacency matrix Ω are defined as $\Omega_{i,j} = w(i, j)$.

Defining the node measure as $\mu(P) = \|\Omega\|_F = \sqrt{\sum_{i,j=1}^{sd} w^2(i,j)}$, the action (5) of the graph Laplacian L can be expressed as a symmetric positive semi-definite matrix given by

$$L = \frac{D - \Omega}{\|\Omega\|_F}, \tag{7}$$

where D is the degree matrix, defined as

$$D_{i,j} = \begin{cases} \sum_{k=1}^{n} \Omega_{i,k} & \text{if } i = j, \\ 0 & \text{otherwise.} \end{cases} \tag{8}$$

The main computational effort in assembling the graph Laplacian operator lies in the construction of the adjacency matrix Ω. Indeed, from the definition of the edge-weight function (6), we have at most $(2R-1)^2$ evaluations of the exponential function for each pixel in the graph. Therefore, the total number of function evaluations required for the construction of the adjacency matrix is at most $(2R-1)^2 n$.

2.2 MM-GKS

In this section, we briefly describe the Majorization–Minimization Generalized Krylov Subspace (MM-GKS) algorithm used to solve the initial problem (2). Further details about this approach can be found in [8].

Since the MM method requires the functional to be minimized to be smooth, as an initial step, we replace the original problem (2) with a smoothed version by substituting the ℓ_1–norm of the regularization term with a smooth approximation. Given $\epsilon > 0$, and defining $\phi_\epsilon(x_i) = \sqrt{x_i^2 + \epsilon^2}$, the ℓ_1–norm can be approximated as $\|\mathbf{x}\|_1 \approx \sum_{i=1}^{n} \phi_\epsilon(x_i)$. Therefore, we consider the smoothed problem

$$\underset{\mathbf{x} \in \mathbb{R}^n}{\arg\min} \underbrace{\frac{1}{2}\|A\mathbf{x} - \mathbf{b}^\delta\|_2^2 + \mu \sum_{i=1}^{n} \phi_\epsilon((B\mathbf{x})_i)}_{\mathcal{J}_\epsilon(\mathbf{x})}. \tag{9}$$

The remainder of the MM strategy involves constructing a sequence that converges to a stationary point of (9). Given an approximation $\mathbf{x}^{(k)}$, we define a quadratic functional $\mathcal{Q}(\mathbf{x}, \mathbf{x}^{(k)})$ that majorizes $\mathcal{J}_\epsilon(\mathbf{x})$ and is tangent to it at $\mathbf{x}^{(k)}$. Specifically, letting $\boldsymbol{\omega}^{(k)} = \mathbf{u}^{(k)}\left(1 - \left(\frac{(\mathbf{u}^{(k)})^2+\epsilon^2}{\epsilon^2}\right)^{-1/2}\right)$, where $\mathbf{u}^{(k)} = (B\mathbf{x}^{(k)})$, we define

$$\mathcal{Q}(\mathbf{x}, \mathbf{x}^{(k)}) = \frac{1}{2}\|A\mathbf{x} - \mathbf{b}^\delta\|_2^2 + \frac{\mu}{2\epsilon}\left(\|B\mathbf{x}\|_2^2 - 2\langle B\mathbf{x}, \boldsymbol{\omega}^{(k)}\rangle\right). \tag{10}$$

The next iterate $\mathbf{x}^{(k+1)}$ is then computed by minimizing $\mathcal{Q}(\mathbf{x}, \mathbf{x}^{(k)})$, which is equivalent to solving the minimization problem

$$\arg\min_{\mathbf{x}\in\mathbb{R}^n} \frac{1}{2}\left\|\begin{bmatrix} A \\ \eta^{1/2}B \end{bmatrix}\mathbf{x} - \begin{bmatrix} \mathbf{b}^\delta \\ \eta^{1/2}\boldsymbol{\omega}^{(k)} \end{bmatrix}\right\|_2^2, \tag{11}$$

where $\eta = \frac{\mu}{\epsilon}$. An approximate solution to (11) can be efficiently computed in a low-dimensional subspace of $\mathbb{R}^n$. Let $V_k \in \mathbb{R}^{n\times k}$ be a matrix with orthonormal columns that span the search subspace. We then look for a solution of the form

$$\mathbf{x}^{(k+1)} = V_k\mathbf{y}^{(k+1)}. \tag{12}$$

At each iteration, the search subspace is expanded by adding the normalized residual to the basis V_k, i.e., $V_{k+1} = [V_k, \mathbf{r}^{(k+1)}/\|\mathbf{r}^{(k+1)}\|_2]$, where

$$\mathbf{r}^{(k+1)} = A^T(AV_k\mathbf{y}^{(k+1)} - \mathbf{b}^\delta) + \eta B^T(BV_k\mathbf{y}^{(k+1)} - \boldsymbol{\omega}^{(k)}). \tag{13}$$

In the numerical experiments, we adopt the same implementation proposed in [5].

2.3 Stability and Convergence

As anticipated in the introduction, given a reasonably accurate approximation of the desired solution $\mathbf{x}^\dagger$, the `graphLa+`Ψ method consists of computing a solution to the problem (3) using the MM-GKS strategy introduced in the previous section. In [4], it has been proven that this method is convergent and stable under very weak assumptions. In this final part, we briefly focus on the theoretical aspects.

Assumption 1. *Firstly, we assume the existence of an element* $\mathbf{x}_0 \in \mathbb{R}^n$ *and a parameter choice rule* $\Theta = \Theta(\delta, \mathbf{b}^\delta)$ *such that*

$$\|\Psi_{\Theta(\delta,\mathbf{b}^\delta)}(\mathbf{b}^\delta) - \mathbf{x}_0\|_2 \to 0 \quad \textit{as } \delta \to 0. \tag{14}$$

Secondly, for $k \to \infty$ *and let* $\Theta_k = \Theta(\delta_k, \mathbf{b}^{\delta_k})$ *be a parameter choice rule. Let also* $\{\delta_k\}$ *and* $\{\mathbf{b}^{\delta_k}\}$ *be two sequences such that* $\delta_k \to \delta$ *and* $\mathbf{b}^{\delta_k} \to \mathbf{b}^\delta$ *, then*

$$\Theta_k \to \Theta \quad \textit{and} \quad \|\Psi_{\Theta_k}(\mathbf{b}^{\delta_k}) - \Psi_\Theta(\mathbf{b}^\delta)\|_2 \to 0 \textit{ for } k \to \infty. \tag{15}$$

Now let $\mathbf{x}_0$ and $\Theta = \Theta(\delta, \mathbf{b})$ be such that $\mathbf{x}_0 := \lim_{\delta\to 0} \Psi_{\Theta(\delta,\mathbf{b}^\delta)}(\mathbf{b}^\delta)$, then we have the following

Definition 3 (Graph minimizing solution). *We define the graph minimizing solution* $\mathbf{x}_{sol}$ *with respect to* $\mathbf{x}_0$ *if*

$$\|L_{\mathbf{x}_0}\mathbf{x}_{sol}\|_1 = \min\{\|L_{\mathbf{x}_0}\mathbf{x}\|_1 : \mathbf{x} \in X \text{ and } A\mathbf{x} = \mathbf{b}\}. \tag{16}$$

The two following results establish convergence and stability of the `graphLa+`Ψ.

Theorem 1 ([4] (Convergence)). *Assume that* $\mu : (0, \infty) \to (0, \infty)$ *satisfies*

$$\begin{aligned} &\lim_{\delta\to 0} \mu(\delta) = 0, \\ &\lim_{\delta\to 0} \mu(\delta) = \frac{\delta^2}{\mu(\delta)} = 0. \end{aligned} \tag{17}$$

Fix a sequence $\{\delta_k\}$ *such that*

$$\lim_{k\to\infty} \delta_k = 0, \quad \|\mathbf{b}^{\delta_k} - \mathbf{b}\|_2 \le \delta_k, \tag{18}$$

and set $\mu_k = \mu(\delta_k)$. *Let* $\Theta = \Theta(\delta, \mathbf{b}^\delta)$ *be the parameter choice rule satisfies the Assumption 1, and set* $\Theta_k = \Theta(\delta_k, \mathbf{b}^{\delta_k})$. *Then every sequence* $\{\mathbf{x}_k\}$ *of elements that minimize the problem (3), with* δ_k *and* Θ_k, *has a converging subsequence. The limit* $\mathbf{x}_{sol}$ *of the converging subsequence* $\{\mathbf{x}_{k'}\}$ *is a graph minimizing solution with respect to* $\mathbf{x}_0$ *and*

$$\lim_{k'\to\infty} \|L\mathbf{x}_{k'}\|_1 = \|L_{\mathbf{x}_0}\mathbf{x}_{sol}\|_1. \tag{19}$$

If $\mathbf{x}_{sol}$ *is unique, then* $\mathbf{x}_k \to \mathbf{x}_{sol}$.

Theorem 2 ([4] (Stability)). *Let now* $\mathbf{b}^\delta$ *be fixed and* $\{\delta_k\}$ *and* $\{\mathbf{b}^{\delta_k}\}$ *be sequences such that* $\delta_k \to \delta$ *and* $\mathbf{b}^{\delta_\mathbf{k}} \to \mathbf{b}^\delta$ *for* $k \to \infty$. *Under Assumption 1 every sequence* $\{\mathbf{x}_k\}$, *such that* $\mathbf{x}_k$ *is a minimizer of the problem*

$$\arg\min_{\mathbf{x}\in\mathbb{R}^n} \frac{1}{2}\|A\mathbf{x} - \mathbf{b}^\delta\|_2^2 + \mu\|L_{\Psi_\Theta^{\delta_k}}\mathbf{x}\|_1, \quad \mu > 0, \tag{20}$$

has a converging subsequence $\{\mathbf{x}_{k'}\}$, *such that its limit is a minimizer for the problem (3).*

Note that Assumption 1 can be satisfied even when Ψ is not a regularization method, i.e., $\mathbf{x}_0$ need not be the minimum norm solution of the least squares problem (1). This happens, for instance, by choosing Ψ as a neural network.

3 GraphLa+Ψ with a Fractional Approach

In this section, we aim to incorporate the fractional strategy proposed in [1] within the `graphLa+`Ψ framework. This involves solving the problem (4) using the MM-GKS algorithm discussed in Subsect. 2.2.

Given $\alpha > 0$, the computation of the fractional exponent of the graph Laplacian operator $L_{\Psi_\Theta^\delta}$ is not as straightforward as it might seem. Since $L_{\Psi_\Theta^\delta}$ is typically a large matrix, directly computing its fractional exponent is computationally prohibitive within a reasonable amount of time. Moreover, even though $L_{\Psi_\Theta^\delta}$ is sparse, its fractional power $L^\alpha_{\Psi_\Theta^\delta}$ results in a dense matrix.

However, the MM–GKS algorithm requires only matrix–vector products, as shown in Eqs. (11)–(13), with $B = L^\alpha_{\Psi_\Theta^\delta}$ and B^T. Thanks to Eq. (7) and the properties of a weighted graph recalled in Definition 1, the graph Laplacian operator $L_{\Psi_\Theta^\delta}$, as well as its fractional extension $L^\alpha_{\Psi_\Theta^\delta}$, is symmetric and positive semidefinite. Therefore, it suffices to compute matrix–vector products with $L^\alpha_{\Psi_\Theta^\delta}$.

Following the approach presented in [13], we address the density issue of the fractional graph Laplacian operator by performing a few preliminary iterations of the Lanczos method. Specifically, let $j \ll n$, we compute

$$L_{\Psi_\Theta^\delta} U_j = U_j T_j + \mathbf{t}_j \mathbf{e}_j^T, \tag{21}$$

where $U_j \in \mathbb{R}^{n \times j}$ has orthonormal columns, $\mathbf{t}_j = (L_{\Psi_\Theta^\delta} - \alpha_j)\mathbf{v}_j - \beta_{j-1}\mathbf{v}_{j-1}$, and

$$T_j = \begin{bmatrix} \alpha_1 & \beta_2 & & & \\ \beta_2 & \alpha_2 & \beta_3 & & \\ & \ddots & \ddots & \ddots & \\ & & \beta_{j-1} & \alpha_{j-1} & \beta_j \\ & & & \beta_j & \alpha_j \end{bmatrix}_{j \times j}. \tag{22}$$

Therefore, we can use the matrix T_j to approximate $L^\alpha_{\Psi_\Theta^\delta}$ in the subspace generated by the columns of U_j, and thus the matrix–vector product $L^\alpha_{\Psi_\Theta^\delta}\mathbf{x}$ can be approximated as

$$L^\alpha_{\Psi_\Theta^\delta}\mathbf{x} \simeq (U_j T_j U_j^T)^\alpha \mathbf{x} = U_j Q_j \Sigma_j^\alpha Q_j^T U_j^T \mathbf{x} \tag{23}$$

In (23), we exploit the symmetry of T_j by considering its eigenvalue decomposition $T_j = Q_j \Sigma_j Q_j^T$, where Q_j is an orthogonal matrix and Σ_j is a diagonal matrix with non–negative entries. In [13], the authors showed that a small subspace, i.e., index j, is sufficient to obtain a good approximation of the matrix–vector product. Thus, the computation of Q_j and Σ_j can be performed with negligible computational cost.

In order to be more precise, we note that the computation of the matrices T_j and U_j using j Lanczos iteration requires O(jn) operations, while the spectral decomposition of the tridiagonal symmetric matrix T_j requires $O(j^3)$ operations. Thus, since we only require a few iterations (i.e., $j \ll n$), the overall computational cost of the previously discussed method for the matrix-vector product is essentially linear in n.

4 Numerical Results

In this section, we present a selection of numerical experiments in 2D computerized tomography, where we integrate the fractional **graphLa+Ψ** method with various reconstructors Ψ. These include Filtered Back Projection (LaFBP), standard Tikhonov (LaTik), Total Variation (LaTV), and a trained neural network (LaNet). For comparison, we will also include the reconstruction achieved by our method using the ground truth image $\mathbf{x}_{gt}$ as a first approximation, labeled as LaTrue. This serves as an upper-bound reference for the effectiveness of each case.

4.1 Experimental Setup

We evaluate the performance of the fractional **graphLa+Ψ** method on the COULE dataset [6], which consists of synthetic images with a resolution of 256×256 pixels. These images depict ellipses and lines of varying gray intensities against a dark background. To simulate the sinogram $\mathbf{b}^\delta$, we consider $n_a = 120$ different angles evenly distributed over the closed interval $[0, 179]$ and $n_d = \lfloor\sqrt{2}n\rfloor$ pixels in the detector. As a result, the sinogram has dimensions $n_d \times n_a$ and is generated using the IRtools toolbox [7]. To simulate real-world conditions, we add white Gaussian noise $\boldsymbol{\xi}$ with an intensity level δ, such that $\mathbf{b}^\delta$ is computed as

$$\mathbf{b}^\delta = \mathbf{b} + \delta \frac{\boldsymbol{\xi}}{\|\boldsymbol{\xi}\|_2}. \tag{24}$$

In the numerical experiments presented below, we consider a noise level $\delta = 2\%$. Figure 1 illustrates both the original image and the corresponding noisy sinogram.

As anticipated, in addition to all possible choices for the reconstructor Ψ, we also consider a Residual U-net (ResU-net) architecture, as previously proposed in [12]. Since the ResU-net architecture is fully convolutional, the network requires an image as input. Consequently, the input sinogram $\mathbf{b}^\delta$ must be pre-processed using a fast algorithm to map the sinogram to a coarse reconstructed image. For our experiments, we employ the FBP algorithm. Lastly, it is worth noting that while we tested the fractional **graphLa+Ψ** algorithm at different noise levels, and despite the known high sensitivity of DNNs to noise, the neural network was trained solely on a noiseless dataset. Nevertheless, the resulting strategy demonstrates stability with respect to the noise level in the sinogram.

Given a reconstructor Ψ, we compute approximate solutions of (3) for different values of the fractional exponent $\alpha \in [0.1, 1.1]$ using the MM–GKS strategy described in Sect. 2.2. For this purpose, we set the maximum number of iterations to 500, the smoothing parameter to $\epsilon = 10^{-1}$, and the number of Lanczos iterations in (21) to $j = 30$. For the graph Laplacian operator, we set $R = 5$ and $\sigma^2 = 10^{-3}$ in the definition of the edge weight function (6).

Finally, to evaluate the quality of the reconstructions, we use two metrics: the Peak Signal-to-Noise Ratio (PSNR) and the Structural Similarity Index Measure (SSIM).

4.2 Numerical Experiments

We evaluate the performance of the fractional `graphLa+`Ψ method for different values of the fractional exponent α. We compare the results obtained using various reconstructors Ψ. In all cases, the regularization parameter is automatically estimated using the discrepancy principle [5].

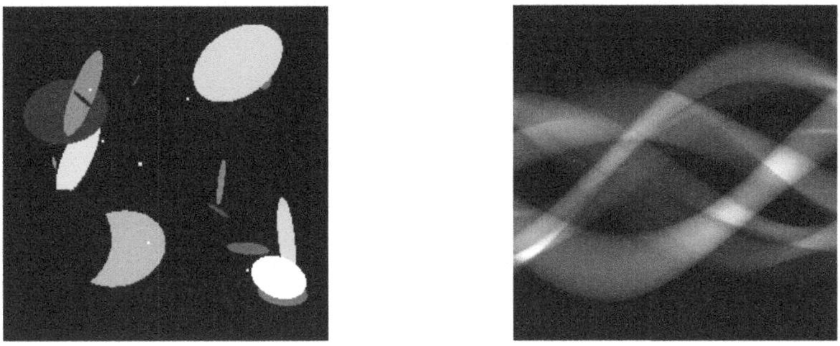

Fig. 1. Ground truth, on the left and its associated sinogram on the right.

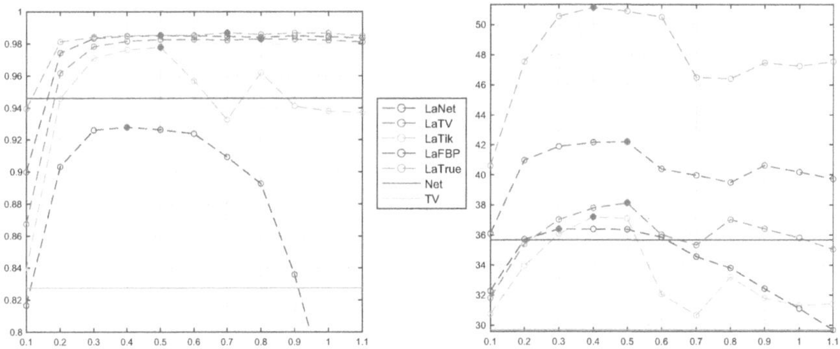

(a) (b)

Fig. 2. (a) SSIM and (b) PSNR for different values of α and various reconstructors Ψ. In both figures, the red marker denotes the optimal α. The noise level is $\delta = 2\%$. (Color figure online)

Figure 2 shows the SSIM and PSNR metrics as functions of the fractional exponent α. Red dots depict the optimal values of α that achieve the highest performance among all the considered choices.

We observe that the parameter α significantly affects the reconstructions. In both metrics, the optimal value of α differs from 1, indicating that the fractional `graphLa+`Ψ method provides better reconstructions compared to the standard

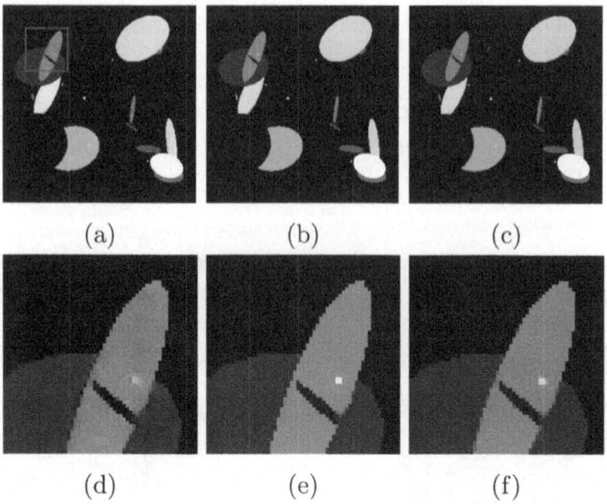

Fig. 3. First row: Reconstructions obtained using the fractional graph Laplacian with (a) as the initial reconstruction provided by the neural network (SSIM = 0.9461, PSNR = 35.6828), (b) $\alpha = 0.5$ (SSIM = 0.9851, PSNR = 42.1872), and (c) $\alpha = 1$ (SSIM = 0.9844, PSNR = 40.1793). Second row: Zoom–in on a detail of the above reconstructions.

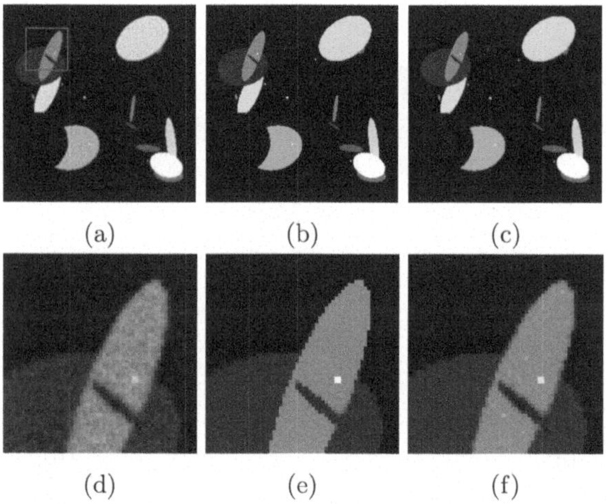

Fig. 4. First row: Reconstructions obtained using the fractional graph Laplacian with (a) as the initial reconstruction obtained solving the Tikhonov model (SSIM = 0.4607, PSNR = 26.4129), (b) $\alpha = 0.4$ (SSIM = 0.9763, PSNR = 37.1978), and (c) $\alpha = 1$ (SSIM = 0.9380, PSNR = 31.3000). Second row: Zoom–in on a detail of the above reconstructions.

case $\alpha = 1$. This is further confirmed in Figs. 3 and 4, where we present the reconstructions obtained for different Ψ and α. Specifically, the first row displays the full reconstructed image, while the second row provides a zoomed-in detail of the reconstruction. This visualization clearly shows that choosing $\alpha \neq 1$ enhances the sharpness of the final reconstruction.

Moreover, Fig. 4 demonstrates that when the initial reconstruction contains a high level of noise, the fractional `graphLa+Ψ` method can effectively address this issue by selecting a smaller value of α, whereas the standard case only slightly reduces the noise intensity in the final reconstruction.

Finally, in Fig. 5, we visualize the absolute value of the action of the graph Laplacian L_{gt} on different reconstructions: Tikhonov ($\mathbf{x}_{Tik}$), Total Variation ($\mathbf{x}_{TV}$), and the neural network ($\mathbf{x}_{NN}$). The first row shows results for the standard graph Laplacian ($\alpha = 1$), while the second row presents the fractional case for varying α. Using $\alpha \neq 1$ increases sparsity and enhances feature distinction, supporting the use of an ℓ^1-based regularizer in models (3) and (4).

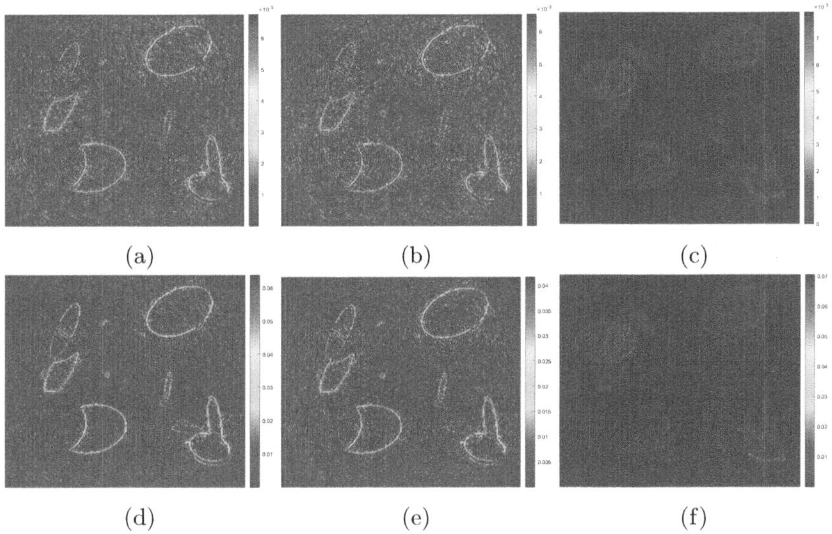

Fig. 5. (a) $|L_{gt}\mathbf{x}_{Tik}|$, (b) $|L_{gt}\mathbf{x}_{TV}|$, (c) $|L_{gt}\mathbf{x}_{NN}|$, (d) $|L_{gt}^{\alpha}\mathbf{x}_{Tik}|$ with $\alpha = 0.4$, (e) $|L_{gt}^{\alpha}\mathbf{x}_{TV}|$ with $\alpha = 0.5$, (f) $|L_{gt}^{\alpha}\mathbf{x}_{NN}|$ with $\alpha = 0.5$.

4.3 Stability and Regularization

As mentioned in Sect. 2.3, the authors in [4] proved the stability and convergence of the `graphLa+Ψ` method. Unfortunately, such theoretical analysis cannot be straightforwardly extended to the fractional version. However, in this section, we aim to experimentally validate the stability of the fractional `graphLa+Ψ` algorithm.

In Fig. 6, we report the evolution of the PSNR and SSIM metrics with respect to the noise level. We observe that, as the noise level approaches 0, the method yields improved reconstructions. Notably, as discussed in [4], integrating the graph Laplacian with a DNN acts as an effective regularization strategy, outperforming the standalone application of the DNN. A similar behavior can be observed in Fig. 6, indicating that, from an experimental perspective, the fractional `graphLa+`Ψ method would seem to behave as a regularization method.

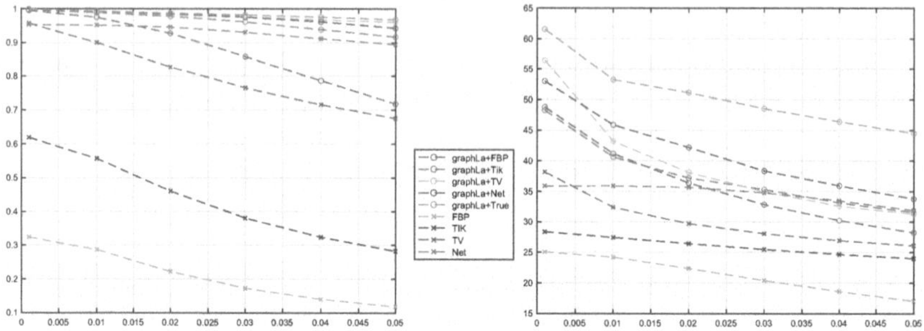

(a) (b)

Fig. 6. (a) SSIM and (b) PSNR for different noise levels. For each case, we reported the performance obtained using the optimal value of α.

Finally, we highlight that as the noise level increases, the performance quality deteriorates. Nonetheless, the method remains stable across all noise levels considered.

5 Conclusions

In this work, we combined the fractional extension of the graph Laplacian proposed in [1] with the `graphLa+`Ψ method introduced in [4]. After a brief introduction on our fractional graph Laplacian method, we provided numerical results demonstrating the performance of the method.

In particular, we showed that using a fractional exponent $\alpha \neq 1$ can significantly improve the quality of the reconstructions, both in terms of performance metrics and visual inspection. Moreover, we observed that when the initial reconstructor Ψ yields poor approximations with a high level of noise, the fractional `graphLa+`Ψ algorithm effectively addresses this issue by selecting a smaller value of α and enhancing the sharpness of the final reconstruction.

Finally, in the last part, we experimentally demonstrated that the fractional extension of the `graphLa+`Ψ method acts as a regularization method and is stable. However, a proper theoretical analysis of this fractional extension is yet to be conducted.

Declarations

Acknowledgments. The first, second, and fourth authors are partially supported by the PRIN project 2022ANC8HL funded by the Italian Ministry of University and Research.

The first and second authors are partially supported by the "INdAM - GNCS Project", code CUP_E53C24001950001. This research was also supported by the Swiss National Science Foundation SNSF via the projects Stress-Based Methods for Variational Inequalities in Solid Mechanics n. 186407 and ExaSolvers n. 162199.

References

1. Aleotti, S., Buccini, A., Donatelli, M.: Fractional graph Laplacian for image reconstruction. Appl. Numer. Math. **200**, 43–57 (2024)
2. Antil, H., Bartels, S.: Spectral approximation of fractional PDEs in image processing and phase field modeling. J. Comput. Methods Appl. Math. **17**(4), 661–678 (2017)
3. Bianchi, D., Buccini, A., Donatelli, M., Randazzo, E.: Graph Laplacian for image deblurring. Electron. Trans. Numer. Anal. **55**, 169–186 (2021)
4. Bianchi, D., Evangelista, D., Aleotti, S., Donatelli, M., Piccolomini, E.L., Li, W.: A data-dependent regularization method based on the graph Laplacian. SIAM J. Sci. Comput. **47**(2), C369–398 (2025)
5. Buccini, A., Reichel, L.: Software for limited memory restarted l^p-l^q minimization methods using generalized krylov subspaces. Electron. Trans. Numer. Anal. **61**, 66–91 (2024)
6. Evangelista, D. Morotti, E., Loli Piccolomini, E.: COULE dataset. https://www.kaggle.com/datasets/loiboresearchgroup/coule-dataset (2021)
7. Gazzola, S., Hansen, P.C., Nagy, J.G.: IR tools: a MATLAB package of iterative regularization methods and large-scale test problems. Numer. Algorithms **81**(3), 773–811 (2019)
8. Huang, G., Lanza, A., Morigi, S., Reichel, L., Sgallari, F.: Majorization-minimization generalized Krylov subspace methods for ℓ_p- ℓ_q optimization applied to image restoration. BIT **57**(2), 351–378 (2017)
9. Keller, M., Lenz, D., Wojciechowski, R.: Graphs and Discrete Dirichlet Spaces, 1st edn. Grundlehren der Mathematischen Wissenschaften, Springer Nature, Cham, Switzerland (2021)
10. Meyer, F.G., Shen, X.: Perturbation of the eigenvectors of the graph Laplacian: application to image denoising. Appl. Comput. Harmon. Anal. **36**(2), 326–334 (2014)
11. Pang, J., Cheung, G.: Graph Laplacian regularization for image denoising: analysis in the continuous domain. IEEE Trans. Image Process. **26**(4), 1770–1785 (2017)
12. Ronneberger, O., Fischer, P., Brox, T.: U-Net: convolutional networks for biomedical image segmentation. In: Lecture Notes in Computer Science, pp. 234–241. Lecture notes in computer science, Springer International Publishing, Cham (2015)
13. Susnjara, A., Perraudin, N., Kressner, D., Vandergheynst, P.: Accelerated filtering on graphs using Lanczos method. arXiv:1509.04537 (2015)
14. Yang, Q., Chen, D., Zhao, T., Chen, Y.: Fractional calculus in image processing: a review. Fract. Calc. Appl. Anal. **19**(5), 1222–1249 (2016)

A Fractional-Order Telegraph Diffusion Model for Multiplicative Noise Removal

Padmaja Yinukula(✉), Mahipal Jetta, Rajkumar Alugunuri, Sarita Panda, and Meraj Alam

Department of Mathematics, Mahindra University, Hyderabad 500043, Telangana, India
devipadmaja1509@gmail.com

Abstract. We introduce a time-fractional telegraph diffusion model to remove high level multiplicative noise while preserving critical features such as texture and edges. The model utilizes α^{th} and β^{th} order time-fractional Caputo derivatives to capture sub-diffusive and super-diffusive behaviors, enhancing noise removal in various images such as synthetic aperture radar (SAR), texture, and natural images. To accurately capture the non-local effects of the Caputo derivatives, we apply first-order and second-order discretization schemes. Extensive experiments demonstrate superior performance of the proposed model over state-of-the-art methods in terms of PSNR, SI, MAE, and MSSIM.

Keywords: Image Denoising · Multiplicative Gamma Noise · Telegraph-Diffusion · Time-fractional Caputo derivative

1 Introduction

Denoising is essential for synthetic aperture radar (SAR), ultrasound, and laser images, where they often suffer from multiplicative noise [17]. This noise severely degrades the edges and fine details of images. In comparison to additive noise, multiplicative noise is more challenging to remove, particularly in texture-rich images, as the oscillatory nature of textures often leads denoising algorithms to misinterpret them as noise. In this context, the observed image f is modeled as $f = v\eta$, where v represents the true image, and η denotes the noise. As in [4], we assume that the noise η follows a Gamma distribution with mean 1 and variance $1/L$:

$$p(\eta) = \frac{L^L}{\Gamma(L)}\eta^{L-1}e^{-L\eta}\mathbf{1}_{\{\eta\geq 0\}}.$$

Here, $\mathbf{1}_{\{\eta\geq 0\}}$ ensures non-negativity, and L is a parameter which controls the noise density.

Various techniques, including filtering [8], wavelet methods [5], and variational approaches [1,10] have been utilized to address multiplicative noise. However, traditional methods often struggle to balance noise reduction with the preservation of important image features, leading to the exploration of alternatives grounded in partial differential equations (PDEs), for example, see [12,16], and the references therein.

T. A. Bubba et al. (Eds.): SSVM 2025, LNCS 15667, pp. 96–107, 2025.
https://doi.org/10.1007/978-3-031-92366-1_8

PDE-based methods, particularly anisotropic diffusion introduced by Perona and Malik (PM) [12], have significantly advanced image denoising. Although the PM model facilitates selective smoothing, it encounters limitations, such as ill-posedness and the 'staircasing' effect, where excessive smoothing occurs in flat regions, resulting in artificial artifacts [12].

Recent extensions of second-order models, such as the Shan model [15] introduces modifications to the diffusion function, enhancing stability and edge preservation, while the Majee model [11] further optimizes performance in images with intricate textures and challenging noise patterns. Although these second-order advancements provide important refinements, they are still restricted by the inherent constraints of integer-order diffusion, limiting their flexibility in adapting to various image characteristics. To enhance smoothing accuracy, researchers have explored fourth-order PDEs, such as the one by You and Kaveh [21], which reduces staircase artifacts but introduces the 'speckle effect'. In contrast, fractional-order partial differential equations (FPDEs) [6,9] offer a powerful alternative to traditional integer-order models in image denoising. These FPDEs leverage non-local information and broader spatial correlations to enable selective diffusion adapted to the image content. This capability makes FPDEs particularly well-suited for processing natural images with complex and diverse features.

Several contributions have significantly advanced fractional-order models for image denoising. Among them, the fully fractional anisotropic diffusion (FFAD) model, proposed by Janev et al. [7], interpolates between parabolic and hyperbolic partial differential equations (PDEs). This model enhances diffusion control, achieving superior image enhancement and edge preservation by combining fractional derivatives with an edge-stopping function. Similarly, Ullah et al. [18] introduced a novel variational approach that integrates a fractional-order total variation (FOTV) filter with a high-order Laplacian norm. This approach excels in preserving edges and mitigating the staircase effect in smooth regions. Additionally, Yin et al. [20] proposed a fractional-order nonlinear anisotropic diffusion model for image restoration, utilizing the p-Laplace norm of the fractional-order gradient. While this model effectively preserves texture details and adapts diffusion based on local geometric features, it struggles with high noise levels, particularly in structure and texture images.

The success of these models, at moderate noise levels, often comes at the expense of high computational efforts and over-smoothing of textures. Motivated by the benefits and limitations of these approaches, this paper proposes a novel method to address high level multiplicative noise using fractional PDEs. Our method emphasizes controlled smoothing while significantly reducing computational overhead.

The remainder of this paper is organized as follows. Section 2 reviews existing image denoising models, emphasizing fractional-order methods and integer-order models. Section 3 introduces the proposed fractional telegraph-diffusion equation, explaining its formulation and the benefits regarding the texture preservation. Section 4 provides a detailed explanation of the numerical scheme employed to solve the equation. Section 5 presents experimental results, comparing the proposed model with existing methods in terms of prevalent quality assessment metrics on various image datasets. Section 6 concludes the paper with key findings and future research directions.

2 Existing PDE Based Models

Perona and Malik [12] proposed a nonlinear filter for removing additive Gaussian noise from an image, governed by the following partial differential equation

$$\begin{cases} \frac{\partial u}{\partial t} = \nabla \cdot (g_1(|\nabla u|)\nabla u)\,, \text{ in } \Omega \times (0,T), \\ \frac{\partial u}{\partial \mathbf{n}} = 0, & \text{on } \partial\Omega \times (0,T), \\ u(\mathbf{x},0) = f(\mathbf{x}), & \text{on } \Omega \times \{t=0\}, \end{cases} \tag{1}$$

where u denotes the restored image, f is the noisy input image, and t represents time. The spatial domain is $\Omega \subset \mathbb{R}^2$, with $\partial\Omega$ representing its boundary. The symbol $\mathbf{n}$ represents unit outward normal to the boundary $\partial\Omega$. The key aspect of this model lies in the function $g_1(|\nabla u|)$, which is smooth and nonincreasing, with $g_1(0) = 1$ and $\lim_{|\nabla u| \to \infty} g_1(|\nabla u|) = 0$. The function $g_1(|\nabla u|)$ is given by

$$g_1(|\nabla u|) = \frac{1}{1 + \left(\frac{|\nabla u|}{k}\right)^2}, \tag{2}$$

where k is a constant, controlling the diffusion behavior. This diffusion process, driven by the local gradient magnitude $|\nabla u|$, effectively preserves edges. However, it may face challenges in handling high noise level, making it less suitable for intensely noisy images.

To address this issue, numerous authors have developed various partial differential equation (PDE)-based models. Among these, Shan et al. [15] adapted the Perona-Malik framework for the removal of multiplicative Gamma noise by modifying the diffusion function to

$$g_2(u_\sigma, |\nabla u_\sigma|) = \left(\frac{|u_\sigma|}{M_\sigma}\right)^{\tilde{\alpha}} \frac{1}{1 + |\nabla u_\sigma|^{\tilde{\beta}}}, \tag{3}$$

where u_σ is the smoothed version of u, obtained by convolving the image u, with a Gaussian kernel G_σ, and $M_\sigma = \max_{\mathbf{x}\in\Omega} |u_\sigma(\mathbf{x},t)|$. The parameters $\tilde{\alpha}$, $\tilde{\beta}$, and σ are positive constants. This modification enables the model to leverage gray level information and edges to control the diffusion process, effectively reducing noise while preserving image structures.

Majee et al. [11] extended the approach of Shan et al. [15] by developing a telegraph-type nonlinear PDE model that combines both diffusion and wave propagation dynamics to achieve enhanced denoising performance. This model takes the following form

$$\begin{cases} \frac{\partial^2 u}{\partial t^2} + \gamma\frac{\partial u}{\partial t} = \nabla \cdot (g_3(u_\sigma, |\nabla u_\sigma|)\nabla u)\,, \text{ in } \Omega \times (0,T), \\ \frac{\partial u}{\partial \mathbf{n}} = 0, & \forall(\mathbf{x},t) \in \partial\Omega \times (0,T), \\ u(\mathbf{x},0) = f(\mathbf{x}), \quad \frac{\partial u}{\partial t}(\mathbf{x},0) = 0, & \forall \mathbf{x} \in \Omega. \end{cases} \tag{4}$$

The diffusion coefficient $g_3(u_\sigma, |\nabla u_\sigma|)$ in this model is defined as

$$g_3(u_\sigma, |\nabla u_\sigma|) = \frac{2|u_\sigma|^\nu}{M_\sigma^\nu + |u_\sigma|^\nu} \cdot \frac{1}{1 + \left(\frac{|\nabla u_\sigma|}{k}\right)^2}, \tag{5}$$

where k, σ, ν, and γ are positive constants. Here M_σ and $u_\sigma$are same as in eq (3). This model proves particularly effective in preserving oscillatory patterns during the denoising process. However, the above two models for multiplicative noise removal often blur edges and fail to preserve fine textures.

Recently, a fractional-order cross-diffusion system [6] has been proposed to remove multiplicative noise while preserving texture details. The model consists of the coupled PDEs

$$\begin{cases} \frac{\partial u_1}{\partial t} = \text{div}^\alpha \left(g_4(u) \left(\cos\theta \nabla^\alpha u_1 - \sin\theta \nabla^\alpha u_2\right)\right), & \text{in } \Omega \times (0, T), \\ \frac{\partial u_2}{\partial t} = \text{div}^\alpha \left(g_4(u) \left(\sin\theta \nabla^\alpha u_1 + \cos\theta \nabla^\alpha u_2\right)\right), & \text{in } \Omega \times (0, T), \\ u_1(\mathbf{x}, t) = 0, \quad u_2(\mathbf{x}, t) = 0, & \text{on } \partial\Omega \times (0, T), \\ u_1(\mathbf{x}, 0) = u_{10}, \quad u_2(\mathbf{x}, 0) = u_{20}, & \text{in } \Omega. \end{cases} \tag{6}$$

Here, the functions $u_1, u_2 : \Omega \times (0, T) \rightarrow \mathbb{R}$ represent the texture and boundary information, respectively, while u_{10} and u_{20} denote the initial texture and boundary information of the grayscale image.The edge-preserving diffusion function $g_4(u)$ is defined as

$$g_4(u) = \left(\frac{|u|}{M}\right)^r \frac{1}{1 + \left(\frac{u_2}{\kappa\theta}\right)^2} + \xi, \tag{7}$$

where $|u| = \sqrt{u_1^2 + u_2^2}$, $M = \sup_{\mathbf{x}\in\Omega} |u(\mathbf{x}, t)|$, and ξ is a small positive constant. The parameters $r > 0$ and $\kappa > 0$ control grayscale range and thresholding, respectively, while $\theta \ll 1$ regulates cross-diffusion.

The small θ assumption ensures $\cos\theta \approx 1$ and $\sin\theta \approx \theta$, simplifying interactions between u_1 and u_2. This choice reduces computational complexity while preserving textures and edges. Larger θ values increase coupling, potentially leading to excessive smoothing. However, tuning these parameters can be challenging and may result in a loss of image quality. While the model performs well at moderate noise levels, its effectiveness diminishes at higher noise intensities and in complex-structured images, such as natural images. Severe noise further impacts its ability to preserve texture and denoise satellite images, limiting its practical applications primarily to scenarios with moderate noise conditions.

To address these limitations, we propose a novel image denoising model that integrates fractional-order dynamics within a telegraph-based diffusion framework. This approach combines the strengths of fractional-order models and telegraph diffusion models [2, 11, 14] to effectively suppress high noise levels while preserving intricate texture patterns. The details of this model are presented in the following section.

3 Proposed Model

In this section, we introduce a novel nonlinear fractional-order telegraph-type partial differential equation (FPDE) model for image denoising.

The mathematical formulation of the proposed model is given by

$$\boldsymbol{D}_C^{\alpha} u + \gamma \boldsymbol{D}_C^{\beta} u = \nabla \cdot \left(g_3(u_\sigma, |\nabla u_\sigma|) \nabla u \right), \quad \forall (\mathbf{x}, t) \in \Omega \times (0, T), \tag{8}$$

subject to the initial and boundary conditions:

$$\begin{aligned} u(\mathbf{x}, 0) &= f(\mathbf{x}), \quad \frac{\partial u}{\partial t}(\mathbf{x}, 0) = 0, \quad \forall \mathbf{x} \in \Omega, \\ \frac{\partial u}{\partial \mathbf{n}} &= 0, \quad \forall (\mathbf{x}, t) \in \partial\Omega \times (0, T). \end{aligned}$$

Here the diffusion function $g_3(u_\sigma, |\nabla u_\sigma|)$ is same as Eq. (5). It consists of two components: an edge indicator, $\frac{1}{1+\left(\frac{|\nabla u_\sigma|}{k}\right)^2}$, which restricts diffusion near edges while allowing stronger diffusion in homogeneous regions for noise suppression, and a gray-level indicator, $\frac{2|u_\sigma|^\nu}{(M_\sigma)^\nu + |u_\sigma|^\nu}$, which adjusts the diffusion rate based on intensity values in the image.

The proposed PDE (8) incorporates two distinct Caputo fractional derivatives: $D_C^{\alpha} u$ and $D_C^{\beta} u$, with orders $\alpha \in (1, 2)$ and $\beta \in (0, 1)$, respectively. These derivatives follow the standard Caputo definition [13] and are used to characterize memory effects and anomalous diffusion. The αth-order term introduces wave-like propagation, preserving structural details and preventing excessive smoothing, while the βth-order term governs diffusion, ensuring smooth and gradual noise removal. By adjusting γ, the model transitions between different behaviors: as $\gamma \to 0$, wave-like effects dominate, helping retain high-frequency features such as edges and textures, whereas for larger γ, the model shifts toward a parabolic regime, prioritizing noise removal while reducing oscillations.

A key advantage of the proposed fractional-order formulation is its inherent memory effect, where the solution evolution depends on its past history rather than just its immediate state. This property enables the model to adaptively balance noise suppression with texture preservation, making it particularly well-suited for denoising images affected by multiplicative noise. Compared to conventional diffusion models, the proposed telegraph-type formulation maintains oscillatory patterns more effectively, reducing the risk of excessive smoothing, particularly in texture-rich images.

4 Numerical Setup

4.1 Numerical Scheme

We introduce a finite difference approximation to discretize the time-fractional derivatives. Let $t_n := n\tau$, where τ is the time step. We use central differences to approximate the second-order spatial derivatives, while the Caputo derivative is used to approximate the fractional time derivative. Together, it provides the numerical solution for a time-fractional PDE. We refer the readers to [19] for more details. For the αth order fractional derivative [22], we have

$$D_C^\alpha u(\mathbf{x}, t_n) = \frac{\partial^\alpha u(\mathbf{x}, t_n)}{\partial t^\alpha} = \frac{1}{\Gamma(2-\alpha)} \int_0^{t_n} \frac{\partial^2 u(\mathbf{x}, s)}{\partial s^2} \frac{ds}{(t_n - s)^{\alpha-1}}, \quad 1 < \alpha < 2.$$

$$\approx \frac{1}{\Gamma(2-\alpha)} \sum_{k=0}^{n} \int_{t_k}^{t_{k+1}} (t_n - s)^{1-\alpha} \left[\frac{u^{k+1} - 2u^k + u^{k-1}}{\tau^2} \right] ds.$$

Thus, the final expression for the approximation of the αth order fractional derivative is,

$$D_C^\alpha u^n = \sigma_1 \left[\left(u^{n+1} - 2u^n + u^{n-1}\right) + \sum_{k=1}^{n} b_k \left(u^{n-k+1} - 2u^{n-k} + u^{n-k-1}\right) \right], \quad (9)$$

where

$$\sigma_1 = \frac{\tau^{-\alpha}}{\Gamma(3-\alpha)}, \quad b_k = (n-k+1)^{2-\alpha} - (n-k)^{2-\alpha}, \text{and} \quad D_C^\alpha u^n \approx D_C^\alpha u(\mathbf{x}, t_n).$$

For the fractional derivative of βth order [3], we have

$$D_C^\beta u(\mathbf{x}, t_n) = \frac{\partial^\beta u(\mathbf{x}, t_n)}{\partial t^\beta} = \frac{1}{\Gamma(1-\beta)} \int_0^{t_n} (t_n - s)^{-\beta} \frac{\partial u(\mathbf{x}, s)}{\partial s} ds, \quad 0 < \beta < 1.$$

$$\approx \frac{1}{\Gamma(1-\beta)} \sum_{k=0}^{n} \left[\frac{u^{k+1} - u^k}{\tau} \right] \int_{t_k}^{t_{k+1}} (t_n - s)^{-\beta} ds.$$

Thus, the final expression for the approximation of the βth order fractional derivative is,

$$D_C^\beta u^n = \sigma_2 \left[\left(u^{n+1} - u^n\right) + \sum_{k=1}^{n} a_k \left(u^{n-k+1} - u^{n-k}\right) \right], \quad (10)$$

where

$$\sigma_2 = \frac{\tau^{-\beta}}{\Gamma(2-\beta)}, \quad a_k = (n-k)^{1-\beta} - (n-k-1)^{1-\beta}, \text{and} \quad D_C^\beta u^n \approx D_C^\beta u(\mathbf{x}, t_n).$$

Substituting the expressions (9) and (10) into Eq. (8) yields

$$\sigma_1 \left[u^{n+1} - 2u^n + u^{n-1} + \sum_{k=1}^{n} b_k \left(u^{n-k+1} - 2u^{n-k} + u^{n-k-1}\right) \right]$$

$$+ \gamma\sigma_2 \left[u^{n+1} - u^n + \sum_{k=1}^{n} a_k \left(u^{n-k+1} - u^{n-k}\right) \right] = \nabla . \left(g_3(|\nabla u_\sigma^n|) \nabla u^n\right).$$

Finally, rearranging the terms leads to

$$u^{n+1} = u^n \left(\frac{2\sigma_1 + \gamma\sigma_2}{\sigma_1 + \gamma\sigma_2}\right) - u^{n-1}\left(\frac{\sigma_1}{\sigma_1 + \gamma\sigma_2}\right) - \left(\frac{\gamma\sigma_2}{\sigma_1 + \gamma\sigma_2}\right)\sum_{k=1}^{n} a_k \left(u^{n-k+1} - u^{n-k}\right)$$
$$- \left(\frac{\sigma_1}{\sigma_1 + \gamma\sigma_2}\right)\sum_{k=1}^{n} b_k \left(u^{n-k+1} - 2u^{n-k} + u^{n-k-1}\right)$$
$$+ \left(\frac{1}{\sigma_1 + \gamma\sigma_2}\right)\left\{\nabla_x \left(g_3^n \left(|\nabla u_\sigma^n|\right)\nabla_x u^n\right) + \nabla_y \left(g_3^n \left(|\nabla u_\sigma^n|\right)\nabla_y u^n\right)\right\}, \quad (11)$$

where $g_3^n(\cdot)$ represents the approximation of Eq. (5) at the nth time-step level. Here, the nonlinear diffusion terms $\nabla_x \left(g_3^n \left(|\nabla u_\sigma^n|\right)\nabla_x u^n\right)$ and $\nabla_y \left(g_3^n \left(|\nabla u_\sigma^n|\right)\nabla_y u^n\right)$ are approximated using the central difference scheme as described in [11]. The discretized equation is solved iteratively using an explicit update scheme (11) with the condition $u^{-1} = u^0 = f$.

4.2 Numerical Experiments

In this section, we evaluate the denoising performance of the proposed model (8), comparing it against established state-of-the-art models. In our experiments, we selected the parameters to ensure a fair comparison with existing models. Specifically, we used fixed values of $\alpha = 1.1$, $\beta = 0.1$, $k = 0.1$, $\nu = 1$ and $\gamma > 0$ across all tested images.

Table 1. Parameters used for different models on various types of Images.

Image	L	FCD Model [6]		Shan Model [15]		Majee Model [11]			Proposed Model
		r	θ	$\tilde{\alpha}$	$\tilde{\beta}$	γ	ν	k	γ
Texture Images									
Texture1	1	0.9	$\frac{\pi}{30}$	0.5	0.5	2	2	1	5
Texture2	1	0.9	$\frac{\pi}{30}$	2.5	1	2	2	1	5
Texture3	1	0.9	$\frac{\pi}{30}$	3	0.5	0.3	1	0.1	0.1
Satellite Images									
Satellite1	1	0.01	$\frac{\pi}{30}$	0.5	0.5	0.5	1	0.1	0.1
Satellite2	1	0.01	$\frac{\pi}{30}$	0.5	0.5	0.5	1	0.1	0.7
Satellite3	1	0.02	$\frac{\pi}{30}$	0.5	0.5	0.5	1	0.1	0.1
Satellite4	1	0.09	$\frac{\pi}{30}$	0.5	0.5	0.5	1	0.1	0.1
Satellite5	1	0.05	$\frac{\pi}{30}$	0.5	0.5	0.5	1	0.1	0.1
Natural, Synthetic, and Ultrasound Images									
Brain	1	0.2	$\frac{\pi}{30}$	2.5	0.5	1	2	0.1	0.1
Circle	1	0.9	$\frac{\pi}{30}$	1	0.5	0.5	1.5	0.5	0.1
Pirate	1	0.2	$\frac{\pi}{30}$	1	0.5	1	1	0.1	0.1
Leaf	1	0.9	$\frac{\pi}{30}$	1	2	1	2	0.1	0.1
House	1	0.2	$\frac{\pi}{30}$	0.5	0.5	1	2	0.1	0.2

Table 2. Comparison of Restoration Performance on Various Images

Satellite Images

Image	Noisy (SI)	FCD Model [6]		Shan Model [15]		Majee Model [11]		Proposed Model	
		SI	CPU Time (s)	SI	CPU Time (s)	SI	CPU Time (s)	SI	CPU Time (s)
satellite1	0.49	0.45	1.73	0.33	8.55	0.29	5.38	**0.25**	**0.62**
satellite2	1.24	0.37	1.71	0.26	3.42	0.20	1.54	**0.14**	**0.51**
satellite3	1.30	0.46	9.30	0.42	14.83	0.38	9.23	**0.31**	**2.49**

Texture Images (L=1)

Image	L	FCD Model [6]			Shan Model [15]			Majee Model [11]			Proposed Model		
		PSNR (idx)	MAE	CPU Time (s)	PSNR (idx)	MAE	CPU Time (s)	PSNR (idx)	MAE	CPU Time (s)	PSNR (idx)	MAE	CPU Time (s)
Texture1	1	11.90 (302)	0.2015	3.26	12.30 (91)	0.2061	5.18	12.49 (47)	0.1961	3.70	**12.65 (12)**	**0.1870**	**1.22**
Texture2	1	15.43 (150)	0.1245	8.37	17.23 (146)	0.1135	10.79	17.52 (114)	0.1047	8.37	**17.62 (5)**	**0.0996**	**1.65**
Texture3	1	17.68 (185)	0.1062	12.79	22.89 (178)	0.1062	89.59	23.25 (117)	0.0865	40.72	**24.54 (12)**	**0.0435**	**5.26**

Natural, Synthetic & Ultrasound Images

Image	L	FCD model [6]		Shan Model [15]		Majee Model [11]		Proposed Model	
		PSNR(idx)	MSSIM(idx)	PSNR(idx)	MSSIM(idx)	PSNR(idx)	MSSIM(idx)	PSNR(idx)	MSSIM(idx)
Brain	1	18.83 (85)	0.61 (147)	24.79 (68)	0.72 (71)	24.97 (72)	0.74 (57)	**25.56 (6)**	**0.82 (6)**
Circle	1	27.00 (8156)	0.49 (9765)	24.64 (231)	0.91 (498)	24.87 (165)	0.92 (279)	**32.97 (26)**	**0.93 (84)**
Pirate	1	19.10 (2927)	0.24 (2190)	22.27 (119)	0.54 (139)	22.54 (203)	0.55 (205)	**22.82 (81)**	**0.56 (81)**
Leaf	1	18.93 (1574)	0.36 (2250)	21.58 (60)	0.76 (198)	21.71 (366)	0.78 (499)	**22.14 (4)**	**0.80 (63)**
House	1	17.79 (2959)	0.22 (3170)	21.24 (84)	0.60 (199)	21.53 (189)	0.60 (176)	**21.92 (8)**	**0.61 (29)**

The evaluation is based on five types of images: texture, satellite, natural, synthetic and ultrasound images, in the presence of high noise level. The proposed fractional-order telegraph diffusion mechanism is designed to handle various levels of noise by adaptively balancing structure preservation and noise suppression. Our experiments primarily focused on Gamma noise with $L = 1$, given its relevance in SAR and texture image analysis. Figures 1 and 2 demonstrate the denoising performance of the proposed model on different images. Additionally, Fig. 3 provides a visual comparison of the performance of various models, focusing on texture preservation and edge sharpness. The denoised images were analyzed with particular attention to key regions, highlighted by red rectangles, to evaluate noise suppression and detail retention. Observations on the Brain(Ultrasound) image reveal that applying the Shan [15] and Majee [11] filters result in blurred outputs, failing to preserve the sharpness of critical features. The Gao model [6], leaves residual noise and fails to completely retain texture details in those areas. In contrast, the proposed model demonstrates superior performance, effectively preserving edges, textures, and fine details while achieving significant noise removal. These results highlight the efficacy of the proposed model in maintaining image quality and structural integrity compared to existing methods.

(a) Noisy Image	(b) Gao [6] PSNR = 11.90	(c) Shan [15] PSNR = 12.30	(d) Majee [11] PSNR = 12.49	(e) Proposed PSNR = **12.65**

(f) Noisy Image	(g) Gao [6] PSNR = 11.80	(h) Shan [15] PSNR = 12.20	(i) Majee [11] PSNR = 12.45	(j) Proposed PSNR = **12.70**

(k) Noisy Image	(l) Gao [6] PSNR = 11.75	(m) Shan [15] PSNR = 12.10	(n) Majee [11] PSNR = 12.35	(o) Proposed PSNR = **12.55**

(p) Noisy Image	(q) Gao [6] PSNR = 17.79	(r) Shan [15] PSNR = 21.24	(s) Majee [11] PSNR = 21.53	(t) Proposed PSNR = **21.92**

(u) Noisy Image	(v) Gao [6] PSNR = 18.93	(w) Shan [15] PSNR = 21.58	(x) Majee [11] PSNR = 21.71	(y) Proposed PSNR = **22.14**

Fig. 1. Comparison of image denoising performance of different models for various texture and natural images with high noise level ($L = 1$).

Numerical experiments were conducted using MATLAB R2023b on a system equipped with an Intel Core i7-7600U CPU @ 2.80GHz. The iterations of the proposed model were terminated once the maximum Peak Signal-to-Noise Ratio (PSNR) was achieved, ensuring both computational efficiency and optimal quality. The proposed model demonstrates improved efficiency and requires fewer iterations compared

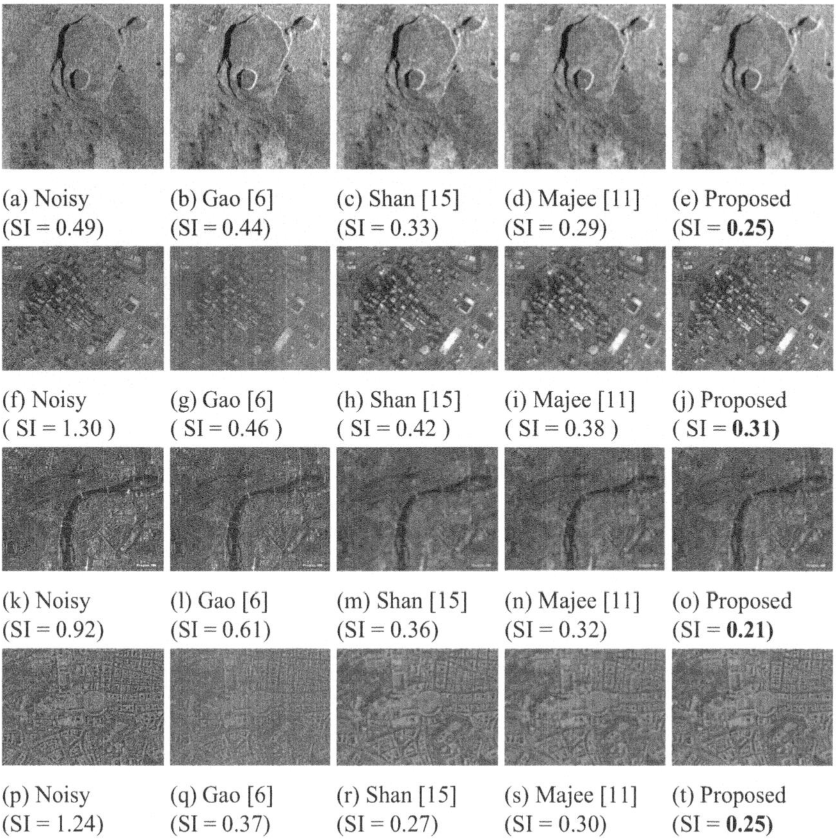

(a) Noisy (SI = 0.49) (b) Gao [6] (SI = 0.44) (c) Shan [15] (SI = 0.33) (d) Majee [11] (SI = 0.29) (e) Proposed (SI = **0.25**)

(f) Noisy (SI = 1.30) (g) Gao [6] (SI = 0.46) (h) Shan [15] (SI = 0.42) (i) Majee [11] (SI = 0.38) (j) Proposed (SI = **0.31**)

(k) Noisy (SI = 0.92) (l) Gao [6] (SI = 0.61) (m) Shan [15] (SI = 0.36) (n) Majee [11] (SI = 0.32) (o) Proposed (SI = **0.21**)

(p) Noisy (SI = 1.24) (q) Gao [6] (SI = 0.37) (r) Shan [15] (SI = 0.27) (s) Majee [11] (SI = 0.30) (t) Proposed (SI = **0.25**)

Fig. 2. Comparison of denoising results on various SAR images using different models

to existing models, as shown in Table 2. The parameter configurations for each model and image type are provided in Table 1. The comparative performance of the models, in terms of the peak signal-to-noise ratio (PSNR), mean structural similarity index measure (MSSIM), mean absolute error (MAE) and Speckle Index (SI) metrics, is summarized in Table 2.

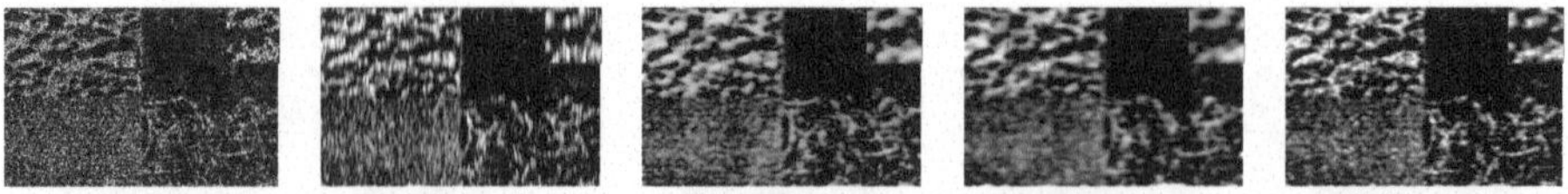

(a) Noisy ($L = 1$) (b) Gao [6] model (c) Shan [15] model (d) Majee [11] model (e) Proposed model

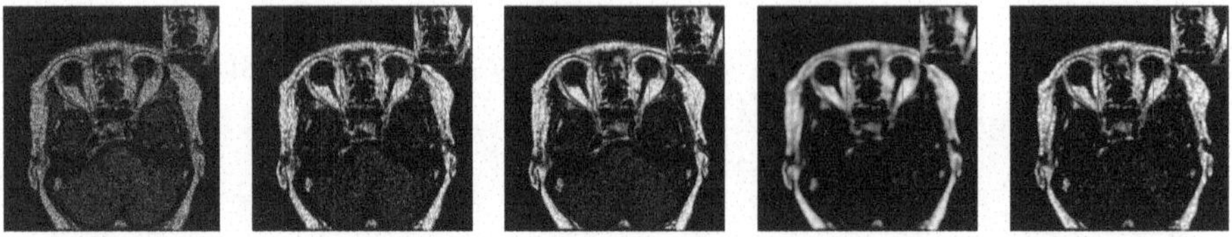

(f) Noisy ($L = 1$) (g) Gao [6] model (h) Shan [15] model (i) Majee [11] model (j) Proposed model

Fig. 3. Multiplicative noise removal results of various models on adversarially attacked texture and ultrasound images.

5 Conclusion

This paper presents a novel fractional-order telegraph diffusion model for image denoising, which effectively balances structure preservation and noise suppression. By incorporating the αth order Caputo fractional derivative, the model generalizes the standard Telegraph equation, introducing memory effects that enhance edge and texture preservation. Meanwhile, the βth order derivative contributes to smooth and artifact-free noise reduction in homogeneous regions. This adaptive diffusion mechanism strikes a balance between preserving structural details and suppressing noise, allowing the model to perform consistently well across diverse image types, including SAR, texture, and natural images. Quantitative metrics (PSNR, MSSIM, MAE, SI) and visual evaluations confirm its robustness, making it a promising approach for advanced image denoising tasks.

Future work will focus on adapting the proposed model to other types of noise and various image processing tasks. Additionally, efforts will be directed towards establishing a rigorous proof of the existence and uniqueness of solution for the model, thereby strengthening its theoretical foundation and ensuring its applicability in a wide range of image restoration problems.

Conflicts of Interest. The authors declare that there are no conflicts of interests.

References

1. Aubert, G., Aujol, J.F.: A variational approach to removing multiplicative noise. SIAM J. Appl. Math. **68**(4), 925–946 (2008)
2. Bai, X., Zhang, D., Shi, S., Yao, W., Guo, Z., Sun, J.: A fractional-order telegraph diffusion model for restoring texture images with multiplicative noise. Fractal Fractional **7**(1), 64 (2023)

3. Ben-Loghfyry, A., Charkaoui, A.: Regularized Perona-Malik model involving Caputo time-fractional derivative with application to image denoising. Chaos, Solitons Fractals **175**, 113925 (2023)
4. Bioucas-Dias, J.M., Figueiredo, M.A.: Multiplicative noise removal using variable splitting and constrained optimization. IEEE Trans. Image Process. **19**(7), 1720–1730 (2010)
5. Bnou, K., Raghay, S., Hakim, A.: A wavelet denoising approach based on unsupervised learning model. EURASIP J. Adv. Sig. Process. **2020**(1), 1–26 (2020). https://doi.org/10.1186/s13634-020-00693-4
6. Gao, J., Sun, J., Shi, S.: Fractional-order cross-diffusion system for multiplicative noise removal. Comput. Math. Appl. **164**, 1–11 (2024)
7. Janev, M., Pilipović, S., Atanacković, T., Obradović, R., Ralević, N.: Fully fractional anisotropic diffusion for image denoising. Math. Comput. Model. **54**(1–2), 729–741 (2011)
8. Kuan, D.T., Sawchuk, A.A., Strand, T.C., Chavel, P.: Adaptive noise smoothing filter for images with signal-dependent noise. IEEE Trans. Pattern Anal. Mach. Intell. **PAMI-7**(2), 165–177 (1985)
9. Li, C., He, C.: Fractional-order diffusion coupled with integer-order diffusion for multiplicative noise removal. Comput. Math. Appl. **136**, 34–43 (2023)
10. Li, Z., Lou, Y., Zeng, T.: Variational multiplicative noise removal by DC programming. J. Sci. Comput. **68**(3), 1200–1216 (2016)
11. Majee, S., Ray, R.K., Majee, A.K.: A gray level indicator-based regularized telegraph diffusion model: application to image despeckling. SIAM J. Imag. Sci. **13**(2), 844–870 (2020)
12. Perona, P., Malik, J.: Scale-space and edge detection using anisotropic diffusion. IEEE Trans. Pattern Anal. Mach. Intell. **12**(7), 629–639 (1990)
13. Podlubny, I.: Fractional differential equations: an introduction to fractional derivatives, fractional differential equations, to methods of their solution and some of their applications, vol. 198. elsevier (1998)
14. Ratner, V., Zeevi, Y.Y.: Image enhancement using elastic manifolds. In: 14th International Conference on Image Analysis and Processing (ICIAP 2007), pp. 769–774. IEEE (2007)
15. Shan, X., Sun, J., Guo, Z.: Multiplicative noise removal based on the smooth diffusion equation. J. Math. Imaging Vis. **61**(6), 763–779 (2019)
16. Shi, J., Osher, S.: A nonlinear inverse scale space method for a convex multiplicative noise model. SIAM J. Imag. Sci. **1**(3), 294–321 (2008)
17. Tur, M., Chin, K.C., Goodman, J.W.: When is speckle noise multiplicative? Appl. Opt. **21**(7), 1157–1159 (1982)
18. Ullah, A., Chen, W., Khan, M.A.: A new variational approach for restoring images with multiplicative noise. Comput. Math. Appl. **71**(10), 2034–2050 (2016)
19. Yamamoto, M.: On time fractional derivatives in fractional Sobolev spaces and applications to fractional ordinary differential equations, pp. 287–308 (2021)
20. Yin, X., Zhou, S., Siddique, M.A.: Fractional nonlinear anisotropic diffusion with p-laplace variation method for image restoration. Multimedia Tools Appl. **75**, 4505–4526 (2016)
21. You, Y.L., Kaveh, M.: Fourth-order partial differential equations for noise removal. IEEE Trans. Image Process. **9**(10), 1723–1730 (2000)
22. Zhang, X., Yao, L.: Numerical approximation of time-dependent fractional convection-diffusion-wave equation by RBF-FD method. Eng. Anal. Boundary Elem. **130**, 1–9 (2021)

Self-supervised Conformal Prediction for Uncertainty Quantification in Imaging Problems

Jasper Marijn Everink[1(✉)], Bernardin Tamo Amougou[2,3], and Marcelo Pereyra[2]

[1] Technical University of Denmark, Kgs. Lyngby, Denmark
jmev@dtu.dk
[2] Heriot-Watt University and Maxwell Institute for Mathematical Sciences, Edinburgh, UK
[3] Université de Paris Cité, Paris, France

Abstract. This paper presents a self-supervised conformal prediction method for uncertainty quantification in imaging problems without ground truth available. Conformal prediction has recently emerged as a flexible framework to equip any estimator with uncertainty quantification capabilities that, by construction, have nearly exact marginal coverage. However, to achieve this, conformal prediction relies on abundant ground truth data for calibration. In image reconstruction problems, reliable ground truth data is often expensive or not possible to acquire. Also, reliance on ground truth data can introduce large biases in situations of distribution shift between calibration and deployment. Our proposed method leverages Stein's Unbiased Risk Estimator to self-calibrate directly from the observed noisy measurements, bypassing the need for ground truth. The method is suitable for any linear inverse problem that is ill-conditioned, and it is especially powerful when used with modern self-supervised imaging techniques that can also be trained directly from measurement data. The proposed approach is demonstrated through image denoising and deblurring experiments, where it delivers results that are remarkably accurate and comparable to those obtained by supervised conformal prediction with ground truth data.

Keywords: Conformal Prediction · High-Dimensional Image Restoration Problems · Stein's Unbiased Risk Estimate

1 Introduction

Image restoration tasks often carry a significant amount of uncertainty, as they admit a wide variety of solutions that are equally in agreement with the observed data. Quantifying and characterizing this uncertainty is important for applications that depend on restored images to inform critical decisions. Several statistical frameworks exist to address uncertainty quantification (UQ) in imaging

T. A. Bubba et al. (Eds.): SSVM 2025, LNCS 15667, pp. 108–118, 2025.
https://doi.org/10.1007/978-3-031-92366-1_9

sciences. Among these, the Bayesian statistical framework has been extensively studied and applied [1], enabling the incorporation of prior knowledge and providing probabilistic interpretations of uncertainty. This framework supports a diverse range of modeling and algorithmic approaches, as demonstrated in works such as [2–5]. Unfortunately, despite significant progress in the field, state-of-the-art Bayesian imaging methods are not yet able to provide accurate UQ on structures that are larger than a few pixels in size [6].

With regards to non-Bayesian approaches to UQ in image restoration, the recently proposed equivariant bootstrapping method [7] offers excellent frequentist accuracy, even for large image structures. This is achieved by exploiting known symmetries in the problem to reduce the bias inherent to bootstrapping. Equivariant bootstrapping is particularly accurate in problems that are highly ill-posed, such as compressive sensing, inpainting and limited angle tomography and radio-interferometry, as in these cases the bias from bootstrapping is almost fully removed by the actions of the symmetry group [7,8]. Conversely, equivariant bootstrapping struggles with problems that are not ill-posed, such as image denoising or mild deblurring, as in this case it is difficult to identify symmetries to remove the bias from bootstrapping (see [7] for details).

Moreover, when sufficient ground truth data are available for calibration, conformal prediction presents another highly flexible strategy for UQ in image restoration. Crucially, conformal prediction can be seamlessly applied to any image restoration technique to deliver UQ results that, by construction, have nearly exact marginal coverage [9]. Also, conformal prediction can be easily combined with other UQ strategies, such as Bayesian imaging strategies (see, e.g., [10]) or equivariant bootstrapping [8], as a correction step.

However, as mentioned previously, in its original form, conformal prediction approaches require access to abundant ground truth data for calibration. In many applications, access to ground truth data is either expensive or impossible. Also, reliance on ground truth data for calibration can lead to poor accuracy in situations of distribution shift between calibration and deployment. To address this limitation of conformal prediction, we propose a self-supervised conformal prediction method that leverages Stein's Unbiased Risk Estimator (SURE) to self-calibrate UQ results directly from the observed noisy measurements, bypassing the need for ground truth.

The remainder of this paper is organized as follows. Section 2 provides an overview of conformal prediction and a formal problem statement. Section 3 introduces the proposed self-supervised conformal prediction method. Section 4 demonstrates the proposed approach through numerical experiments on image denoising and non-blind deblurring tasks and by considering model-based as well as learning-based estimators. Conclusions and perspectives for future work are finally reported in Sect. 5.

2 Problem Statement

We consider the estimation of a set of plausible values for an unknown image of interest $x^\star \in \mathbb{R}^n$, from a noisy degraded measurement $y \in \mathbb{R}^m$. We assume that $x^\star$ is a realization of a random variable X, and y is a realization of the conditional random variable $Y|X = x^\star$. A point estimator for $x^\star$, derived from Y, is henceforth denoted by $\hat{x}(Y)$. We focus on the case where observations follow a Gaussian noise model:

$$(Y|X = x^\star) \sim \mathcal{N}(Ax^\star, \sigma^2 \mathbb{I}_m),$$

where $A \in \mathbb{R}^{m \times n}$ models deterministic instrumental aspects of the image restoration problem, $\sigma^2 > 0$ is the noise variance, and $\mathbb{I}_m$ is the $m \times m$ identity matrix. Throughout the paper we assume A is a full-rank but potentially highly ill-conditioned linear operator.

Our goal is to construct a region $C(Y) \subset \mathbb{R}^n$ in the solution space such that

$$\mathbb{P}_{(X,Y)}\left(X \in C(Y)\right) \geq 1 - \alpha\,, \tag{1}$$

where the probability is taken with respect to the joint distribution of (X, Y), and $\alpha \in [0, 1]$ specifies the desired confidence level.

To illustrate, suppose that $x^\star$ is a high-resolution MRI scan of an adult brain. The random variable X characterizes the distribution of brain MRI scans for a generic individual within the population, as obtained by an ideal noise-free and resolution-perfect MRI scanner. The specific image $x^\star$ corresponds to an MRI scan of a particular individual, while y represents the noisy, degraded measurement acquired in practice. The estimator $\hat{x}(y)$ produces an estimate of $x^\star$. The region $C(y)$ encapsulates a set of plausible solutions, rather than a single estimate, and satisfies the guarantee in (1). This means that if the procedure is repeated across a large number of individuals from the population, the constructed regions C will contain the respective true images $x^\star$ in at least $1 - \alpha$ of the cases.

Conformal prediction provides a general framework for constructing sets C with the desired probabilistic guarantee [9]. This is achieved using a *non-conformity measure* $s : \mathbb{R}^n \times \mathbb{R}^m \to \mathbb{R}$. By computing the top $(1 - \alpha)$-quantile q_α of the statistic $s(X, Y)$, the set $C(y)$ is defined as:

$$C(y) := \{x \in \mathbb{R}^n \mid s(x, y) \leq q_\alpha\} \quad \text{for all } y \in \mathbb{R}^m.$$

By construction, this set satisfies:

$$\mathbb{P}_{(X,Y)}\big(X \in C(Y)\big) = \mathbb{P}_{(X,Y)}\big(s(X, Y) \leq q_\alpha\big) \geq 1 - \alpha,$$

for any confidence level $\alpha \in (0, 1)$. With a sufficiently large sample $\{x_i, y_i\}_{i=1}^M$ to calibrate q_α, any suitable function s can be used to construct a set C that contains $x^\star$ with high probability under the joint distribution of (X, Y).

A popular specific implementation of this framework is *split conformal prediction*. Given a training sample $\{X_i, Y_i\}_{i=1}^M$ of independent (or exchangeable) realizations of (X, Y), the method estimates the top $\frac{\lceil (M+1)(1-\alpha) \rceil}{M}$-quantile $\hat{Q}_\alpha$ of $\{s(X_i, Y_i)\}_{i=1}^M$. For a new observation $(X_{\text{new}}, Y_{\text{new}})$, the prediction set is then:

$$\hat{C}(Y_{\text{new}}) := \{X_{\text{new}} \in \mathbb{R}^n \mid s(X_{\text{new}}, Y_{\text{new}}) \leq \hat{Q}_\alpha\}.$$

This set satisfies the guarantee:

$$\mathbb{P}_{(X,Y)^{M+1}}\left(X_{\text{new}} \in \hat{C}(Y_{\text{new}})\right) \geq 1 - \alpha\,, \tag{2}$$

where the probability accounts for the joint distribution of the M training samples and the new observation. Notably, this formulation includes a finite-sample correction because $\hat{Q}_\alpha$ is derived from the $\frac{\lceil (M+1)(1-\alpha) \rceil}{M}$-quantile. This correction becomes negligible as $M \to \infty$. For an excellent introduction to conformal prediction, please see [9].

While the conformal prediction framework is highly flexible, not all non-conformity measures $s(x, y)$ deliver prediction regions that are equally useful in practice. Indeed, all prediction sets $\hat{C}(y)$ take the form of a sub-level set of the function $x \mapsto s(x, y)$, which can be arbitrarily chosen. As a result, it is possible to construct infinitely many regions in $\mathbb{R}^n$ that satisfy the guarantee of containing the true solution $x^\star$ with probability at least $1 - \alpha$. However, many of these regions may be excessively large, especially in high-dimensional settings, where poorly designed score functions $s(x, y)$ can lead to regions $\hat{C}(y)$ that are overly conservative and cover most of the support of X.

Carefully designing $s(x, y)$ allows obtaining conformal prediction sets that are compact and informative, even in high dimension. In particular, it is essential that $s(x, y)$ is constructed in a way that reduces the variability of $s(X, Y)$. Of particular interest are normalized non-conformity measures of the form [11]:

$$s(x, y) = \|x - \hat{x}(y)\|^2_{\Sigma(y)} = (x - \hat{x}(y))^\top \Sigma(y)\,(x - \hat{x}(y))\,, \tag{3}$$

where $\Sigma(y)$ is a positive definite matrix of size $n \times n$. A well-chosen $\Sigma(y)$ reduces variability in $s(X, Y)$, leading to prediction sets that are well-centered around $\hat{x}(y)$, compact, and highly informative. In practice, $\Sigma(y)$ is often chosen as an approximation of the inverse-covariance matrix of the error $X - \hat{x}(Y)$ [11].

A fundamental obstacle to applying conformal prediction to image restoration problems is the need for paired samples $\{x_i, y_i\}$, as obtaining the ground truth image x_i is precisely the goal of solving the imaging problem in the first place. This can be partially mitigated by relying on a training dataset, at the risk of delivering poor results in situations of distribution shift (e.g., returning to our illustrative example related to MRI imaging, when the population encountered in deployment differs significantly from the population used for calibration). In this paper, we propose a greatly more robust and deployable approach to conformal prediction that relies solely on the observed measurements $\{y_i\}_{i=1}^M$.

3 Proposed Method

Our proposed self-supervised conformal prediction method circumvents the need for ground truth data by leveraging Stein's unbiased risk estimate (SURE) [12].

We begin by pooling together M exchangeable imaging problems, where each problem involves an unknown image $x_i^\star$ and an observation y_i which we consider to be a realisation of the conditional random variable $(Y|X = x_i^\star) \sim \mathcal{N}(Ax_i^\star, \sigma^2\mathbb{I}_m)$. To specify the non-conformity measure, we consider $\Sigma(y) = A^\top A$ which is a natural choice for approximation for the error inverse-covariance when $(Y|X = x_i^\star) \sim \mathcal{N}(Ax_i^\star, \sigma^2\mathbb{I}_m)$, as we expect $\hat{x}(Y)$ to be accurate along the leading eigenvectors of $A^\top A$ and the estimation error to concentrate along weak eigenvectors of $A^\top A$. This leads to the non-conformity measure

$$s(x, y) = \frac{1}{m}\|Ax - A\hat{x}(y)\|_2^2, \tag{4}$$

where we recall that A is assumed full-rank, but potentially very poorly conditioned . We require A to be full rank as otherwise $\Sigma(y)$ is only positive semidefinite, implying that the corresponding prediction set can be unbounded.

To calibrate without ground truth data, instead of relying on a sample quantile of $\{s(x_i, y_i)\}_{i=1}^M$, we rely on SURE to provide unbiased estimates of $\{s(x_i, y_i)\}_{i=1}^M$ from the observed measurements $\{y_i\}_{i=1}^M$. We then use those noisy quantile estimates for calibration, at the expense of a small amount of bias.

More precisely, assuming that the estimator $\hat{x}$ is differentiable almost everywhere, the SURE estimate of (4) is given by

$$\text{SURE}(y) = \frac{1}{m}\|y - A\hat{x}(y)\|_2^2 - \sigma^2 + \frac{2\sigma^2}{m}\text{div}\left(A\hat{x}(y)\right), \tag{5}$$

where $\text{div}(\cdot)$ denotes the divergence operator [12]. It is easy to show that $\text{SURE}(Y)$ provides an estimate of $s(X, Y)$ that is unbiased [12]. Crucially, from the law of large numbers, the estimate provided by SURE is not only unbiased but also often very accurate when $m = \dim(y)$ is large (see [12,13] for a theoretical analysis of the variance of SURE and [14] for an empirical analysis in an imaging setting). As a consequence, we expect that the conformal calibration quantiles obtained from SURE will be in close agreement with the true quantiles of $s(X, Y)$, ensuring that the resulting conformal prediction remain nearly exact.

With regards to the evaluation of SURE, for some model-based estimators it is possible to identify a closed-form expression [15]. Otherwise, computing SURE requires a numerical approximation of the divergence term. A common approach is the Monte Carlo SURE (MC-SURE) algorithm [16], which is estimator-agnostic. We use the so-called Hutchinson's stochastic trace approximation method [17], which is more computationally efficient than MC-SURE and does not require hyper-parameter fine-tuning. More precisely, the divergence $\text{div}(A\hat{x}(y))$ is approximated as follows:

$$\text{div}(A\hat{x}(y)) = \text{trace}(\boldsymbol{J}_{h(y)}) = \mathbb{E}[\tilde{n}^\top \boldsymbol{J}_{h(y)}\tilde{n}], \tag{6}$$

$$\approx \frac{1}{K}\sum_{i=1}^{K} \tilde{n}_i^\top \boldsymbol{J}_{h(y)}\tilde{n}_i, \tag{7}$$

$$\approx \frac{1}{K}\tilde{n}^\top \boldsymbol{J}_{\tilde{n}^\top h(y)}, \tag{8}$$

where $\{\tilde{n}_i\}_{i=1}^K$ are a K i.i.d. standard normal random vectors, $\boldsymbol{J}_{h(y)}$ is the Jacobian matrix of the predicted measurements $h(y) = A\hat{x}(y)$ with respect to y, and (7) corresponds to Hutchinson's method, which for computationally efficiency we implement (8) using automatic differentiation, as suggested in [17]. This allows obtaining accurate SURE estimates in a highly efficient manner, even in very large problems.

Adopting a split-conformal strategy, after computing $\hat{x}(y_i)$ and $\text{SURE}(y_i)$, for each $i = \{1, \ldots, M\}$ we construct the $(1-\alpha)$-prediction set $\hat{C}(y_i)$ as:

$$\hat{C}(y_i) = \{x \in \mathbb{R}^n : \|Ax - A\hat{x}(y_i)\|_2^2/m \leq \hat{Q}_\alpha^{(i)}\},$$

where $\hat{Q}_\alpha^{(i)}$ is the top $\frac{\lceil M(1-\alpha) \rceil}{M-1}$-quantile of the sample $\{\text{SURE}(y_j)\}_{j=1}^M$ with the ith element, $\text{SURE}(y_i)$, removed. Note that while the proposed approach has some bias due to the estimation error introduced by SURE, in our experience the bias is small and arguably significantly smaller than the bias that is likely to be incurred due to distribution shift in deployment. It is also worth mentioning that the estimates $\text{SURE}(y_i)$ can be computed in parallel. The proposed method is summarised in Algorithm 1 below.

Algorithm 1. SURE-based conformal prediction

Input: Forward operator A, noise variance σ^2, estimator $\hat{x}$, measurement y, samples $\{y_i\}_{i=1}^M$, precision level $1 - \alpha \in (0, 1)$.

1: **for** $i \leftarrow 1$ to M **do**
2: $S_i \leftarrow \text{SURE}(y_i)$ using (5) and (8)
3: **end for**
4: $\hat{Q}_\alpha \leftarrow$ top $\frac{\lceil (M+1)(1-\alpha) \rceil}{M}$-quantile of $\{S_i\}_{i=1}^M$
5: $m \leftarrow dim(y)$
6: $\hat{C}(y) \leftarrow \{x \in \mathbb{R}^n \mid \|Ax - A\hat{x}(y)\|_2^2/m \leq \hat{Q}_\alpha\}$

Output: Prediction set $\hat{C}(y)$

4 Experiments

We demonstrate the proposed self-supervised conformal prediction approach by applying it to two image restoration problems: image denoising and image deblurring. To showcase the versatility of the method, for image denoising we construct conformal prediction sets by using a learning-based image restoration technique trained in a self-supervised end-to-end manner, whereas for image deblurring we use the model-driven technique. For each experiment, we implement our method by using Algorithm 1 to compute conformal prediction sets. We consider a fine grid of values for the confidence parameter $\alpha \in (0, 1)$, ranging from 0% to 100%. The corresponding prediction sets should cover the solution space with a probability of approximately $1 - \alpha$. We evaluate the accuracy of these prediction sets by calculating the empirical coverage probabilities on a test set. Specifically, we compute the proportion of test images that lie within the conformal prediction sets for various confidence levels. In all experiments, the calibration sample size M is chosen to be sufficiently large to ensure that the variability of the sample quantiles caused by the finite-sample correction is negligible. Furthermore, for comparison, in each experiment we also report results by supervised conformal prediction (i.e., using ground truth data for calibration, rather than SURE). The experiments are implemented using the Deep Inverse library[1] for imaging with deep learning using PyTorch.

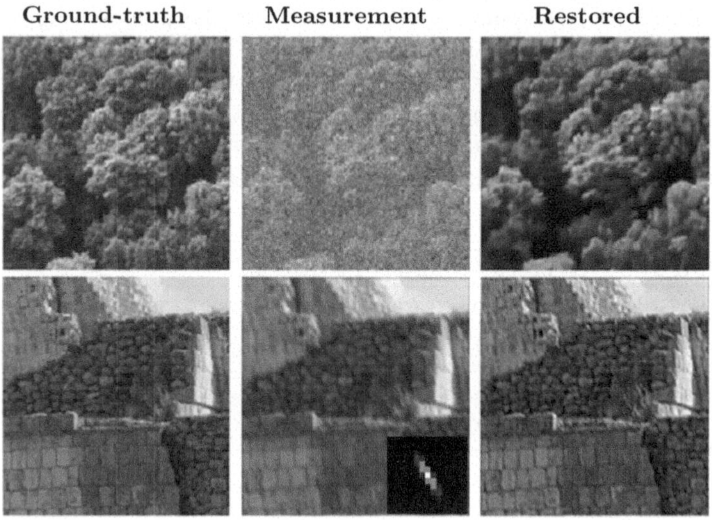

Fig. 1. Image reconstruction results for various imaging problems. Top: Gaussian noise on DIV2K. Bottom: Noisy Gaussian blur on DIV2K

4.1 Image Denoising

For the image denoising experiment, we consider colour images of size 128×128 pixels obtained by cropping images from the DIV2K dataset [18], which we artificially degrade by adding white Gaussian noise with a standard deviation of $\sigma = 0.1$. As image restoration method, we use a DRUNet model [19] trained in a self-supervised manner by using the SURE loss [20]. The training data consists of 900 noisy measurements, we do not use any form of ground truth for training or for conformal prediction. See the top row of Fig. 1 for an example of a clean image, its noisy measurement, and the estimated reconstruction.

[1] https://deepinv.github.io/deepinv/.

We then use these same noisy images to compute our proposed self-supervised conformal prediction sets, and assess their accuracy empirically by using 200 measurement-truth pairs from the test dataset. The results are reported in Fig. 2 below, together with the results obtained by using the equivalent supervised conformal prediction approach that relies on ground truth data for calibration. We observe that our method delivers prediction sets that are remarkably accurate and in close agreement with the results obtained by using supervised conformal prediction, demonstrating that the bias stemming from using SURE instead of ground truth data is negligible in this case. For completeness, Fig. 4 (left) below shows the empirical distribution of the non-similarity function $s(x, y)$ for the supervised conformal prediction based on the MSE, and the proposed self-supervised conformal prediction based on a SURE estimate of the MSE. Again, we observe close agreement between these distributions, with the SURE distribution being slightly more spread due to the random error inherent to SURE.

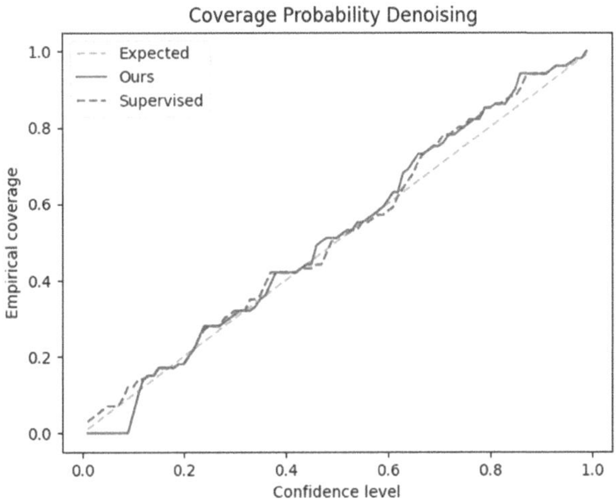

Fig. 2. Image denoising experiment: desired confidence level vs empirical coverage. Both supervised and the proposed self-supervised conformal prediction methods deliver prediction sets with near perfect coverage.

4.2 Image Deblurring

We now consider a non-blind image deblurring experiment with colour images of size 256 × 256 pixels, derived from the DIV2K dataset [18] and artificially degraded with a diagonal Gaussian blur of major bandwidth $\varsigma_0 = 2$, $\varsigma_1 = 0.3$ along the minor axis, and an inclination of $\pi/6$ degrees, as well as additive white Gaussian noise with a standard deviation $\sigma = 0.01$. As image restoration technique, we use the recently proposed model-based Polyblur technique [21], a highly efficient restoration method for removing mild blur from natural images based on a truncated polynomial approximation of the inverse of the blur kernel. See the bottom row of Fig. 1 for an example of a clean image, its noisy measurement, and the estimated reconstruction.

We use 900 blurred and noisy images to compute our proposed self-supervised conformal prediction sets, and assess their accuracy empirically by using 200 measurement-truth pairs from the test dataset. The results are reported in Fig. 3 below, together with the results obtained by using the equivalent supervised conformal prediction approach that relies on ground truth data for calibration. Again, we observe that our method delivers prediction sets that are accurate and remarkably close to the results obtained by using supervised conformal prediction, demonstrating that the bias stemming from using SURE instead of

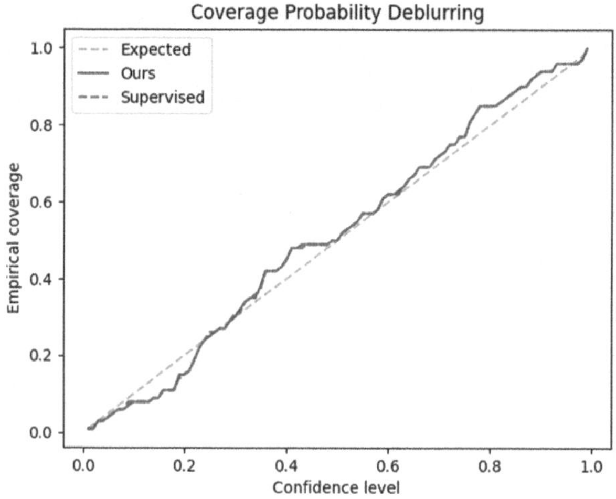

Fig. 3. Image deblurring experiment: desired confidence level vs empirical coverage. Both supervised and the proposed self-supervised conformal prediction methods deliver prediction sets with near perfect coverage.

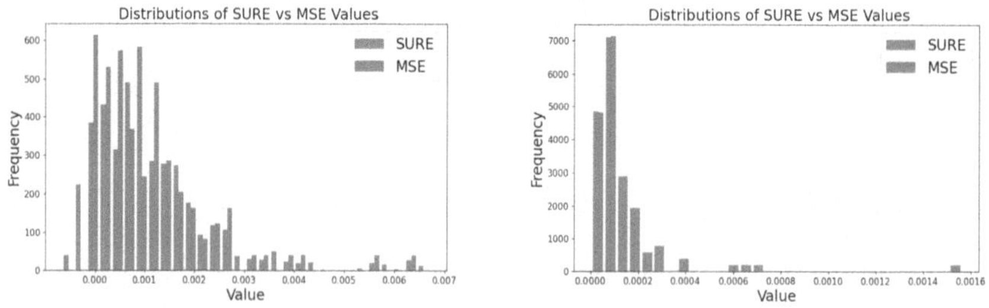

(a) Image denoising

(b) Image deblurring

Fig. 4. Calibration histograms (empirical distribution of the non-similarity function $s(x, y)$) for the supervised case (MSE) and the self-supervised case based on a SURE estimate of the MSE, for the denoising and deblurring experiments.

ground truth data is again negligible in this case. For completeness, Fig. 4 (right) below shows the empirical distribution of the non-similarity function $s(x, y)$ for the supervised conformal prediction based on the MSE, and the proposed self-supervised conformal prediction based on a SURE estimate of the MSE. Again, we observe close agreement between these distributions.

5 Discussion and Conclusion

This paper presented a self-supervised approach for constructing conformal prediction regions for linear imaging inverse problems that are ill-posed. Unlike previous conformal prediction methods, the proposed approach does not require any form of ground truth data. This is achieved by leveraging Stein's unbiased risk estimator and by pooling together a group of exchangeable imaging problems. This allows delivering conformal prediction sets in situations where there is no ground truth data available for calibration, and provides robustness to distribution shift. Additionally, the proposed framework is estimator-agnostic, as it uses a Monte Carlo implementation of SURE that does not require any explicit knowledge of the image restoration algorithm used. This flexibility allows the framework to be straightforwardly applied to model-driven as well as data-driven image restoration techniques, including self-supervised data-driven techniques that are also trained directly from the measurement data. Moreover, the method is computationally efficient as computations can be performed in parallel. We demonstrated the effectiveness of the proposed method through image denoising and deblurring experiments, where we observed that our method delivers extremely accurate prediction sets.

Future work will study the theoretical underpinning of the proposed method and the biases induced by SURE. Also, the generalization of the proposed approach to other noise models, such as Poisson and Poisson-Gaussian, as well as to problems in which the parameters of the noise model are unknown [20] could be explored. Another important perspective for future work is to extend this approach to problems in which the forward A is not full rank, for example by leveraging equivariance properties. In particular, it would be interesting to study the integration of our proposed method and the equivariant bootstrap [7].

References

1. Kaipio, J., Somersalo, E.: Statistical and computational inverse problems. Springer, Germany (2004). https://doi.org/10.1007/b138659
2. Durmus, A., Moulines, E., Pereyra, M.: Efficient Bayesian computation by proximal Markov chain Monte Carlo: when Langevin meets Moreau. SIAM J. Imag. Sci. **11**(1), 473–506 (2018)
3. Pereyra, M.: Maximum-a-posteriori estimation with Bayesian confidence regions. SIAM J. Imag. Sci. **10**(1), 285–302 (2017)
4. Laumont, R., Bortoli, V.D., Almansa, A., Delon, J., Durmus, A., Pereyra, M.: Bayesian imaging using plug & play priors: when Langevin meets Tweedie. SIAM J. Imag. Sci. **15**(2), 701–737 (2022)

5. Holden, M., Pereyra, M., Zygalakis, K.C.: Bayesian imaging with data-driven priors encoded by neural networks. SIAM J. Imag. Sci. **15**(2), 892–924 (2022)
6. Thong, D.Y.W., Mbakam, C.K., Pereyra, M.: Do Bayesian imaging methods report trustworthy probabilities? (2024). https://arxiv.org/abs/2405.08179
7. Tachella, J., Pereyra, M.: Equivariant bootstrapping for uncertainty quantification in imaging inverse problems. Proc. Mach. Learn. Res. **238**, 4141–4149 (2024)
8. Cherif, M., Liaudat, T.I., Kern, J., Kervazo, C., Bobin, J.: Uncertainty quantification for fast reconstruction methods using augmented equivariant bootstrap: application to radio interferometry (2024). https://arxiv.org/abs/2410.23178
9. Angelopoulos, A.N., Bates, S.: A gentle introduction to conformal prediction and distribution-free uncertainty quantification. arXiv preprint arXiv:2107.07511 (2021)
10. Narnhofer, D., Habring, A., Holler, M., Pock, T.: Posterior-variance-based error quantification for inverse problems in imaging. SIAM J. Imag. Sci. **17**(1), 301–333 (2024)
11. Johnstone, C., Cox, B.: Conformal uncertainty sets for robust optimization. In: Conformal and Probabilistic Prediction and Applications, pp. 72–90. PMLR (2021)
12. Stein, C.M.: Estimation of the mean of a multivariate normal distribution. Ann. Stat. **9**, 1135–1151 (1981)
13. Bellec, P.C., Zhang, C.-H.: Second-order stein: sure for sure and other applications in high-dimensional inference. Ann. Stat. **49**(4), 1864–1903 (2021)
14. Candès, E.J., Sing-Long, C.A., Trzasko, J.D.: Unbiased risk estimates for singular value thresholding and spectral estimators. IEEE Trans. Signal Process. **61**(19), 4643–4657 (2013)
15. Tibshirani, R.J., Taylor, J.: Degrees of freedom in lasso problems. Ann. Stat. **40**(2), 1198–1232 (2012)
16. Ramani, S., Blu, T., Unser, M.: Monte-Carlo SURE: a black-box optimization of regularization parameters for general denoising algorithms. IEEE Trans. Image Process. **17**(9), 1540–1554 (2008)
17. Soltanayev, S., Giryes, R., Chun, S.Y., Eldar, Y.C.: On divergence approximations for unsupervised training of deep denoisers based on stein's unbiased risk estimator. In: ICASSP 2020 - 2020 IEEE International Conference on Acoustics, Speech and Signal Processing (ICASSP), pp. 3592–3596 (2020)
18. Agustsson, E., Timofte, R.: NTIRE 2017 challenge on single image super-resolution: Dataset and study. In: The IEEE Conference on Computer Vision and Pattern Recognition (CVPR) Workshops (2017)
19. Zhang, K., Li, Y., Zuo, W., Zhang, L., Van Gool, L., Timofte, R.: Plug-and-play image restoration with deep denoiser prior. IEEE Trans. Pattern Anal. Mach. Intell. **44**(10), 6360–6376 (2021)
20. Tachella, J., Davies, M., Jacques, L.: Unsure: unknown noise level stein's unbiased risk estimator. arXiv preprint arXiv:2409.01985 (2024)
21. Delbracio, M., Garcia-Dorado, I., Choi, S., Kelly, D., Milanfar, P.: Polyblur: removing mild blur by polynomial reblurring. IEEE Trans. Comput. Imaging **7**, 837–848 (2021)

Real-Time Scene Recovery from Image Scale Space and Perceptual Hue Similarity

Roy Y. He(✉) and Han Wang

City University of Hong Kong, Kowloon Tong, Hong Kong
royhe2@cityu.edu.hk, hwang474-c@my.cityu.edu.hk

Abstract. Chromatic degradation is often caused by multiple factors that are difficult to retrieve after images have been acquired, making scene recovery a highly ill-posed inverse problem. In this paper, we leverage the nature of human vision and propose an effective algorithm to achieve robust and visually appealing reconstructions. Specifically, we develop a reliable method for estimating chromatic bias by averaging colors from large, blurred regions identified through scale-space analysis. We then utilize CIELAB perceptual hue similarity to determine the degree of degradation caused by transmission and backscattering. By combining these estimators, we recover the scene by adaptively discounting the bias across the image domain. We conduct numerical experiments and comparative studies to demonstrate the effectiveness of our method.

Keywords: Color image reconstruction · Image scale space · CIELAB perceptual color space

1 Introduction

Image formation is a complex process that involves transformations of the spectrum of incident light through filters composed of object surfaces and scattering media such as dust particles and water [26,31]. During transmission, the light reflected from the scene mixes with ambient light. Depending on the distance between the scene and the sensor, different wavelengths are attenuated at varying rates. This process can be simplified [31,32] as

$$\mathbf{I}(\mathbf{x}) = \mathbf{J}(\mathbf{x}) \odot (\mathbf{1} - \mathbf{t}(\mathbf{x})) + \mathbf{A} \odot \mathbf{t}(\mathbf{x}), \quad \mathbf{x} \in \Omega, \tag{1}$$

where the operation $\odot$ denotes the Hadamard product, $\mathbf{I} = (I^r, I^g, I^b) : \Omega \to \mathbb{R}^3$ is the acquired wideband image defined on the rectangular domain $\Omega \subset \mathbb{R}^2$; $\mathbf{J} : \Omega \to \mathbb{R}^3$ is the corresponding wideband field near the scene surface; $\mathbf{A} \in \mathbb{R}^3$ represents the globally uniform *ambient light*; and $\mathbf{t} : \Omega \to [0,1]^3$ is the *transmission map*, which depends on the distance from the scene to the camera through the Beer-Lambert law [36]. Model (1) indicates that the obtained image $\mathbf{I}$ is a channel-wise convex combination of the scene $\mathbf{J}$ and the ambient light $\mathbf{A}$, weighted by $\mathbf{t}$. More sophisticated models are available [2].

T. A. Bubba et al. (Eds.): SSVM 2025, LNCS 15667, pp. 119–131, 2025.
https://doi.org/10.1007/978-3-031-92366-1_10

Fig. 1. Left column: original images. Right column: proposed results.

The objective of scene recovery is to estimate **J** from **I** without knowledge of **t** or **A**, making it a highly ill-posed inverse problem. Many priors have been proposed to establish correlations among these variables for estimation, and they can be classified into three major categories. The first category relies on *image intensity features*, including the Dark Channel Prior (DCP) [16], Red Channel Prior (RCP) [14], haze lines [8], and color lines [13]. The second category relies on *image structure features*, such as edges [20], blurriness [27], and patch recurrences [5]. Some of these are integrated into frameworks like retinex algorithms [29]. The third category is data-driven, where priors are learned from sets of images [11,19,23,34]. More comprehensive reviews can be found in [1,15,38].

In this paper, we take a different perspective and interpret **t** as the channel-wise *degree of degradation* perceived by an observer. Although **t** is determined by physical quantities such as scene depth or the sizes of scattering particles, human eyes rarely conduct such measurements in practice. Recent studies reveal that image colors are correlated with depth in natural images [33,35], suggesting the possibility of deducing spatial information about the scene from colors. Furthermore, consistent with classical Gestalt principles [10], evidence [28] shows that grouped patches with similar colors influence the perception of illuminants, and the area occupied by such groups plays a critical role.

Based on these guiding principles, we propose an effective algorithm for recovering a color-balanced scene from a degraded image (Fig. 1). Specifically, we leverage the scale-space [21] induced by **I** to identify large smooth regions, where the ambient light **A** is estimated by averaging. Since dark areas often have low contrast, they are easily classified as large-scale objects that lack sufficient information about the ambient light. To obtain a more reliable estimation, we refine the regions for estimating **A** by focusing on subsets with brighter pixels. We then utilize the perceptually uniform CIELAB color space [17] to measure the hue similarity between the estimated ambient light and the image colors, deducing the degree of degradation. Notably, the estimated **t** is adaptive to individual images. Thus, without adjusting model parameters, our method produces satisfying results for degraded images acquired under various conditions, such as haze, dust, or underwater environments. Additionally, we introduce a subjective parameter that can be adjusted by the user to achieve a desired enhancement level. The effectiveness of our method are validated through qualitative and quantitative experiments.

This work makes the following contributions. First, we develop an effective, real-time, non-learning-based algorithm for recovering color-balanced scenes from degraded images. Second, we propose novel methods for reliable ambient light estimation based on scale-space analysis and perceptually consistent approximation of the degree of degradation using CIELAB hue similarity. Third, we conduct systematic experiments and comparative studies to demonstrate the model's behavior and validate its effectiveness.

The paper is organized as follows. In Sect. 2, we provide an overview of our method. In Sect. 3, we present our approach for estimating ambient light based on scale-space analysis. In Sect. 4, we describe the estimation of the degree of degradation through hue comparison in the CIELAB color space. In Sect. 5, we present qualitative and quantitative experiments to validate and compare our method with several benchmarks. Finally, we conclude the paper in Sect. 6.

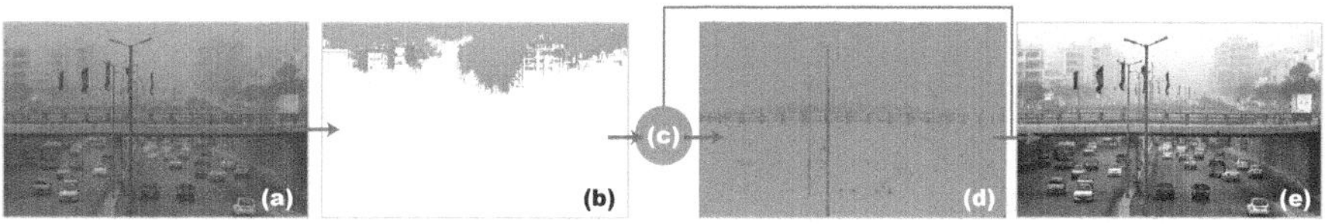

Fig. 2. Outline of the proposed algorithm. Starting from (a) the input image, we detect (b) large-scale blur regions (Sect. 3), where the average color (c) is used to estimate the ambient light. Next, the degree of degradation (d) is estimated (Sect. 4). These two estimations are combined to recover the scene in (e).

2 Overview of the Proposed Method

Our proposed algorithm consists of the following two major components:

1. **Reliable ambient light estimation.** Estimate the ambient light $\mathbf{A} \in \mathbb{R}^3$ in (1) from large and homogeneous regions detected using the scale-space induced by the luminance of **I**. Details are provided in Sect. 3.
2. **Perceptual estimation of the degree of degradation.** Estimate the degree of degradation $\mathbf{t} : \Omega \to \mathbb{R}^3$ in (1) by leveraging its influence on hue similarity between **I** and **A**. Details are provided in Sect. 4.

Algorithm 1. Proposed algorithm

Input: RGB image $\mathbf{I} : \Omega \to \mathbb{R}^3$ defined on a rectangular domain Ω; $\zeta > 0$ brightness threshold; $\varepsilon > 0$ constant to avoid division-by-zero.
Output: Color-balanced estimation $\widehat{\mathbf{J}} : \Omega \to \mathbb{R}^3$.
1: Compute the scalar field S (4) from the gray-scale image $|\mathbf{I}|$.
2: Estimate the background light $\widehat{\mathbf{A}}$ from (6) and (7).
3: Estimate the transmission map $\widehat{\mathbf{t}}$ from (11) in CIELAB space.
4: Reconstruct $\widehat{\mathbf{J}}$ via (2).

Once we obtain the estimated ambient light $\widehat{\mathbf{A}}$ and degree of degradation $\widehat{\mathbf{t}}$, we recover the color-balanced image by inverting the relation in (1). Specifically,

$$\widehat{\mathbf{J}} = \frac{\mathbf{I}(\mathbf{x}) - \widehat{\mathbf{A}} \odot \widehat{\mathbf{t}}(\mathbf{x})}{\max(\mathbf{1} - \widehat{\mathbf{t}}(\mathbf{x}), \varepsilon)}, \tag{2}$$

where the division is performed element-wise, and $\varepsilon > 0$ is a small constant to avoid division by zero. Figure 2 outlines our proposed method, and Algorithm 1 provides the pseudocode. Since the computational complexity of each step is $\mathcal{O}(n)$, where n denotes the number of pixels, our algorithm runs in real-time.

3 Ambient Light from Image Scale-Space

Fig. 3. Illustration of the scale detection function. (a) Input image. (b) and (c) display D_σ (3) for $\sigma = 8$ and $\sigma = 32$, respectively. (d) shows $\overline{S}(\mathbf{x})$ (4), where brighter pixels indicate image regions with larger scales.

3.1 Detection of Large Smooth Regions

We propose to detect large smooth regions in $\mathbf{I}$ using its associated scale-space [21]. Let $|\mathbf{I}|$ denote the luminance of $\mathbf{I}$[1]. The scale-space induced by $|\mathbf{I}|$ is obtained by applying Gaussian blurring with a sequence of increasing kernel sizes. Since image features with scale σ' are smoothed out in $G_\sigma * |\mathbf{I}|$ when $\sigma > \sigma'$, the difference

$$D_\sigma(\mathbf{x}) := \big||\mathbf{I}|(\mathbf{x}) - G_\sigma * |\mathbf{I}|(\mathbf{x})\big| \tag{3}$$

[1] The scalar $|\mathbf{I}|(\mathbf{x}) = 0.299 \times I^r(\mathbf{x}) + 0.587 \times I^g(\mathbf{x}) + 0.114 \times I^b(\mathbf{x})$ is computed for every $\mathbf{x} \in \Omega$.

captures image content with scales greater than σ around each $\mathbf{x} \in \Omega$. Specifically, larger values of $D_\sigma(\mathbf{x})$ indicate the presence of small-scale patterns or object boundaries near $\mathbf{x}$, while smaller values of $D_\sigma(\mathbf{x})$ suggest that the region around $\mathbf{x}$ is smooth. To account for multiple scales in the image, we introduce

$$\overline{S}(\mathbf{x}) = \frac{S(\mathbf{x}) - \min_{\mathbf{y}\in\Omega} S(\mathbf{y})}{\max_{\mathbf{y}\in\Omega} S(\mathbf{y}) - \min_{\mathbf{y}\in\Omega} S(\mathbf{y})}, \quad \text{where} \quad S(\mathbf{x}) = 1 - \frac{1}{H}\sum_{i=1}^{H} D_{\sigma_i}(\mathbf{x}), \tag{4}$$

for a sequence of increasing scale parameters $\sigma_1 < \sigma_2 < \cdots < \sigma_H$. See Fig. 3 for illustrations of D_σ with different values of σ and $\overline{S}$. Except for a thin tubular region covering object boundaries, $\overline{S}(\mathbf{x})$ is close to 1 if $\mathbf{x}$ lies within a homogeneous region of relatively large scale in the given image.

We estimate the ambient light $\mathbf{A}$ in (1) by averaging colors from the interiors of large smooth regions, which are determined by thresholding:

$$\mathcal{A}_\zeta = \{\mathbf{x} \in \Omega : \overline{S}(\mathbf{x}) > \zeta\}, \tag{5}$$

where $\zeta \in [0, 1)$ is a hyperparameter adjusting the scale of interest. The area of $\mathcal{A}_\zeta$ is a nonincreasing function of ζ. This criterion, based on scale, can be further improved by incorporating illumination information. We observe that darker regions are often identified as large, smooth areas due to their low contrast. These regions typically represent surfaces that do not receive ambient light or where the reflected light is directed away from the viewer's perspective. Therefore, we focus only on sub-regions of $\mathcal{A}_\zeta$ where the luminance of $\mathbf{I}$ is sufficiently high to reduce the influence of dark pixels.

Based on the above discussion, we propose iterating the following process:

$$\mathcal{A}_\zeta^k = \left\{\mathbf{x} \in \mathcal{A}_\zeta^{k-1} : |\mathbf{I}|(\mathbf{x}) > \frac{\eta}{|\mathcal{A}_\zeta^{k-1}|}\int_{\mathcal{A}_\zeta^{k-1}} |\mathbf{I}|(\mathbf{y})\, d\mathbf{y}\right\}, \quad \mathcal{A}_\zeta^0 = \mathcal{A}_\zeta, \tag{6}$$

for $k = 1, \ldots, K$, where $K \geq 0$ denotes the number of iterations, and $\eta \in [0, 1]$ is a model parameter. During each iteration of (6), pixels in $\mathcal{A}_\zeta^{k-1}$ with luminance greater than η times the average luminance of $\mathcal{A}_\zeta^{k-1}$ are retained. We propose using

$$\widehat{\mathbf{A}} = \frac{1}{|\mathcal{A}_\zeta^K|}\int_{\mathcal{A}_\zeta^K} \mathbf{I}(\mathbf{y})\, d\mathbf{y} \tag{7}$$

as the estimated ambient light. Further details are provided in Sect. 5.

4 Degree of Degradation from Hue Similarity

4.1 Hue Similarity in CIELAB

The CIELAB color space provides a coordinate system for color specification and measurements. Specifically, each color is parameterized by a point $(L^*, a^*, b^*) \in \mathbb{R}^3$. The first coordinate, $L^* \in [0, 100]$, represents the lightness channel; $a^* \in$

$[-128, 127]$ represents the red-green channel; and $b^* \in [-128, 127]$ represents the blue-yellow channel. The CIELAB space relates to the RGB space through the device-independent XYZ space [9] via the following equations:

$$L^* = 116 f(Y/Y_n) - 16,$$
$$a^* = 500\left(f(X/X_n) - f(Y/Y_n)\right),$$
$$b^* = 200\left(f(Y/Y_n) - f(Z/Z_n)\right),$$

where

$$f(z) = \begin{cases} z^{1/3} & \text{if } z > \left(\frac{6}{29}\right)^3, \\ \frac{841}{108} z + \frac{16}{116} & \text{otherwise.} \end{cases} \tag{8}$$

The inverse formula is also available[2]. The non-linearity in (8) allows for quantifying the perceptive difference between colors in CIELAB via their Euclidean distance. In other words, CIELAB is designed to be perceptually uniform.

Focusing on the chromatic component, i.e., the projections of CIELAB colors onto their respective second and third components—the *hue similarity* between two colors can be conveniently computed by:

$$s_{\text{hue}}((L_1^*, a_1^*, b_1^*), (L_2^*, a_2^*, b_2^*)) = \frac{a_1^* a_2^* + b_1^* b_2^*}{\max(\varepsilon', \sqrt{(a_1^*)^2 + (b_1^*)^2}\sqrt{(a_2^*)^2 + (b_2^*)^2})}. \tag{9}$$

This represents the cosine similarity between (a_1^*, b_1^*) and (a_2^*, b_2^*), where ε' is used to avoid division by zero. Specifically, when (9) is close to 1, the two compared colors are perceived as similar, with possibly different saturation or lightness.

4.2 Estimation of the Degree of Degradation

To estimate the degree of degradation, we first compute the hue similarity score

$$\overline{s}(\mathbf{x}) = \frac{s_{\text{hue}}(\mathbf{I}_{\text{LAB}}(\mathbf{x}), \widehat{\mathbf{A}}_{\text{LAB}}) - m}{M - m}, \tag{10}$$

where m and M are the minimum and maximum values of $s_{\text{hue}}(\mathbf{I}_{\text{LAB}}(\mathbf{y}), \widehat{\mathbf{A}}_{\text{LAB}})$ for all $\mathbf{y} \in \Omega$; and $\mathbf{I}_{\text{LAB}}(\mathbf{y})$ and $\widehat{\mathbf{A}}_{\text{LAB}}$ are the CIELAB coordinates for the RGB representations $\mathbf{I}(\mathbf{x})$ and $\widehat{\mathbf{A}}$, respectively. The normalization in (10) ensures that $\overline{s}(\mathbf{x})$ varies between 0 and 1. When $\mathbf{I}_{\text{LAB}}(\mathbf{x})$ has a similar hue to the estimated ambient light, $\overline{s}(\mathbf{x})$ is close to 0. In contrast, when $\mathbf{I}_{\text{LAB}}(\mathbf{x})$ has a different hue from the estimated background light, $\overline{s}(\mathbf{x})$ can approach 1. After obtaining $\overline{s}$, we convolve it with a Gaussian kernel to account for noise.

[2] A comprehensive list of conversion formulas among major color spaces can be found at http://www.brucelindbloom.com/index.html?Eqn_RGB_to_XYZ.html.

As a consequence of the wavelength-dependent attenuation, the intensity of $\widehat{\mathbf{A}}$ strongly correlates with the rate of degradation in the corresponding channels. Higher values of the ambient light in certain channels reflect more severe chromatic bias. Therefore, it is natural to incorporate $\widehat{\mathbf{A}}$ (7) to estimate the degree of degradation as follows:

$$\widehat{\mathbf{t}}(\mathbf{x}) = \omega \overline{s}(\mathbf{x}) \cdot \widehat{\mathbf{A}}, \quad \mathbf{x} \in \Omega, \tag{11}$$

where $\omega \in (0, 1)$ is a parameter that can be easily adjusted by the user to achieve different levels of enhancement. In particular, larger values of ω mean higher sensitivity to chromatic deviation, and thus a stronger correction will be enforced by our model. This will be confirmed in the numerical section.

5 Numerical Experiments

This section presents numerical results to demonstrate and compare our method's performance with other algorithms.

Parameter Settings. We fix the following set of parameters for our method. For the sequence of increasing scale parameters used for detecting smooth regions (4), we use $\sigma_i = 2^{i+2}$ for $i = 1, 2, 3, 4$; the threshold for generating the mask (5) is $\zeta = 0.94$, and that for the refinement (6) is $\eta = 90\%$; the sensitivity parameter for computing the degree of degradation (11) is $\omega = 0.9$; and the number of iterations K for estimating the ambient light in (6) is 10. To avoid division by zero, $\varepsilon = 1 \times 10^{-8}$ in (2) and $\varepsilon' = 5$ in (9) are used to achieve the best performance.

General Performance. In Fig. 4, we test the proposed algorithm using two images acquired under different environments. In particular, (a) shows an underwater image, which is typically greenish or bluish due to the more substantial attenuation of red light; hence, colors such as yellow, brown, or red are severely distorted. As shown in (b), our method successfully recovers the fish's colors and the rocks in the background. In (c), we test our method with an image affected by a sandstorm, which yields a yellowish cast. Our method effectively corrects the chromatic bias and renders a colorful reconstruction in (d).

Qualitative Comparison. In Fig. 5, we visually compare our method with some benchmark algorithms when the images are affected by haze, sandstorms, or underwater environments. For different categories, we select methods for general purposes, e.g., Multiscale Retinex [29], and methods tailored for specific scenarios. For Rank1+ [22], appropriate sets of parameters are used to adjust for different settings. For our model, the parameters are fixed but can be fine-tuned to achieve better results. We observe that for these challenging images from [3,4,6,12], many methods either fail to recover colors from faded scenes or produce oversaturated, unnatural colors. The learning-based method [23] does not yield consistent results across different inputs. The Multiscale Retinex [29]

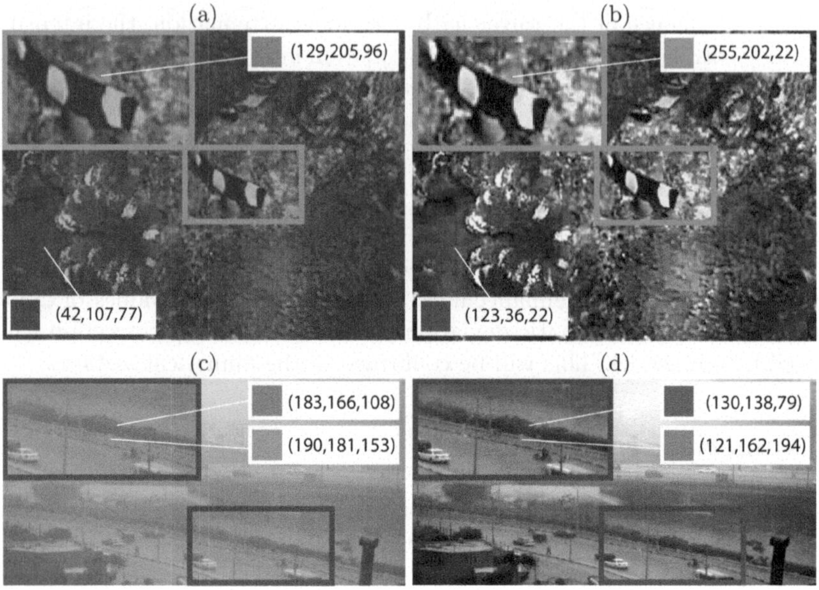

Fig. 4. General performance of the proposed algorithm. (a) Input underwater image. (b) Reconstruction of (a). (c) Input sandstorm image. (d) Reconstruction of (c). Our algorithm improves the colorfulness of the degraded images while effectively reducing the chromatic bias in different scenarios.

achieves good white balancing but loses some desirable colors. Compared to Rank1+ [22], our method produces similar degrees of white balancing but with more appealing colorfulness. Specifically, we note that Rank1+ suffers from overexposure, whereas our results appear more natural.

Quantitative Comparison. To quantitatively compare our method with the aforementioned algorithms, we evaluate them over the RUSH dataset [22] and report the average performance in Table 1 using non-reference metrics, including PIQE [37], NIQE [25], BRISQUE [24], and MetaIQA [39]. We note that there is no single method that excels in all metrics, and our method achieves either first or second place under most of these criteria.

Regions for Ambient Light Estimation. We compare regions used for ambient light estimation. Figure 6(a) shows the input image, while (b), (c), and (d) display the regions produced by DCP [16], Rank1+ [22], and our method, respectively. In (b), we selected the top 10% of the lightest pixels in the dark channel; however, this approach focuses on a small portion of pixels, making it difficult to generalize to other regions. In (c), Rank1+ [22] includes both bright and dark regions, which may lead to dimmer ambient light estimations. In (d), our method generates large and blurry regions with brighter colors, whose average is closer to the perceived chromatic bias.

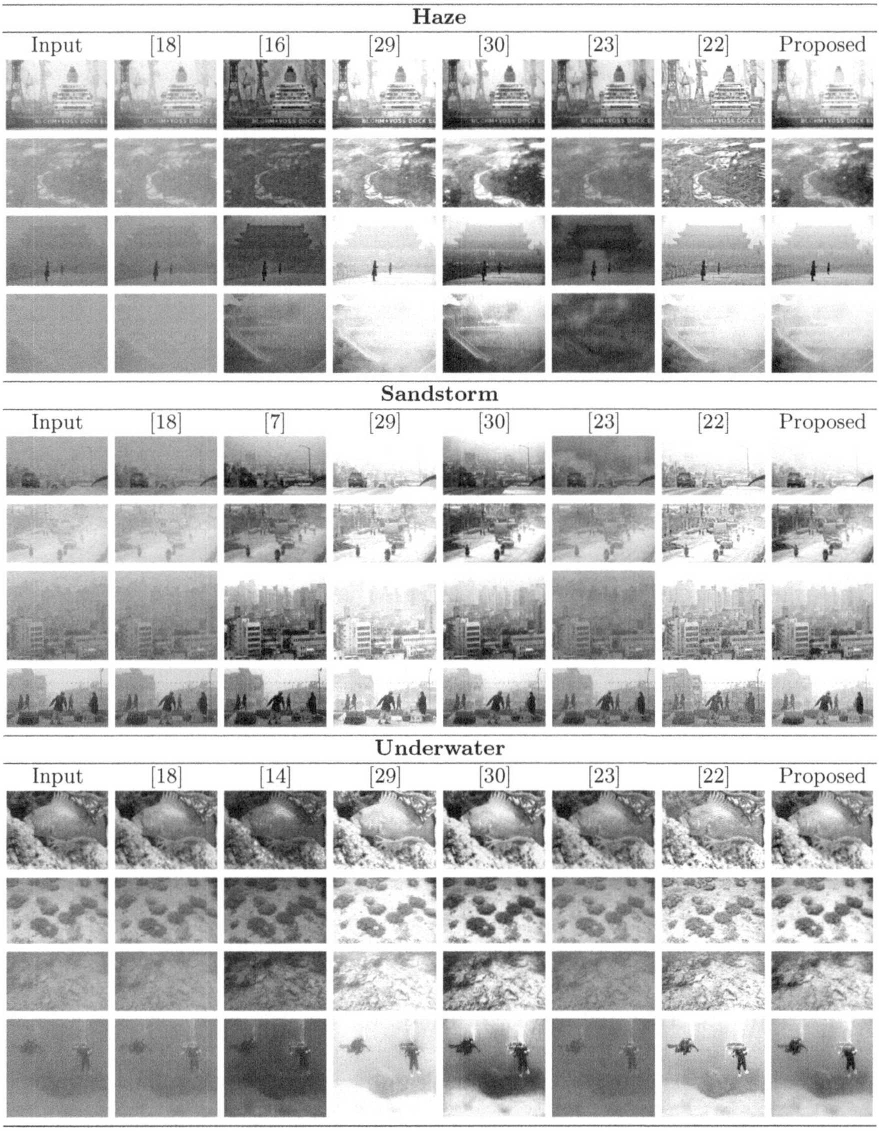

Fig. 5. Qualitative comparisons with benchmark methods, including Gray World Prior [18], Multiscale Retinex [29], Adaptive Histogram Equalization [30], and Rank1+ [22] for general purposes; Red Channel Prior [14] for underwater images; Dark Channel Prior [16], Non-local Dehazing [7], and GridDehazeNet [23] for hazy or dusty images. (Color figure online)

Effects of Parameters ω and ζ. We investigate the effects of our model parameters. By adjusting the parameter ω in (11), our model achieves varying levels of color enhancement. Figure 7 presents zoomed-in results for (a) $\omega = 0.3$, (b)

Table 1. Quantitative comparison. Methods are compared using average non-reference metrics over the RUSH dataset [22]. The symbol ↓ means smaller values indicate better quality; ↑ means the opposite; bold numbers indicate the best results; and † denotes the second best. Parameters of other methods are tuned using their respective default settings.

Haze							
	[18]	[30]	[29]	[16]	[23]	[22]	Proposed Method
piqe ↓	47.90	48.13	45.56	45.86	**41.27**	48.38	45.55†
niqe ↓	3.13	3.08	2.90†	2.95	2.97	2.91	**2.90**
brisque ↓	34.91	30.29	31.96	29.22†	**28.73**	32.33	29.81
metaIQA ↑	0.37	**0.45**	0.35	0.32	0.39	0.39	0.39†
Sandstorm							
	[18]	[30]	[29]	[7]	[23]	[22]	Proposed Method
piqe ↓	44.48	49.62	43.88†	45.77	**42.04**	46.53	44.44
niqe ↓	3.08	3.01	2.92	2.93	2.94	**2.86**	2.88†
brisque ↓	34.82	31.25	31.47	30.93	**27.62**	32.36	28.74†
metaIQA ↑	**0.40**	0.36	0.33	0.34	0.34	0.37	0.37†
Underwater							
	[18]	[30]	[29]	[14]	[23]	[22]	Proposed Method
piqe ↓	33.48	31.43	28.35	32.77	29.93†	31.76	**29.42**
niqe ↓	2.91	**2.65**	2.89	2.78	2.88	2.69†	2.73
brisque ↓	30.51	**24.69**	25.90†	27.87	28.44	27.38	26.14
metaIQA ↑	**0.26**	0.26†	0.23	0.22	0.25	0.25	0.24

$\omega = 0.6$, and (c) $\omega = 0.9$. As ω increases, the colors of the recovered scene become more saturated, demonstrating the flexibility of our proposed method. Another key parameter in our method is ζ, which determines the regions used for ambient light estimation (7). Figure 8 (b)–(d) illustrates the mask $\mathcal{A}_\zeta$ (5) for $\zeta = 0.9$, 0.94, and 0.98, respectively, with the input shown in (a). As ζ increases, more regions are excluded from the mask, indicating that the ambient light is

Fig. 6. Comparison of masks generated for estimating the ambient light. (a) Input image. (b) The mask generated DCP [16], (c) Rank1+ [22] and (d) the proposed method.

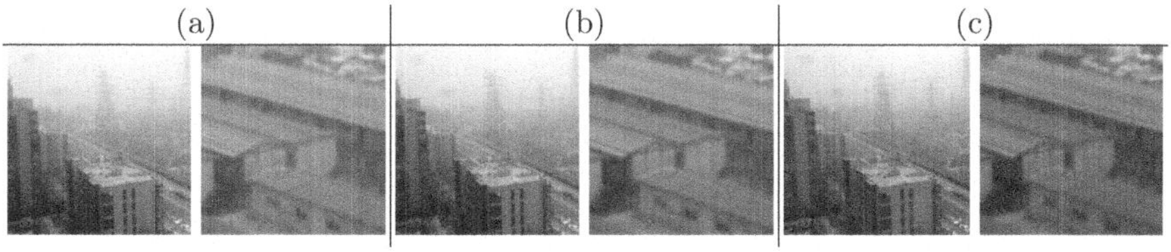

Fig. 7. Effect of the parameter ω (11). We present zoomed-in results for (a) $\omega = 0.3$; (b) $\omega = 0.6$; and (c) $\omega = 0.9$. Higher values of ω result in more saturated colors.

Fig. 8. Effect of the parameter ζ on the mask $\mathcal{A}_\zeta$ (5) used for estimating the ambient light (7). (a) Input image. The white regions in (b), (c), and (d) represent the mask for $\zeta = 0.9$, 0.94, and 0.98, respectively. Higher values of ζ result in smaller regions in the mask, thereby focusing more on the blurry areas.

estimated primarily from the blurry areas. Based on these results, we select $\zeta = 0.94$ as the default value.

6 Conclusion

In this paper, we proposed a simple yet effective algorithm for color image reconstruction. Our method leverages the nature of human vision, which infers ambient light from large-scale homogeneous regions and depth from hue comparisons. We implement these principles through scale-space analysis and perceptual hue similarity in the CIELAB color space. The proposed algorithm is applicable in real-time and can be extended to process video streams. Its effectiveness is validated both qualitatively and quantitatively through a series of numerical experiments and comparative studies.

Acknowledgments. Roy Y. He was partially supported by CityU grant 7200779.

References

1. Agrawal, S.C., Jalal, A.S.: A comprehensive review on analysis and implementation of recent image dehazing methods. Arch. Comput. Methods Eng. **29**(7), 4799–4850 (2022)
2. Akkaynak, D., Treibitz, T.: A revised underwater image formation model. In: Proceedings of the IEEE Conference on Computer Vision and Pattern Recognition, pp. 6723–6732 (2018)

3. Ancuti, C.O., Ancuti, C., Sbert, M., Timofte, R.: Dense haze: a benchmark for image dehazing with dense-haze and haze-free images. In: IEEE International Conference on Image Processing (ICIP). IEEE ICIP 2019 (2019)
4. Ancuti, C.O., Ancuti, C., Timofte, R., Gool, L.V., Zhang, L., Yang, M.H.: NTIRE 2019 Image Dehazing Challenge Report. In: Proceedings of the IEEE Conference on Computer Vision and Pattern Recognition Workshops. IEEE CVPR 2019 (2019)
5. Bahat, Y., Irani, M.: Blind dehazing using internal patch recurrence. In: 2016 IEEE International Conference on Computational Photography (ICCP), pp. 1–9. IEEE (2016)
6. Bartani, A., Abdollahpouri, A., Ramezani, M., Tab, F.A.: An adaptive optic-physic based dust removal method using optimized air-light and transfer function. Multimedia Appl. **81**, 1–27 (2022)
7. Berman, D., Avidan, S.: Non-local image dehazing. In: Proceedings of the IEEE Conference on Computer Vision and Pattern Recognition, pp. 1674–1682 (2016)
8. Berman, D., Treibitz, T., Avidan, S.: Single image dehazing using haze-lines. IEEE Trans. Pattern Anal. Mach. Intell. **42**(3), 720–734 (2018)
9. Connolly, C., Fleiss, T.: A study of efficiency and accuracy in the transformation from RGB to CIELAB color space. IEEE Trans. Image Process. **6**(7), 1046–1048 (1997)
10. Desolneux, A., Moisan, L., Morel, J.M.: From Gestalt Theory to Image Analysis: A Probabilistic Approach, vol. 34. Springer Science & Business Media (2007). https://doi.org/10.1007/978-0-387-74378-3
11. Engin, D., Genç, A., Kemal Ekenel, H.: Cycle-dehaze: enhanced cyclegan for single image dehazing. In: Proceedings of the IEEE Conference on Computer Vision and Pattern Recognition Workshops, pp. 825–833 (2018)
12. Fabbri, C., Islam, M.J., Sattar, J.: Enhancing underwater imagery using generative adversarial networks. In: 2018 IEEE International Conference on Robotics and Automation (ICRA), pp. 7159–7165. IEEE (2018)
13. Fattal, R.: Dehazing using color-lines. ACM Trans. Graph. (TOG) **34**(1), 1–14 (2014)
14. Galdran, A., Pardo, D., Picón, A., Alvarez-Gila, A.: Automatic red-channel underwater image restoration. J. Vis. Commun. Image Represent. **26**, 132–145 (2015)
15. Gui, J., et al.: A comprehensive survey and taxonomy on single image dehazing based on deep learning. ACM Comput. Surv. **55**(13s), 1–37 (2023)
16. He, K., Sun, J., Tang, X.: Single image haze removal using dark channel prior. IEEE Trans. Pattern Anal. Mach. Intell. **33**(12), 2341–2353 (2010)
17. Jain, A.K.: Fundamentals of Digital Image Processing. Prentice-Hall, Inc (1989)
18. Kwok, N.M., Wang, D., Jia, X., Chen, S., Fang, G., Ha, Q.P.: Gray world based color correction and intensity preservation for image enhancement. In: 2011 4th International Congress on Image and Signal Processing, vol. 2, pp. 994–998. IEEE (2011)
19. Li, C., Anwar, S., Porikli, F.: Underwater scene prior inspired deep underwater image and video enhancement. Pattern Recogn. **98**, 107038 (2020)
20. Li, Z., Zheng, J.: Edge-preserving decomposition-based single image haze removal. IEEE Trans. Image Process. **24**(12), 5432–5441 (2015)
21. Lindeberg, T.: Scale-space theory: a basic tool for analyzing structures at different scales. J. Appl. Stat. **21**(1–2), 225–270 (1994)
22. Liu, J., Liu, R.W., Sun, J., Zeng, T.: Rank-one prior: real-time scene recovery. IEEE Trans. Pattern Anal. Mach. Intell. **45**(7), 8845–8860 (2022)

23. Liu, X., Ma, Y., Shi, Z., Chen, J.: GriddeHazenet: attention-based multi-scale network for image dehazing. In: Proceedings of the IEEE/CVF International Conference on Computer Vision, pp. 7314–7323 (2019)
24. Mittal, A., Moorthy, A.K., Bovik, A.C.: Blind/referenceless image spatial quality evaluator. In: 2011 Conference Record of the Forty Fifth Asilomar Conference on Signals, Systems and Computers (ASILOMAR), pp. 723–727. IEEE (2011)
25. Mittal, A., Soundararajan, R., Bovik, A.C.: Making a "completely blind" image quality analyzer. IEEE Sign. Process. Lett. **20**(3), 209–212 (2012)
26. Nayar, S.K., Narasimhan, S.G.: Vision in bad weather. In: Proceedings of the Seventh IEEE International Conference on Computer Vision, vol. 2, pp. 820–827. IEEE (1999)
27. Peng, Y.T., Cosman, P.C.: Underwater image restoration based on image blurriness and light absorption. IEEE Trans. Image Process. **26**(4), 1579–1594 (2017)
28. Perkins, K.R., Schirillo, J.A.: Three-dimensional spatial grouping affects estimates of the illuminant. JOSA A **20**(12), 2246–2253 (2003)
29. Petro, A.B., Sbert, C., Morel, J.M.: Multiscale retinex. In: Image Processing on Line, pp. 71–88 (2014)
30. Pizer, S.M., et al.: Adaptive histogram equalization and its variations. Comput. Vis. Graph. Image Process. **39**(3), 355–368 (1987)
31. Schechner, Y.Y., Karpel, N.: Recovery of underwater visibility and structure by polarization analysis. IEEE J. Oceanic Eng. **30**(3), 570–587 (2005)
32. Schechner, Y.Y., Narasimhan, S.G., Nayar, S.K.: Instant dehazing of images using polarization. In: Proceedings of the 2001 IEEE Computer Society Conference on Computer Vision and Pattern Recognition. CVPR 2001, vol. 1, pp. I. IEEE (2001)
33. Schirillo, J.A., Shevell, S.K.: Lightness and brightness judgments of coplanar retinally noncontiguous surfaces. JOSA A **10**(12), 2442–2452 (1993)
34. Song, Y., He, Z., Qian, H., Du, X.: Vision transformers for single image dehazing. IEEE Trans. Image Process. **32**, 1927–1941 (2023)
35. Su, C.C., Cormack, L.K., Bovik, A.C.: Color and depth priors in natural images. IEEE Trans. Image Process. **22**(6), 2259–2274 (2013)
36. Swinehart, D.F.: The Beer-Lambert law. J. Chem. Educ. **39**(7), 333 (1962)
37. Venkatanath, N., Praneeth, D., Bh, M.C., Channappayya, S.S., Medasani, S.S.: Blind image quality evaluation using perception based features. In: 2015 Twenty First National Conference on Communications (NCC), pp. 1–6. IEEE (2015)
38. Xu, S., Zhang, M., Song, W., Mei, H., He, Q., Liotta, A.: A systematic review and analysis of deep learning-based underwater object detection. Neurocomputing **527**, 204–232 (2023)
39. Zhu, H., Li, L., Wu, J., Dong, W., Shi, G.: MetaIQA: deep meta-learning for no-reference image quality assessment. In: Proceedings of the IEEE/CVF Conference on Computer Vision and pattern recognition, pp. 14143–14152 (2020)

A Novel Interpretation of the Radon Transform's Ray and Pixel-Driven Discretizations Under Balanced Resolutions

Richard Huber(✉)

Technical University of Denmark, Department of Applied Mathematics and Computer Science, 2800 Kongens Lyngby, Denmark
richu@dtu.dk

Abstract. Tomographic investigations are a central tool in medical applications, allowing doctors to image the interior of patients. The corresponding measurement process is commonly modeled by the Radon transform. In practice, the solution of the tomographic problem requires discretization of the Radon transform and its adjoint (called the backprojection). There are various discretization schemes; often structured around three discretization parameters: spatial-, detector-, and angular resolutions. The most widespread approach uses the ray-driven Radon transform and the pixel-driven backprojection in a balanced resolution setting, i.e., the spatial resolution roughly equals the detector resolution. The use of these particular discretization approaches is based on anecdotal reports of their approximation performance, but there is little rigorous analysis of these methods' approximation errors. This paper presents a novel interpretation of ray-driven and pixel-driven methods as convolutional discretizations, illustrating that from an abstract perspective these methods are similar. Moreover, we announce statements concerning the convergence of the ray-driven Radon transform and the pixel-driven backprojection under balanced resolutions. Our considerations are supported by numerical experiments highlighting certain aspects of the discussed methods.

Keywords: Radon Transform · Computed Tomography · Discretization Errors · Numerical Analysis · X-Ray Transform

1 Introduction

Computed Tomography (CT) is a crucial tool in medicine, allowing the investigation of the interior of patients' bodies [8]. The measurement process consists of emitting radiation that travels through the patient, the intensity of which is then measured by a detector on the other side. This paper considers planar geometries (i.e., ignoring the third spatial dimension) with the patient's body located inside the two-dimensional unit ball $\Omega := B(0,1) \subset \mathbb{R}^2$ (called the spatial domain). The radiation is assumed to move along straight lines (beams)

$$L_{\phi,s} := \{ s\vartheta_\phi + t\vartheta_\phi^\perp \in \mathbb{R}^2 \,|\, t \in \mathbb{R} \} \tag{1}$$

T. A. Bubba et al. (Eds.): SSVM 2025, LNCS 15667, pp. 132–145, 2025.
https://doi.org/10.1007/978-3-031-92366-1_11

for $(\phi, s) \in \mathcal{S} := [0, \pi[\times]\text{-}1,1[$ (called the sinogram domain, having an angular and a detector component), where $\vartheta_\phi := (\cos(\phi), \sin(\phi)) \in \mathbb{R}^2$ is the unit vector associated with the angle ϕ and $\vartheta_\phi^\perp := (-\sin(\phi), \cos(\phi)) \in \mathbb{R}^2$ is the direction rotated by 90° counterclockwise; see Fig. 1. The beam's loss of intensity is the accumulation of losses caused by the beam's interaction with the patient's mass within the body. This accumulation is commonly modeled by the Radon transform (in this context also called forward projection) [4,15], $\mathcal{R}\colon L^2(\Omega) \to L^2(\mathcal{S})$ according to

$$[\mathcal{R}f](\phi, s) := \int_\Omega f(x)\, \mathrm{d}\mathcal{H}^1 \llcorner L_{\phi,s}(x) = \int_{-\sqrt{1-s^2}}^{\sqrt{1-s^2}} f(s\vartheta_\phi + t\vartheta_\phi^\perp)\, \mathrm{d}t \tag{2}$$

for $f \in L^2(\Omega)$ and $(\phi, s) \in \mathcal{S}$ (where $\mathcal{H}^1 \llcorner L_{\phi,s}$ denotes the one-dimensional Hausdorff measure restricted to $L_{\phi,s}$), i.e., $\mathcal{R}$ is a collection of line integrals. Thus, the tomographic reconstruction corresponds to the solution of the inverse problem $\mathcal{R}f = g$ for known measurements g and unknown mass density distributions f.

While there is a direct inversion called the filtered backprojection [15], many more evolved reconstruction techniques are iterative, e.g., iterative algebraic reconstruction algorithms [1,7,18] and solution algorithms for convex optimization problems (e.g., total variation regularized reconstructions) [5,12,19]. These iterative methods also involve the adjoint operator $\mathcal{R}^*\colon L^2(\mathcal{S}) \to L^2(\Omega)$ (called the backprojection [15]), which, given $g \in L^2(\mathcal{S})$, reads

$$[\mathcal{R}^* g](x) := \int_0^\pi g(\phi, x \cdot \vartheta_\phi)\, \mathrm{d}\phi \qquad \text{for } x \in \Omega. \tag{3}$$

Only finite amounts of data can be measured and processed in practical applications. Thus, proper discretization $\mathcal{R}_\delta$ for $\mathcal{R}$ (and the adjoints) is imperative, where δ describes the degree of discretization. For us, $\delta = (\delta_x, \delta_\phi, \delta_s)$ denotes the spatial resolution of reconstructions, and the angular and detector resolutions of measured data (we think of both reconstructions and measurements as images). The expectation is that with ever finer resolution ($\delta \to 0$), also the approximation gets arbitrarily good (i.e., $\mathcal{R}_\delta \to \mathcal{R}$ in some sense). The most widespread discretization framework uses the ray-driven Radon transform $\mathcal{R}_\delta^{\mathrm{rd}}$ [6,20,21] and the pixel-driven backprojection $\mathcal{R}_\delta^{\mathrm{pd}^*}$ [2,16,17,23] (we call it an rd-pd* approach) with balanced resolutions, i.e., $\delta_x \approx \delta_s$. Note that usually δ_ϕ and δ_s are determined from the measurements, while δ_x is somewhat selectable. The ray-driven approach discretizes line integrals by splitting them into sections associated with the intersections of the line with pixels, while the pixel-driven backprojection is based on a linear interpolation approximation of (3); see Fig. 3. There is also a ray-driven backprojection and a pixel-driven Radon transform as the adjoint operators of the aforementioned operators. However, they do not see much use due to being supposed poor approximations that create artifacts [13]. Not using the forward and backprojections from the same framework (e.g., rd-rd* or pd-pd*) can potentially harm iterative solvers [14,22].

However, this danger seems to be outweighed in practice by the supposed better approximation performance of rd-pd* methods. There is little rigorous analysis of approximation errors, and anecdotal knowledge of performance is much more prevalent. The author's article [2] rigorously discussed approximation errors for pixel-driven methods in the case that the spatial resolution δ_x is asymptotically smaller than the detector resolution δ_s, finding convergence in the operator norm. Thus, a pd-pd* approach is justified when $\frac{\delta_x}{\delta_s} \to 0$ (both avoiding the unmatched operator issue, while yielding proper approximations). However, in practice, it is much more common to use balanced resolutions (i.e., $\delta_x \approx \delta_s$). This paper will partially justify the use of rd-pd* approaches for balanced resolutions by describing a novel interpretation of ray-driven methods that allows for much more structured analysis. In particular, based on these novel interpretations, we announce convergence results in the strong operator topology in Theorem 1, thus substantiating heuristic notions of approximation performance. The rigorous proof of the statement is available as a preprint in [10], with a corresponding journal article in preparation. Some of these results were already presented in the doctoral thesis [9].

2 Discretization Approaches for the Radon Transform

We start by discretizing the spatial domain $\Omega = B(0,1) \subset \mathbb{R}^2$ and the sinogram domain $\mathcal{S} = [0,\pi[\times]\text{-}1,1[$ into 'pixels'; one can think of data and reconstructions as digital two-dimensional images; see Fig. 1. We fix $N_x \in \mathbb{N}$ and set $\delta_x = \frac{2}{N_x}$. We define the spatial pixel centers $x_{ij} := (\frac{2i+1}{N_x} - 1, \frac{2j+1}{N_x} - 1) = ((i+\frac{1}{2})\,\delta_x - 1, (j+\frac{1}{2})\,\delta_x - 1)$ for $i,j \in [N_x] := \{0,\dots,N_x-1\}$ and $X_{ij} := x_{ij} + [-\frac{\delta_x}{2}, \frac{\delta_x}{2}]^2$ denotes the corresponding squared spatial pixel with side-length δ_x. We consider a finite number $N_\phi \in \mathbb{N}$ of angles $\phi_0 < \cdots < \phi_{N_\phi-1} \in [0,\pi[$ and associate them with the angular pixels $\Phi_q := \left[\frac{\phi_{q-1}+\phi_q}{2}, \frac{\phi_{q+1}+\phi_q}{2}\right[$ for $q \in [N_\phi]$, where we understand $\phi_{-1} = \phi_{N_\phi-1} - \pi$ and $\phi_{N_\phi} = \phi_0 + \pi$. Correspondingly, we set $\delta_\phi := \max_{q\in[N_\phi]} |\Phi_q|$. For the sake of readability, we write ϑ_q when we mean ϑ_{ϕ_q} for the unit vector associated with ϕ_q. We assume a fixed number $N_s \in \mathbb{N}$ of detector pixels and set $\delta_s = \frac{2}{N_s}$. The associated (equispaced) detector pixels are $S_p := s_p + [-\frac{\delta_s}{2}, \frac{\delta_s}{2}[$ for $p \in [N_s]$ with centers $s_p := \frac{2p+1}{N_s} - 1 = (p+\frac{1}{2})\delta_s - 1$. Hence, we have discretized the domain Ω into a Cartesian $N_x \times N_x$ grid with resolution δ_x, while the sinogram space $\mathcal{S}$ contains $N_\phi \times N_s$ rectangular pixels $\Phi_q \times S_p$ for $q \in [N_\phi]$ and $p \in [N_s]$, i.e., with angular resolution δ_ϕ and detector resolution δ_s; see Fig. 1. We notationally combine all these resolutions to $\delta = (\delta_x, \delta_\phi, \delta_s) \in \mathbb{R}^+ \times \mathbb{R}^+ \times \mathbb{R}^+$, and N_x, N_ϕ and N_s are tacitly chosen accordingly.

One can naturally associate pixels values of an image representing $f \in L^2(\Omega)$ with average values $f_{ij} := \frac{1}{\delta_x^2} \int_{X_{ij}} f(x)\,\mathrm{d}x$. Moreover, one can consider the associated (pixelized) functions $f_\delta = \sum_{i,j=0}^{N_x-1} f_{ij} \chi_{X_{ij}}$ in $U_\delta := \operatorname{span}\{\chi_{X_{ij}}\} \widehat{=} \mathbb{R}^{N_x^2}$, where $\chi_M(x)$ equals 1 if $x \in M$ and zero otherwise. Analogously we define $V_\delta := \operatorname{span}\{\chi_{\Phi_q \times S_p}\}$ for discrete sinograms. Discretizations of $\mathcal{R}$ translate to

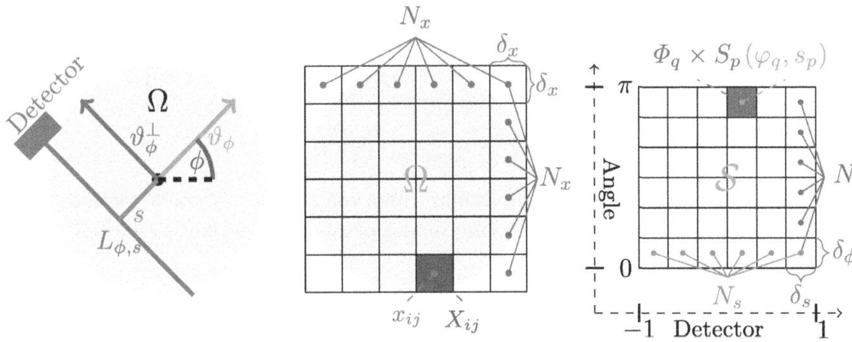

Fig. 1. On the left is the geometry of the Radon transform; in the middle is the discretization of the spatial domain Ω into pixels X_{ij} with width $\delta_x \times \delta_x$; on the right is the discretization of the sinogram domain $\mathcal{S}$ into pixels $\Phi_q \times S_p$ of width $|\Phi_q| \times \delta_s$.

a matrix-vector multiplication with $A \in \mathbb{R}^{(N_\phi \cdot N_s) \times N_x^2}$, where A_{qpij} denotes the weight attributed to a pixel X_{ij} in the calculation of L_{ϕ_q,s_p} (q and p determine a row in A, while i and j determine a column). Interpreting A as a mapping from $U_\delta \subset L^2(\Omega)$ to $V_\delta \subset L^2(\mathcal{S})$, we can understand the discretizations as finite rank operators between L^2 spaces, thus allowing for a proper comparison with $\mathcal{R}$.

In order to compute $[\mathcal{R}f](\phi, s)$, one has to take relatively few values (the values of f along $L_{\phi,s}$) into account; therefore also the A_{qpij} should be non-zero only for pixels X_{ij} that are close to L_{ϕ_q,s_p}. To this end, we will set the weights A_{qpij} using the following weight functions.

Definition 1. *Given δ and $\phi \in [0, \pi[$, we set $\overline{s}(\phi) := \frac{\delta_x}{2}(|\cos(\phi)| + |\sin(\phi)|)$, $\underline{s}(\phi) := \frac{\delta_x}{2}(\big||\cos(\phi)| - |\sin(\phi)|\big|)$ and $\kappa(\phi) := \min\left\{\frac{1}{|\cos(\phi)|}, \frac{1}{|\sin(\phi)|}\right\}$ (with $\frac{1}{0} = \infty$). We define the ray-driven weight function for $t \in \mathbb{R}$ according to*

$$\omega_\delta^{\mathrm{rd}}(\phi, t) := \frac{1}{\delta_x} \begin{cases} \kappa(\phi) & \text{if } |t| < \underline{s}(\phi), \\ \frac{\overline{s}(\phi) - |t|}{\delta_x |\cos(\phi)\sin(\phi)|} & \text{if } |t| \in [\underline{s}(\phi), \overline{s}(\phi)[, \\ \frac{1}{2} & \text{if } \phi \in \{0, \frac{\pi}{2}\} \text{ and } |t| = \overline{s}(\phi), \\ 0 & \text{else.} \end{cases} \tag{4}$$

Moreover, we define the pixel-driven weight function to be

$$\omega_\delta^{\mathrm{pd}}(\phi, t) = \omega_\delta^{\mathrm{pd}}(t) := \frac{1}{\delta_s^2} \max\{\delta_s - |t|, 0\} \qquad \text{for } \phi \in [0, \pi\,[,\ t \in \mathbb{R}\,. \tag{5}$$

For the ray-driven method, one typically uses weights $A_{qpij} := \mathcal{H}^1(X_{ij} \cap L_{\phi_q,s_p})$ (i.e., the intersection length of line and pixel) computed in an iterative manner following the ray [6]. The novel function $\delta_x^2 \omega_\delta^{\mathrm{rd}}(\phi_q, x_{ij} \cdot \vartheta_q - s_p)$ is a closed form expression of this weight, allowing more structured analysis (Fig. 2).

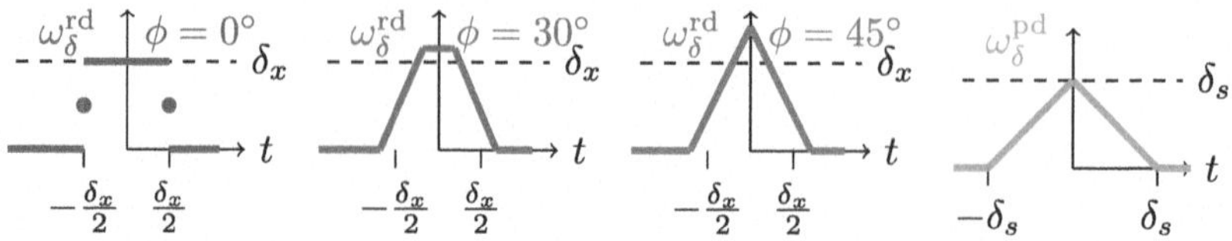

Fig. 2. Depiction of the ray-driven weight function $t \mapsto \delta_x^2 \omega_\delta^{\mathrm{rd}}(\phi, t)$ for fixed $\phi \in \{0°, 30°, 45°\}$. On the right, the depiction of the pixel-driven weight $t \mapsto \delta_s^2 \omega_\delta^{\mathrm{pd}}(t)$.

Definition 2. *Given a (weight) function $\omega\colon [0, \pi[\times \mathbb{R} \to \mathbb{R}$, we define the convolutional Radon transform $\mathcal{R}_\omega\colon L^2(\Omega) \to L^2(\mathcal{S})$, such that, for a function $f \in L^2(\Omega)$,*

$$[\mathcal{R}_\omega f](\phi, s) := \sum_{q=0}^{N_\phi - 1} \sum_{p=0}^{N_s - 1} \chi_{\Phi_q \times S_p}(\phi, s) \sum_{i,j=0}^{N_x - 1} \omega(\phi_q, x_{ij} \cdot \vartheta_q - s_p) \int_{X_{ij}} f(x)\, \mathrm{d}x. \quad (6)$$

The ray-driven Radon transform $\mathcal{R}_\delta^{\mathrm{rd}}$ and pixel-driven Radon transform $\mathcal{R}_\delta^{\mathrm{pd}}$ are defined as special cases of $\mathcal{R}_\omega$ with $\omega = \omega_\delta^{\mathrm{rd}}$ or $\omega = \omega_\delta^{\mathrm{pd}}$, respectively. The corresponding ray-driven or pixel-driven backprojections $\mathcal{R}_\delta^{\mathrm{rd}^*}$ and $\mathcal{R}_\delta^{\mathrm{pd}^*}$ are special cases of the convolutional backprojection $\mathcal{R}_\omega^*\colon L^2(\mathcal{S}) \to L^2(\Omega)$ according to

$$[\mathcal{R}_\omega^* g](x) = \sum_{i,j=0}^{N_x - 1} \chi_{X_{ij}}(x) \sum_{q=0}^{N_\phi - 1} \sum_{p=0}^{N_s - 1} \omega(\phi_q, x_{ij} \cdot \vartheta_q - s_p) \int_{\Phi_q \times S_p} g(\phi, s)\, \mathrm{d}(\phi, s) \quad (7)$$

for $g \in L^2(\mathcal{S})$ when setting ω to $\omega_\delta^{\mathrm{rd}}$ or $\omega_\delta^{\mathrm{pd}}$, respectively.

Note that these definitions are operators between L^2 spaces, whose corresponding matrices A coincide with the classical interpretation of these methods in the literature as intersection lengths for weights and linear interpolation, respectively; see Fig. 3. Also, (6) and (7) show that from a certain point of view, the ray-driven and pixel-driven methods are actually quite similar. Therefore, they can also be implemented using structurally analogous algorithms with similar degrees of parallelization.

3 Theoretical Results

Next, we describe some relevant analytical properties of weight functions and the corresponding discretization operators (Lemma 1), as well as describe the main theoretical result (Theorem 1) concerning convergence for ever finer resolutions.

Lemma 1. *Given δ, and $f_\delta \in U_\delta$, for all $\hat{q} \in [N_\phi]$ and $\hat{p} \in [N_s]$, we have*

$$[\mathcal{R} f_\delta](\phi_{\hat{q}}, s_{\hat{p}}) = [\mathcal{R}_\delta^{\mathrm{rd}} f_\delta](\phi_{\hat{q}}, s_{\hat{p}}). \quad (\mathrm{exact}^{\mathrm{rd}})$$

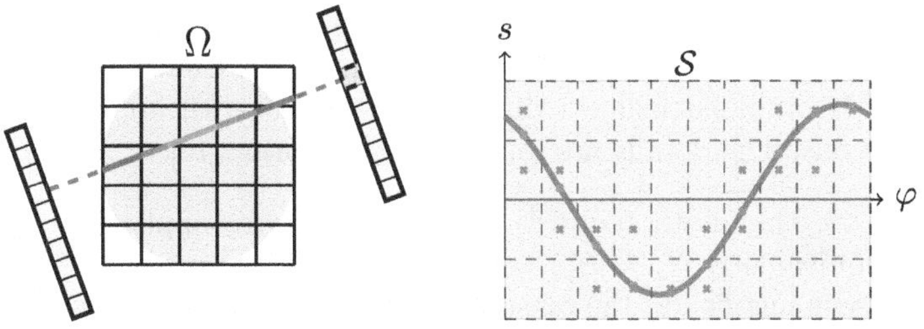

Fig. 3. Illustration of the ray-driven Radon transform (left) and pixel-driven backprojection (right). In the former, integration along a straight line is split into the sum of integrals along the intersections with pixels. The pixel-driven backprojection discretizes the angular integral (3) (along the blue curve $(x \cdot \vartheta_\phi, \phi)$) by a finite sum of angular evaluations $(x \cdot \vartheta_q, \phi_q)$ (red dots), with linear interpolation in the detector dimension (using the points represented by the green dots). (Color figure online)

Moreover, for fixed $\hat{i}, \hat{j} \in [N_x]$ such that $x_{\hat{i}\hat{j}} \cdot \vartheta_{\hat{q}} \in [s_0, s_{N_s-1}]$, we have

$$\sum_{p=0}^{N_s-1} \omega_\delta^{\mathrm{pd}}(x_{\hat{i}\hat{j}} \cdot \vartheta_{\hat{q}} - s_p) = \frac{1}{\delta_s} \qquad (\mathrm{intpol}^{\mathrm{pd}})$$

with exactly two non-zero summands $\hat{p}$ and $\hat{p}+1$ if $x_{\hat{i}\hat{j}} \cdot \vartheta_{\hat{q}} \in]s_{\hat{p}}, s_{\hat{p}+1}[$ and a single non-zero summand $\hat{p}$ if $x_{\hat{i}\hat{j}} \cdot \vartheta_{\hat{q}} = s_{\hat{p}}$.

One can interpret (exactrd) as the ray-driven forward being precise for piecewise constant functions (images) in the detector pixel centers. The ray-driven and pixel-driven discretizations have been known for decades, but no rigorous convergence results were presented. Theorem 1 complements anecdotal reports on the performance of these discretization approaches, announcing convergence in the strong operator topology.

Theorem 1. *Let $(\delta^n)_{n\in\mathbb{N}} = (\delta_x^n, \delta_\phi^n, \delta_s^n)_{n\in\mathbb{N}}$ be a sequence of discretization parameters with $\delta^n \overset{n\to\infty}{\to} 0$ (componentwise) and let $c > 0$ be a constant. If $\frac{\delta_s^n}{\delta_x^n} < c$ for all $n \in \mathbb{N}$, then, for any $f \in L^2(\Omega)$, we have*

$$\lim_{n\to\infty} \|\mathcal{R}f - \mathcal{R}_{\delta^n}^{\mathrm{rd}} f\|_{L^2(\mathcal{S})} = 0. \qquad (\mathrm{conv}^{\mathrm{rd}})$$

If the sequence $(\delta^n)_{n\in\mathbb{N}}$ satisfies $\frac{\delta_s^n}{\delta_x^n} \overset{n\to\infty}{\to} 0$, then, for each $g \in L^2(\mathcal{S})$, we have

$$\lim_{n\to\infty} \|\mathcal{R}^* g - {\mathcal{R}_{\delta^n}^{\mathrm{rd}}}^* g\|_{L^2(\Omega)} = 0. \qquad (\mathrm{conv}^{\mathrm{rd}*})$$

If $\frac{\delta_x^n}{\delta_s^n} \leq c$ for all $n \in \mathbb{N}$, then, for each $g \in L^2(\mathcal{S})$, we have

$$\lim_{n\to\infty} \|\mathcal{R}^* g - {\mathcal{R}_{\delta^n}^{\mathrm{pd}}}^* g\|_{L^2(\Omega)} = 0. \qquad (\mathrm{conv}^{\mathrm{pd}*})$$

The proof involves showing convergence for smooth functions, using Taylor's theorem and suitable properties of the weight functions. Convergence for general L^2 functions is obtained using a diagonal argument with approximations via smooth functions. The details are found in the preprint [10].

Remark 1. Note that both (conv^{rd}) and ($\text{conv}^{\text{pd}*}$) are applicable in the case $\delta_x^n \approx \delta_s^n$ (or equivalently $N_x \approx N_s$), so using the rd-pd* approach in the balanced resolution case is justified in the sense that we have pointwise convergence of the operators. Note that the convergence described in Theorem 1 is not necessarily uniform, i.e., the convergence speed could depend on the specific f and g.

4 Insights into Weight Functions via Simulations

Throughout this section, we use small numerical experiments to highlight certain properties of the weight functions and show how they impact the discretized operators. All calculations were implemented in Python using PyOpenCL with the pixel-driven methods from Gratopy [3], and with a custom ray-driven implementation in an analogous style. All calculations were executed on a 12th Gen Intel(R) Core(TM) i7-12650H in parallel with single precision. The code is available as a GitHub repository at [11]. We consider angles $\phi_q = \pi \frac{q}{N_\phi}$ for $q \in [N_\phi]$. In our illustrations below, we only visualize $x, x_{ij} \in B(0, 0.9)$ to avoid technical issues occurring at the boundary $\partial\Omega$ that are not central to our observations.

Example 1

We consider $f(x) = \chi_E(x)$ with the ellipse $E := \{x \in \mathbb{R}^2 \mid a^2(x \cdot \vartheta_\theta)^2 + b^2(x \cdot \vartheta_\theta^\perp)^2 < 1\}$ with $a = 4$, $b = 2$ and $\theta = \frac{\pi}{4}$; see Fig. 4 a). We consider the bijective transform $T\colon E \to B(0,1)$ according to $Tx := a(x \cdot \vartheta_\theta)\vartheta_\theta + b(x \cdot \vartheta_\theta^\perp)\vartheta_\theta^\perp$. One calculates

$$[\mathcal{R}\chi_E](\phi, s) = \eta(\phi)[\mathcal{R}\chi_{B(0,1)}](\alpha(\phi), \zeta(\phi)s) = \eta(\phi)\tilde{g}(\zeta(\phi)s) \tag{8}$$

with $\alpha(\phi) := \arg(T\vartheta_\phi^\perp) - \frac{\pi}{2}$, $\eta(\phi) := \frac{1}{\|T\vartheta_\phi^\perp\|}$ and $\zeta(\phi) := (T\vartheta_\phi) \cdot \vartheta_{\alpha(\phi)}$, and $\tilde{g}(t) = 2\sqrt{1-t^2}$ if $|t| < 1$ and zero otherwise.

We can explicitly calculate the $L^2([-1,1])$ difference between a function g^δ that is piecewise constant on the detector pixels S_p, and the function $g(s) = \eta(\phi)\tilde{g}(\zeta(\phi)s)$, using $\|g^\delta - g\|^2_{L^2([-1,1])} = \|g^\delta\|^2_{L^2} + \|g\|^2_{L^2} - 2\langle g^\delta, g\rangle$. Using $P(s) := \min(\max(s,-1),1)$, we have

$$\|g^\delta\|^2_{L^2} = \sum_{p=0}^{N_s-1} \delta_s |g_p^\delta|^2, \qquad \|g\|^2_{L^2} = \frac{\eta(\phi)^2}{\zeta(\phi)} \sum_{p=0}^{N_s-1} \left[P(s) - \frac{P(s)^3}{3}\right]_{s=(s_p-\frac{\delta_s}{2})\zeta(\phi)}^{s=(s_p+\frac{\delta_s}{2})\zeta(\phi)},$$

$$\langle g, g^\delta\rangle = \frac{\eta(\phi)}{\zeta(\phi)} \sum_{p=0}^{N_s-1} g_p^\delta \left(G(\zeta(\phi)(s_p+\delta_s)) - G(\zeta(\phi)(s_p-\delta_s))\right) \tag{9}$$

with $G(t) := \frac{1}{2}\big(P(t)\sqrt{1-P(t)^2} + \arcsin(P(t))\big)$. Hence, for fixed angle ϕ_q, we can calculate $\|[\mathcal{R}f - \mathcal{R}_\delta^{\mathrm{rd}}](\phi_q,\cdot)\|_{L^2([-1,1])}$, and we refer below with the L^2 error to $\sqrt{\sum_{q=0}^{N_\phi-1} |\Phi_q| \|[\mathcal{R}f - \mathcal{R}_\delta^{\mathrm{rd}}](\phi_q,\cdot)\|^2_{L^2([-1,1])}}$.

We numerically evaluated the ray-driven and pixel-driven Radon transforms (see Fig. 4 b and c)) and considered the evaluation of errors under increasing balanced resolutions in Fig. 4 d). In the latter, for the pixel-driven case, the error appears to reach a plateau (or at least slows down significantly) far before reaching zero, while the ray-driven error reduces much further. This is in line with Theorem 1. Moreover, the pixel-driven error appears to reduce with the number of angles. The reason is that there are some angles for which the approximation is very poor (e.g., $\phi = 45°$, see Fig. 4 e)), but whose overall contribution reduces if there are more projections in total. In contrast, the error induced by the ray-driven method appears virtually independent of the number of angles, suggesting the error is similar for all projections.

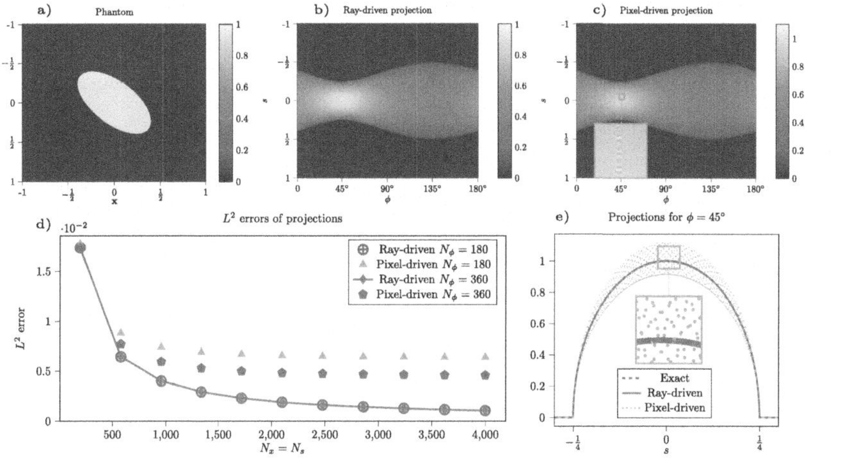

Fig. 4. Illustration of Example 1, showing the phantom used in a), the corresponding ray-driven b) and pixel-driven c) forward projections with $N_s = N_x = 580$ and $N_\phi = 360$. Figure d) plots the L^2 errors for varying balanced resolutions $N_s = N_x$ and changing N_ϕ. Figure e) shows the projection for $\phi = 45°$ (with $N_x = N_s = 4000$), showing hugely oscillating values for the pixel-driven method.

Example 2

We consider the constant sinogram $\mathbf{1}(\phi, s) = 1$ for all $(\phi, s) \in \mathcal{S}$. Since the integrand in (3) is constant, the backprojection satisfies $[\mathcal{R}^*\mathbf{1}](x) = \pi$ for $x \in \Omega$. A discretization method will be precise iff

$$\sum_{q=0}^{N_\phi-1} \sum_{p=0}^{N_s-1} \delta_s |\phi_q| \omega(\phi_q, x_{ij}\cdot\vartheta_q - s_p) \overset{\text{per}}{\underset{\text{def}}{=}} [\mathcal{R}_\omega^*\mathbf{1}](x_{ij}) \overset{!}{=} [\mathcal{R}^*\mathbf{1}](x_{ij}) = \pi. \quad (10)$$

(Requesting (10) to hold for all $x \in \Omega$ seems quite natural then.) The pixel-driven backprojection (i.e., with $\omega = \omega_\delta^{\mathrm{pd}}$) satisfies (10) due to (intpol^{pd}) (for $\|x\| \leq |s_0|$) and since $\sum_{q=0}^{N_\phi-1} |\Phi_q| = \pi$. Hence, the pixel-driven $\mathcal{R}_\delta^{\mathrm{pd}^*}\mathbf{1}$ is correct up to machine precision ($\approx 10^{-6}$), which numerical experiments confirm.

In contrast, for the ray-driven method, it is unclear whether (10) holds. Figures 5 a) and b) depict the error of simulations using the ray-driven backprojections, revealing some geometric patterns whose amplitude and L^2 error do not decrease with increased balanced resolutions $N_x = N_s$. The relative error is roughly one percent, i.e., not huge. Still, this illustrates a potential issue that appears to not be solved by increasing resolutions. Note that the structured nature of the error suggests this has a deeper cause in the weight function $\omega_\delta^{\mathrm{rd}}$.

According to (conv$^{\mathrm{rd}*}$), if $\delta_x \ll \delta_s$ the ray-driven backprojection should converge. And indeed, as depicted in Fig. 5 c), the error reduces significantly (by a factor of 10) when considering $N_x = 1000$ and $N_s = 4000$. It might seem counter-intuitive that using coarser spatial resolution (compared to Fig. 5 b)) yields a significantly better approximation, but this is in line with (conv$^{\mathrm{rd}*}$).

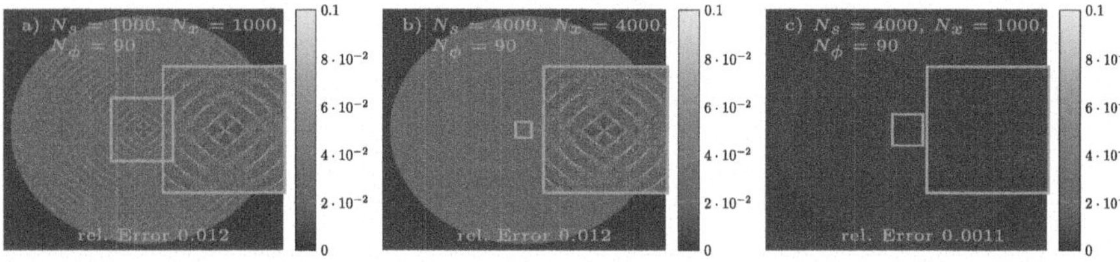

Fig. 5. Illustration of the error incurred by ray-driven backprojections $\mathcal{R}_\delta^{\mathrm{rd}^*}\mathbf{1}$ for Example 2. In a) and b), the backprojections for balanced resolution with $N_\phi = 90$ and $N_s = N_x = 1000$ and $N_s = N_x = 4000$, respectively, are shown. (In both, we truncated very few values to make the patterns better visible.) In c), we depict the error when reducing the spatial resolution to $N_x = 1000$ while keeping $N_s = 4000$ detector pixels.

In Fig. 6, the error of $\mathcal{R}_\delta^{\mathrm{rd}^*}\mathbf{1}$ for balanced resolution and increasing N_ϕ is shown. The error appears to reduce with increasing N_ϕ, though not very quickly. As illustrated in the pixel-driven backprojections' precision, the angular resolution should not be a natural limiting factor for this example. Despite that, it appears so for the ray-driven method. This can be understood as follows: (10) is a mean value of $\phi \mapsto \sum_{p=0}^{N_s-1} \delta_s \omega(\phi, x_{ij} \cdot \vartheta_\phi - s_p)$, and while this does not have a fixed value for all angles ϕ or points x_{ij}, it still averages out to 1. Thus, the more angles are considered, the closer is (10) to be satisfied.

To emphasize this point, we fix some $\hat{q} \in [N_\phi]$ and consider $g_{\hat{q}}(\phi, s) = \frac{1}{|\Phi_{\hat{q}}|}$ for $s \in]-1, 1[$ and $\phi \in \Phi_{\hat{q}}$ and zero otherwise, i.e., the sinogram has only very small angular support. We have $[\mathcal{R}^* g_{\hat{q}}](x) = 1$ for all $x \in \Omega$. Hence, the convolutional approximations $\mathcal{R}_\omega^* g_{\hat{q}}$ are precise iff

$$\delta_s \sum_{p=0}^{N_s-1} \omega(\phi_{\hat{q}}, x_{ij} \cdot \vartheta_{\hat{q}} - s_p) = 1. \tag{11}$$

($\hat{q}$ was arbitrary, so we would want (11) to hold for all angles.) For the pixel-driven method, (11) holds via (intpol^{pd}), and therefore, the pixel-driven method is again precise (up to machine precision). The averaging with respect to the angles described above that moderated (10) does not apply here. Therefore, the issue seen in $\mathcal{R}_\delta^{\text{rd}^*}\mathbf{1}$ becomes much more severe for $\mathcal{R}_\delta^{\text{rd}^*} g_{\hat{q}}$. In fact, a key step in proving convergence of the ray-driven backprojection in (conv$^{\text{rd}*}$) is to show that ray-driven weight functions asymptotically satisfy (11) when $\delta_s \ll \delta_x$.

Figure 6 also shows the errors for $\mathcal{R}_\delta^{\text{rd}^*} g_{\hat{q}}$ (with $\hat{q}$ such that $\phi_{\hat{q}} = \frac{\pi}{4}$). While the error for $\mathbf{1}$ reduces with increasing N_ϕ under balanced resolution, for $g_{\hat{q}}$, the balanced resolutions cause relative L^2 errors of 19% independent of the number of projections used. In the case $N_\phi = 90$, choosing $N_x = 1000$ significantly smaller than $N_s = 4000$ reduces the relative L^2 error for $\mathcal{R}_\delta^{\text{rd}^*} g_{\hat{q}}$ significantly (by a factor of roughly 30) to 0.00625.

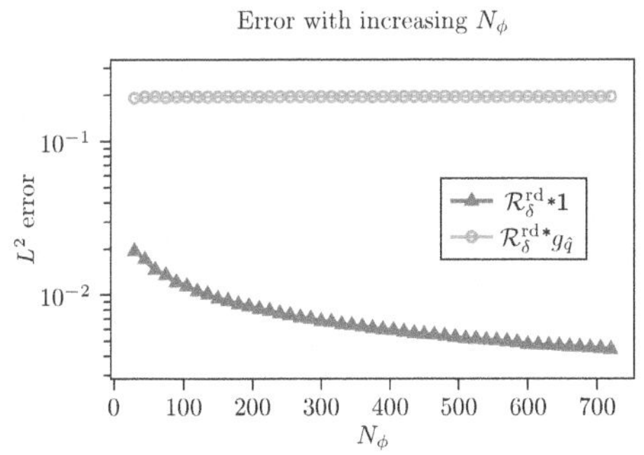

Fig. 6. Illustration of L^2 errors of the ray-driven backprojections $\mathcal{R}_\delta^{\text{rd}^*}\mathbf{1}$ and $\mathcal{R}_\delta^{\text{rd}^*}\mathbf{g}_{\hat{q}}$ of Example 2 for increasing N_ϕ in balanced resolution $N_x = N_s = 2000$.

It might be possible to translate these considerations into the construction of a function g with $\|\mathcal{R}^* g - \mathcal{R}_\delta^{\text{rd}^*} g\|_{L^2(\Omega)} \geq c\|g\|_{L^2(\mathcal{S})}$ for some constant $c > 0$ and all $\delta_x \approx \delta_s \to 0$ and $\delta_\phi \to 0$; then, the ray-driven Radon transforms could not converge in the operator norm. Consequently, for any δ, there would be a function $f = f(\delta)$ such that $\|\mathcal{R}f - \mathcal{R}_\delta^{\text{rd}} f\|_{L^2(\mathcal{S})} \geq c\|f\|_{L^2(\Omega)}$, and no matter how fine the resolution, there would always be functions for which the ray-driven Radon transform is a poor approximation.

Example 3

Next, we consider the sinogram $g(\phi, s) = s$ for all $\phi \in [0, \pi[$ and $s \in [-1, 1]$ with

$$[\mathcal{R}^* g](x) = \int_0^\pi \mathbf{x}\cos(\phi) + \mathbf{y}\sin(\phi)\,\mathrm{d}\phi = 2\mathbf{y}, \tag{12}$$

where $x = (\mathbf{x}, \mathbf{y})$. Since the pixel-driven method uses linear interpolation (see ($\text{intpol}^{\text{pd}}$)), and g is linear in the detector dimension and constant in the angular dimension, one might suspect that $\mathcal{R}_\delta^{\text{pd}^*}$ is again accurate up to machine precision. However, this would be a fallacy. One can explicitly calculate

$$[\mathcal{R}_\delta^{\text{pd}^*} g](x_{ij}) = \sum_{q=0}^{N_\phi - 1} |\Phi_q| \Big(\mathbf{x}_i \cos(\phi_q) + \mathbf{y}_j \sin(\phi_q)\Big), \tag{13}$$

where $x_{ij} = (\mathbf{x}_i, \mathbf{y}_j)$. Hence, the error that the pixel-driven backprojection suffers is the error of the Riemann sum of the integrals in (12), i.e.,

$$[\mathcal{R}^* g](x_{ij}) - [\mathcal{R}_\delta^{\text{pd}} g](x_{ij}) = -\mathbf{x}_i \sum_{q=0}^{N_\phi - 1} |\Phi_q| \cos(\phi_q) + \mathbf{y}_j \left(2 - \sum_{q=0}^{N_\phi - 1} |\Phi_q| \sin(\phi_q)\right).$$

Thus, the pixel-driven backprojection can lead to highly structured errors related to how well these Riemann sums approximate the cosine and sine integrals. In particular, although g does not depend on ϕ, δ_ϕ can have an impact here.

For our numerical experiments, we started with the naive choice $\phi_0 = 0°$, $\phi_1 = 0.5°, \ldots, \phi_{359} = 179.5°$. Then, the error of the Riemann sum approximating the cosine integral is roughly $\pi/360 \approx 0.0087$, while the Riemann sum of the sine terms equals roughly 1.99998. Hence, when computing the pixel-driven method, the error is dominated by the linear $0.0087\,\mathbf{x}_i$, see Fig. 7 a).

If we shift the considered angles by 0.25°, i.e., $\phi_0 = 0.25°$, $\phi_1 = 0.75°$, and so on, by symmetry the cosine Riemann sum is zero, while the sine Riemann sum's error is roughly $6\,10^{-6}$. Hence, the error is dominated by $6\mathbf{y}_j 10^{-6}$, which overall reduces the approximation error substantially (roughly by a factor of 10^{-3}); see Fig. 7 b). Note that neither of the presented backprojections is terrible, however, their difference and the structure of the errors is unexpected.

Naturally, such considerations cannot be taken into account for real data. Nonetheless, this gives some interesting insight into the significant impact of the angular discretization. Theorem 1 suggests that with an increasing number of angles, this issue reduces (irrespective of the specific angles). In contrast, in the ray-driven backprojection, the error is structured differently (see Fig. 7 c)), suggesting that factors other than the Riemann sum drive this error.

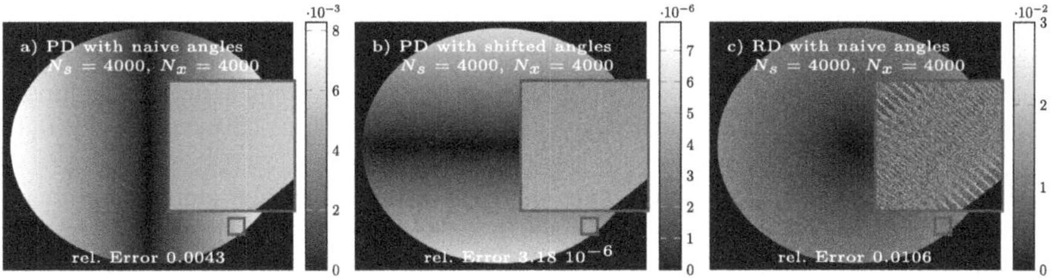

Fig. 7. Illustration of the error of backprojections for Example 3 with $N_x = N_s = 4000$ and $N_\phi = 360$. In a) the pixel-driven backprojections' errors with the naive angles, and in b) with the described angle shift. The ray-driven backprojection's error is depicted in c). Note the different colormaps used and corresponding different relative errors.

5 Conclusion and Outlook

This paper presented a novel interpretation of ray-driven and pixel-driven methods as 'convolutional discretizations', suggesting that in some sense the methods are more alike than previously thought. Moreover, we announced novel convergence statements in the strong operator topology in Theorem 1. These results give a theoretical foundation for the widespread use of ray-driven forward and pixel-driven backprojection operators under balanced resolutions. Anecdotes also suggest that ray-driven backprojections and pixel-driven forward projections do not converge, which our simulations support. Hence, the presented results might suggest that the combination of ray-driven and pixel-driven methods is better in terms of approximation for the balanced resolution setting than the matched rd-rd* and pd-pd* approaches. This work did not address the issue of unmatched operators, however, using unmatched operators may be preferable to using non-approximating discretizations.

The convergence results are probably extendable to other types of tomography (with fanbeam or conebeam operators), and might be the topic of future investigations.

The convergence result (conv$^{\text{rd}*}$) shows that while the ray-driven backprojection might be unsuitable for balanced resolutions, it approximates the backprojection if $\frac{\delta_s}{\delta_x} \to 0$. Thus, future works might investigate if convergence in the operator norm is achieved in that setting. Moreover, Example 2 might yield a way to show that ray-driven methods do not converge in the operator norm when using balanced resolutions, and thus might also merit future investigations.

Acknowledgments. This work was supported by The Villum Foundation (Grant No.25893) and is in part based on work supported by the International Research Training Group "Optimization and Numerical Analysis for Partial Differential Equations with Nonsmooth Structures", funded by the German Research Council (DFG) and Austrian Science Fund (FWF) grant W1244.

References

1. Andersen, A.H., Kak, A.C.: Simultaneous algebraic reconstruction technique (SART): a superior implementation of the ART algorithm. Ultrason. Imag. **6**(1), 81–94 (1984)
2. Bredies, K., Huber, R.: Convergence analysis of pixel-driven Radon and fanbeam transforms. SIAM J. Num. Anal. **59**(3), 1399–1432 (2021). https://doi.org/10.1137/20M1326635
3. Bredies, K., Huber, R.: Gratopy 0.1 release candidate 1 [software] (2021). https://doi.org/10.5281/zenodo.5221443
4. Deans, S.R.: The Radon Transform and Some of its Applications. Krieger Publishing Company (1993)
5. Dong, B., Li, J., Shen, Z.: X-ray CT image reconstruction via wavelet frame based regularization and Radon domain inpainting. J. Sci. Comput. **54**(2), 333–349 (2013). https://doi.org/10.1007/s10915-012-9579-6
6. Gao, H.: Fast parallel algorithms for the X-ray transform and its adjoint. Med. Phys. **39**(11), 7110–7120 (2012). https://doi.org/10.1118/1.4761867
7. Gilbert, P.: Iterative methods for the three-dimensional reconstruction of an object from projections. J. Theor. Biol. **36**(1), 105–117 (1972)
8. Hsieh, J.: Computed Tomography: Principles, Design, Artifacts, and Recent Advances. WA: SPIE — The International Society for Optical Engineering (2009)
9. Huber, R.: Pixel-driven projection methods' approximation properties and applications in electron tomography. PhD thesis, University of Graz, Austria (2022)
10. Huber, R.: Convergence of ray- and pixel-driven discretization frameworks in the strong operator topology (2025). https://arxiv.org/abs/2503.03069
11. Huber, R.: Github repository with simulations accompanying 'A novel interpretation of the Radon transform's ray- and pixel-driven discretizations under balanced resolutions'. https://github.com/huberr92/Simulations_SSVM_2025 (2025)
12. Huber, R., Haberfehlner, G., Holler, M., Kothleitner, G., Bredies, K.: Total generalized variation regularization for multi-modal electron tomography. Nanoscale **11**, 5617–5632 (2019). https://doi.org/10.1039/C8NR09058K
13. Liu, R., Fu, L., De Man, B., Yu, H.: GPU-based branchless distance-driven projection and backprojection. IEEE Trans. Comput. Imaging **3**(4), 617–632 (2017). https://doi.org/10.1109/TCI.2017.2675705
14. Lorenz, D., Schneppe, F.: Chambolle-pock's primal-dual method with mismatched adjoint. Appl. Math. Optim. **87**(2) (2023). https://doi.org/10.1007/s00245-022-09933-5
15. Natterer, F.: The Mathematics of Computerized Tomography. Society for Industrial and Applied Mathematics, Philadelphia (2001)
16. Peters, T.M.: Algorithms for fast back- and re-projection in computed tomography. IEEE Trans. Nucl. Sci. **28**(4), 3641–3647 (1981). https://doi.org/10.1109/TNS.1981.4331812
17. Qiao, Z., Redler, G., Gui, Z., Qian, Y., Epel, B., Halpern, H.: Three novel accurate pixel-driven projection methods for 2D CT and 3D EPR imaging. J. Xray Sci. Technol. **26**(1), 83–102 (2017)
18. Scales, J.: Tomographic inversion via the conjugate gradient method. Geophysics **52**, 179–185 (1987). https://doi.org/10.1190/1.1442293
19. Scherzer, O., Grasmair, M., Grossauer, H., Haltmeier, M., Lenzen, F.: Variational Methods in Imaging. Springer, 1 edn. (2008). https://doi.org/10.1007/978-0-387-69277-7

20. Siddon, R.L.: Fast calculation of the exact radiological path for a three-dimensional CT array. Med. Phys. **12**(2), 252–255 (1985)
21. Sundermann, E., Jacobs, F., Christiaens, M., De Sutter, B., Lemahieu, I.: A fast algorithm to calculate the exact radiological path through a pixel or voxel space. J. Comput. Inf. Technol. **6**, 89–94 (1998)
22. Zeng, G.L., Gullberg, G.T.: Unmatched projector/backprojector pairs in an iterative reconstruction algorithm. IEEE Trans. Med. Imaging **19**(5), 548–555 (2000). https://doi.org/10.1109/42.870265
23. Zhuang, W., Gopal, S.S., Hebert, T.J.: Numerical evaluation of methods for computing tomographic projections. IEEE Trans. Nucl. Sci. **41**(4), 1660–1665 (1994). https://doi.org/10.1109/23.322963

Product of Gaussian Mixture Diffusion Model for Non-linear MRI Inversion

Laurenz Nagler[1(✉)], Martin Zach[2,3], and Thomas Pock[1]

[1] Institute of Visual Computing, Graz University of Technology, 8010 Graz, Austria
lnagler@student.tugraz.at, thomas.pock@tugraz.at

[2] Biomedical Imaging Group, École Polytechnique Fédérale de Lausanne, 1015 Lausanne, Switzerland
martin.zach@epfl.ch

[3] Center for Biomedical Imaging, 1015 Lausanne, Switzerland

Abstract. Diffusion models have recently shown remarkable results in magnetic resonance imaging reconstruction. However, the employed networks typically are black-box estimators of the (smoothed) prior score with tens of millions of parameters, restricting interpretability and increasing reconstruction time. Furthermore, parallel imaging reconstruction algorithms either rely on off-line coil sensitivity estimation, which is prone to misalignment and restricting sampling trajectories or perform per-coil reconstruction, making the computational cost proportional to the number of coils. To overcome this, we jointly reconstruct the image and the coil sensitivities using the lightweight, parameter-efficient, and interpretable product of Gaussian mixture diffusion model as an image prior and a classical smoothness priors on the coil sensitivities. The proposed method delivers promising results while allowing for fast inference and demonstrating robustness to contrast out-of-distribution data and sampling trajectories, comparable to classical variational penalties such as total variation. Finally, the probabilistic formulation allows the calculation of the posterior expectation and pixel-wise variance.

Keywords: Diffusion Model · MRI Reconstruction · Non-linear Inverse Problem

1 Introduction

Magnetic resonance imaging (MRI) is one of the pillars of modern medicine, exhibiting high spatial resolution and contrast flexibility. However, these strengths are limited by low temporal resolution, long scan times and artefacts induced by patient movement.

This can be overcome by acquiring less data, resulting in an ill-posed reconstruction problem. One way to obtain solutions to the problem is parallel imaging (PI), where multiple spatially structured receiver coils are used. The individual coil data is then combined in either the image domain [22] or the frequency

T. A. Bubba et al. (Eds.): SSVM 2025, LNCS 15667, pp. 146–158, 2025.
https://doi.org/10.1007/978-3-031-92366-1_12

domain [10]. Another way of tackling this problem is through variational methods or compressed sensing (CS) theory, which leverages sparse representations of the data in a specific domain and incoherent sampling to enable image reconstruction [1,19].

In recent years, data-driven methods have achieved remarkable results in MRI image reconstruction [7,13,33]. These models can be categorized by whether they follow a Bayesian separation of likelihood and prior, i.e. whether the learned components account for the measurement setup or not. In the context of MRI, the Bayesian separation obviates the need for multi-coil data for learning [33] and enables the learned model to be used for reconstructing images from arbitrary k-space sampling trajectories. Methods pursuing this Bayesian separation in MRI include works based on generative adversarial networks (GANs) [20], energy based models (EBMs) [33], and, more recently, diffusion models [7,13,18,27]. Diffusion models deliver state-of-the-art results but have shortcomings: The networks are, in general, not the gradient of a potential [24], and the time conditioning is opaque, seemingly unrelated to the underlying stochastic differential equation. Furthermore, models with millions of parameters are often accompanied by long inference times, a lack of interpretability and the need for heuristics to be applied to data of arbitrary size [7,9]. Existing methods for PI such as [13,18] typically require off-line coil sensitivity estimation, e.g. with ESPIRiT [31]. This is sensitive to patient motion and hard to adapt to arbitrary k-space sampling trajectories [14]. In a different line of work, [7] overcomes this by reconstructing individual coil images, making the computational cost proportional to the number of coils, which is unacceptably large when paired with the typical large networks used to model the score. In [4], the authors propose to tackle blind inverse problems by additionally learning a diffusion model on the parameters of the forward operator. In the context of PI, this translates to learning a diffusion model on the coil sensitivities, again making the computational cost proportional to the number of coils and introducing the engineering challenge of training and tuning the additional diffusion model. In [11], the authors propose an algorithm for PI, where the coil sensitivities are assumed to be sufficiently close to an initial guess, again requiring off-line estimation.

Recently, Zach et al. [34] introduced the product of Gaussian mixture diffusion model (PoGMDM),

$$p_\theta(x,t) \propto \prod_{i,j=1}^{n,m} \prod_{k=1}^{o} \psi_k((K_k x)_{i,j}, w_k, t), \tag{1}$$

a diffusion model with fields-of-experts-type structure and relations to classical shrinkage-type regularization. Here, to model the density of $n \times m$ images, for all $k = 1, \ldots, o$,

$$\psi_k(x,w,t) = \sum_{i=1}^{L} w_i / \sqrt{2\pi\sigma(t)^2} \exp\big((x-\mu_i)^2/(2\sigma_k^2(t))\big) \tag{2}$$

is a Gaussian mixture model (GMM) with weights w_k, $K_k : \mathbb{R}^{n\times m} \to \mathbb{R}^{n\times m}$ is a convolution operator, and the time conditioning adapts the variance σ_k^2 in the

GMM in accordance with the stochastic differential equation (SDE) based on characteristics of K_k. This model is analytically tractable and interpretable, has few learnable parameters and allows for fast inference. In their work, they only considered denoising due to the close relations between generative modelling and denoising via Tweedies identity. In this work, we demonstrate the feasibility of PoGMDM as an image prior in joint non-linear MRI reconstruction by pairing it with simple smoothness priors on the coil sensitivities.

2 Background

2.1 Generative Modeling and Diffusion

In generative modeling, we are interested in learning a parametric model approximating a reference distribution from a given dataset. Let X be the random variable of the reference distribution with associated density p_X. Learning a model p_θ approximating p_X is hard as the high dimensional density space is sparsely populated. One way to ease learning is to smooth (thus filling low-density regions) and approximate the density at different scales [26].

Song et al. generalized this idea to SDEs [28]. In this work, we consider the SDE

$$dY_t = \sqrt{2}dw_t, \quad Y_0 = X \tag{3}$$

where w is the standard Wiener process and is a random variable Y_t with density $p_Y(\cdot, t)$. The Fokker-Planck equation [23] links Y_t to its density, which is given as the partial differential equation (PDE) $\partial p_Y(\cdot, t)/\partial t = \Delta p_Y(\cdot, t)$ with $p_Y(\cdot, 0) = p_X$. This is the classic heat diffusion with the solution $p_Y(\cdot, t) = \mathcal{N}(0, 2tI) * p_X$, thus constructing a scale space in the space of probability [34]. In practice, a model $p_\theta(\cdot, t) \approx p_Y(\cdot, t)$ can be learned by optimizing the objective function

$$\min_\theta \int_0^\infty \mathbb{E}_{(x,y_t)\sim p_{X,Y_t}} \left[\|x - y_t - 2t\nabla p_\theta(y_t, t)\|_2^2\right] dt \tag{4}$$

known as denoising score matching [28,32]. With this configuration, sampling amounts to running the diffusion process,

$$d\bar{Y}_t = -2\nabla \log p_Y(\bar{Y}_t, t)dt + \sqrt{2}d\bar{w}_t, \quad \bar{Y}_T = Y_T, \tag{5}$$

where dt is a negative infinitesimal time step and $\bar{w}$ is the reverse time Wiener process starting, from a time T to zero. The only unknown quantity in (5) is the gradient of the log density p_Y (score) at each time t, that we model with $p_\theta(\cdot, t)$.

2.2 Gaussian Mixture Diffusion Models

Gaussian mixture models provide a tractable and flexible approach to modelling probability distributions and have recently been studied in the diffusion framework [25,34]. In this work, we model the time-dependent distribution of images according to the PoGMDM specified in (1) and (2). Here, the closure property of Gaussians enables explicit diffusion under the SDE in (3) and has an analytical solution. Specifically, Zach et al. [34] showed that convolving $p_\theta(x,t)$ with $\mathcal{N}(0, 2tI)$ can be achieved by adapting the variances of the individual experts $\psi_k(\cdot, w, t)$ as

$$\sigma_k^2(t) = \sigma_0^2 + \xi_k^2 2t, \tag{6}$$

where ξ_k is the spectral magnitude of the convolutional filters implemented by K_k on their support, and $\sigma_0^2 > 0$ is the initially chosen variance of the experts. They further showed that for (6) to hold, the spectra of the filters implemented by the convolutional operator K_k must have a constant magnitude on their support and be non-overlapping. These assumptions are met by the nonseparable shearlets introduced in [17]. Here, the frequency plane is divided into non-overlapping cones that are specified by shearing and scaling parameters; see [16,17] for more details.

2.3 Inverse Problems and Diffusion Priors

Solving an inverse problem amounts to recovering an unknown signal x from a set of measurements $z = \mathcal{A}(x) + \varepsilon$, where $\mathcal{A}$ models the acquisition and ε summarizes additive measurement noise. The probabilistic treatment of the recovery problem amounts to constructing a posterior density $p_{X|Z}(\cdot, z)$ from the likelihood $p_{Z|X}$, which makes use of the known acquisition physics encoded in $\mathcal{A}$, and prior information about the reconstruction, encoded in p_X; see [8] for more details. The diffusion framework can account for this by conditioning the stochastic process (5) on the measurements, resulting in

$$d\bar{Y}_t = -2\nabla \log p_{Y|Z}(\bar{Y}_t, z, t)dt + \sqrt{2}d\bar{w}_t, \quad \bar{Y}_T = Y_T, \tag{7}$$

where $\log p_{Y|Z}(y_t, z, t) \propto \log p_{Z|Y}(y_t, z, t) + \log p_Y(y_t, t)$. The relationship between Y_t and Z is usually only known at $t = 0$, necessitating approximations for $t > 0$ [5,13,27].

3 Methods

The conditional reverse SDE (7) is valid for general, linear and nonlinear, inverse problems. In this section, we state our acquisition model for PI MRI and formulate our approximations. The resulting reconstruction algorithm is summarized in Algorithm 1.

3.1 Image Reconstruction Algorithm

We view the PI recovery problem as a non-linear inverse problem of jointly reconstructing the spin density $x \in \mathbb{C}^{n\times m}$ as well as the coil sensitivities $\sigma = (\sigma_1, \ldots, \sigma_c) \in \mathbb{C}^{n\times m\times c}$ from measured data $z = (z_1, \ldots, z_c) \in \mathbb{C}^{f\times c}$ where $f \in \mathbb{N}$

is the number of measured spatial frequencies. The relation between the data and the variables is given as

$$z = \mathcal{A}(x, \sigma) + \varepsilon = \begin{pmatrix} MF(\sigma_1 \odot x) \\ \vdots \\ MF(\sigma_c \odot x) \end{pmatrix} + \varepsilon \tag{8}$$

where $M : \mathbb{C}^{n \times m} \to \mathbb{C}^f$ is a binary sampling operator, $F : \mathbb{C}^{n \times m} \to \mathbb{C}^{n \times m}$ is the discrete Fourier transform and $\varepsilon \in \mathbb{C}^{f \times c}$ is additive Gaussian noise and $\odot$ denotes element-wise multiplication. Recovering (x, σ) from z is hard even in the fully sampled case [30]. To approach this, we introduce priors on the spin density and the sensitivities: Motivated by the bilinear form of (8), we view the recovery problem as a blind inverse problem [4,11] and utilize the approximation

$$p_{X,\Sigma|Z}(x_t, \sigma_t, z, t) \propto p_{Z|X,\Sigma}(x_t, \sigma_t, z, t) \cdot p_\theta(x_t, t) \cdot p_\Sigma(\sigma_t, t). \tag{9}$$

i.e. that the distribution of (X, Σ) factorizes. We choose a Gaussian likelihood of the form $p_{Z|X,\Sigma}(x_t, \sigma_t, z, t) \propto \exp\left(-\frac{1}{2}\|\mathcal{A}(x_t, \sigma_t) - z\|_2^2\right)$. This is a popular and simple choice originating from [13], but only correct for $t = 0$ in the diffusion process [7]. $p_\theta(\cdot, t)$ is the PoGMDM (1) trained on reference images, see the details in Sect. 3.2. Following [33] we chose, the classical smoothness prior $p_\Sigma(\sigma, 0) \propto \exp(-s(\sigma))$, with

$$s : \mathbb{C}^{n \times m \times c} \to \mathbb{R} : \sigma \mapsto \frac{1}{2} \sum_{i=1}^{c} \left(\|D_{\mathrm{D}}\mathrm{Re}(\sigma_i)\|_2^2 + \|D_{\mathrm{D}}\mathrm{Im}(\sigma_i)\|_2^2\right), \tag{10}$$

where $D_{\mathrm{D}} : \mathbb{R}^{n \times m} \to \mathbb{R}^{n \times m \times 2}$ is a forward finite differences operator with Dirichlet boundary conditions. We discretize the reverse SDE with an exponential schedule $\zeta(t) = \zeta_{\min}(\zeta_{\max}/\zeta_{\min})^t$ and use an adapted version of the predictor-corrector algorithm proposed in [9] for posterior sampling. The likelihood is incorporated by doing a gradient descent step in each iteration, where the gradient of the log-likelihood w.r.t. x_t is

$$\nabla_{x_t} \log p_{Z|X,\Sigma}(x_t, \sigma_t, y, t) = \sum_{i=1}^{c} \bar{\sigma}_{t,i} \odot (F^* M^*(MF(\sigma_{t,i} \odot x_t) - z_i)), \tag{11}$$

with $\bar{\sigma}_{t,i}$ being the complex conjugate sensitivity of the coil indexed by i. We found it sufficient to fix $p_\Sigma(\,\cdot\,, t) = p_\Sigma(\,\cdot\,, 0)$ for all $t > 0$, thus depart from the diffusion framework and perform proximal gradient descent steps on σ_t. The proximal map of $\log p_\Sigma$ has a closed-form solution

$$\mathrm{prox}_{\mu s}(\sigma) = \begin{pmatrix} Q_\mu(\mathrm{Re}(\sigma_1)) + iQ_\mu(\mathrm{Im}(\sigma_1)) \\ \vdots \\ Q_\mu(\mathrm{Re}(\sigma_c)) + iQ_\mu(\mathrm{Im}(\sigma_c)) \end{pmatrix}, \tag{12}$$

with $\mu \in \mathbb{R}_{>0}$ and $Q_\mu : \mathbb{R}^{n \times m} \to \mathbb{R}^{n \times m}; x \mapsto \mathcal{S}^*\left(\mathcal{S}(\mu x) \odot (\tau + \mu)^{-1}\right)$. Here, $\mathcal{S}$ is the two-dimensional discrete sine transform, and τ are the eigenvalues of the two-dimensional discrete Laplace operator [33]. The gradient of the log-likelihood w.r.t. σ_t is

$$\nabla_{\sigma_t} \log p_{Z|X,\Sigma}(x_t, \sigma_t, y, t) = \begin{pmatrix} x_t \odot (F^* M^* (MF(\sigma_{t,1} \odot x_t) - z_1)) \\ \vdots \\ x_t \odot (F^* M^* (MF(\sigma_{t,c} \odot x_t) - z_c)) \end{pmatrix}. \tag{13}$$

Algorithm 1: Joint Reconstruction Algorithm

Output: x^0, σ^0

$x^N \sim \mathcal{N}(0, \zeta_{max}^2 I)$
for $i \in [N-1, 0]$ **do**
  for $k \in [\mathrm{Re}, \mathrm{Im}]$ **do**
    $x^{k,i} = x^{k,i+1} + (\zeta_{i+1}^2 - \zeta_i^2)\nabla \log p_\theta(x^{k,i+1}, \zeta_{i+1}^2) + \sqrt{\zeta_{i+1}^2 - \zeta_i^2}\,\xi_i^k$
  $x^i = x^{\mathrm{Re},i} + \mathrm{i}x^{\mathrm{Im},i}$
  $x^i = x^i + \lambda \nabla_{x^i} \log p_{Z|X,\Sigma}(x^i, \sigma^{i+1}, z, t)$
  for $j \in [0, M]$ **do**
    for $k \in [\mathrm{Re}, \mathrm{Im}]$ **do**
      $x^{k,i} = x^{k,i} + \epsilon_i \nabla \log p_\theta(x^{k,i}, \zeta_i^2) + \sqrt{2\epsilon_i}\,\xi_j^k$
    $x^i = x^{\mathrm{Re},i} + \mathrm{i}x^{\mathrm{Im},i}$
    $x^i = x^i + \lambda \nabla_{x^i} \log p_{Z|X,\Sigma}(x^i, \sigma^{i+1}, z, t)$
  $\sigma^i = \mathrm{prox}_{\mu s}(\sigma^{i+1} - \mu \nabla_{\sigma^{i+1}} \log p_{Z|X,\Sigma}(x^i, \sigma^{i+1}, z))$

Similar to [7], we apply the prior, trained on magnitude images, to the real- and imaginary parts individually. We summarize the joint reconstruction in Algorithm 1, where ξ stores independent samples of a standard complex-normal distribution, $\lambda, \mu \in \mathbb{R}_{\geq 0}$ are hyperparameters balancing the data consistency and $\{\epsilon_i\}_{i=0}^{N-1}$ is a stepsize sequence.

3.2 Model Architecture and Implementational Details

We largely use the same model parameterization as in [34]: The convolution operators $\{K_k = \gamma_k \tilde{K}_k\}_{k=1}^{o}$ are nonseparable shearlets $\tilde{K}_k$ comprising two scales with five shearings for each of the two (horizontal and vertical) cones, yielding $o = 20$, each endowed with a learnable weight $\gamma_k \geq 0$. Following [2,34], we learn the construction blocks of the shearlet system, a one-dimensional low-pass-filter $h \in \mathbb{R}^9$ and a two-dimensional directional high-pass filter $P \in \mathbb{R}^{17\times 17}$. For each $k = 1, \ldots, o$, we chose $\psi_k : \mathbb{R} \times \triangle^L \times \mathbb{R}_{\geq 0} \to \mathbb{R}_{\geq 0}$ as an $L = 125$ component GMM whose weights w_k are constrained to the L-dimensional unit simplex $\triangle^L$. This results in $9 + 17^2 + o\lceil L/2 \rceil + o = 1578$ (as in [34] the number of learnable weights is half the number of components due to symmetry) learnable parameters that are optimized for (4) with projected AdaBelief [37] for 100 000 steps, where we implement the same constraints as in [34]. Furthermore, we use exponential moving average (EMA), with a momentum of 0.999.

For posterior sampling we choose $N = 1000$, $M = 1$, $\lambda = 1$, $\mu = 10$ and the noise schedule $\zeta(t) = \zeta_{\min}(\zeta_{\max}/\zeta_{\min})^t$ with $\zeta_{\min} = 0.01$ and $\zeta_{\max} = 10$. We calculate ϵ_i according to [28]. Furthermore, we find optimal hyperparameters for each sampling pattern with grid search. The coil sensitivities σ are initialized with the zero-filled coil images normalized by the initial root sum of squares (RSS) reconstruction. Finally, we accelerate sampling by using the algorithm proposed in [6] with vanilla initialization.

3.3 Experimental Data

We use the fastMRI knee dataset [15] for model training, hyper-parameter search, and evaluation. We use the central eleven slices of each scan in the coronal proton density (CORPD) training split, resulting in 5324 images of size 320 × 320. For the hyper-parameter search and testing, we divide the validation split into 30 validation files and 58 test files, excluding k-space data with width different from 368 and 372 for simplicity of implementation, again taking the central eleven slices. For out of distribution (OOD) experiments we use the coronal proton density fat suppressed (CORPDFS) validation dataset, again excluding based on the width and using the eleven central slices. Training images are normalized to a maximum of 1 via $x \mapsto x/\|x\|_\infty$.

3.4 Comparison and Evaluation

Our evaluation is twofold: Firstly, we conduct a synthetic single coil experiment, where we retrospectively sample the k-space data of CORPD reference RSS images $x \in \mathbb{R}^{320\times 320}$. Secondly, a PI experiments with joint reconstruction on the CORPD/FS datasets. In both experiments, we compare against the Charbonnier smoothed isotropic total variation (TV) as a classical variational penalty in the joint nonlinear inversion algorithm proposed in [33]. ScoreMRI [7] (single coil), the fastMRI baseline [35] (single coil), and the end-to-end variational network (VN) [29] (PI) serve as state-of-the-art references, where the latter two primarily serve as examples of possible pitfalls in signal recovery without proper Bayesian separation. We did not include ScoreMRI in the PI experiments as it can only handle fixed-size data (320 × 320).

We estimate the minimum mean squared error (MMSE) by averaging ten posterior samples and, similar to [18], find the maximum a posteriori (MAP) by performing 250 accelerated gradient descent [21] steps with a fixed σ and a step size of 0.001 after the full reverse diffusion. For quantitative comparison, we calculate the peak signal-to-noise ratio (PSNR), structural similarity (SSIM), and normalized mean squared error (NMSE) with respect to the fully-samples RSS reconstruction. To quantitatively compare the reconstruction to other RSS reconstructions, we weigh the reconstruction by the RSS of the coils as described in [30], and we follow [33] and fit a spline curve to match the reconstructed and reference intensities.

4 Results

We demonstrate the interpretability of the image prior by showing the learned filters $\tilde{K}_k$, weights w_k, and potentials $-\log \psi_k(\cdot, w_k, t)$ for different t in Fig. 1. Due to the overcompleteness of the model, the potentials significantly differ from the leptokurtic filter responses observed in [12], instead showing multiple minima different from zero, allowing the enhancement of certain image structures. This matches the observations from [3,34,36].

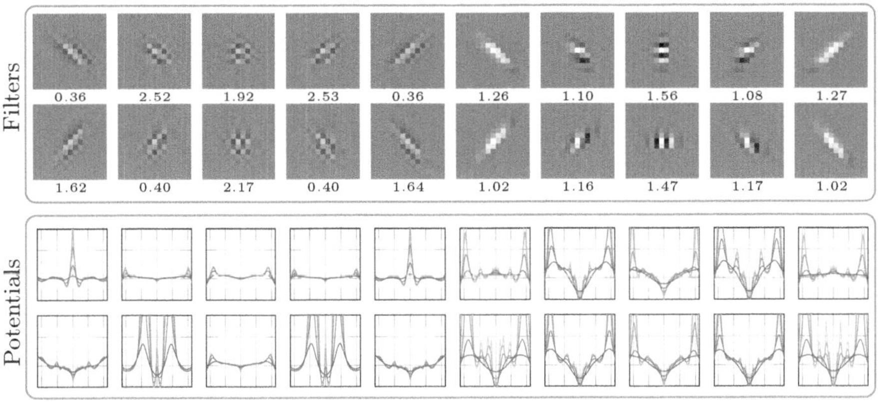

Fig. 1. The learned overcomplete model with Gaussian mixture experts. Below each filter the weight γ_k is shown. The first row shows the vertical cone and the second row is the horizontal cone of the shearlet system. The first five entries correspond to the shearing for the first scale and the second five for the second scale. The colours indicate the diffusion time $\sqrt{2t} = 0$ ___, 0.025 ___, 0.05 ___, 0.1 ___, 0.2 ___; .

4.1 Reconstruction

Table 1 shows the quantitative results for the synthetic single coil experiment. As expected from the large, resource-intensive model, ScoreMRI [7] performs best across all sampling trajectories. The performance of the UNet, which does not follow a strictly Bayesian separation, significantly degrades for sampling trajectories different from the training setup (Cartesian with an acceleration of four). In contrast to other learning-based methods, our model has relatively few learnable parameters while achieving satisfactory results, outperforming TV across all sampling trajectories, except for the radial pattern, where it performs equally well.

The quantitative results for the joint nonlinear inversion in Table 2 demonstrate that our method outperforms TV on CORPD data on all sampling patterns except for Gaussian, where the performance is comparable. On CORPDFS data, our method is second to TV for most sampling trajectories, which we

Table 1. Quantitative results for the single-coil experiment. Bold numbers represent the best results, and underlined numbers are the second-best. The NMSE is scaled by 10^2.

Pattern	A	Metric	ZF	TV	UNet	ScoreMRI	Ours (MMSE)
Cartesian	4	PSNR	24.03	30.2	33.85	**34.74**	31.66
		SSIM	0.69	0.81	**0.87**	**0.87**	0.83
		NMSE	6.32	1.23	0.48	**0.39**	0.84
Spiral	≈ 5		21.15	31.32	27.81	**35.67**	32.99
			0.63	0.84	0.78	**0.88**	0.86
			10.61	0.91	1.91	**0.31**	0.63
Radial	≈ 6		24.6	31.34	29.85	**33.62**	31.34
			0.68	0.82	0.79	**0.84**	0.82
			5.33	0.89	1.19	**0.50**	0.89
2D Gaussian	8		21.77	30.40	24.34	**34.22**	33.19
			0.65	0.85	0.73	0.85	**0.86**
			7.90	1.38	4.35	**0.44**	0.64
Learnable parameters			–	–	4.9×10^8	6.7×10^7	1578

A: Acceleration

Table 2. Quantitative results for PI experiments with different sampling trajectories. The rows in each pattern alternate between PSNR, SSIM and NMSE. The NMSE is scaled by 10^2.

Pattern	A	ACL	In-distribution (CORPD)					Out-of-distribution (CORPDFS)				
			ZF	TV	VN	Ours		ZF	TV	VN	Ours	
						MAP	MMSE				MAP	MMSE
C	4	8%	27.18	33.07	**36.92**	33.22	33.15	26.24	**31.42**	30.05	31.30	31.33
			0.73	0.83	**0.91**	0.86	0.86	0.67	0.73	**0.76**	0.74	0.74
			2.23	0.56	**0.24**	0.61	0.62	5.17	**1.56**	2.38	1.63	1.61
		8%[1]	31.12	33.65	24.72	**35.38**	35.46	26.52	32.00	28.65	**32.04**	31.91
			0.81	0.83	0.67	0.89	**0.90**	0.70	0.74	0.70	**0.77**	0.76
			0.93	0.50	4.01	**0.33**	**0.33**	5.09	**1.40**	2.91	**1.40**	1.43
R	11	–	28.76	33.21	20.55	33.63	**33.65**	25.11	**31.50**	26.38	31.22	31.16
			0.74	0.82	0.69	**0.85**	**0.85**	0.61	0.72	0.68	**0.73**	0.72
			1.56	0.55	10.12	**0.50**	**0.50**	7.13	**1.54**	4.90	1.65	1.66
G	8	–	32.14	**34.26**	23.52	34.07	34.00	26.66	**31.99**	28.20	31.74	31.58
			0.83	0.85	0.72	**0.87**	**0.87**	0.69	0.74	0.71	**0.75**	0.74
			0.73	**0.43**	5.20	0.45	0.46	5.40	**1.38**	3.33	1.48	1.52
Learnable params.			–	–	3×10^7	1578		–	–	3×10^7	1578	

[1] horizontal; C: Cartesian, R: Radial, G: 2D Gaussian, A: Acceleration

believe is due to the noise in the reference images, impairing quantitative comparison. This is supported by the qualitative results in Fig. 2, where the error maps in the insets show that TV introduces errors in the anatomy, e.g. removing the cartilage from the tibia in the CORPDFS reconstruction, or the blood vessels in the lateral region. In contrast, our reconstruction is able to retain these features while delivering an artifact-free reconstruction. As in the single coil case, the end-to-end VN only performs well in its training configuration.

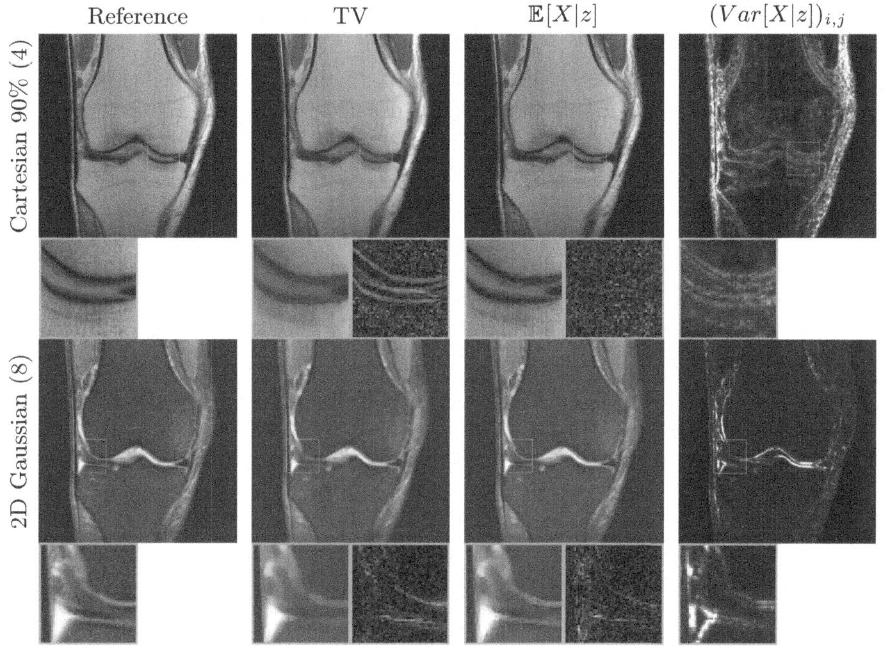

Fig. 2. Qualitative comparison for PI against TV. The first row shows results on ID data, the second on OOD data. The zoom shows an image region and the corresponding absolute error and the pixel-wise variance in the right column (0 ▭ 0.25).

Without adding any extra computational cost to the MMSE estimate, the diffusion framework provides valuable insights into reconstruction uncertainty by delivering pixel-wise marginal variances, as shown in 2. Reconstructing an image of size 640×372 with 16 coils takes twenty seconds with our algorithm. In contrast, ScoreMRI [7] takes approximately five minutes to reconstruct an image of size 320×320 from single-coil data on the same hardware.

Finally, we compare the estimated sensitivity maps qualitatively to ESPIRiT [31] by showing the RSS null-space residuals in Fig. 3, where any residual signal points to a suboptimal estimation. We observe that [31] produces slightly better estimates than our method when more ACLs are available but falls off as the calibration region gets smaller. In contrast, our method still gives a robust estimate of the sensitivities in all cases.

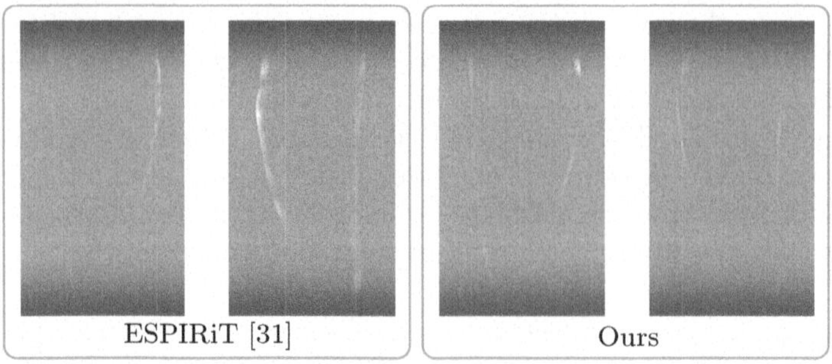

Fig. 3. RSS nullspace residuals for Cartesian subsampling. In each block, the measured data has 8% ACLs (left) and 4% ACLs (right) respectively.

5 Conclusion

In this paper, we pair an analytically tractable product of Gaussian mixture diffusion model with classic smoothness penalties to tackle the non-linear MRI reconstruction problem in the diffusion framework. The proposed method delivers satisfactory results on single- and multi-coil data, allows for fast inference, and demonstrates robustness to changes in the forward model and contrast. A drawback of the algorithm is that it requires the tuning of many hyperparameters. Interesting directions for future work include developing a diffusion prior for $p_{X,\Sigma}$, not assuming a factorization, consequently aligning the method better with recent algorithms for posterior sampling with diffusion models. Additionally, the proposed method could be extended to other inverse problems such as blind deconvolution by adapting the forward operator and p_Σ appropriately.

References

1. Block, K.T., Uecker, M., Frahm, J.: Undersampled radial MRI with multiple coils. iterative image reconstruction using a total variation constraint. Magn. Reson. Med. **57**(6), 1086–1098 (2007)
2. Bogensperger, L., Chambolle, A., Pock, T.: Convergence of a piggyback-style method for the differentiation of solutions of standard saddle-point problems. SIAM J. Math. Data Sci. **4**(3), 1003–1030 (2022)
3. Chen, Y., Pock, T.: Trainable nonlinear reaction diffusion: a flexible framework for fast and effective image restoration. IEEE Trans. Pattern Anal. Mach. Intell. **39**(6), 1256–1272 (2017)
4. Chung, H., Kim, J., Kim, S., Ye, J.C.: Parallel diffusion models of operator and image for blind inverse problems. In: IEEE/CVF Conference on Computer Vision and Pattern Recognition (2023)
5. Chung, H., Kim, J., Mccann, M.T., Klasky, M.L., Ye, J.C.: Diffusion posterior sampling for general noisy inverse problems. In: International Conference on Learning Representations (2023)

6. Chung, H., Sim, B., Ye, J.C.: Come-closer-diffuse-faster: accelerating conditional diffusion models for inverse problems through stochastic contraction. In: 2022 IEEE/CVF Conference on Computer Vision and Pattern Recognition (CVPR), pp. 12403–12412 (2021)
7. Chung, H., Ye, J.C.: Score-based diffusion models for accelerated MRI. Med. Image Anal. **80**, 102479 (2022)
8. Dashti, M., Stuart, A.M.: The bayesian approach to inverse problems. In: Ghanem, R., Higdon, D., Owhadi, H. (eds.) Handbook of Uncertainty Quantification, pp. 311–428. Springer, Cham (2017). https://doi.org/10.1007/978-3-319-12385-1_7
9. Erlacher, M., Zach, M.: Joint non-linear mri inversion with diffusion priors (2023). https://arxiv.org/abs/2310.14842
10. Griswold, M.A., et al.: Generalized autocalibrating partially parallel acquisitions (grappa). Magn. Reson. Med. **47**(6), 1202–1210 (2002)
11. Hu, Y., Peng, A., Gan, W., Kamilov, U.S.: Adobi: adaptive diffusion bridge for blind inverse problems with application to mri reconstruction (2024)
12. Huang, J., Mumford, D.: Statistics of natural images and models. In: Proceedings. 1999 IEEE Computer Society Conference on Computer Vision and Pattern Recognition (Cat. No PR00149), vol. 1, pp. 541–547 (1999)
13. Jalal, A., Arvinte, M., Daras, G., Price, E., Dimakis, A.G., Tamir, J.: Robust compressed sensing MRI with deep generative priors. Adv. Neural. Inf. Process. Syst. **34**, 14938–14954 (2021)
14. Knoll, F., Bredies, K., Pock, T., Stollberger, R.: Second order total generalized variation (TGV) for MRI. Magn. Reson. Med. **65**(2), 480–491 (2011)
15. Knoll, F., et al.: fastmri: a publicly available raw k-space and dicom dataset of knee images for accelerated mr image reconstruction using machine learning. Radiol. Artif. Intell. **2**, e190007 (2020)
16. Kutyniok, G., Labate, D.: Introduction to Shearlets, pp. 1–38. Birkhäuser Boston (2012)
17. Lim, W.Q.: Nonseparable shearlet transform. IEEE Trans. Image Process. **22**(5), 2056–2065 (2013). https://doi.org/10.1109/TIP.2013.2244223
18. Luo, G., Heide, M., Uecker, M.: Mri reconstruction via data driven markov chain with joint uncertainty estimation. arXiv preprint arXiv:2202.01479 (2022)
19. Lustig, M., Donoho, D., Pauly, J.M.: Sparse MRI: the application of compressed sensing for rapid MR imaging. Magn. Reson. Med. **58**(6), 1182–1195 (2007)
20. Narnhofer, D., Hammernik, K., Knoll, F., Pock, T.: Inverse gans for accelerated MRI reconstruction. In: Wavelets and Sparsity XVIII (2019), sPIE Optics+Photonics 2019, 23–27 Aug 2019 (2019)
21. Nesterov, Y.: A method for solving the convex programming problem with convergence rate $o(1/k^2)$. Proc. USSR Acad. Sci. **269**, 543–547 (1983)
22. Pruessmann, K.P., Weiger, M., Scheidegger, M.B., Boesiger, P.: Sense: sensitivity encoding for fast MRI. Magn. Reson. Med. **42**(5), 952–962 (1999)
23. Risken, H.: The Fokker-Planck Equation: Methods of Solution and Applications. Springer, Heidelberg (1996)
24. Salimans, T., Ho, J.: Should EBMs model the energy or the score? In: Energy Based Models Workshop - ICLR 2021 (2021)
25. Shah, K., Chen, S., Klivans, A.R.: Learning mixtures of gaussians using the ddpm objective (2023)
26. Song, Y., Ermon, S.: Generative modeling by estimating gradients of the data distribution. Adv. Neural Inf. Process. Syst., 11895–11907 (2019)

27. Song, Y., Shen, L., Xing, L., Ermon, S.: Solving inverse problems in medical imaging with score-based generative models. In: International Conference on Learning Representations (2022)
28. Song, Y., Sohl-Dickstein, J., Kingma, D.P., Kumar, A., Ermon, S., Poole, B.: Score-based generative modeling through stochastic differential equations. In: International Conference on Learning Representations (2021)
29. Sriram, A., et al.: End-to-End variational networks for accelerated MRI reconstruction. In: Martel, A.L., et al. (eds.) MICCAI 2020. LNCS, vol. 12262, pp. 64–73. Springer, Cham (2020). https://doi.org/10.1007/978-3-030-59713-9_7
30. Uecker, M., Hohage, T., Block, K.T., Frahm, J.: Image reconstruction by regularized nonlinear inversion-joint estimation of coil sensitivities and image content. Magn. Reson. Med. **60**(3), 674–682 (2008)
31. Uecker, M., et al.: Espirit an eigenvalue approach to autocalibrating parallel MRI: where sense meets grappa. Magn. Reson. Med. **71**(3), 990–1001 (2014)
32. Vincent, P.: A connection between score matching and denoising autoencoders. Neural Comput. **23**, 1661–1674 (2011)
33. Zach, M., Knoll, F., Pock, T.: Stable deep MRI reconstruction using generative priors. IEEE Trans. Med. Imaging (2023)
34. Zach, M., Kobler, E., Chambolle, A., Pock, T.: Product of gaussian mixture diffusion models. J. Math. Imaging Vision **66**, 1–25 (2024)
35. Zbontar, J., et al.: fastmri: an open dataset and benchmarks for accelerated MRI (2019)
36. Zhu, S., Mumford, D.: Prior learning and Gibbs reaction-diffusion. IEEE Trans. Pattern Anal. Mach. Intell. **19**, 1236–1250 (1997)
37. Zhuang, J., et al.: Adabelief optimizer: adapting stepsizes by the belief in observed gradients. In: Conference on Neural Information Processing Systems (2020)

Whiteness-Based Bilevel Estimation of Weighted TV Parameter Maps for Image Denoising

Monica Pragliola[1(✉)], Luca Calatroni[2], and Alessandro Lanza[3]

[1] Department of Mathematics and Applications, University of Naples Federico II, Naples, Italy
monica.pragliola@unina.it

[2] MaLGa Center, DIBRIS, Universitá di Genova, Istituto Italiano di Tecnologia, Genoa, Italy
luca.calatroni@unige.it

[3] Department of Mathematics, University of Bologna, Bologna, Italy
alessandro.lanza2@unibo.it

Abstract. We consider a bilevel optimisation strategy based on normalised residual whiteness loss for estimating the weighted total variation parameter maps for denoising images corrupted by additive white Gaussian noise. Compared to supervised and semi-supervised approaches relying on prior knowledge of (approximate) reference data and/or information on the noise magnitude, the proposal is fully unsupervised. To avoid noise overfitting an early stopping strategy is used, relying on simple statistics of optimal performances on a set of natural images. Numerical results comparing the supervised/unsupervised procedures for scalar/pixel-dependent parameter maps are shown.

Keywords: Adaptive regularisation · Total variation image denoising · Unsupervised bilevel optimisation · Residual whiteness

1 Introduction

Variational methods are a popular paradigm to overcome the difficulties related to inverse imaging problems: they seek an estimate of the unknown image by minimizing a functional encoding information on the noise and prior beliefs on the image. For image denoising, denoting by $\mathbf{y} \in \mathbb{R}^n$ the (vectorised) data corrupted by Additive White Gaussian Noise (AWGN), variational methods solve

$$\mathbf{x}^*(\boldsymbol{\lambda}) \in \arg\min_{\mathbf{x}\in\mathbb{R}^n} \left\{ \frac{1}{2}\|\mathbf{x}-\mathbf{y}\|_2^2 + R(\mathbf{x};\boldsymbol{\lambda}) \right\},$$

where $\mathbf{x}^*$ depends on a certain number of parameters $\boldsymbol{\lambda} \in \Lambda$ possibly involved in the so-called regularisation term R. In general, flexible models are characterised by a large number of parameters for which robust selection strategies need to be

T. A. Bubba et al. (Eds.): SSVM 2025, LNCS 15667, pp. 159–172, 2025.
https://doi.org/10.1007/978-3-031-92366-1_13

introduced. How to properly select parameters in variational models is an issue that has attracted the interest of many researchers working on inverse problems.

Strategies for the selection of $\boldsymbol{\lambda}$ maps in variational models can be divided into supervised, semi-supervised and unsupervised. Supervised methods rely on the existence of datasets of groundtruth and corrupted data that allow to customise the regularisation on a specific class of images [12]. Semi-supervised methods exploit the availability of partial information, such as the noise statistics - see, e.g. the Discrepancy Principle (DP) [16] or its bilevel counterpart [9], and the Stein's Unbiased Risk Estimator (SURE), all requiring to know the noise level. Unsupervised methods solely rely on quantities depending on the data $\mathbf{y}$. Empirical approaches for selecting a single regularisation parameter $\boldsymbol{\lambda} = \lambda$, such as the L-curve [4] or the Generalised Cross Validation [6], belong to this class, which also includes strategies detecting a general number of parameters - see the recent Unknown Noise level SURE (UNSURE) [21] and references therein.

Among the unsupervised approaches, in the last years attention has been given to methods observing that a good reconstruction is such that the residual image $\boldsymbol{r}^*(\boldsymbol{\lambda}) := \mathbf{y} - \mathbf{x}^*(\boldsymbol{\lambda})$ is close to the noise realisation degrading the data in terms of its spectrum; this can be done, e.g., enforcing whiteness of the residual, whenever applicable [1]. In [13], the authors introduced the Residual Whiteness Principle (RWP) in presence of AWGN for setting a scalar regularisation parameter λ. In [20], for the same problem, the RWP has been formulated as a bilevel task, where the upper loss measures the whiteness of the residual, and compared to other supervised or semi-supervised losses. Recently, a general bilevel formulation of the RWP, called Generalised Whiteness Principle (GWP), applicable for the estimation of a general number of parameters under the corruption of non white but whitenable noises, has been proposed [2]. Nonetheless, the GWP has been only applied to set a small number of parameters and the bilevel task is solved approximately due to the presence of non-smooth lower cost functionals.

In this work, our goal is to extend the bilevel framework introduced in [20] for the estimation of a single parameter λ to the case in which the variational model involves a large number of unknown parameters $\boldsymbol{\lambda}$. We are exploring the potential of the whiteness loss while mitigating some of the drawbacks that the GWP formulated in [2] brings along when applied to such challenging scenario. Here, our interest is focused on the selection of the parameters defining a space-variant regularisation term [18] for the denoising problem under the action of AWGN; more specifically, we are considering the following variational model

$$\mathbf{x}^*(\boldsymbol{\lambda}) \in \arg\min_{\mathbf{x}\in\mathbb{R}^n} \left\{ \frac{1}{2}\|\mathbf{x}-\mathbf{y}\|_2^2 + \mathrm{WTV}(\mathbf{x};\boldsymbol{\lambda}) \right\}, \quad \mathrm{WTV}(\mathbf{x};\boldsymbol{\lambda}) = \sum_{i=1}^{n} \lambda_i \|(\mathbf{D}\mathbf{x})_i\|_2, \quad (1)$$

where the Weighted Total Variation (WTV) term is a weighted version of the TV regulariser [19], and allows to tune the amount of regularisation pixelwise. Numerical tests show that the whiteness principle returns informative $\boldsymbol{\lambda}$-maps, and suggest that its employment in highly-parametrised settings could be beneficial.

2 Bilevel Estimation of WTV Parameter Maps

We consider a bilevel optimisation approach to compute the optimal parameters $\widehat{\boldsymbol{\lambda}}$ for the WTV regulariser in (1) given a noisy image $\mathbf{y} \in \mathbb{R}^n$, i.e. we consider

$$\widehat{\boldsymbol{\lambda}} \in \underset{\boldsymbol{\lambda} \in \Lambda}{\arg\min}\ \mathcal{Q}(\mathbf{x}^*(\boldsymbol{\lambda})) := \frac{1}{2}\|\boldsymbol{\rho}(\mathbf{x}^*(\boldsymbol{\lambda}))\|_2^2 \tag{2}$$

$$\text{s.t.}\quad \mathbf{x}^*(\boldsymbol{\lambda}) \in \underset{\mathbf{x} \in \mathbb{R}^n}{\arg\min}\ \frac{1}{2}\|\mathbf{x} - \mathbf{y}\|_2^2 + \mathrm{WTV}(\mathbf{x}; \boldsymbol{\lambda}), \tag{3}$$

where $\mathcal{Q}$ is a metric assessing the quality of the reconstruction $\mathbf{x}^*(\boldsymbol{\lambda})$ depending on either the availability of reference data and/or some knowledge on the noise/problem statistics, with $\boldsymbol{\rho}(\cdot)$ being chosen accordingly. Note that although several approximation techniques have been considered to address bilevel problems with non-smooth lower-level constraints (see, e.g., [17]), we restrict in the following to the more simple case where in (2)–(3) smoothness is assumed for the lower-level variational constraint, so as to apply implicit differentiation w.r.t. $\boldsymbol{\lambda}$ by means of the implicit function theorem, see, e.g. [10,12]. We will thus consider a smoothed version of (1) depending on a smoothing parameter $0 < \epsilon \ll 1$,

$$\mathrm{WTV}_\epsilon(\mathbf{x}; \boldsymbol{\lambda}) = \sum_{i=1}^{n} \lambda_i\, h_\epsilon((\mathbf{D}\mathbf{x})_i)\,, \quad h_\epsilon(\boldsymbol{v}) = \begin{cases} \frac{3}{4\epsilon}\|\boldsymbol{v}\|_2^2 - \frac{1}{8\epsilon^3}\|\boldsymbol{v}\|_2^4 & \text{if } \|\boldsymbol{v}\|_2 < \epsilon \\ \|\boldsymbol{v}\|_2 - \frac{3\epsilon}{8} & \text{if } \|\boldsymbol{v}\|_2 \geq \epsilon \end{cases},$$

with $h_\epsilon : \mathbb{R}^2 \to \mathbb{R}_{\geq 0}$ a C^2-Huber smoothing function. The smoothed functional

$$F_\epsilon(\mathbf{x}; \mathbf{y}, \boldsymbol{\lambda}) := \frac{1}{2}\|\mathbf{x} - \mathbf{y}\|_2^2 + \mathrm{WTV}_\epsilon(\mathbf{x}; \boldsymbol{\lambda}) \tag{4}$$

is thus C^2 and 1-strongly convex. We report in the following its main properties whose proof follows easily from [20, Proposition III.1]. In the following proposition, $\odot$ denotes element-wise multiplication while $\boldsymbol{\Lambda} := [\boldsymbol{\lambda}\,;\boldsymbol{\lambda}] \in \mathbb{R}^{2n}_{>0}$.

Proposition 1. *The functional F_ϵ defined in (4) is twice continuously differentiable and 1-strongly convex on $\mathbb{R}^n$, hence it admits a unique minimiser. Its gradient $\boldsymbol{\nabla}_{\mathbf{x}} F_\epsilon \in \mathbb{R}^n$ and Hessian $\boldsymbol{\nabla}^2_{\mathbf{x}} F_\epsilon \in \mathbb{R}^{n \times n}$ are given by*

$$\begin{aligned} \boldsymbol{\nabla}_{\mathbf{x}}\, F_\epsilon(\mathbf{x}; \mathbf{y}, \boldsymbol{\lambda}) &= (\mathbf{x} - \mathbf{y}) + \boldsymbol{\lambda} \odot\ \mathbf{D}^{\mathrm{T}}\, \boldsymbol{\nabla}_{\mathbf{x}} H_\epsilon(\mathbf{D}\mathbf{x}), \\ \boldsymbol{\nabla}^2_{\mathbf{x}}\, F_\epsilon(\mathbf{x}; \mathbf{y}, \boldsymbol{\lambda}) &= \mathbf{I}_n + \mathbf{D}^{\mathrm{T}}\, \mathrm{diag}(\boldsymbol{\Lambda}) \boldsymbol{\nabla}^2 H_\epsilon(\mathbf{D}\mathbf{x})\, \mathbf{D}, \end{aligned} \tag{5}$$

where $\boldsymbol{\nabla} H_\epsilon$ and $\boldsymbol{\nabla}^2 H_\epsilon$ denote the vector (respectively, matrix) of first-(respectively, second-)order derivatives of $H_\epsilon(\boldsymbol{z}) := \sum_{i=1}^n h_\epsilon(z_i)$ w.r.t. $\boldsymbol{z} = \mathbf{D}\mathbf{x}$ and $\mathrm{diag}(\boldsymbol{\Lambda}) \in \mathbb{R}^{2n \times 2n}_{>0}$ is the diagonal matrix having $\boldsymbol{\Lambda}$ on the diagonal. By defining $\lambda_{\mathrm{MAX}} := \|\boldsymbol{\lambda}\|_\infty$, the gradient $\boldsymbol{\nabla}_{\mathbf{x}} F_\epsilon$ is thus $L_{\lambda_{\mathrm{MAX}},\epsilon}$-Lipschitz continuous, and

$$L_{\lambda_{\mathrm{MAX}},\epsilon} = \max_{\mathbf{x} \in \mathbb{R}^n} \left\|\boldsymbol{\nabla}^2_{\mathbf{x}} F_\epsilon(\mathbf{x}; \mathbf{y}, \boldsymbol{\lambda})\right\|_2 \leq 1 + \frac{12\,\lambda_{\mathrm{MAX}}}{\epsilon} =: \overline{L}_{\lambda_{\mathrm{MAX}},\epsilon}. \tag{6}$$

In the context of supervised approaches, the upper loss may be chosen as (half) the mean squared error (MSE) between the reference image $\overline{\mathbf{x}}$ and the reconstruction $\mathbf{x}^*(\boldsymbol{\lambda})$ [7,12], so as to maximise the Signal-to-Noise-Ratio (SNR):

$$\mathcal{Q}_{\text{MSE}}(\mathbf{x}^*(\boldsymbol{\lambda});\overline{\mathbf{x}}) = \frac{1}{2}\|\boldsymbol{\rho}(\mathbf{x}^*(\boldsymbol{\lambda}))\|_2^2, \quad \boldsymbol{\rho}(\mathbf{x}^*(\boldsymbol{\lambda})) := \frac{1}{\sqrt{n}}(\mathbf{x}^*(\boldsymbol{\lambda}) - \overline{\mathbf{x}}). \tag{7}$$

In our settings, the MSE loss is employed for the computation of benchmark results but its adoption is unfeasible in practice. Assuming that no information is available on the noise level and on the reference image, a natural choice is to consider the bilevel counterpart of the RWP, formally introduced in [20], that amounts to maximise the whiteness of the residual image $\boldsymbol{r}^*(\boldsymbol{\lambda})$ via minimisation of (half) the squared norm of its *sample normalised auto-correlation* $\boldsymbol{\gamma}(\boldsymbol{r}^*(\boldsymbol{\lambda}))$:

$$\mathcal{Q}_{\text{white}}(\mathbf{x}^*(\boldsymbol{\lambda});\mathbf{y}) = \frac{1}{2}\|\boldsymbol{\rho}(\mathbf{x}^*(\boldsymbol{\lambda}))\|_2^2, \ \boldsymbol{\rho}(\mathbf{x}^*(\boldsymbol{\lambda})) := \boldsymbol{\gamma}(\boldsymbol{r}^*(\boldsymbol{\lambda})) = \frac{\boldsymbol{r}^*(\boldsymbol{\lambda}) \circledast \boldsymbol{r}^*(\boldsymbol{\lambda})}{\|\boldsymbol{r}^*(\boldsymbol{\lambda})\|_2^2}, \tag{8}$$

where the normalisation term eliminates the dependence on σ, and where $\mathbf{x}_1 \circledast \mathbf{x}_2$ denotes the discrete circular cross-correlation between the matrices $\boldsymbol{X}_1, \boldsymbol{X}_2 \in \mathbb{R}^{n_1 \times n_2}$ such that $\text{vec}(\boldsymbol{X}_d) = \mathbf{x}_d \in \mathbb{R}^n$, $d = 1, 2$ and $n_1 n_2 = n$, which is defined for $(j_1, j_2) \in \{0, \ldots, n_1 - 1\} \times \{0, \ldots, n_2 - 1\}$ by

$$(\boldsymbol{X}_1 \circledast \boldsymbol{X}_2)_{j_1,j_2} = \sum_{k_1=0}^{n_1-1} \sum_{k_2=0}^{n_2-1} (\boldsymbol{X}_1)_{k_1,k_2} (\boldsymbol{X}_2)_{(j_1+k_1) \text{mod } n_1, (j_2+k_2) \text{mod } n_2}.$$

The sample normalised auto-correlation $\boldsymbol{\gamma}$ in (8) is clearly not defined for a null residual $\boldsymbol{r}^*(\boldsymbol{\lambda}) = \mathbf{0}_n$ which, however, is obtained only in the limit/degenerate case $\boldsymbol{\lambda} = \mathbf{0}_n$, of no practical interest. Moreover, $\boldsymbol{\gamma}$ in (8) has the properties [13]:

(P_1) $\gamma_{0,0} = 1, \ \gamma_{j_1,j_2} \in [-1, 1] \ \forall (j_1, j_2) \neq (0, 0)$;
(P_2) $\lim_{n \to \infty} \gamma_{j_1,j_2} = 0 \ \forall (j_1, j_2) \neq (0, 0)$ if $\boldsymbol{r}^*(\boldsymbol{\lambda})$ is a realisation of AWGN.

It follows from (P_1) that the metric $\mathcal{Q}_{\text{white}}$ in (8) has a positive lower bound, $\mathcal{Q}_{\text{white}}(\mathbf{x}^*(\boldsymbol{\lambda}); \mathbf{y}) \geq 1/2$. The asymptotic property (P_2), however, does not imply that the lower bound $1/2$ is reached when solving (even exactly) the bilevel problem (2)–(3) with $\mathcal{Q} = \mathcal{Q}_{\text{white}}$ and a generic lower-level optimisation model, for two reasons. Firstly, the space of all possible solutions $\mathbf{x}^*(\boldsymbol{\lambda})$ to the lower problem, parameterised by $\boldsymbol{\lambda}$, rarely contains the perfect reconstruction $\mathbf{x}^*(\boldsymbol{\lambda}) = \overline{\mathbf{x}}$, and it is very unlikely that the space of the associated residual images $\boldsymbol{r}^*(\boldsymbol{\lambda})$ contains the AWGN realisation actually corrupting $\overline{\mathbf{x}}$. Roughly, the probability that the $\boldsymbol{r}^*(\boldsymbol{\lambda})$-space contains the actual AWGN realisation is higher the greater the dimensionality of $\boldsymbol{\lambda}$. Secondly, even when the estimated residual image $\boldsymbol{r}^*(\boldsymbol{\lambda})$ is a realisation of AWGN - that is $\mathbf{x}^*(\boldsymbol{\lambda}) = \overline{\mathbf{x}}$ - and even when the size $n = n_1 n_2$ of the image is very large - so that (P_2) can be assumed to "almost hold true" - $\mathcal{Q}_{\text{white}}$ does not even approach the lower bound $1/2$. This is the effect, when increasing the size n, of the simultaneous decrease of the values $|\gamma_{j_1,j_2}|$ and growth of the number n of γ_{j_1,j_2} terms normed in (8). The results of some Montecarlo simulations (not reported for shortness) suggest that the $\mathcal{Q}_{\text{white}}$ metric of realisations of white Gaussian random noises assume values between 0.9 and 1.1. When the $\mathcal{Q}_{\text{white}}$ metric has been coupled with lower optimisation models with few free parameters (one or two), the solution of the bilevel problem was characterised by $\mathcal{Q}_{\text{white}}$ values slightly greater than 1, and the reconstruction $\mathbf{x}^*(\boldsymbol{\lambda})$ was of good quality (see, e.g. Figs. 6.2,6.4 in [13] and Figs. 4,7 in [2]).

However, this is the first time that the $\mathcal{Q}_{\text{white}}$ upper loss is coupled to a highly-parameterised lower model, with the number of free parameters equal to the number of pixels in the image to reconstruct. Although on one hand we can expect that the (very large) solution space $\mathbf{x}^*(\boldsymbol{\lambda})$ of the lower model contains images very close to the true one $\overline{\mathbf{x}}$ - hence, associated residual images very close to the AWGN realisation - on the other hand there is a real risk that the space contains also bad-quality reconstructions associated to $\mathcal{Q}_{\text{white}}$ values less than the range $0.9 - 1.1$ characterising white noise realisations. In Sect. 3.3, especially for the latter issue, we will show evidences of the expected behavior.

3 Optimisation Details

In this section, we detail the optimisation algorithms used to solve both the lower- and the upper-level problems in (2)–(3). First, we enforce strict positivity of the desired regularisation map by the (element-wise) change of variables $\boldsymbol{\lambda} = \exp(\boldsymbol{\beta})$ for $\boldsymbol{\beta} \in \mathbb{R}^n$, thus considering the optimisation over the unconstrained variable $\boldsymbol{\beta}$ so that $\mathcal{Q}(\mathbf{x}^*(\boldsymbol{\lambda})) = \mathcal{Q}(\mathbf{x}^*(\exp(\boldsymbol{\beta})))$. Minimising the upper level term in (2) requires the computation of the gradient of the loss functional w.r.t. $\boldsymbol{\beta} \in \mathbb{R}^n$, which we perform by the chain rule and implicit differentiation as follows:

$$
\begin{aligned}
\boldsymbol{\nabla}_{\beta}\, \mathcal{Q}(\mathbf{x}^*) &= (\boldsymbol{J}_{\beta}\, \mathbf{x}^*)^T\, \boldsymbol{\nabla}_{\mathbf{x}} \mathcal{Q}(\mathbf{x}^*) \\
&= -\boldsymbol{J}_{\beta}\, \exp(\boldsymbol{\beta}) \boldsymbol{J}_{\boldsymbol{\lambda}}\, \boldsymbol{\nabla}_{\mathbf{x}} F_{\epsilon}(\mathbf{x}^*) \left(\boldsymbol{\nabla}^2_{\mathbf{x}}\, F_{\epsilon}(\mathbf{x}^*)\right)^{-1} \boldsymbol{\nabla}_{\mathbf{x}} \mathcal{Q}(\mathbf{x}^*),
\end{aligned} \tag{9}
$$

where we dropped the dependence on $\boldsymbol{\beta}$ from $\mathbf{x}^*$ and on $(\mathbf{y}, \exp(\boldsymbol{\beta}))$ from F_{ϵ}. Formula (9) is typically used in the design of gradient- and Newton-type methods for solving (2); the particular expression of $\boldsymbol{\nabla}_{\mathbf{x}} \mathcal{Q}(\cdot)$ depends on the specific choice of the loss function considered, such as $\mathcal{Q}_{\text{MSE}}$ in (7) or $\mathcal{Q}_{\text{White}}$ in (8).
Next, we outline the solvers used to compute at each outer iteration an estimate $\mathbf{x}^*$ of the lower-level problem and to compute a solution $\hat{\boldsymbol{\beta}}$ of the bilevel problem.

3.1 Strongly-Convex Accelerated Gradient Descent Lower-Level Solver

We compute the minimiser $\mathbf{x}^*(\exp(\boldsymbol{\beta}))$ of the functional F_{ϵ} in (4) via a strongly-convex variant of Nesterov's accelerated gradient-descent algorithm described, e.g., in [3,5], see Algorithm 1, which enjoys linear convergence rate. Due to the C^2-regularity of F_{ϵ}, (quasi) Newton-type techniques could have been used for fast convergence, similarly as in [8,12]. This, however, may suffer from high computational costs required for the inversion/approximation of the Hessian along the iterations. On the other hand, the proposed Algorithm is computationally

Algorithm 1. Strongly-convex Nesterov AGD, $\texttt{SC\_AGD}(\mathbf{x}_0, L_i, \boldsymbol{\lambda}_i, \texttt{tol}, \mu = 1)$

Initialise: $\theta_0 = 1, \tau = 1/L_i,\ \mathbf{x}_{-1} = \mathbf{x}_0, t = 0,\ \kappa = \mu/L_i$
while $\|\mathbf{x}_{t+1} - \mathbf{x}_t\|_2 > \texttt{tol}$ **do**
  $\theta_{t+1} = \frac{1-\kappa\theta_t^2+\sqrt{(1-\kappa\theta_t^2)^2+4\theta_t^2}}{2};\quad \beta_{t+1} = \frac{\theta_t - 1}{\theta_{t+1}} \frac{1-\theta_{t+1}\mu\tau}{1-\tau\mu}$
  $z_{t+1} = \mathbf{x}_t + \beta_{t+1}(\mathbf{x}_t - \mathbf{x}_{t-1});\quad \mathbf{x}_{t+1} = z_{t+1} - \tau\nabla_{\mathbf{x}} F_\epsilon(z_{t+1}; \mathbf{y}, \boldsymbol{\lambda}_i)$
  $t = t + 1$
end while
return $\mathbf{x}^* = \mathbf{x}^*(\boldsymbol{\lambda}_i)$

Algorithm 2. Gradient-descent for solving (9), $\texttt{GD\_BIL}(\boldsymbol{\beta}^0, \eta, \varepsilon)$

Initialise: $i = 0,\ \mathbf{x}^*(\exp(\boldsymbol{\beta}^{-1})) = \mathbf{y}$
while $\|\boldsymbol{\beta}^{i+1} - \boldsymbol{\beta}^i\|_2 > \varepsilon$ **do**
  compute $\mathbf{x}^*(\exp(\boldsymbol{\beta}^i))=\texttt{SC\_AGD}(\mathbf{x}^*(\exp(\boldsymbol{\beta}^{i-1})), L_i, \exp(\boldsymbol{\beta}^i), \texttt{tol}, \mu = 1)$
  compute $\boldsymbol{\beta}^{i+1} = \boldsymbol{\beta}^i - \eta\nabla_{\boldsymbol{\beta}}\, \mathcal{Q}(\mathbf{x}^*(\exp(\boldsymbol{\beta}^i)))$
  $i = i + 1$
end while
return $\hat{\boldsymbol{\lambda}} = \exp(\hat{\boldsymbol{\beta}})$

lighter. To get a good approximation of $\mathbf{x}^*$ a small relative tolerance parameter $\texttt{tol}$ is employed and at each iteration $i \geq 1$ of the outer optimisation solver a fixed step-size $1/L_i$, $L_i = \overline{L}_{\lambda^i_{\text{MAX}},\epsilon}$, in (6) depending on the current λ^i_{MAX} is used.

3.2 Gradient-Descent Upper-Level Solver

We use standard gradient-descent for solving the non-convex bilevel problem (2)–(3), see Algorithm 2, that takes as input an initial value $\boldsymbol{\beta}^0$, a step-size $\eta > 0$ and a tolerance ε. While most of the existing approaches make use of quasi-Newton updates for solving efficiently the nested problem [7,9,12], here we focused on a plain first-order scheme. Based on (7), (8), we first observe that:

$$\nabla_{\boldsymbol{\beta}}\mathcal{Q}(\mathbf{x}^*) = (\boldsymbol{J}_{\boldsymbol{\beta}}\, \mathbf{x}^*)^T (\boldsymbol{J}_{\boldsymbol{x}}\, \boldsymbol{\rho})^T \boldsymbol{\rho}(\mathbf{x}^*),$$

where we can proceed as in (9) to compute the Jacobian of $\boldsymbol{x}^*$ w.r.t. $\boldsymbol{\beta}$ and of $\boldsymbol{\rho}$ w.r.t. $\boldsymbol{x}$ for the specific choice of $\boldsymbol{\rho}(\cdot)$ considered. This can be further exploited to design, e.g., Gauss-Newton solvers which could improve speed of convergence.

3.3 Early Stopping

While in the supervised setting Algorithm 2 can be run until convergence, we observed that running the unsupervised whiteness-based approach until convergence allows in most cases to achieve values of the $\mathcal{Q}_{\text{white}}$ loss significantly

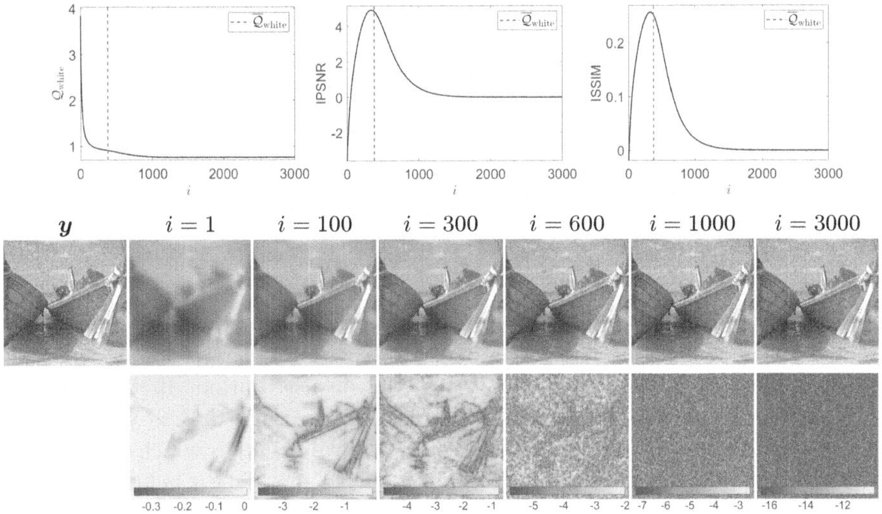

Fig. 1. Behavior of $\mathcal{Q}$, IPSNR and ISSIM values (top), and of the reconstructions and $\boldsymbol{\beta}$-maps (bottom) along the iterations of Algorithm 2 when $\mathcal{Q} = \mathcal{Q}_{\text{white}}$.

less than the expected interval $0.9 - 1.1$ and possibly close to the theoretical lower bound $1/2$, which typically corresponds to meaningless parameters $\boldsymbol{\lambda}$ and solutions $\mathbf{x}^*(\boldsymbol{\lambda})$.

In most of the experiments performed, however, we observed an improvement in the quality metrics during the early iterations of the outer process, which degrade to suboptimal values in later ones. To show such behavior, Algorithm 2 has been run for 3000 iterations with $\mathcal{Q} = \mathcal{Q}_{\text{White}}$ for recovering image #0 shown in Fig. 2 starting from a noisy data $\boldsymbol{y}$ degraded by AWGN with $\sigma = 0.05$ and shown in the left bottom panel of Fig. 1. Along the iterations, we monitored the $\mathcal{Q}$-values as well as the Improved Peak-SNR (IPSNR) and the Improved Structural Similarity Index (ISSIM), measuring the improvement of the output w.r.t. $\mathbf{y}$ in terms of PSNR and SSIM, respectively. The quantities are displayed in the top of Fig. 1: note that the whiteness loss takes values below 0.9, while the IPSNR and ISSIM, after an initial increase, start decreasing and stabilise around 0, i.e. no improvement with respect to the data is obtained. In the bottom of Fig. 1 we show the reconstruction along the iterations of Algorithm 2: one can notice that at the final iteration the output image resembles the data $\mathbf{y}$.

By analysing the weights computed along the iterations - see bottom of Fig. 1 - we observe that at first informative $\boldsymbol{\beta}$-values are recovered, which enforce the regularisation on smooth regions of the image while preserving the edges of the structures. However, eventually the $\boldsymbol{\beta}$ entries tend to $-\infty$, i.e. the $\boldsymbol{\lambda}$-values tend to 0, and the regularisation effect vanishes.

In order to prevent such overfitting, we thus enforced in the optimisation procedure an early stopping regularisation based on statistics of optimal performances estimated off-line. The proposed strategy consists in running the unsupervised procedure until the highest IPSNR value is achieved for a set of $N = 50$ natural images from the BSD400 repository [15] corrupted by AWGN with three different standard deviations $\sigma = 0.01, 0.05, 0.1$ and then to compute the values $\mathcal{Q}_{\text{white}}(\mathbf{x}_n^*)$ for all the computed reconstructions $n = 1\ldots, N$. By considering the empirical average of such values, an estimate $\overline{\mathcal{Q}}_{\text{White}}$ of $\mathbb{E}_{\mathbf{x}}(\mathcal{Q}_{\text{white}}(\mathbf{x}))$ is found and can be used to stop Algorithm 2 before overfitting, by replacing the stopping criterion in the **while** there with a condition checking whether

$$\|\boldsymbol{\beta}^{i+1} - \boldsymbol{\beta}^i\|_2 > \varepsilon \quad \wedge \quad \mathcal{Q}_{\text{white}}(\mathbf{x}^*(\exp(\boldsymbol{\beta}^{i+1}))) \geq \overline{\mathcal{Q}}_{\text{white}}.$$

The estimated stopping value is $\overline{\mathcal{Q}}_{\text{White}} = 0.9081$. The procedure performs well in practice and does not alter the unsupervised feature of the proposed approach, since inference is performed on an image out-of-distribution. In the previous example, the novel stopping criterion can be employed: the dashed vertical lines in the top panels of Fig. 1 represent the estimated value $\overline{\mathcal{Q}}_{\text{white}}$ which corresponds to values of quality metrics almost coincident with the highest achievable.

4 Numerical Examples

In this section, we assess the performance of the whiteness-based procedure for the bilevel problem (2)–(3) when the early stopping criterion outlined in Sect. 3.3 is adopted. Our goal is twofold: on one hand we aim to highlight how the unsupervised whiteness loss is capable of estimating a very large number of parameters arising in the variational model while returning high quality results. Moreover, we want to emphasise the potential of the WTV_ϵ regulariser w.r.t the TV_ϵ term, showing how the bilevel problem (2)–(3) in presence of the supervised MSE loss can boost the performance of the TV_ϵ-based regularisation.

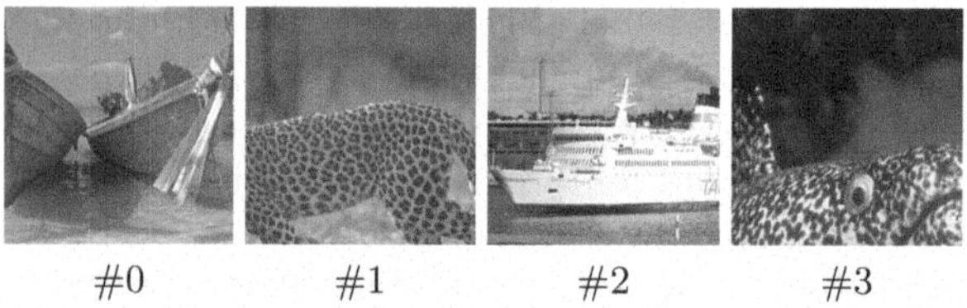

Fig. 2. Original test images of size 180×180 from the BSD400 repository.

Hence, our unsupervised loss is compared to the supervised MSE loss in (2). The two losses are also compared in presence of a TV_ϵ-regularised lower problem. The four methods are tested on the denoising of test images #1-3 from the BSD400 dataset, shown in Fig. 2, corrupted by AWGN with different standard deviation $\sigma = 0.03, 0.06, 0.09$ not used in the computation of the stopping value. The smoothing parameters in the smoothed TV_ϵ and WTV_ϵ terms have been selected so as to maximise the performance of the regularised models. More

Table 1. IPSNR and ISSIM values achieved by the considered models.

Image	σ	supervised				unsupervised			
		TV		WTV		TV		WTV	
		IPSNR	ISSIM	IPSNR	ISSIM	IPSNR	ISSIM	IPSNR	ISSIM
#1	0.03	2.848	0.147	**6.438**	**0.194**	2.729	0.129	**3.176**	**0.176**
	0.06	4.257	0.266	**8.354**	**0.372**	4.210	0.247	**4.546**	**0.322**
	0.09	5.345	0.314	**9.461**	**0.471**	5.332	0.303	**5.462**	**0.370**
#2	0.03	2.896	0.129	**6.689**	**0.179**	2.805	0.117	**3.265**	**0.150**
	0.06	4.467	0.258	**8.707**	**0.362**	4.447	0.248	**4.817**	**0.286**
	0.09	5.624	0.319	**10.016**	**0.465**	5.617	0.313	**5.891**	**0.346**
#3	0.03	3.309	0.193	**7.471**	**0.241**	3.289	0.199	**3.881**	**0.224**
	0.06	4.604	0.331	**9.063**	**0.434**	4.589	0.343	**4.951**	**0.386**
	0.09	5.554	0.376	**10.018**	**0.525**	5.529	0.395	**5.794**	**0.439**

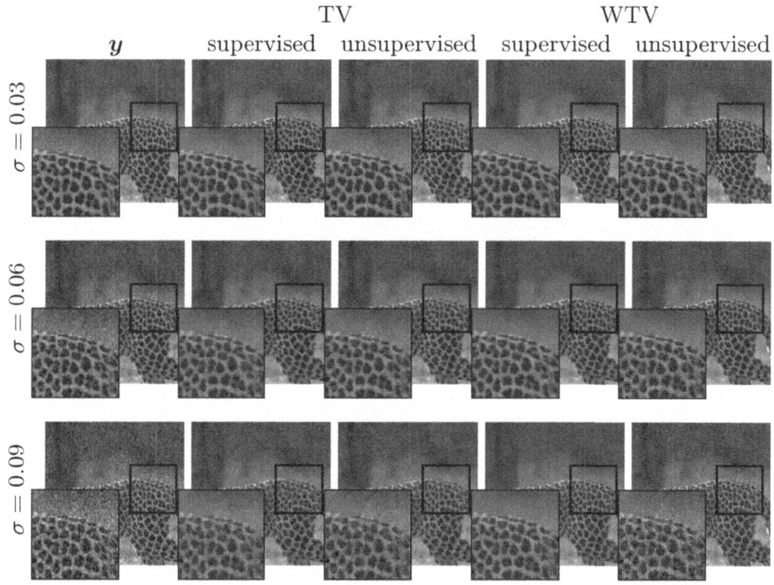

Fig. 3. Output reconstructions for test image #1.

specifically, $\epsilon = 10^{-2}$ in the TV_ϵ term and $\epsilon = 10^{-1}$ in the WTV_ϵ term. The initial guess $\boldsymbol{\beta}_0$ is (a vector of) one(s), while $\eta = 1000$ when $\mathcal{Q} = \mathcal{Q}_{\text{white}}$ and $\eta = 100$ when $\mathcal{Q} = \mathcal{Q}_{\text{MSE}}$. Also, to avoid degenerate configurations, we impose an upper bound on $\boldsymbol{\beta}$, i.e. on $\boldsymbol{\lambda}$, $\lambda_i \leq 5$. Finally, `tol` and ε are set equal to 10^{-6}.

In Table 1, for each method we show the IPSNR and ISSIM values obtained, with in bold the best achieved in the sub-classes of supervised and unsupervised approaches. The metrics are extremely high when solving (2)–(3) with the

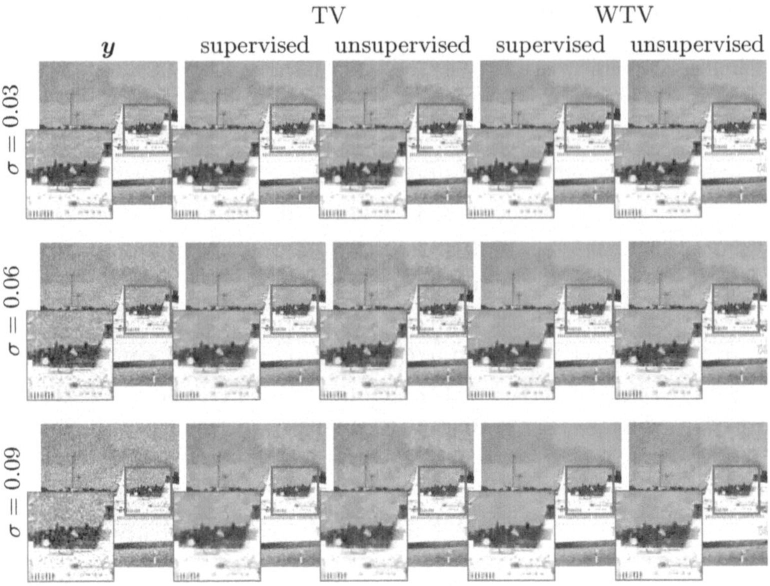

Fig. 4. Output reconstructions for test image #2.

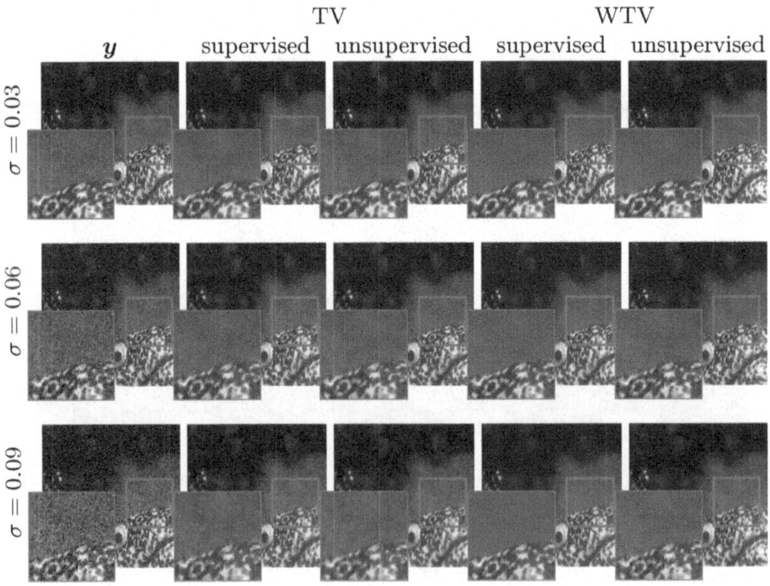

Fig. 5. Output reconstructions for test image #3.

$\mathcal{Q}_{\mathrm{MSE}}$; also, while the unsupervised method upon the selection of the TV_ϵ term is comparable to the supervised approach, the performance of the WTV_ϵ term in

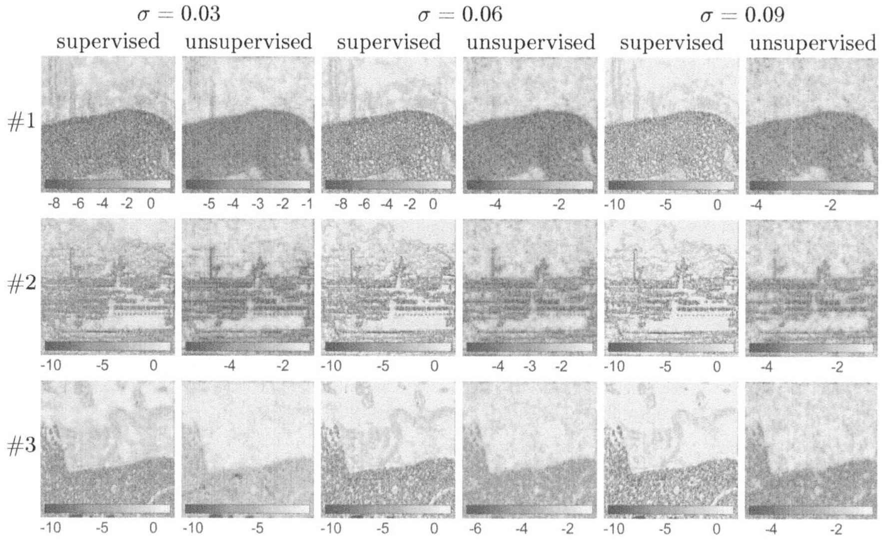

Fig. 6. Output parameters for the supervised $\mathcal{Q}_{\text{MSE}}$ and the unsupervised $\mathcal{Q}_{\text{white}}$.

the supervised case does not seem to be reproducible with the unsupervised loss. Nonetheless, in the latter scenario the results outperform the performance of the TV_ϵ term, both in the supervised and unsupervised settings, thus confirming the potentials of the WTV_ϵ term and assessing the robustness of the whiteness loss when moving from a single to a general high number of parameters.

The numeric results reported in Table 1 are confirmed by the output images obtained by the four methods for the different noise levels - see Figs. 3, 4 and 5.

Finally, in Fig. 6 we show the maps of the output $\boldsymbol{\beta}$ for the WTV_ϵ regulariser in the supervised and unsupervised scenario; for rendering purposes, a lower bound of -10 has been considered in the visualisation of the maps. The supervised and unsupervised approach are equally good at differentiating between smooth regions, characterised by larger values of the parameters, and textured parts of the image, for which smaller values are more suitable. However the scale of the maps in the two cases are significantly different, this possibly being responsible for the loss of quality observed in the unsupervised case.

The unsupervised whiteness-based strategy with the WTV regulariser relies on a training on the BSD400 natural images dataset. To prove the flexibility of the approach, we also perform a training on the COCO dataset of common objects [14]. From $N = 50$ images of the set, we extract a 180×180 patch and consider the same degradation of Sect. 3.3. The learned stopping value is $\widetilde{\mathcal{Q}}_{\text{white}} = 0.9190$, very close to the value $\bar{\mathcal{Q}}_{\text{white}}$ employed in the experimentation so far. In Fig. 7, for image #2 and the three noise levels, we plot the IPSNRs achieved along a fixed number of iterations of Algorithm 2 for solving the whiteness-based bilevel problem with the WTV regulariser. The vertical blue and red lines indicate the stopping values returned by the training on BSD400 and COCO, respectively. One can notice that they correspond to very close IPSNRs.

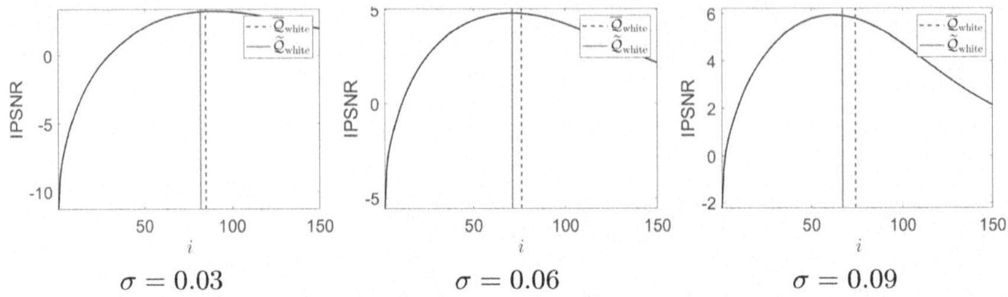

Fig. 7. IPSNR along the iterations of Algorithm 2 for different noise levels. (Color figure online)

Finally, we consider a signal **x** corrupted by additive white uniform noise distribution, and solve the bilevel problem in (2)–(3) under the same settings detailed above for Gaussian data. The resulting IPSNRs/ISSIMs for image #3 are summarised in Table 2, and show the same behavior of those reported in Table 1, suggesting the robustness of the unsupervised whiteness loss also when coupled to lower cost functionals that do not properly model the noise at hand.

Table 2. IPSNR and ISSIM values achieved by the considered models in case of additive white uniform noise corruption.

Image	σ	supervised				unsupervised			
		TV		WTV		TV		WTV	
		IPSNR	ISSIM	IPSNR	ISSIM	IPSNR	ISSIM	IPSNR	ISSIM
#3	0.03	3.419	0.199	**7.377**	**0.243**	3.354	0.209	**3.815**	**0.225**
	0.06	4.712	0.342	**8.900**	**0.436**	4.629	0.365	**5.007**	**0.392**
	0.09	5.697	0.387	**9.887**	**0.524**	5.585	0.423	**5.831**	**0.427**

5 Conclusions

We addressed the estimation of spatially-adaptive weighted TV_ϵ regularisation maps through a bilevel optimisation framework with an unsupervised upper loss defined in terms of a residual whiteness measure. This approach enables the simultaneous recovery of both the denoised image and the optimal regularisation maps for data corrupted by AWGN. The approach requires the employment of an early stopping criterion based on simple statistics of optimal performances computed on a set of natural images, and returns maps morphologically similar to the ones obtained by employing supervised MSE loss, while outperforming the standard (global) TV_ϵ term, both in supervised and unsupervised settings. Our tests are encouraging in terms of the expected performance of the whiteness loss in highly-parametrised settings. Further work should address the use of the whiteness loss in proper learning settings where, e.g., the $\boldsymbol{\lambda}$ map is parametrised as the output of a Convolutional Neural Network (CNN), as in [11] where a similar strategy is performed for MRI image reconstruction in a supervised setting.

Acknowledgments. This work is based upon work supported by the ANR JCJC project TASKABILE ANR-22-CE48-0010 and by the ERC grant MALIN under the European Union's Horizon Europe programme (grant agreement No. 101117133). Views and opinions expressed are however those of the author(s) only and do not necessarily reflect those of the European Union or the European Research Council Executive Agency. Neither the European Union nor the granting authority can be held responsible for them. MP acknowledges the GNCS-INdAM funding and the 2022 research program FRA of the University of Naples Federico II.

References

1. Almeida, M., Figueiredo, M.: Parameter estimation for blind and non-blind deblurring using residual whiteness measures. IEEE Trans. Signal Process. **22**, 2751–2763 (2013)
2. Bevilacqua, F., Lanza, A., Pragliola, M., Sgallari, F.: A general framework for whiteness-based parameters selection in variational models. Comput. Optim. Appl. (2024)
3. Calatroni, L., Chambolle, A.: Backtracking strategies for accelerated descent methods with smooth composite objectives. SIAM J. Optim. **29**(3), 1772–1798 (2019)
4. Calvetti, D., Lewis, B., Reichel, L.: L-curve for the MINRES method. In: Advanced Signal Processing Algorithms, Architectures, and Implementations X, vol. 4116, pp. 385–395 (2000)
5. Chambolle, A., Pock, T.: An introduction to continuous optimization for imaging. Acta Numer. **25**, 161–319 (2016)
6. Craven, P.G., Wahba, G.: Smoothing noisy data with spline functions: estimating the correct degree of smoothing by the method of generalized cross-validation. Numer. Math. **31**, 377–403 (1978)
7. De los Reyes, J.C., Schönlieb, C.-B., Valkonen, T.: Bilevel parameter learning for higher-order total variation regularisation models. J. Math. Imaging Vis. **57**(1), 1–25 (2017)
8. De los Reyes, J.C.: Numerical PDE-Constrained Optimization. SpringerBriefs in Optimization. Springer (2015)
9. Fehrenbach, J., Nikolova, M., Steidl, G., Weiss, P.: Bilevel image denoising using gaussianity tests. In: SSVM, pp. 117–128. Springer, Cham (2015)
10. Haber, E., Tenorio, L.: Learning regularization functionals: a supervised training approach. Inverse Probl. **19**(3), 611–626 (2003)
11. Kofler, A., Altekrüger, F., Antarou Ba, F., et al.: Learning regularization parameter-maps for variational image reconstruction using deep neural networks and algorithm unrolling. SIAM J. Imaging Sci. **16**(4), 2202–2246 (2023)

12. Kunisch, K., Pock, T.: A bilevel optimization approach for parameter learning in variational models. SIAM J. Imaging Sci. **6**(2), 938–983 (2013)
13. Lanza, A., Pragliola, M., Sgallari, F.: Residual whiteness principle for parameter-free image restoration. Electron. Trans. Numer. Anal. **53**, 329–351 (2020)
14. Lin, T.Y., et al.: Microsoft COCO: common objects in context. CoRR, abs/1405.0312 (2014)
15. Martin, D.R., Fowlkes, C.C., Tal, D., Malik, J.: A database of human segmented natural images and its application to evaluating segmentation algorithms and measuring ecological statistics. In: ICCV 2001, vol. 2, pp. 416–423 (2001)
16. Morozov, V.A.: On the solution of functional equations by the method of regularization. Dokl. Math. **7**, 414–417 (1966)
17. Ochs, P., Ranftl, R., Brox, T., Pock, T.: Techniques for gradient-based bilevel optimization with non-smooth lower level problems. J. Math. Imaging Vis. **56**(2), 175–194 (2016)
18. Pragliola, M., Calatroni, L., Lanza, A., Sgallari, F.: On and beyond total variation regularization in imaging: the role of space variance. SIAM Rev. **65**(3), 601–685 (2023)
19. Rudin, L.I., Osher, S., Fatemi, E.: Nonlinear total variation based noise removal algorithms. Physica D: Nonlinear Phen. **60**(1), 259–268 (1992)
20. Santambrogio, C., Pragliola, M., Lanza, A., Donatelli, M., Calatroni, L.: Whiteness-based bilevel learning of regularization parameters in imaging. In: 2024 32nd European Signal Processing Conference (EUSIPCO), pp. 1801–1805 (2024)
21. Tachella, J., Davies, M., Jacques, L.: UNSURE: self-supervised learning with unknown noise level and Stein's Unbiased Risk Estimate. In: The Thirteenth International Conference on Learning Representations (2025)

Direct Atomistic Reconstruction in Homogeneous Cryo-EM Using Protein Geometry Regularization

Jonathan Krook[1(✉)], Willem Diepeveen[2], and Ozan Öktem[1]

[1] KTH, Stockholm, Sweden
{jkroo,ozan}@kth.se
[2] University of California, Los Angeles, Los Angeles, CA 90095, USA
wdiepeveen@math.ucla.edu

Abstract. Direct reconstruction of macromolecular structures from cryogenic electron microscopy (Cryo-EM) data has shown to be a challenge both in the homogeneous and heterogeneous setting. In this work we propose a new direct reconstruction method based on a combination of recent developments on protein geometry and orientation estimation. Even though this method is set up for the homogeneous setting, we aim to gain insight into challenges that atomistic methods have been facing for the heterogeneous case. In numerical experiments we observe that the method is able to recover the structure to almost inter-atomic resolution from as few as 100 2D Cryo-EM images due to the strong bias the regularizer gives. We conclude this work with a discussion on how the obtained results indicate possibilities and challenges for the generalization to the heterogeneous case.

Keywords: Cryo-EM · Atomistic reconstruction · Regularization

1 Introduction

Single particle cryogenic-electron microscopy (Cryo-EM) is a tomographic imaging method that can be used to recover the 3D structure of a biomolecule (see Fig. 1). This is highly valuable for downstream applications, such as drug development [2], and for advancing structural biology as a whole. This technique involves rapidly freezing multiple copies of the same molecule in a thin layer of vitreous ice, preserving them in a nearly physiological state. Hence, the specimen consists of isolated particles that represent randomly distributed and oriented exemplars of the biomolecule. The ice is usually thin enough to ensure particles cannot stack on top of each other along the optical axis. The electron microscope captures a 2D image (micrograph) of this ice slab, reflecting the squared intensity of the electron wave after it has scattered against the electrostatic potential of the particles and then focused onto the detector by the microscope's optics [15]. The micrograph is then cropped into several small 2D images, each representing

T. A. Bubba et al. (Eds.): SSVM 2025, LNCS 15667, pp. 173–184, 2025.
https://doi.org/10.1007/978-3-031-92366-1_14

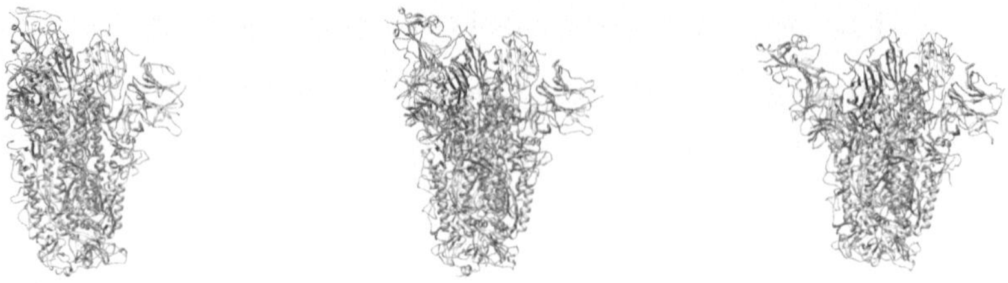

(a) (b) (c)

Fig. 1. Conformations of the delta variant of the SARS-CoV-2 spike protein.

(a) (b) (c)

Fig. 2. Noisy Cryo-EM images of the delta variant of the SARS-CoV-2 spike protein.

a very noisy and low contrast projection image of a single particle (see Fig. 2), which are used to reconstruct the 3D structure of the biomolecule.

Traditionally, the standard approach has been to first reconstruct the electrostatic potential of the biomolecule – also known as the *3D map* or *volume* – and then use this for assembling its 3D structure (*model building*). This approach is particularly efficient when the biomolecule is rigid, so all particles have identical structure (homogeneous particles) and their 3D maps differ only by rigid body transformations [18]. However, although it has become clear that many biomolecules do not meet this rigidity assumption, many modern deep learning-based solvers have continued to use this intermediate step of 3D map reconstruction before performing model building in a now heterogeneous set of 3D maps [6,20].

Nonetheless, recent advancements indicate that recovering the 3D structure directly from 2D Cryo-EM images present several benefits. In particular, thanks to AlphaFold [11] one can generate an initial template structure. Combined with recent work in protein geometry [4], this allows us to formulate a robust regularizer that penalizes non-physical deformations of such a template, which is challenging to enforce in (intermediate) 3D map reconstruction. Furthermore, improvements in orientation estimation [5] enable the integration of the entire process into a unified approach that is both robust against noise and data-consistent – an issue that has remained a challenge even for homogeneous 3D map reconstruction until recently.

Our goal in this work is less ambitious than solving the full heterogeneous reconstruction problem. Instead, we aim to take an initial step by combining the above ideas into a new method for the homogeneous case to gain insights for tackling the heterogeneous scenario. Specifically, we are focused on constructing an effective regularizer, understanding the effect of adding orientation estimation, and determining how much data from sufficiently similar conformations is needed to use this method for a realistic noise level. This information will be crucial when adapting this approach to the heterogeneous case.

1.1 Related Work

The main reason for focusing on the homogeneous case in this work is twofold: (i) the literature has shown mixed results on using (pseudo-)atomic models, or more generally point cloud-based models, for single particle analysis in heterogeneous setting, and (ii) there is a scientific gap between the dominant lines of approaches that might be too ambitious to bridge right away. In other words, directly proposing a point cloud-based heterogeneous reconstruction scheme composed of just the above-mentioned parts [4,5,11] without further consideration is expected to be overly ambitious.

To give the reader a better overview of the challenges faced by these recent approaches and the scientific gap we try to close, we review several methods for heterogeneous reconstruction below.

Normal Mode Analysis-Based Latent Space. Normal mode analysis (NMA) [19] is a technique for estimating the most likely directions a given conformation can morph into based on linearisation of a physics-based potential. Typically, only a small amount of these normal modes can be used to model a (local) linear approximation of the conformation landscape. This gives in principle a very interpretable low-dimensional latent space representation of the heterogeneous reconstruction problem, which partially deals with the ill-posedness inherent to Cryo-EM reconstruction. Also in practice, this idea seems to be yielding promising results for reconstructing conformations that originate from medium-scale deformations when integrated into a reconstruction pipeline [9,10,14,21]. However, realistic conformational changes are often large-scale and non-linear, which is something that cannot be expected to be captured from local energy landscape linearization.

Autoencoder-based latent space. Autoencoders are a neural networks with a bottleneck in their architectures to enforce a low-dimensional latent space representation. Especially Variational Autoencoders (VAE) [12], which also underlie successful approaches for recovering the 3D map, like CryoDRGN [22], have also been tried in a point cloud-based setting [3,17,23]. While these approaches are not restricted by the linearity of the latent space as in NMA-based approaches, there is by default no physical prior in the reconstruction on large-scale deformations besides typical penalization of the degradation of bond lengths. On top of that, the latent space of an autoencoder does not offer the same interpretability as NMA, since even distances in latent space cannot be expected to carry any meaning regarding the (global) geometry of the data. In practice, the results by these approaches as presented in [3,17,23] have been less impressive than for NMA due to too restrictive simplifications such as assuming known rotations, unrealistically low noise levels or only 1-dimensional latent spaces. Having that said, the same seems to be true for a proposed approach by [16] that does incorporate the non-linear physics-based prior from [4].

Manifold Learning-Based Latent Space. Finally, we would like to mention the work [7], in which the authors use spectral-based manifold learning [13] to learn (low-dimensional) deformation coefficients after which their proposed algorithm tries to solve for a latent space basis with non-linearity take care of by modeling the deformations through dihedral angles. This method showcases promising results and the learning step that does the heavy lifting can be sped up [8]. However, this first step also requires known orientations.

1.2 Outline

The paper is organized as follows. Section 2 reviews some recent developments that underlie the proposed method for homogeneous reconstruction. Subsequently, our method is lined out in Sect. 3. Section 4 discusses several numerical experiments from which we aim to find out constructing an effective regularizer, understanding the effect of adding orientation estimation, and determining how much data from sufficiently similar conformations is needed to use this method for a realistic noise level. We conclude and discuss prospects for the heterogeneous case in Sect. 5.

2 Theory

Before introducing the algorithm, we outline its foundational components, which are essential for developing a variational model to address the homogeneous reconstruction problem. Understanding these elements individually is crucial before employing alternating minimization techniques to achieve a solution. Both subproblems (conformation and pose estimation) in this context are inherently ill-posed. For protein conformation, the challenge lies in the lack of sufficient information to determine a conformation, which could lead to non-physical outcomes if parts of the chain are improperly adjusted. To prevent this, a regularizer is necessary. In the case of the pose, direct minimization often produces poor results due to the high noise levels. Although averaging can mitigate these noise effects, it can be computationally expensive once passed to the conformation update step, particularly when computing gradients. Therefore, it is preferable to identify a single orientation that minimizes the computational burden.

In the following, we recapitulate recent theoretical developments that tackle these problems.

2.1 Regularizing Conformation Estimation

In this work, we approximate the 3D structure of a protein by a discrete curve, or chain, made out of the C_α atoms of the backbone as in [7]. If there are n atoms in the chain, this means we can describe the 3D structure by a matrix in $\mathbb{R}^{3\times n}$, where the i:th column is the 3D coordinates of the i:th C_α atom in the chain.

Given a physically realistic template with accurate adjacent bond lengths[1], it is important to employ a regularizer that prevents non-physical deformations of the atomic structure.

Recently, a physics-based Riemannian geometry was introduced [4], featuring an approximate closed form Riemannian distance[2] whose square is given by

$$w^\delta(\mathbf{X},\mathbf{Y})^2 := \sum_i \sum_{j>i} \left(\log\left(\frac{\|\mathbf{x}_i - \mathbf{x}_j\|_2}{\|\mathbf{y}_i - \mathbf{y}_j\|_2} \right) \right)^2 + \delta \left(\log\left(\frac{\det(\sum_i (\mathbf{x}_i - \frac{1}{n}\mathbf{X}\mathbf{1}_n) \otimes (\mathbf{x}_i - \frac{1}{n}\mathbf{X}\mathbf{1}_n))}{\det(\sum_i (\mathbf{y}_i - \frac{1}{n}\mathbf{Y}\mathbf{1}_n) \otimes (\mathbf{y}_i - \frac{1}{n}\mathbf{Y}\mathbf{1}_n))} \right) \right)^2, \quad (1)$$

for some $\delta > 0$, and where $\mathbf{1}_n = (1,\dots,1)^t \in \mathbb{R}^n$. This distance function is designed to preserve local structures, ensuring that small-scale deformations are penalized more heavily than large-scale ones, which aligns well with the principles underlying well-known models, like the Lennard-Jones potential.

2.2 Regularizing Pose Estimation

Next, we seek a global optimization scheme for pose estimation that incorporates regularization to ensure stable and meaningful solutions. The recently proposed method of ellipsoidal support lifting (ESL) [5] offers a promising approach to achieve this goal. ESL works by relaxing the problem of minimising an objective function $F : \mathcal{M} \to \mathbb{R}$ over a smooth manifold $\mathcal{M}$ into one defined over measures, similar to an averaging-based methods [18]:

$$\mathbf{p}^* \in \operatorname*{argmin}_{\mathbf{p}\in\mathcal{M}} F(\mathbf{p}) \quad \Longleftrightarrow \quad \delta_{\mathbf{p}^*} \in \operatorname*{argmin}_{\mu\in\mathbb{P}(\mathcal{M})} \int_{\mathcal{M}} F(\mathbf{p})\,\mathrm{d}\mu. \quad (2)$$

To prevent this measure from degenerating into a single point, regularization must be applied, ensuring that the measure remains distributed. If the resulting measure is sufficiently concentrated, it can be approximated by its barycentre

$$\int_{\mathcal{M}} F(\mathbf{p})\,\mathrm{d}\mu(\mathbf{p}) \approx F(\mathbf{p}_\mu), \quad \text{where} \quad \mathbf{p}_\mu \in \operatorname*{argmin}_{\mathbf{p}'\in\mathcal{M}} \int_{\mathcal{M}} d_{\mathcal{M}}(\mathbf{p}',\mathbf{p})^2\,\mathrm{d}\mu(\mathbf{p}). \quad (3)$$

[1] For example a template obtained from AlphaFold [11].

[2] That is, on the space $P(3,n) \subset \mathbb{R}^{3\times n}/\mathbb{E}(3)$ as defined in [4].

In practice, ESL involves selecting a sufficiently dense sampling set $\mathcal{X} \subset \mathcal{M}$ from the underlying manifold – in our case, we use rotation matrices to represent poses, so the manifold is the special orthogonal group SO(3). The problem is then reduced to solving a variational problem on the unit simplex, incorporating regularization parameters to control the distribution of the sampling points:

$$\beta^* := \underset{\beta \in \mathbb{R}^{|\mathcal{X}|}}{\operatorname{argmin}} \Big\{ F_{\mathcal{X}} \cdot \beta + \chi_{\Delta^{|\mathcal{X}|}}(\beta) + \frac{\gamma}{2|\mathcal{X}|^{\eta}} \|\beta\|_2^2 \Big\} \quad \text{for } \gamma > 0 \text{ and } \eta \in \mathbb{R}, \tag{4}$$

where $F_{\mathcal{X}} := (F(\mathbf{p}^i))_{i=1}^{|\mathcal{X}|} \in \mathbb{R}^{|\mathcal{X}|}$ and $\chi_{\Delta^{|\mathcal{X}|}} : \mathbb{R}^{\mathcal{X}} \to \mathbb{R} \cup \{\infty\}$ is the indicator function on the unit simplex $\Delta^{|\mathcal{X}|} \subset \mathbb{R}^{|\mathcal{X}|}$. Finally, the optimal measure is projected onto a single point by computing the barycentre

$$\mathbf{p}^* \in \underset{\mathbf{p} \in \mathbb{R}^d}{\operatorname{argmin}} \Big\{ \frac{1}{2|\mathcal{X}|} \sum_{i=1}^{|\mathcal{X}|} \beta_i^* d_{\mathcal{M}}(\mathbf{p}, \mathbf{p}^i)^2 \Big\}. \tag{5}$$

Here, the locality of the measure and control of the barycentre is ensured through careful selection of the sampling set and appropriate parameter choices [5, Theorems 3.6 & 3.7].

3 Method Description

In this section we describe how the ideas presented in the previous section can be combined in order to solve the homogeneous 3D reconstruction problem.

3.1 Homogeneous Reconstruction Problem

First we want to pose the reconstruction problem as a variational problem. As in the previous section, the 3D structure of a protein is represented by a matrix $\mathbf{X} \in \mathbb{R}^{3 \times n}$, where the columns are atomic coordinates.

Definition 1. *The Cryo-EM image formation operator* $\mathcal{A}\colon \mathbb{R}^{3\times n} \to L^2(\mathbb{R}^2)$, *mapping a 3D structure to an image, is constructed by first generating a 3D volume by placing an isotropic Gaussian with covariance* $\zeta^2 I$ *at each point in* $\mathbf{X} \in \mathbb{R}^{3\times n}$, *and then applying a parallel beam ray transform on this 3D volume, thus producing a 2D image in* $L^2(\mathbb{R}^2)$.

The nonlinear operator in Definition 1 maps an atomic structure $\mathbf{X} \in \mathbb{R}^{3\times n}$ for a protein (given as a point cloud representing the backbone conformation) onto a 2D grayscale image recorded in a transmission electron microscope. For computational reasons, one can evaluate $\mathcal{A}$ by first geometrically projecting the atomic coordinates along the optical axis followed by placing 2D Gaussians at the projected coordinates in the detector plane. It is easy to show that this is the same as computing the ray transform of the 3D Gaussians defined by the point cloud.

Remark 1. The actual model for image formation in Cryo-EM is more complicated than the one in Definition 1. As an example, the model in Definition 1 disregards the blurring due to the optics (defocus) in the microscope and detector response, see for example [15, chapter 4] for further details.

Given a set of Cryo-EM data $\{f\}_{i=1}^{N}$ consisting of N 2D images, we initially consider the following variational approach to homogeneous reconstruction: Find a minimizer of

$$\mathbb{R}^{3\times n} \ni \mathbf{X} \mapsto \sum_i \left\|\mathcal{A}(R_i\mathbf{X}) - f_i\right\|_2^2 + \lambda w^{\delta,\mathbf{A}}(\mathbf{X}, \mathbf{X}^0)^2. \tag{6}$$

Here, $R_i \in \mathrm{SO}(3)$ represents the orientation of the particle that is imaged in f_i, $\mathbf{X}^0 \in \mathbb{R}^{3\times n}$ is a template structure, $\lambda > 0$ is a regularization parameter, and $w^{\delta,\mathbf{A}}\colon \mathbb{R}^{3\times n} \to \mathbb{R}_{\geq 0} \cup \{\infty\}$ is for some $\delta > 0$ and $\mathbf{A} \in \mathbb{R}_{\geq 0}^{n\times n}$ defined as

$$\begin{aligned} w^{\delta,\mathbf{A}}(\mathbf{X},\mathbf{Y})^2 := \sum_i \sum_{j>i} \mathbf{A}_{ij}\left(\log\left(\frac{\|\mathbf{x}_i - \mathbf{x}_j\|_2}{\|\mathbf{y}_i - \mathbf{y}_j\|_2}\right)\right)^2 \\ + \delta\left(\log\left(\frac{\det\left(\sum_i (\mathbf{x}_i - \frac{1}{n}\mathbf{X}\mathbf{1}_n) \otimes (\mathbf{x}_i - \frac{1}{n}\mathbf{X}\mathbf{1}_n)\right)}{\det\left(\sum_i (\mathbf{y}_i - \frac{1}{n}\mathbf{Y}\mathbf{1}_n) \otimes (\mathbf{y}_i - \frac{1}{n}\mathbf{Y}\mathbf{1}_n)\right)}\right)\right)^2. \end{aligned}$$

We assume that the Cryo-EM images are centered and that particles are not reflected so that the orientation can be described by a rotation, and not a general element in E(3). The regularization term $w^{\delta,\mathbf{A}}$ depends on the choice of δ and $\mathbf{A}$. In order to reduce computational costs, we choose $\delta = 0$. As for $\mathbf{A}$, our idea is to construct the weights $\mathbf{A}_{ij}$ so that preservation of interatomic distances is more strongly enforced for atoms that are close together along the chain. Thus, we put $\mathbf{A}_{ij} = e^{-p|i-j|}$, where $p \geq 0$ is some parameter to be determined.

3.2 Algorithm

We now describe how we solve the variational problem (6). The orientations R_i are assumed to be unknown, so our first step is to find orientation estimates R_i^0. This is done using the method described in Sect. 2.2, with the template $\mathbf{X}^0$ and data $\{f_i\}$ as input. At step $k+1$, we assume that we have orientation estimates R_i^k, which are used to formulate the variational problem

$$\inf_{\mathbf{X}\in\mathbb{R}^{3\times n}} \sum_i \left\|\mathcal{A}(R_i^k\mathbf{X}) - f_i\right\|_2^2 + \lambda w^{\delta,\mathbf{A}}(\mathbf{X}, \mathbf{X}^0)^2. \tag{7}$$

The minimizer $\mathbf{X}^{k+1}$ is then be used to obtained updated guesses R_i^{k+1}, and so on. The algorithm can be stopped after a fixed number of iterations or for example when $w^{\delta,\mathbf{A}}(\mathbf{X}^k, \mathbf{X}^{k+1}) < \varepsilon$ for some threshold $\varepsilon > 0$, at which point it outputs the reconstruction $\mathbf{Y}$.

4 Experiments

In this section we present numerical results of our method. We use the gradient descent based algorithm ADAM implemented in pytorch with autograd for solving optimization problem (7).

4.1 Generating Synthetic Data

In order to test the algorithm, we generate synthetic data. For this we pick out the C_α atoms of a simulated molecular dynamics trajectory [1], giving us a series of matrices in $\mathbb{R}^{3\times n}$. As the objective $\mathbf{X} \in \mathbb{R}^{3\times n}$, that is the structure that we aim to reconstruct, we choose the final frame of the trajectory, and for template $\mathbf{X}^0 \in \mathbb{R}^{3\times n}$, we choose the first frame of the trajectory. Let $(R_i)_{i=1}^N$ be a list of uniformly random generated orientations. The 2D Cryo-EM image corresponding to the structure $\mathbf{X}$ in orientation R_i is modeled by $f_i = h * \mathcal{A}(R_i\mathbf{X}) + \epsilon_i$ where we have noise $\epsilon_i \in \mathcal{N}(0, \sigma^2 I)$ and a point spread function $h : \mathbb{R}^2 \to \mathbb{R}$. Our synthetic Cryo-EM dataset is then $(f_i)_{i=1}^N$. A formula for the point spread function is found in [7]. To measure the level of noise in our synthetic data, we use the signal to noise ratio (SNR). The SNR in the image f is given by $\text{SNR} = \|f - \epsilon\|_2^2 / \|\epsilon\|_2^2$. See Fig. 3 for a small sample of the data.

Throughout our experiments we fix $\lambda = 10^{-6}$, $\zeta = 2.4$ Å, and use a detector with effective pixel size of 1 Å. The γ and η parameters for the orientation estimation are chosen according to [5, Theorem 3.6 & 3.7].

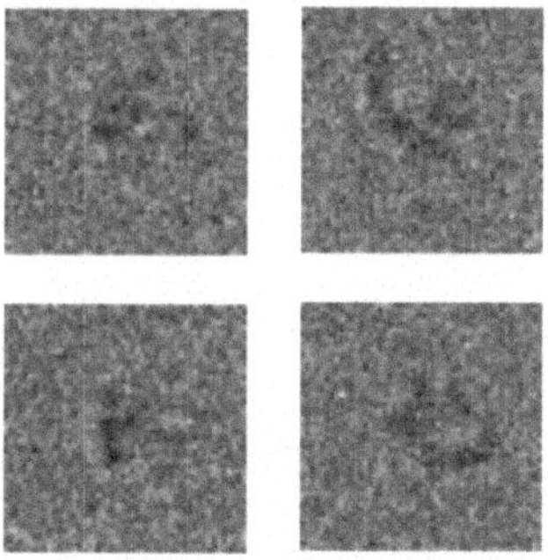

Fig. 3. Sample of four simulated Cryo-EM images with SNR = 0.1.

4.2 Choosing Weights

Our first step is to choose the weights $\mathbf{A}$ that define the regularization term in (7). Recall that we let $\mathbf{A}_{ij} = e^{-p|i-j|}$, where p is some parameter to be determined. The idea is to run the reconstruction algorithm with known orientations and without noise for different values of p, and then choosing for future experiments the value giving best results.

As a measure for how good our reconstruction is, we take the root mean squared deviation (RMSD). The RMSD between structures $\mathbf{X}$ and $\mathbf{Y}$ is defined as

$$\mathrm{RMSD} = \sqrt{\frac{1}{n}\sum_i \|\mathbf{x}_i - \mathbf{y}_i\|_2^2}. \tag{8}$$

When computing the RMSD between the true structure $\mathbf{X}$ and the reconstruction, we first find a rotation that aligns the reconstruction with $\mathbf{X}$, and then take the RMSD between the aligned reconstruction and $\mathbf{X}$.

From Fig. 4, we conclude that $p = 0.5$ should be a reasonable choice.

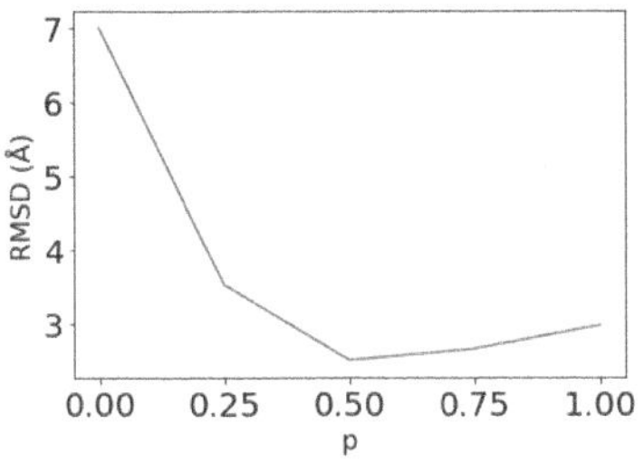

Fig. 4. Reconstruction RMSD after 200 iterations as a function of the parameter p, displaying a local minimum at $p = 0.5$. We used known orientations, no noise, and simulated data without using a point spread function. We had $N = 100$ Cryo-EM images in the data.

4.3 Reconstruction Results

Having determined the weights in $\mathbf{A}$, we now tackle the homogeneous reconstruction problem using the full method described in the previous section.

Alongside the RMSD, we are also interested in how well we are able to recover the true orientations. For this we want a measure of how close two rotation matrices $a, b \in \mathrm{SO}(3)$ are to each other. We do this by measuring the rotation angle θ of the rotation $a^\top \cdot b$, which is given as

$$\theta = \arccos\Big(\big(\mathrm{tr}(a^\top \cdot b) - 1\big)/2\Big). \tag{9}$$

Hence, if a and b are close, then θ is small. We note that, up to scalar multiplication, θ is the geodesic distance between a and b in the unique bi-invariant metric on the compact Lie group SO(3). The plots in Fig. 5 show how the orientation error (the distance θ between true and estimated orientations) and RMSD between the true structure and reconstruction evolve during optimization.

Figure 7 shows how well individual atomic positions are reconstructed depending on the number of images in the data. A view of the 3D-reconstruction is found in Fig. 6.

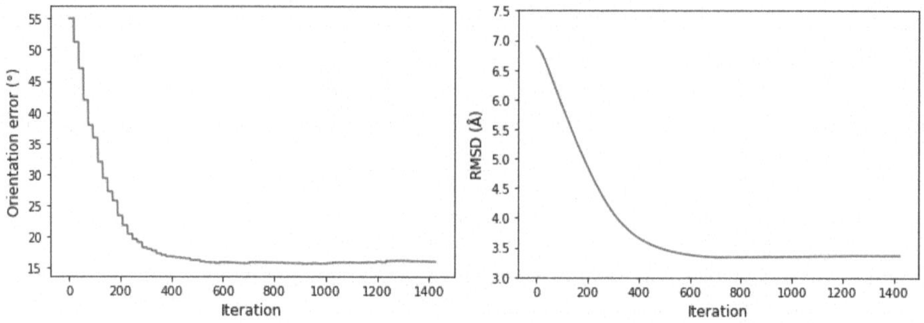

(a) (b)

Fig. 5. Accuracy of orientation estimates (a) and RMSD (b) as functions of the number of iterations. Here we had data with SNR = 0.1 and $N = 100$.

(a) Template $\mathbf{X}^0$ (b) True structure (c) Reconstruction

Fig. 6. 3D reconstruction after 1500 iterations with $N = 100$ and using data with SNR = 0.1. Reconstruction RMSD = 3.40 Å. As a reference point, the distance between adjacent C_α atoms is roughly 3.85 Å.

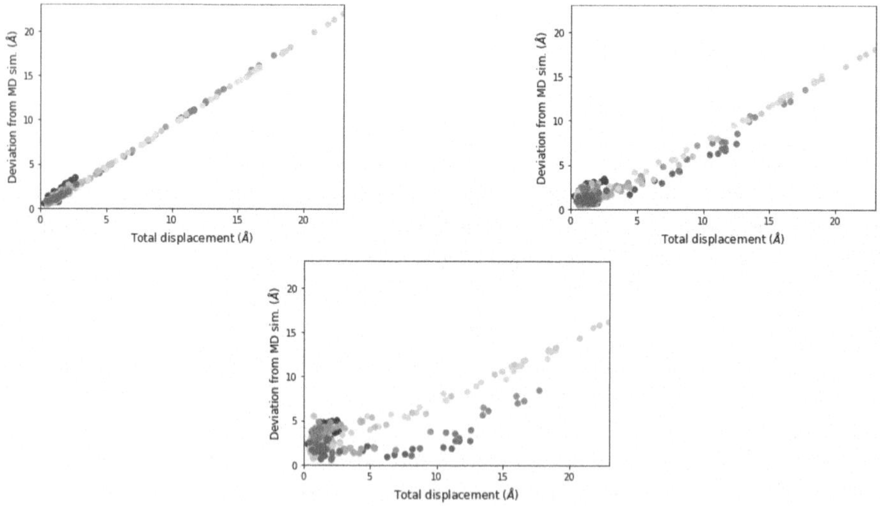

Fig. 7. Atomic displacements of reconstruction after 1500 iterations using data with SNR = 0.1 for $N = 10$ (left), $N = 50$ (middle), and $N = 100$ (right).

5 Conclusions

In this work, we set out to take an initial step towards direct heterogeneous Cryo-EM reconstruction of macromolecular structures through combining recent advancements into a new method for the homogeneous case. In particular, we aimed to gain insights for tackling the heterogeneous scenario from in the homogeneous case.

We posed the reconstruction problem as a variational problem in (6), using a regularization term from [4]. We found in Fig. 4 that reconstruction results were improved by modifying the regularisation term by more strongly enforcing the preservation of interatomic distances of atoms that are close to each other along the backbone chain. Further, we found in Fig. 6 that if given enough data, we are able to reconstruct the transition from closed to open state. However, the results in Fig. 7 show that we need more than order 10 images for a good reconstruction – 10 images do not even give a better reconstruciton than the template. The number of images of a single conformation required could prove a difficulty when considering the heterogeneous reconstruction problem. For future work, the main challenge will be to reduce the number of images necessary for reconstruction of a specific conformation for realistic noise levels. At this point, 100 images are needed to get close to inter-atomic resolution for noise levels that are on the edge of being realistic. Regularization will be key as there is simply not enough information in a single projection image to find a reconstruction. A potentially promising approach could be to learn or optimize over the weight matrix $\mathbf{A}$ rather than hand-craft it, which might allow us to move towards the heterogeneous case. This setup could be particularly promising as it has been observed that dynamics of proteins look low-dimensional under the Riemannian structure generated by the regularizer used in this work [4].

Acknowledgments. Krook and Öktem acknowledges supported from Swedish Research Council grant 2020-03107.

References

1. Beckstein, O., Seyler, S.L., Kumar, A.: Simulated trajectory ensembles for the closed-to-open transition of adenylate kinase from DIMS MD and FRODA (2018)
2. Castells-Graells, R., et al.: Cryo-EM structure determination of small therapeutic protein targets at 3 Å-resolution using a rigid imaging scaffold. Proc. Natl. Acad. Sci. **120**(37), e2305494120 (2023)
3. Chen, M., Ludtke, S.J.: Deep learning-based mixed-dimensional gaussian mixture model for characterizing variability in cryo-EM. Nat. Methods **18**(8), 930–936 (2021)
4. Diepeveen, W., Esteve-Yagüe, C., Lellmann, J., Öktem, O., Schönlieb, C.-B.: Riemannian geometry for efficient analysis of protein dynamics data. Proc. Natl. Acad. Sci. **121**(33), e2318951121 (2024)

5. Diepeveen, W., Lellmann, J., Öktem, O., Schönlieb, C.-B.: Regularizing orientation estimation in cryogenic electron microscopy three-dimensional map refinement through measure-based lifting over riemannian manifolds. SIAM J. Imag. Sci. **16**(3), 1440–1490 (2023)
6. Donnat, C., Levy, A., Poitevin, F., Zhong, E.D., Miolane, N.: Deep generative modeling for volume reconstruction in cryo-electron microscopy. J. Struct. Biol. **214**(4), 107920 (2022)
7. Esteve-Yagüe, C., Diepeveen, W., Öktem, O., Schönlieb, C.-B.: Spectral decomposition of atomic structures in heterogeneous cryo-EM. Inverse Prob. **39**(3), 034003 (2023)
8. Gilles, M.A., Singer, A: A bayesian framework for cryo-EM heterogeneity analysis using regularized covariance estimation. BioRxiv (2023)
9. Hamitouche, I., Jonic, S.: DeepHEMNMA: ResNet-based hybrid analysis of continuous conformational heterogeneity in cryo-EM single particle images. Front. Mol. Biosci. **9**, 965645 (2022)
10. Jin, Q., et al.: Iterative elastic 3D-to-2D alignment method using normal modes for studying structural dynamics of large macromolecular complexes. Structure **22**(3), 496–506 (2014)
11. Jumper, J., et al.: Highly accurate protein structure prediction with AlphaFold. Nature **596**(7873), 583–589 (2021)
12. Kingma, D.P.: Auto-encoding variational Bayes. arXiv preprint arXiv:1312.6114 (2013)
13. Moscovich, A., Halevi, A., Andén, J., Singer, A.: Cryo-EM reconstruction of continuous heterogeneity by Laplacian spectral volumes. Inverse Prob. **36**(2), 024003 (2020)
14. Nashed, Y., et al.: Heterogeneous reconstruction of deformable atomic models in Cryo-EM. arXiv preprint arXiv:2209.15121 (2022)
15. Öktem, O.: Mathematics of electron tomography (2015)
16. Prins, V., Diepeveen, W., Bekkers, E.J., Öktem, O.: Physics-informed geometric regularization of heterogeneous reconstructions in cryo-EM. In: ICLR 2024 Workshop on Generative and Experimental Perspectives for Biomolecular Design (2024)
17. Rosenbaum, D., et al.: Inferring a continuous distribution of atom coordinates from cryo-EM images using VAEs. arXiv preprint arXiv:2106.14108 (2021)
18. Scheres, S.: RELION: implementation of a bayesian approach to cryo-EM structure determination. J. Struct. Biol. **180**(3), 519–530 (2012)
19. Tirion, M.M.: Large amplitude elastic motions in proteins from a single-parameter, atomic analysis. Phys. Rev. Lett. **77**(9), 1905 (1996)
20. Toader, B., Sigworth, F.J., Lederman, R.R.: Methods for cryo-EM single particle reconstruction of macromolecules having continuous heterogeneity. J. Mol. Biol. **435**(9), 168020 (2023)
21. Woollard, G., Shekarforoush, S., Wood, F., Brubaker, M.A., Duc, K.D.: Physics aware inference for the cryo-EM inverse problem: anisotropic network model heterogeneity, global pose and microscope defocus. In: NeurIPS Workshop on Machine Learning for Structural Biology (2022)
22. Zhong, E.D., Bepler, T., Berger, B., Davis, J.H.: CryoDRGN: reconstruction of heterogeneous cryo-EM structures using neural networks. Nat. Methods **18**(2), 176–185 (2021)
23. Zhong,E.D., Lerer, A., Davis, J.H., Berger, B.: Exploring generative atomic models in cryo-EM reconstruction. arXiv preprint arXiv:2107.01331 (2021)

Enhanced Denoising and Convergent Regularisation Using Tweedie Scaling

Naïl Khelifa[1(✉)], Carola-Bibiane Schönlieb[2], and Ferdia Sherry[2]

[1] ENSAE Paris, Paris, France
nail.khelifa@ensae.fr
[2] DAMTP, University of Cambridge, Cambridge, UK
{cbs31,fs436}@cam.ac.uk

Abstract. The inherent ill-posed nature of image reconstruction problems, due to limitations in the physical acquisition process, is typically addressed by introducing a regularisation term that incorporates prior knowledge about the underlying image. The iterative framework of Plug-and-Play methods, specifically designed for tackling such inverse problems, achieves state-of-the-art performance by replacing the regularisation with a generic denoiser, which may be parametrised by a neural network architecture. However, these deep learning approaches suffer from a critical limitation: the absence of a control parameter to modulate the regularisation strength, which complicates the design of a convergent regularisation. To address this issue, this work introduces a novel scaling method that explicitly integrates and adjusts the strength of regularisation. The scaling parameter enhances interpretability by reflecting the quality of the denoiser's learning process, and also systematically improves its optimisation. Furthermore, the proposed approach ensures that the resulting family of regularisations is provably stable and convergent.

Keywords: Image Reconstruction · Plug-and-Play · Convergent Regularisation

1 Introduction

Image Reconstruction. In various fields, physical devices capture real-world scenes and convert them into digital form. The image reconstruction (IR) problem involves reversing this process by estimating a realistic input image that best matches the observed data, which is typically a degraded version of the original scene due to limitations in the acquisition system. Mathematically, if $\mathcal{X} = \mathbb{R}^n$ denotes the space of images and $\mathcal{Y} = \mathbb{R}^m$ that of measurements, the physics behind a measurement $y \in \mathcal{Y}$ and its original image $x \in \mathcal{X}$ is commonly modeled by a linear operation,

$$y = \mathrm{A}x + \xi, \tag{1}$$

Work completed during a research visit to DAMTP at the University of Cambridge.

T. A. Bubba et al. (Eds.): SSVM 2025, LNCS 15667, pp. 185–197, 2025.
https://doi.org/10.1007/978-3-031-92366-1_15

where $A : \mathcal{X} \to \mathcal{Y}$ is the *forward operator* and ξ is an additive white Gaussian noise (AWGN). The reconstruction problem writes as the minimisation over $\mathcal{X}$ of a data-fidelity term, often and here taken to be $\frac{1}{2}\|Ax - y\|^2$. However, this problem is commonly *ill-posed*, meaning that its solution is either (or both) not unique or highly sensitive to changes in the initial data. To address this issue, a standard approach is to incorporate regularisation by reformulating the IR problem as

$$\min_{x \in \mathcal{X}} \left\{ \frac{1}{2}\|Ax - y\|^2 + \lambda g(x) \right\}, \tag{2}$$

where $g : \mathcal{X} \to \mathbb{R}$ is the regularisation term and $\lambda \geq 0$ is a term controlling its intensity. By viewing x as the realization of a random variable X with prior distribution p_X and y as the realization of a random variable Y with a conditional density $p_{Y|X}$, the *maximum a posteriori* (MAP) estimate is defined by,

$$\max_{x \in \mathcal{X}} p_{X|Y}(x|y) = \min_{x \in \mathcal{X}} \left\{ -\log p_{Y|X}(y|x) - \log p_X(x) \right\}. \tag{3}$$

Identifying problems (2) and (3), the MAP links the data-fidelity term to the likelihood and the regularisation to the prior.

Plug-and-Play. In most cases, problem (2) does not have an analytical solution, necessitating the use of iterative optimisation algorithms. First-order proximal splitting methods [5] such as Proximal Gradient Descent (PGD, [15]), Half-Quadratic Splitting (HQS, [9]) and Alternating Direction Method of Multipliers (ADMM, [3]) are a common tool for minimising the sum of two functions. These methods use the proximal operator of a functional f, defined as,

$$\operatorname{prox}_f(x) = \operatorname*{argmin}_z \left\{ \frac{1}{2}\|x - z\|^2 + f(z) \right\}.$$

In the context of IR, the Plug-and-Play (PnP) framework [22] extends these methods by substituting the proximal operator of the regularisation term with a generic Gaussian denoiser, i.e. an operator designed to solve the Gaussian denoising task that corresponds to the physics (1) with $A = I_n$ and ξ an AWGN. Initially applied with non-deep denoisers such as BM3D [4] or pseudo-linear denoisers [8,14], more recent approaches [1,13,19,20,24] rely on denoisers parametrised with deep convolutional neural networks such as DnCNN [26], IR-CNN [27] or DRUNet [25]. The PnP framework has been applied in conjunction with first-order proximal splitting algorithms yielding PnP-HQS [27], PnP-ADMM [16,17], and PnP-PGD [21].

Problem. While these deep learning methods yield impressive restoration results, they come with notable limitations. In particular, their interpretation as regularisation terms becomes problematic within the framework of convergent regularisation. These models are typically trained at fixed noise levels, yet their penalising role in a PnP setting, along with the requirement for convergent regularisation, calls for adjustable control over the regularisation strength. Thus, it is desirable to devise a general approach that not only introduces an interpretable intensity control to a pre-trained denoiser but that also leads to convergent regularisation—the method introduced in [23] does not, see Sect. 5.3.

Related Works. In [23], the authors propose a methodology for adjusting the regularisation strength of a denoiser by applying a scaling, which is motivated by interpreting the denoiser as the proximal operator of a 1-homogeneous functional. In that work, the scaling is shown to improve IR performance, but the question of convergent regularisation is not studied, and in fact, as we show later in this work, it does not generally give convergent regularisation. The problem of showing convergent regularisation for PnP methods has received some attention recently [6,10]. In particular, the work in [6] introduces a theoretical framework for constructing a convergent regularisation for the PnP-PGD; however, the authors do not propose a concrete method to implement this construction with generic denoisers. Let us take the chance to contrast this line of work with a different but related line of work on proving that PnP methods define convergent iterations, a topic which has received considerably more attention from various perspectives recently [11,14,17]: having convergent iterations is necessary, but not sufficient, to make sense of convergent regularisation in our setting.

Main Contributions. This paper bridges the gap between the two contributions in [6,23] by introducing a novel framework for constructing convergent regularisation through a scaling transformation of a single pre-trained denoiser, aligning with the setting proposed in [6]. Additionally, it is shown that the introduced scale parameter has a highly interesting interpretation. Indeed, a specific value of the scale parameter consistently enhances the performance of deep denoisers trained with the $\mathcal{L}^2$ loss and acts as a quality indicator for the training.

Notations. Throughout this work, for simplicity and seeing images and measurements as high-dimensional tensors, let $\mathcal{X} = \mathbb{R}^n$ and $\mathcal{Y} = \mathbb{R}^m$. The noise level is denoted $\sigma > 0$. Let ξ be a random variable distributed according to $\mathcal{N}(0, \mathrm{I}_n)$ and let $\mathbf{D} : \mathcal{Y} \to \mathcal{X}$ denote a Gaussian denoiser trained to minimise the $\mathcal{L}^2$ loss at a noise level σ. Let also p_X denote the prior distribution of clean images over $\mathcal{X}$, G_σ the density of the Gaussian distribution $\mathcal{N}(0, \sigma^2 \mathrm{I}_n)$, and $p_\sigma = p_X * G_\sigma$ the distribution of noisy images at noise level σ over $\mathcal{Y}$. The minimum mean square error (MMSE) denoiser, is denoted and defined by

$$\mathbf{D}^{\mathrm{MMSE}} := \underset{\mathbf{D}\ \text{measurable}}{\operatorname{argmin}}\ \mathbb{E}_{X,Y}[\|\mathbf{D}(Y) - X\|^2] = \mathbb{E}[X|Y].$$

For any denoiser $\mathbf{D}$, the mean squared error of Gaussian denoising is denoted,

$$\mathcal{L}^2(\mathbf{D}) = \mathbb{E}_{X,\xi}[\|\mathbf{D}(X + \sigma\xi) - X\|^2].$$

Outline. The first section presents the proposed scaling-based regularisation method, providing a justification for its theoretical foundation and underlying intuition. The second section explores the interpretation of the scaling parameter that governs the introduced regularisation. Finally, the last section demonstrates that the resulting family of regularisations forms a convergent regularisation.

2 Construction

In this section, the aim is twofold: introduce a scaling-based control parameter for a pre-trained denoiser and provide insights into the meaning of the parameter.

Homogeneous Scaling. In [23], the scaling approach is motivated by the observation that, if $h : \mathcal{X} \to \mathbb{R}$ is a 1-homogeneous function (i.e. $h(\tau u) = \tau h(u)$ for all $\tau > 0, u \in \mathcal{X}$) whose proximal operator is well-defined, then for all $x \in \mathcal{X}$,

$$\operatorname{prox}_{\tau h}(x) = \tau \operatorname{prox}_h \left(\frac{x}{\tau}\right).$$

This implies that if the initial denoiser is given by $\mathbf{D} = \operatorname{prox}_h$ for some 1-homogeneous h, then, for all $\delta > 0$ and for all $x \in \mathcal{X}$,

$$\frac{1}{\delta}\mathbf{D}(\delta x) = \frac{1}{\delta} \operatorname{prox}_h(\delta x) = \operatorname{prox}_{\frac{h}{\delta}}(x).$$

From this approach, henceforth referred to as *homogeneous scaling*, we aim to retain the concept of a scaling transformation while discarding the strong assumption on the proximal expression of $\mathbf{D}$.

Proposed Approach. The proposed approach builds upon Tweedie's identity [7], which states,

$$\mathbf{D}^{\mathrm{MMSE}} = \operatorname{Id} + \sigma^2 \nabla \log p_\sigma. \tag{4}$$

The objective is to define a method ensuring convergent regularisation. This will be achieved by studying the behaviour of the identity as σ approaches zero. The idea here is to provide an intuition about the proposed method. Assuming p_X is sufficiently smooth ($\mathcal{C}^3$ suffices, for example), p_σ satisfies the heat equation with time variable $t = \sigma^2$,

$$\partial_{\sigma^2} p_\sigma = \Delta p_\sigma, \qquad \partial_{\sigma^2} \nabla p_\sigma = \nabla(\Delta p_\sigma). \tag{5}$$

Using the first-order Taylor expansion of p_σ and ∇p_σ around $\sigma^2 = 0$, and incorporating (5), the following expressions are obtained,

$$\begin{cases} p_\sigma = p_X + \partial_{\sigma^2} p_\sigma\big|_{\sigma^2=0} \sigma^2 + \mathcal{O}_0(\sigma^4) = p_X + \sigma^2 \Delta p_X + \mathcal{O}_0(\sigma^4), \\ \nabla p_\sigma = \nabla p_X + \partial_{\sigma^2} \nabla p_\sigma\big|_{\sigma^2=0} \sigma^2 + \mathcal{O}_0(\sigma^4) = \nabla p_X + \sigma^2 \nabla \Delta p_X + \mathcal{O}_0(\sigma^4). \end{cases}$$

As $\sigma \to 0$, neglecting higher-order terms in the expansion yields,

$$\nabla \log p_\sigma = \frac{\nabla p_\sigma}{p_\sigma} = \frac{\nabla p_X + \sigma^2 \nabla \Delta p_X}{p_X + \sigma^2 \Delta p_X} = \frac{\nabla p_X}{p_X} + \mathcal{O}_0(\sigma^2) \sim_{\sigma \to 0} \nabla \log p_X.$$

Thus, for Tweedie's identity, it follows that,

$$\operatorname{Id} + \sigma^2 \nabla \log p_\sigma \sim_{\sigma \to 0} \operatorname{Id} + \sigma^2 \nabla \log p_X.$$

This leads to the introduction of a scaling parameter to control regularisation, expressed in Tweedie's identity as,

$$\mathrm{Id} + \frac{\sigma^2}{\delta^2}\nabla \log p_X = \mathrm{Id} + \frac{1}{\delta^2}\left(\mathrm{Id} + \sigma^2 \nabla \log p_X - \mathrm{Id}\right) \sim_{\sigma \to 0} \mathrm{Id} + \frac{\mathbf{D}^{\mathrm{MMSE}} - \mathrm{Id}}{\delta^2}.$$

The key observation is that training a denoiser $\mathbf{D}$ to minimise the $\mathcal{L}^2$-loss for a noise level σ effectively approximates $\mathbf{D}^{\mathrm{MMSE}}$. By substituting $\mathbf{D}$ for $\mathbf{D}^{\mathrm{MMSE}}$ in the above expression, the *Tweedie Scaling Method* is defined as,

$$\mathbf{D}_\delta = \mathrm{Id} + \frac{\mathbf{D} - \mathrm{Id}}{\delta^2}.$$

This formulation assumes no specific structure for $\mathbf{D}$ and relies on two key approximations: (1) the approximation of $\mathbf{D}$ as $\mathbf{D}^{\mathrm{MMSE}}$ through proper $\mathcal{L}^2$-training, and (2) the validity of expansions as $\sigma \to 0$. In fact, as we will show in Sect. 4.2, this scaling approach can result in convergent regularisation even in a setting in which we do not assume that the denoiser is close to being optimal.

3 Enhanced Denoising and Interpretation of δ

To interpret δ, recall that the construction of $\mathbf{D}_\delta$ in the previous section started with an approximation of $\mathbf{D}^{\mathrm{MMSE}}$ by $\mathbf{D}$, where the accuracy was related to the performance of $\mathbf{D}$ in minimizing the $\mathcal{L}^2$ loss, or equivalently, to the tightness of the following bound—which holds by definition of $\mathbf{D}^{\mathrm{MMSE}}$:

$$\mathcal{L}^2(\mathbf{D}_{\mathrm{MMSE}}) \leq \mathcal{L}^2(\mathbf{D}) \tag{6}$$

To proceed with the interpretation, this bound is refined by the following result,

Proposition 1. *For*

$$\delta^2_{\mathrm{opt}} = -\frac{\mathbb{E}[\|\mathbf{D}(X+\sigma\xi) - (X+\sigma\xi)\|^2]}{\mathbb{E}[\langle \sigma\xi, \mathbf{D}(X+\sigma\xi) - (X+\sigma\xi)\rangle]},$$

it holds that,

$$\mathcal{L}^2(\mathbf{D}_{\mathrm{MMSE}}) \leq \mathcal{L}^2(\mathbf{D}_{\delta_{\mathrm{opt}}}) \leq \mathcal{L}^2(\mathbf{D}) \tag{7}$$

Proof. The proof directly follows from studying the function $\delta \mapsto \mathcal{L}^2(\mathbf{D}_\delta)$.

Now, If equality holds in (6), perfect training of $\mathbf{D}$ follows, implying:

$$\mathcal{L}^2(\mathbf{D}_{\mathrm{MMSE}}) = \mathcal{L}^2(\mathbf{D}_{\delta_{\mathrm{opt}}}) = \mathcal{L}^2(\mathbf{D}) \implies \mathbf{D}_{\delta_{\mathrm{opt}}} = \mathbf{D} \iff \delta^2_{\mathrm{opt}} = 1$$

Thus, the value of δ_{opt} reveals the quality of the initial denoiser's training. It is noteworthy that, when $\mathbf{D}$ is not perfectly trained, it is possible to construct a denoiser with a lower theoretical training error simply by choosing the appropriate value of the transformation parameter.

4 Plug-and-Play Properties of $(\mathbf{D}_\delta)_\delta$

Attention is now directed towards the behaviour of $(\mathbf{D}_\delta)_\delta$ within a Plug-and-Play framework. Specifically, it is first demonstrated that the Tweedie scaling method (1) defines convergent iterations for PnP-PGD algorithms under mild assumptions and (2) establishes a convergent regularisation for the original problem, again under mild assumptions.

4.1 Convergence of Iterations

The focus is on the convergence of iterations within the PnP-PGD [21] algorithm which, for a mean square data-fidelity term and considering $\mathbf{D}_\delta$, iterates as:

$$x^{(i+1)} = \mathbf{D}_\delta(x^{(i)} - \tau \mathrm{A}^\top(\mathrm{A}x^{(i)} - y)) := \mathbf{D}_\delta(\mathbf{G}(x^{(i)})) := \mathbf{T}_\delta(x^{(i)}) \qquad (8)$$

where $\mathbf{G}(x) := x - \tau \mathrm{A}^\top(\mathrm{A}x - y)$ denotes the gradient-step operator and $\mathbf{T}_\delta(x) := \left(1 - 1/\delta^2\right)\mathbf{G}(x) + (\mathbf{D} \circ \mathbf{G})(x)/\delta^2$. In this setting, we have the following result:

Proposition 2 ($\mathbf{T}_\delta$ defines convergent iterations). *Assume that the step size τ is chosen with $0 < \tau \leq 1/\|\mathrm{A}^\top \mathrm{A}\|$ and that $\mathbf{D}$ is non-expansive. Then, $\mathbf{T}_\delta$ is $\frac{\delta^2}{2\delta^2-1}$-averaged. If $\mathbf{T}_\delta$ has at least one fixed point and $\delta^2 > 1$, then the iterations (8) converge to a fixed point of $\mathbf{T}_\delta$.*

Proof. The proof follows in three steps:

1. Recalling that an operator $\mathbf{O}$ is said to be θ-averaged for $\theta \in (0,1)$ if it can be expressed as $\mathbf{O} = \theta \mathbf{R} + (1-\theta)\,\mathrm{Id}$, where $\mathbf{R}$ is non-expansive, and recalling that $\mathbf{D}_\delta = \left(1 - 1/\delta^2\right)\mathrm{Id} + \mathbf{D}/\delta^2$, one directly gets that, assuming $\mathbf{D}$ to be non-expansive, implies that $\mathbf{D}_\delta$ is $1/\delta^2$-averaged.
2. Moreover, it is recalled from [2] that if $\mathbf{O}_1$ and $\mathbf{O}_2$ are θ_1-averaged and θ_2-averaged operators, respectively, then their composition $\mathbf{O}_1 \circ \mathbf{O}_2$ is θ-averaged with
$$\theta = \frac{\theta_1 + \theta_2 - 2\theta_1\theta_2}{1 - \theta_1\theta_2}.$$
Combining this, the $1/\delta^2$-averagedness of $\mathbf{D}_\delta$ and, the result from [2] that states that under the stepsize condition $0 < \tau \leq 1/\|\mathrm{A}^\top \mathrm{A}\|$, $\mathbf{G}$ is firmly non-expansive, i.e. 1/2-averaged and the fact that $\mathbf{T}_\delta = \mathbf{D}_\delta \circ \mathbf{G}$ it follows that $\mathbf{T}_\delta$ is $\delta^2/(2\delta^2 - 1)$-averaged.
3. According to a corollary of the Krasnosel'skii-Mann theorem [2], if $\mathbf{T}_\delta$ admits at least one fixed point and if $\delta^2/(2\delta^2 - 1) \in (0,1)$-a condition equivalent to $\delta^2 > 1$-then the sequence of iterations (8) converges to a fixed point.

The central assumption underlying this result is the non-expansiveness of $\mathbf{D}$, which is influenced by the architecture of $\mathbf{D}$. In the experiments, we will use a standard neural network denoiser, the DRUNet (see Sect. 5 for more details). As demonstrated in Table 1, the maximum value of $\|\mathbf{D}(y_1) - \mathbf{D}(y_2)\|/\|y_1 - y_2\|$ for two images y_1, y_2 from the (noisy) CBSD68 patch-extend dataset is below one for sufficiently large noise levels, supporting the assumption of non-expansiveness.

Table 1. $\max\{\frac{\|\mathbf{D}(y_1)-\mathbf{D}(y_2)\|_2}{\|y_1-y_2\|_2} | y_1, y_2 \in \mathcal{D}_{\text{CBSD68}}(\sigma^2)\}$ as the noise level σ^2 varies.

	$\sigma = 0.01$	$\sigma = 0.03$	$\sigma = 0.05$	$\sigma = 0.1$	$\sigma = 0.15$	$\sigma = 0.2$	$\sigma = 0.25$	$\sigma = 0.3$
max	1.005	0.997	0.994	0.985	0.966	0.950	0.919	0.905

4.2 Convergent Regularisation Properties

The properties of convergent regularisation are now considered. In [6], a theoretical framework for studying convergent regularisation in PnP methods is introduced, but no general method is provided for constructing it from a pre-trained denoiser $\mathbf{D}$. This work demonstrates that the proposed Tweedie scaling method defines a convergent regularisation that satisfies the framework in [6]. For $\delta > 0$ and noisy measurements y, the set of fixed points of the iterations of the PnP-PGD algorithm is defined as follows,

$$\text{PnP}(\delta, y) := \text{Fix}(\mathbf{D}_\delta \circ (\text{Id} - \tau \text{A}^\top(\text{A}(\cdot) - y))), \tag{9}$$

where $\text{Fix}(\mathbf{T}) = \{x \in \mathcal{X} \mid \mathbf{T}(x) = x\}$. By adapting the result of Definition 2.5 in [6] to our case, $(\mathbf{D}_\delta)_\delta$ define a convergent regularisation for the problem $\text{argmin}_x \frac{1}{2}\|\text{A}x - y\|^2$ over $E \subseteq \mathcal{X}$, if the following conditions are satisfied,

(a) **Stability:** For all $\delta > 0$, $\mathbf{D}_\delta$ is continuous.
(b) **Convergence:** For all $y \in E$, $(1/\delta_k^2)_{k\in\mathbb{N}} \subset (0,\infty)^{\mathbb{N}}$ converging to 0 and all $y_k \in \mathcal{Y}$ with $\|y - y_k\| \leq 1/\delta_k^2$, the sequence $(\text{PnP}(1/\delta_k^2, y_k))_{k\in\mathbb{N}}$ has a convergent subsequence and the limit of every convergent subsequence of $(\text{PnP}(1/\delta_k^2, y_k))_{k\in\mathbb{N}}$ is a minimiser of $x \mapsto \frac{1}{2}\|\text{A}x - y\|^2$.

Stability. To demonstrate stability, Theorem 3.10 and Corollary 3.11 from [6] are considered. In the statement of these results, four conditions are imposed: a contractivity condition on $\mathbf{D}_\delta$ and three conditions on the data-fidelity function. It has been shown in [6] that the data-fidelity between a clean image x and its noisy measurements y defined by $\frac{1}{2}\|\text{A}x - y\|^2$ satisfies these conditions. Thus, the only assumption that remains to be verified is the contractivity of $\mathbf{D}_\delta$ for all $\delta > 0$, which is equivalent to the contractivity of the original denoiser $\mathbf{D}$. By directly applying Corollary 3.11 from [6], it follows that

Assuming that $\mathbf{D}_\delta$ *is contractive,* $x \mapsto \text{PnP}(\delta, x)$ *is continuous for all* $\delta > 0$.

Convergence. To prove convergence, Theorem 3.14 from [6] is applied, requiring that $(\mathbf{D}_\delta)_\delta$ be an admissible family of denoisers, defined by,

(A1) $\forall \delta > 0$, $\mathbf{D}_\delta(\cdot) : \mathbb{R}^n \to \mathbb{R}^n$ is a contraction.
(A2) $\mathbf{D}_\delta(\cdot) \longrightarrow \text{Id}_{\mathbb{R}^n}$ strongly, pointwise as $\delta \to +\infty$.
(A3) $\mathbf{D}_\delta(\cdot) \longrightarrow \text{Id}_{\mathbb{R}^n}$ weakly, uniformly over bounded sets as $\delta \to +\infty$.
(A4) $\exists E \subseteq \mathbb{R}^n\ \forall x \in E : \|\mathbf{D}_\delta(x) - x\| = \mathcal{O}(1 - \text{Lip}(\mathbf{D}_\delta))$.

The first condition is overly restrictive regarding the architecture defining $\mathbf{D}$, and it is desirable to move beyond it. In [6], a rescaling method is proposed to relax the contractivity condition **(A1)** to non-expansiveness. More precisely, it is suggested that if the family is non-expansive and satisfies **(A2)** and **(A3)**, then it is always possible to construct a family of admissible denoisers by considering $(\gamma(\frac{1}{\delta^2})\mathbf{D}_\delta)_{\delta>0}$, where $\gamma : [0,\infty) \to (0,1]$ is a strictly decreasing and continuous function at zero, satisfying $\gamma(0)=1$, and where $\|\mathbf{D}_\delta(x)-x\| = \mathcal{O}(1-\gamma(\frac{1}{\delta^2}))$ as $\frac{1}{\delta^2} \to 0$. For instance, a valid choice for γ is $\gamma\left(1/\delta^2\right) = \delta^2/(1+\delta^2)$. It is noted in [6] that if conditions **(A2)** and **(A3)** are satisfied by $(\mathbf{D}_\delta)_{\delta>0}$, then they are also satisfied by $(\gamma(\frac{1}{\delta^2})\mathbf{D}_\delta)_{\delta>0}$, and in fact this rescaled family then satisfies **(A4)**. Consequently, in what follows, by a slight abuse of notation, $\mathbf{D}_\delta$ will be identified with this rescaled version. Thus, only conditions **(A2)** and **(A3)** need to be verified, since it was shown in the previous section that condition **(A1)**, now relaxed to non-expansiveness, is empirically satisfied. Since we have assumed to be in a finite-dimensional setting, weak and strong convergence coincide, so it is clear that **(A3)** implies **(A2)**, and we only need to prove **(A3)**:

(A3): Let $B \subseteq \mathbb{R}^n$ be a bounded set with $R = \sup_{x\in B}\|x\|$. For any $x \in B$,

$$\|\mathbf{D}_\delta(x)-x\| = \left\|\frac{\mathbf{D}(x)-x}{\delta^2}\right\| \leq \frac{\|\mathbf{D}-\mathrm{Id}\,\|}{\delta^2}\|x\| \leq \frac{R\|\mathbf{D}-\mathrm{Id}\,\|}{\delta^2}$$

Now, the bound on the right-hand side is uniform over all $x \in B$ and tends to zero as $\delta \to \infty$, proving uniform convergence on bounded sets.

Hence, $(\mathbf{D}_\delta)_\delta$ is admissible, completing the proof that the proposed family defines a convergent regularisation.

5 Experiments

In all conducted experiments, the denoisers correspond to a DRUNet [25] architecture with 5 blocks, trained to denoise images corrupted by noise levels σ. To explicitly control the noise level, a dataset is constructed by augmenting the Berkeley segmentation dataset CBSD68 [12] through random erasing and random cropping operations, resulting in 8,000 images. Specifically, for each image $x \in \mathcal{D}_{\text{CBSD68}}$, its noisy measurements are generated as $y = \mathrm{A}x + \sigma\xi$, where $\xi \sim \mathcal{N}(0, \mathrm{I}_n)$ and A depends on the typical task. Unless stated otherwise, as in the section on δ^2_{opt} where the noise level σ varies, it is set to 0.1.

5.1 Optimal Value δ_{opt}

In this section, the results of Sect. 3 concerning the optimal scaling parameter δ_{opt} will be studied experimentally, and its interpretation as an indicator of the training quality of the denoiser $\mathbf{D}$ will be validated. Since the denoising task becomes harder to learn at lower noise levels, given fixed training conditions (architecture, data, and training hyperparameters), a denoiser trained on lower noise levels is expected to be less effectively trained. In this experiment, we consider four noise levels, $\sigma_1 < \sigma_2 < \sigma_3 < \sigma_4$, and four denoisers, $\mathbf{D}_1, \mathbf{D}_2, \mathbf{D}_3, \mathbf{D}_4$, trained under the fixed procedure described earlier. As such, $\mathbf{D}_1$ is the least well-trained, while $\mathbf{D}_4$ is the best-trained. According to the intuition outlined in Sect. 3, where the value of δ_{opt} reflects the training quality, with $\delta_{\text{opt}} = 1$ indicating perfect training, we anticipate the relationship $1 < \delta_{\text{opt},4} < \delta_{\text{opt},3} < \delta_{\text{opt},2} < \delta_{\text{opt},1}$. To observe this, the function $\delta \mapsto \mathbb{E}_{X,\xi}\left[\|\mathbf{D}_\delta(x + \sigma\xi) - x\|^2\right]$ (computed as the empirical mean of the norm over the augmented dataset) is plotted for $\sigma \in \{0.01, 0.07, 0.10, 0.12\}$. The experimental results align with and confirm this intuition as shown in Fig. 1.

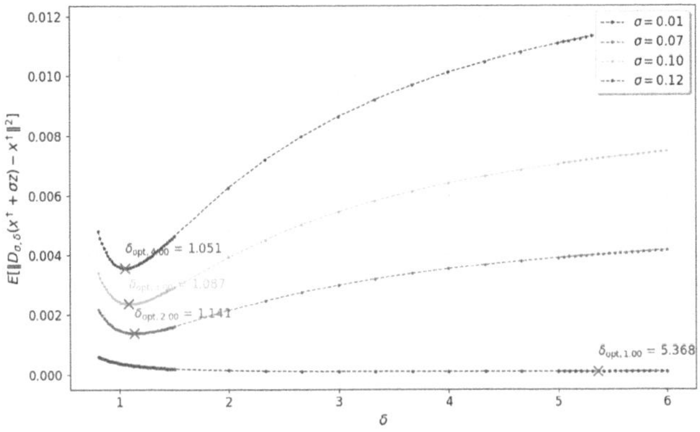

Fig. 1. $\delta \mapsto \mathbb{E}_{X,\xi}\left[\|\mathbf{D}_\delta(x + \sigma\xi) - x\|^2\right]$ for $\sigma \in \{0.01, 0.07, 0.10, 0.12\}$.

5.2 Stability

In the previous section, the following stability result was stated:

Assuming that $\mathbf{D}_\delta$ *is contractive,* $x \mapsto \text{PnP}(\delta, x)$ *is continuous for all* $\delta > 0$.

This section provides an empirical evaluation of the continuity of the operator $x \mapsto \text{PnP}(\delta, x)$. A sequence $(y_k)_k$ satisfying $\|y_k - y\| \to 0$ as $k \to +\infty$ is considered. More precisely, a clean image x is given, and $y = \text{A}x$ is considered, where A corresponds to a physics operator for an inpainting task, where 20% of the pixels in the image are masked. For all $k \in \mathbb{N}^*$, y_k is defined as $y_k = y + \frac{\sigma}{k}\xi$, where $\xi \sim \mathcal{N}(0, \text{I}_n)$. δ is here fixed to its optimal value for $\sigma = 0.1$, i.e. as before to 1.09. The convergence of $(\text{PnP}(\delta, y_k))_k$ towards $\text{PnP}(\delta, y)$ is then illustrated in Fig. 2.

5.3 Empirical Evidence of Convergent Regularisation

This section aims to (1) empirically demonstrate that the proposed Tweedie scaling method defines a convergent regularisation for the most basic tasks in PnP-PGD, and (2) show that this is not the case for the method proposed in [23]. Empirical results show that in the basic cases of denoising ($\text{A} = \text{I}_n$) and of inpainting, no convergence is observed for the homogeneous scaling approach,

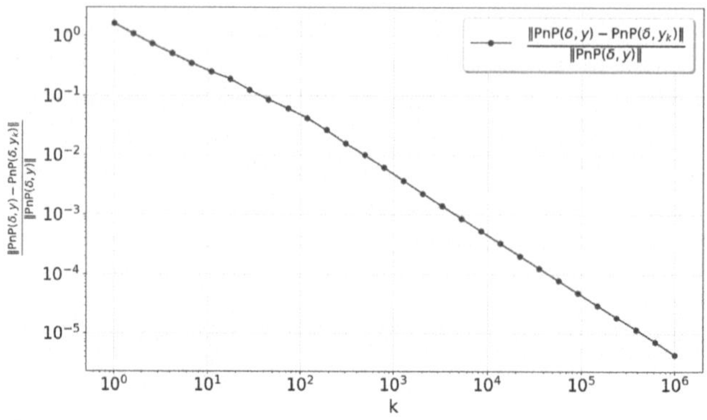

Fig. 2. Convergence of $(\text{PnP}(\delta, y_k))_k$ towards $\text{PnP}(\delta, y)$.

whereas convergence is observed for Tweedie scaling. The experiment consists of taking a clean image $x^\dagger$ in the CBSD68 dataset and setting $y_0 = \text{A}x^\dagger$. A logarithmic range of parameters δ is then considered, with a step size $\tau = 1$, and y_0 is perturbed to $y_\delta = y_0 + \frac{\sigma}{\delta}\xi$, where $\xi \sim \mathcal{N}(0, \text{I}_m)$. For each level of penalisation intensity δ, $x_\delta^\dagger$ is computed as the output of the PnP-PGD algorithm using the Tweedie transformed denoiser $\mathbf{D}_\delta$ and a fixed number of 300 iterations. The initial image x_0 is a tensor with all values equal to 0 (dark image). Two quantities are plotted in Fig. 3.

1. *Data consistency*: the convergence of $\|\text{A}x_\delta^\dagger - \text{A}x^\dagger\| / \|\text{A}x^\dagger\| \to 0$ as $\delta \to +\infty$, to show that the limit of the sequence $(x_\delta^\dagger)_\delta$ satisfies $\text{A}x_\delta^\dagger = \text{A}x^\dagger$.
2. *Convergence*: the convergence of $(x_\delta^\dagger)_\delta$ towards a limit x^* to show that there is a notion of convergence.

Regarding the second point, note that the limit x^* need not be the same as the clean image $x^\dagger$ when the measurements are incomplete, i.e. $\ker(\text{A}) \supsetneq \{0\}$. Indeed, just as in variational regularisation, where there is the notion of a g-minimising solution associated with (2) [18], the regularisation can overcome the non-uniqueness of the solution to the inverse problem, but can only be expected to recover the true clean image if this image is represented well by the prior information encoded in the regularisation.

The fundamental premise underlying the construction of the homogeneous scaling method is the existence of a 1-homogeneous functional h such that $\mathbf{D} = \text{prox}_h$. Under this assumption, it is evident (by variational regularisation theory) that the homogeneous scaling family $(\mathbf{D}_\delta)_\delta$ defines a convergent regularisation. Consequently, if convergence of this regularisation is not observed, it implies that the assumption does not hold for the particular $\mathbf{D}$ being trained, despite its otherwise generic nature. This observation highlights the limitations inherent in the homogeneous scaling approach.

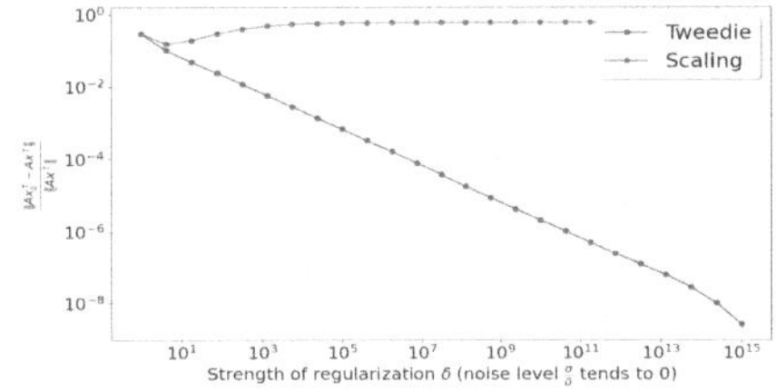

(a) Inpainting: data consistency

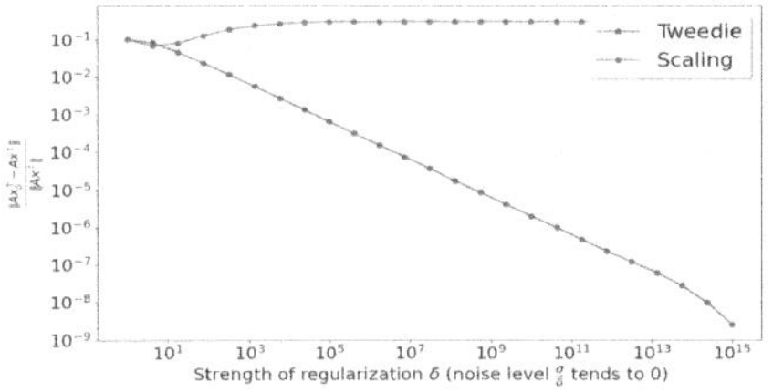

(c) Denoising: data consistency

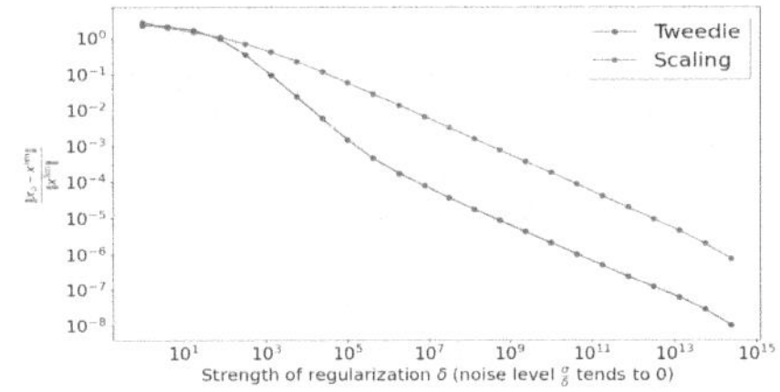

(b) Inpainting: convergence

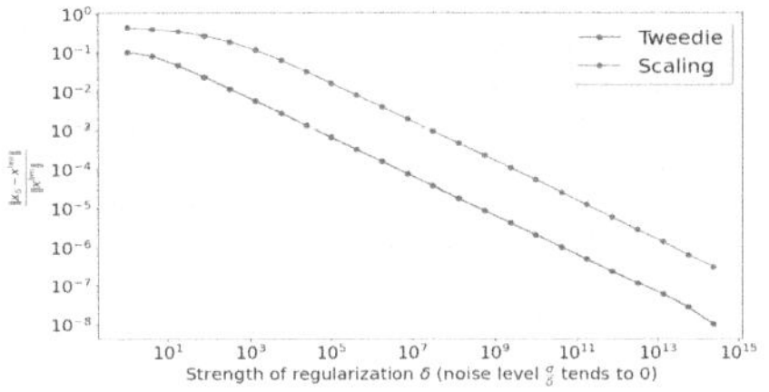

(d) Denoising: convergence

Fig. 3. Data consistency and convergence (or lack thereof) for the Tweedie scaling method (blue) and homogeneous scaling (green) in Inpainting and Denoising. (Color figure online)

6 Conclusion

The proposed Tweedie scaling method bridges the theoretical framework for convergent regularisation developed in [6] by introducing a strongly interpretable scaling parameter. More specifically, it enables principled and explicit modulation of the regularisation intensity in pre-trained denoisers, without relying on restrictive assumptions about the form of the denoiser as in [23]. The scaling parameter is shown to provide a meaningful characterization of training quality in deep learning-based denoisers. Empirical evidence confirms the framework's ability to guarantee convergence in PnP algorithms. A future direction could involve proposing new scaling techniques to improve the proposed method, including a study of the rate of convergence of the regularisation, investigating whether the property of convergent regularisation holds for other PnP algorithms, and adapting or optimising the scaling parameter according to the algorithms. Other potential research avenues include that of relaxing the assumption of non-expansiveness of the initial denoiser, and considering the infinite-dimensional setting, as is usually the case in the theory of inverse problems.

References

1. Ahmad, R., et al.: Plug-and-play methods for magnetic resonance imaging: Using denoisers for image recovery. IEEE SPM (2020)
2. Bauschke, H.H., Combettes, P.L.: Convex Analysis and Monotone Operator Theory in Hilbert Spaces. Springer (2011)

3. Boyd, S., Parikh, N., Chu, E., Peleato, B., Eckstein, J.: Distributed optimization and statistical learning via the alternating direction method of multipliers. Found. Trends Optimization (2011)
4. Chan, S.H., Wang, X., Elgendy, O.A.: Plug-and-play ADMM for image restoration: fixed-point convergence and applications. IEEE TCI (2016)
5. Combettes, P.L., Pesquet, J.C.: Proximal Splitting Methods in Signal Processing. Springer NY (2011)
6. Ebner, A., Haltmeier, M.: Plug-and-play image reconstruction is a convergent regularization method. IEEE TIP (2024)
7. Efron, B.: Tweedie's formula and selection bias. JASA (2011)
8. Gavaskar, R.G., Athalye, C.D., Chaudhury, K.N.: On plug-and-play regularization using linear denoisers. IEEE TIP (2021)
9. Geman, D., Yang, C.: Nonlinear image recovery with half-quadratic regularization. IEEE TIP (1995)
10. Hauptmann, A., Mukherjee, S., Schönlieb, C.B., Sherry, F.: Convergent regularization in inverse problems and linear plug-and-play denoisers. FoCM (2024)
11. Hurault, S., Chambolle, A., Leclaire, A., Papadakis, N.: A relaxed proximal gradient descent algorithm for convergent plug-and-play with proximal denoiser. In: SSVM (2023)
12. Martin, D., Fowlkes, C., Tal, D., Malik, J.: A database of human segmented natural images and its application to evaluating segmentation algorithms and measuring ecological statistics. In: ICCV (2001)
13. Meinhardt, T., Möller, M., Hazirbas, C., Cremers, D.: Learning proximal operators: using denoising networks for regularizing inverse imaging problems. In: ICCV (2017)
14. Nair, P., Gavaskar, R.G., Chaudhury, K.N.: Fixed-point and objective convergence of plug-and-play algorithms. IEEE TCI (2021)
15. Passty, G.B.: Ergodic convergence to a zero of the sum of monotone operators in Hilbert space. J. Math. Anal. Appl. (1979)
16. Romano, Milanfar, E.: The little engine that could: regularization by denoising (RED). SIIMS (2017)
17. Ryu, E., Liu, J., Wang, S., Chen, X., Wang, Z., Yin, W.: Plug-and-play methods provably converge with properly trained denoisers. In: ICML (2019)
18. Scherzer, O., Grasmair, M., Grossauer, H., Haltmeier, M., Lenzen, F.: Variational Methods in Imaging. Springer (2009)
19. Sun, Y., Liu, J., Kamilov, U.S.: Block coordinate regularization by denoising. IEEE TCI (2020)
20. Sun, Y., Wu, Z., Xu, X., Wohlberg, B., Kamilov, U.S.: Scalable plug-and-play ADMM with convergence guarantees. IEEE TCI (2021)
21. Terris, M., Repetti, A., Pesquet, J.C., Wiaux, Y.: Building firmly nonexpansive convolutional neural networks. In: ICASSP (2020)
22. Venkatakrishnan, Wohlberg, B.: Plug-and-play priors for model based reconstruction. In: GlobalSIP (2013)
23. Xu, X., Liu, J., Sun, Y., Wohlberg, B., Kamilov, U.S.: Boosting the performance of plug-and-play priors via denoiser scaling. In: ACSSC (2020)
24. Yuan, X., Liu, Y., Suo, J., Dai, Q.: Plug-and-play algorithms for large-scale snapshot compressive imaging. In: CVPR (2020)
25. Zhang, K., Li, Y., Zuo, W., Zhang, L., Van Gool, L., Timofte, R.: Plug-and-play image restoration with deep denoiser prior. IEEE TPAMI (2022)

26. Zhang, K., Zuo, W., Chen, Y., Meng, D., Zhang, L.: Beyond a gaussian denoiser: residual learning of deep CNN for image denoising. IEEE TIP (2017)
27. Zhang, K., Zuo, W., Gu, S., Zhang, L.: Learning deep CNN denoiser prior for image restoration. In: CVPR (2017)

Machine and Deep Learning in Imaging

Max-Sparsity Atomic Autoencoders with Application to Inverse Problems

Ali Joundi[1(✉)], Yann Traonmilin[1], and Alasdair Newson[2]

[1] University of Bordeaux, CNRS, Bordeaux INP, IMB, UMR 5251, 33400 Talence, France
ali.joundi@u-bordeaux.fr
[2] ISIR, Sorbonne Université, Paris, France

Abstract. An atomic autoencoder is a neural network architecture that decomposes an image as a sum of low dimensional atoms. While it is efficient for image datasets which are well represented by this summation model, it is more limited for the representation of more generic images. In this article, we propose a new atomic model, the max-sparsity model to better represent images. We study some theoretical properties of this model and implement the corresponding atomic autoencoder. We show experimentally that it leads to a sparse decomposition of input images with interpretable low-level visual features. With this new architecture, we solve a super resolution inverse problem via a projected gradient descent that uses the trained network as a projection operator. The resulting estimation shows improved robustness compared to previous architectures.

Keywords: Atomic autoencoders · Sparse models · Image super-resolution

1 Introduction

An autoencoder is a neural network that approximates the identity function on a given dataset. By reducing the size of hidden layers, autoencoders produce low dimensional latent representations of elements of the dataset, an important property in the context of image processing. Formally, an autoencoder is the composition $f = f_D \circ f_E : \mathbb{R}^n \to \mathbb{R}^n$ of an encoder $f_E : \mathbb{R}^n \to \mathbb{R}^d$ that projects inputs into the latent space, and a decoder $f_D : \mathbb{R}^d \to \mathbb{R}^n$ that recovers them from their latent representations. To train an autoencoder on a dataset X, we minimize the loss function $\mathcal{L}$ defined by:

$$\mathcal{L}_X(f) = \mathcal{L}_X(f_D \circ f_E) := \sum_{x \in X} \|f_D \circ f_E(x) - x\|_2^2. \tag{1}$$

The benefits of accessing an explicit latent space are multiple. Autoencoders can generate new elements of the induced low dimensional space by decoding

T. A. Bubba et al. (Eds.): SSVM 2025, LNCS 15667, pp. 201–213, 2025.
https://doi.org/10.1007/978-3-031-92366-1_16

new latent vectors. If enough structure of the latent space is guaranteed, it allows to manipulate, edit or interpolate input images efficiently. In addition, autoencoders enable, by providing a low dimensional prior, the solving of ill-posed imaging problems such as super-resolution or deconvolution problems. However, training autoencoders with the sole constraint that it needs to recover the input can lead to a latent set of image representations with little structure, due to its non-convexity. This heavily limits the effectiveness of autoencoders for these tasks.

Adding more structure to the latent space has been an important subject of research in the field of generative models. In particular, atomic autoencoders, proposed in [11], are custom autoencoders (see Fig. 1) designed to decompose images into sparse representations, i.e. a sum of atoms from a given continuous dictionary. Given an image x, an atomic autoencoder encodes x into a latent vector $f_E(x) = \theta \in \mathbb{R}^{kd_0}$. The latent vector θ is divided into k atomic latent codes $\theta_i \in \mathbb{R}^{d_0}$ such that $\theta = (\theta_i)_{i=1}^k$. These codes are merged with a decoder having the form $f_D(\theta) = \sum_{i=1}^k g(\theta_i)$, where g is the atomic decoder modelling the continuous dictionary. It was shown in [11] that these atoms represent decorrelated low-level features of images. This property of neural networks is called *atomic disentanglement* of the latent space (not to be confused with the classical disentanglement notion that aims to represent high level interpretable features). However, on some image datasets, while giving promising decomposition of shapes, atomic autoencoders show some limits for providing a precise representation of images. One observation is that the sum of localized atoms fail to represent accurately typical image sets. Indeed, instead of being sums of objects, natural images can be seen as a set of objects occluding one another, which was e.g. statistically described by the dead leaves model [4].

Contribution. In this paper, we propose an improvement of atomic autoencoders for image processing through a new sparsity model that models images as the maximum pixel wise of atoms in a given dictionary. This way of merging atoms aims to better represent natural images where objects may occlude one another. The resulting atomic autoencoder encodes the input image into a latent vector divided into blocks that are decoded separately to elementary images – or atoms. We obtain the output by computing the maximum pixel wise over these atoms instead of summing them.

In Sect. 2, we define the max-sparsity model and describe some of its structural properties in relation to a classical sparsity model. We show in particular that it may be included within a classical sparsity model under certain assumptions, which indicates greater structure induced by this model.

In Sect. 2.3, we propose a modification of the atomic autoencoder from [11] (SUM-AAE) to perform the maximum pixel wise operation (MAX-AAE) over the obtained atoms instead of summing them. We compare the two atomic autoencoders and the decomposition they achieve over two different datasets: Sect. 3.1 shows experimentally that SUM-AAE and MAX-AAE achieve similar reconstruction performance. However, our proposed architecture MAX-AAE achieves a decomposition with sharper and more disjoint atoms compared to SUM-AAE. Section 3.2 shows that the decomposition achieved by MAX-AAE yields sparser representations of the considered images.

In Sect. 4, we use atomic autoencoders to solve a super resolution inverse problem. In this setting, they are used as a projection step for the Projected Gradient Descent algorithm (PGD). This is motivated by the fact that these networks, when trained properly, project the input into a low dimensional space of features captured during training. Experiments in Sect. 4 prove that even if they are more constrained, using MAX-AAE in the PGD can recover original images as good as when we use a simple autoencoder instead and with improved performances compared to SUM-AAE.

Related Work. Atomic autoencoders [11] are closely related to dictionary learning. Given a dictionary $\mathcal{D}$ and a weight vector w (often considered sparse), an image x is modelled as $x = \mathcal{D}w$. This provides a prior in optimization problems that forces the recovered solution to have a desired structure (see [10] for an overview). To estimate the underlying dictionary $\mathcal{D}$ of a dataset, many algorithms have been proposed using for instance Non Negative Matrix factorization (NMF) [9], Singular Value Decomposition – as k-SVD [1] or even deep neural networks for continuous versions [6,15]. However each of them supposes a linear relation between the learned dictionary and the original images.

Other natural images models have been explored and do not directly rely on a linear dictionary. The dead leaves model [4] supposes that each image consists of the superposition of several independent and different objects that partly occlude one another. It is built over a random process that generates objects to recover the final image. However, it does not provide a fixed dictionary that can be used to solve optimization problems. The idea of using a maximum element-wise function with atomic autoencoders is directly inspired by the dead leaves model, with the goal of modelling the superposition of objects through obtaining a dictionary.

By decomposing images into atoms, low level features are obtained. A plethora of works aims to discover underlying features of images. This operation is referred to as *disentanglement*. Previous autoencoder architectures have been built to disentangle latent variables by modifying the loss to decorrelate them [3]. Non deterministic approaches were proposed: β-VAE [5] generalizes variational autoencoders with a weighted Kullback-Leibler loss. It achieves a better disentanglement at the expense of the reconstruction quality. Apart from autoencoders, Generative Adversarial Networks (GANs) have proven to be efficient in providing disentangled latent variables [2]. In this work, we specifically focus on deterministic latent representations (i.e. with an explicit encoder).

In terms of applications, generative models are mainly used to generate new data, but they are also useful in solving inverse problems. [13] gives an overview of several techniques with some associated theoretical results. To solve the underlying optimization problem, it is possible, in particular, to use generative neural networks in a projected gradient descent. The generative function projects inputs into the natural images set at each iteration. For example, a basic autoencoder is used in [12] whereas [14] uses GANs to perform this projection.

2 The Max-Sparsity Model

In the classical dictionary approach, an image x is modelled as a sum of elementary images within a dictionary $\mathcal{D}$ weighted by a weight matrix w *i.e.* $x = \mathcal{D}w$. Instead, we propose to consider each image as the superposition of different independent objects that occlude one another. To model such occlusions with a dictionary approach, we propose to perform a maximum pixel wise operation over the weighted elements of $\mathcal{D}$ to form the output. We call this model a *max-sparsity model* as opposed to the classical *sum-sparsity models.*

We define in this section the *max-sparsity model* and overview some of its properties in relation to the *sum-sparsity model.*

2.1 Definition

We denote by $\mathcal{D}$ a finite or infinite dictionary of vectors in $\mathbb{R}^n$. In the finite case, we can write $\mathcal{D} = \{a_i \in \mathbb{R}^n, i \leq \#\mathcal{D}\}$. We call $a_{i,j}$ the j-th coordinate of the i-th atom in $\mathcal{D}$ and x_i the i-th component of vector $x \in \mathbb{R}^n$.

We define a generalized sparsity model Σ as

$$\Sigma = \{x = \sigma(\lambda_1 a_1, \ldots, \lambda_k a_k), a_i \in \mathcal{D}, \lambda_i \in \mathbb{R}\} \tag{2}$$

where σ is an arbitrary aggregation function : $\mathbb{R}^{n \times k} \longrightarrow \mathbb{R}^n$ and k is the sparsity of the image in $\mathcal{D}$. Through this generalized sparsity model we define:

- the classical sparsity model Σ_k [10] (which we call *sum-sparsity model*) when the aggregation function σ is a sum, i.e.:

$$\Sigma_k := \left\{ x = \sum_{i=1}^{k} \lambda_i a_i, a_i \in \mathcal{D}, \lambda_i \in \mathbb{R} \right\}; \tag{3}$$

- the *max-sparsity* model $\Sigma_k^{\max}$ when the aggregation function $\sigma(\cdot) = \mathrm{maxel}(\cdot)$ is the maximum element wise operation, i.e.:

$$\Sigma_k^{\max} := \{x = \mathrm{maxel}(\lambda_1 a_1, \ldots, \lambda_k a_k), a_i \in \mathcal{D}, \lambda_i \in \mathbb{R}\}; \tag{4}$$

where $v = \mathrm{maxel}(u_1, \ldots, u_k)$ is defined by $(v_j)_{j \leq n} = (\max(u_{1,j}, \ldots, u_{k,j}))_{j \leq n}$.

As the representation is generally supposed to be sparse, we suppose that $k < n$. While we focus on the structural properties of the *max-sparsity* model in the following, this general formulation opens the broader question of the choice of aggregation function for a given dataset.

2.2 Properties of $\Sigma_k^{\max}$

In inverse problems, the model structure is crucial to provide guarantees. For example, positively homogeneous or homogeneous models have different recovery guarantees [16]. All proofs can be found in the extended version of the paper [8]

Definition 1 (Homogeneity). *A set Σ is positively homogeneous (called a cone) if for all $\lambda \in \mathbb{R}^+$ and $x \in \Sigma$, $\lambda x \in \Sigma$. The set Σ is homogeneous (called a union of subspaces) if for all $\lambda \in \mathbb{R}$ and $x \in \Sigma$, $\lambda x \in \Sigma$.*

Proposition 1. *The max-sparsity model set $\Sigma_k^{\max}$ is a cone. The set $\Sigma_k^{\max}$ is not a homogeneous model in general.*

This shows that $\Sigma_k^{\max}$ is not a union of linear subspaces. Thus, it has in general less structure than the sum-sparsity model. However, with more constraints on the atoms, it is included in the sum-sparsity model.

Proposition 2. *If the dictionary $\mathcal{D}$ verifies the two following assumptions:*

1. *for all $i \neq j$, $\operatorname{supp}(a_i) \cap \operatorname{supp}(a_j) = \varnothing$*
2. *for all i and $j \leq n$, $k \leq n$ $a_{i,j} \cdot a_{i,k} \geq 0$*

then $\Sigma_k^{\max} \subsetneq \Sigma_k$. In general, $\Sigma_k^{\max} \not\subset \Sigma_k$.

For example, if $\mathcal{D}$ is a subset of the canonical basis of $\mathbb{R}^n$ then $\Sigma_k^{\max} \subsetneq \Sigma_k$. However, it is not ensured that for every dictionary $\mathcal{D}$ the inclusion is still valid.

Given an underdetermined linear measurement operator $\mathbf{A}$, the problem of recovering $x \in \Sigma$ from $\mathbf{A}x$ has a unique solution if and only if $\ker(\mathbf{A}) \cap (\Sigma - \Sigma) = 0$ where $\Sigma - \Sigma$ is the set $\{z \in \mathbb{R}^n : \exists\, x, y \in \Sigma, z = x - y\}$ [17].

If $\Sigma_k^{\max} \subset \Sigma_k$, then, if a matrix $\mathbf{A}$ verifies the property $\ker(\mathbf{A}) \cap (\Sigma_k - \Sigma_k) = \{0\}$, then $\ker(\mathbf{A}) \cap (\Sigma_k^{\max} - \Sigma_k^{\max}) = \{0\}$. This shows that, when atoms have separated supports, the recovery of elements of Σ_k implies the recovery of elements of $\Sigma_k^{\max}$, but the converse is not generally true. Hence, for a fixed dictionary, if $\Sigma_k^{\max} \subset \Sigma_k$, recovering $\Sigma_k^{\max}$ is "easier" than recovering Σ_k. Given this conclusion, we compare the recovery of images by atomic autoencoders that enforce them to be in Σ_k and in $\Sigma_k^{\max}$ in Sect. 4. We will indeed observe that max-sparse atomic autoencoders provide better solutions to the image super-resolution problem.

2.3 Atomic Autoencoders with Max-Sparsity Model (MAX-AAE)

In [11], the sum-sparsity model is used to build the atomic autoencoder $f = f_D \circ f_E$ such that: $\theta = f_E(x)$ and $\tilde{x} = f_D(\theta) = \sum_{i=1}^{k} g(\theta_i)$, where f_E and g (defining f_D) are implemented with deep Leaky ReLU networks – Fig. 1. In this work, we propose to modify the decoder f_D to define a max-sparsity model.

Definition 2 (Max-sparsity atomic autoencoder). *A max-sparsity atomic autoencoder is a function* $f = f_D \circ f_E$ *that is the composition of an encoder* $f_E : \mathbb{R}^n \to \mathbb{R}^{kd_0}$ *and a decoder* $f_D : \mathbb{R}^d \to \mathbb{R}^n$ *defined by*

$$f_D(\theta) = \text{maxel}(g(\theta_1), \ldots, g(\theta_k)) \tag{5}$$

where $g : \mathbb{R}^{d_0} \to \mathbb{R}^n$ *is the function decoding a single atomic latent code – see Fig. 1.*

Note that using a maximum pixel wise operation adds a non-linearity to the decoder acting as a ReLU function because $\max(a, b) = \texttt{ReLU}(a - b) + b$.

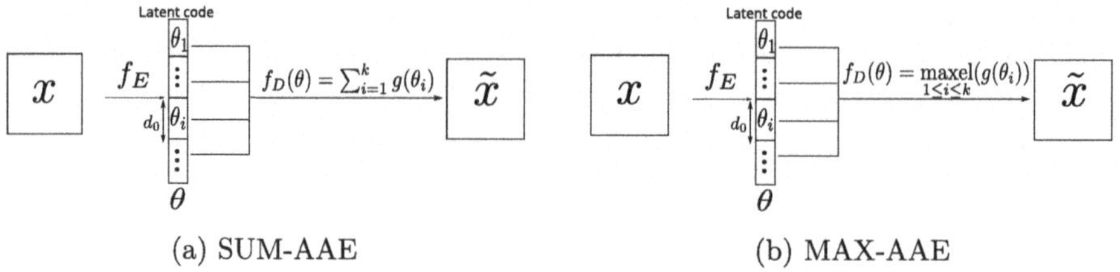

Fig. 1. Architecture of sum-sparsity and max-sparsity atomic autoencoders.

By constraining the decoder, the autoencoder output is falling in a potentially smaller space. This is related to the sparsity of the decomposition obtained by an atomic autoencoder. To assess this sparsity in the context of atomic autoencoders, we define the notion of *activated atoms*:

Definition 3 (Activated atoms). *Given an image* $x \in \mathbb{R}^n$ *and a vector* $\theta \in \mathbb{R}^{kd}$ *such that* $f_D(\theta) = x$, *an atom* $a_i = g(\theta_i)$ *is activated with respect to a threshold* s *if* $\frac{\|a_i\|_2}{\|x\|_2} \geq s$

We will observe in Sect. 3.1 that max-sparsity atomic autoencoders (MAX-AAE) yield sparser representations than sum-sparsity atomic autoencoders (SUM-AAE) thus giving an interpretation of the improved performance of MAX-AAE in the context of inverse imaging problem.

3 Application to Image Decomposition

We compare the sum-sparsity atomic autoencoder (SUM-AAE) and the max-sparsity atomic autoencoder (MAX-AAE), defined in 2.3. We train both of them on two different datasets MNIST and Fashion-MNIST of size 30000. They have the same architecture: an encoder with two convolutional layers followed by two fully connected layers and a latent code composed of 20 blocks of size 10. Each block is decoded via the same decoder. It consists of four fully connected layers and three convolutional layers yielding 20 images (atoms) before applying the aggregation function. SUM-AAE sums the atoms whereas MAX-AAE applies the maximum pixel wise operation. We use Leaky ReLU as an activation function for all hidden layers. As a baseline, we train a simple autoencoder (AE) with a latent space of size 200 and having roughly the same number of parameters. Every comparison in this paper is achieved over subsets of size 600 of MNIST and Fashion MNIST test sets. We compare the recovered images through Peak signal-to-noise ratio (PSNR) metric. Furthermore, we assess the sparsity of autoencoded images through the mean number of *activated atoms* per image (defined in 3). The chosen threshold is 0.05. The code to reproduce experiments is available in [7].

3.1 Image Decomposition with Max-Sparsity Atomic Autoencoder

We train two atomic autoencoders SUM-AAE and MAX-AAE over MNIST, a 28×28 digit images dataset and Fashion MNIST a 28×28 clothe images dataset. We keep the same structure for both of them but increased the number of filters and the size of the layers for Fashion MNIST.

Figure 2 compares the autoencoding of some images from the 2 datasets and their associated PSNRs. Table 1 gives the average PSNR over the test sets. According to these figures, MAX-AAE performs better than SUM-AAE and AE over MNIST. For Fashion MNIST, textured images are slightly degraded with MAX-AAE. However, we will observe in the context of image super-resolution that this slight degradation is a positive feature as MAX-AAE provides a better prior for image reconstruction and limits hallucinations (compared to AE and SUM-AAE).

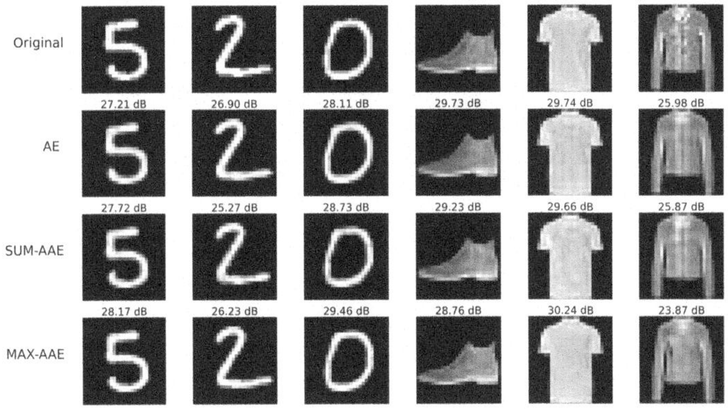

Fig. 2. Reconstruction of digits (MNIST) and clothes (Fashion MNIST) images through the autoencoders AE, SUM-AAE and MAX-AAE. The performances are similar for MNIST. For Fashion MNIST, MAX-AAE show a slight degradation for textured images.

Table 1. Average PSNR of the autoencoded images. We compare the autoencoders MAX-AAE, SUM-AAE and AE over two test sets of size 600.

	MNIST	Fashion MNIST
AE	26.72 ± 2.66 dB	$\mathbf{25.46 \pm 3.76}$ dB
SUM-AAE	27.35 ± 2.34 dB	25.28 ± 3.57 dB
MAX-AAE	$\mathbf{28.42 \pm 2.74}$ dB	24.25 ± 3.27 dB

Figure 3 shows the decompositions performed by SUM-AAE and MAX-AAE over MNIST and FashionMNIST. In both cases, MAX-AAE provides a decomposition with sharp distinct atoms whereas in SUM-AAE many atoms are blurry and scattered despite the mean PSNR of MAX-AAE being lower. We also observe that the support of atoms in the decomposition are mostly separated in MAX-AAE, indicating an inclusion in a sum-sparsity model (see Proposition 2).

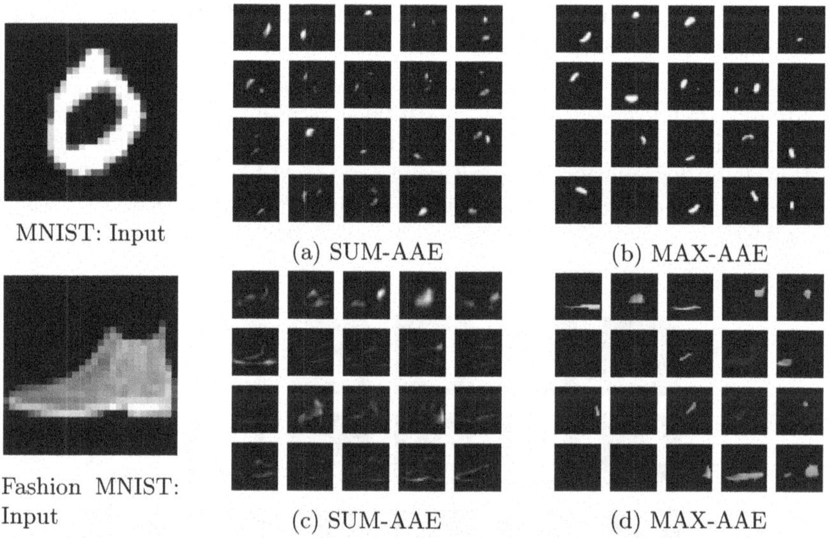

Fig. 3. Decomposition of an input from MNIST and Fashion MNIST through the two atomic autoencoders. SUM-AAE decomposes into scattered, blurry atoms while MAX-AAE produces sharp, disjoint atoms with a sparser decomposition.

In addition to these improved decompositions, the latent representations provided by MAX-AAE are sparser. We represent in Fig. 4 the histogram of the number of activated atoms for a test set of size 600. To complement it, we show the average evolution of the norm of each atoms after ordering them in decreasing order. It shows that MAX-AAE activates less atoms than SUM-AAE as its related histogram is shifted to the left. The norm of the atoms decreases to 0 from the 16-th block for MAX-AAE when none of them is zero in SUM-AAE despite the faster decrease. Consequently, MAX-AAE produces a sparser representation. It must be noted that the training is performed without any regularization of the latent set in the loss. Hence, we blindly estimate the intrinsic dimensions of these two datasets with the atomic autoencoder architecture.

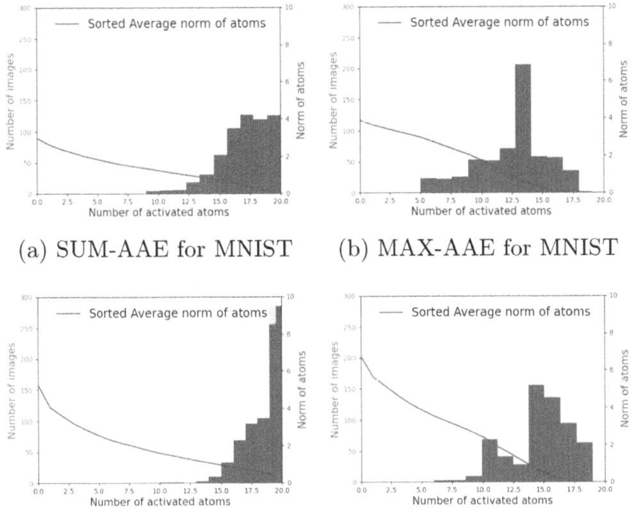

(a) SUM-AAE for MNIST (b) MAX-AAE for MNIST

(c) SUM-AAE for Fashion MNIST (d) MAX-AAE for Fashion MNIST

Fig. 4. Histogram of activated atoms for 600 images of the test set (right y-axis) and the average norm of each atom in descending order (left y-axis). For MAX-AAE, the histogram is shifted to the left which shows that on average less atoms are activated and the norm of the atoms shatters to 0 earlier than SUM-AAE.

3.2 Atomic Disentanglement of Latent Variables

As defined in [11], an atomic autoencoder achieves *atomic disentanglement* if it decomposes the input into atoms, each of them associated to a low-level feature. Modifying a latent block must act over the sole feature it represents. It is shown in ([11]) that SUM-AAE provides atomic disentanglement of its variables through two examples: MNIST Dataset and off-the-grid spikes. We focus here on Fashion MNIST. We show experimentally that even though it is a more complex dataset with more shape diversity, MAX-AAE achieves an atomic disentanglement of its latent set. Figure 3 shows that MAX-AAE produces sharp atoms. In particular, for Fashion MNIST, it is possible to assign each block to a particular low level feature. The shoe for instance has precise activated blocks that represent specific parts of the shoe. In Fig. 3, we see that the lower part of the sole of the shoe is represented by the first block e.g. the first image in MAX-AAE decomposition.

A way of showing atomic disentanglement is to modify values from the latent vector and see how it influences the output image. To do so, after choosing a dimension in a block, we compute the two extreme values along this dimension over all the test set. We, then, interpolate with a constant step between these two values. In other words, for a dataset X, if $\theta \in \mathbb{R}^{kd}$, $n \in \mathbb{N}$, $i \in [\![0, n]\!]$, $p \in [\![1, k]\!]$ and $q \in [\![1, d]\!]$, we modify θ and decode via f_D as follows:

$$\hat{x}_i = f_D(\theta^i) = f_D(\theta_{1,1} \dots | \dots \theta^i_{p,q} \dots | \dots \theta_{k,d}), \text{ where } \theta^i_{p,q} = \theta^{\min}_{p,q} + i\frac{\theta^{\max}_{p,q} - \theta^{\min}_{p,q}}{n}$$
$$\theta^{\max}_{p,q} = \max_{\theta \in f_E(X)} \theta_{p,q} \text{ and } \theta^{\min}_{p,q} = \min_{\theta \in f_E(X)} \theta_{p,q}, \text{ and } \hat{x}^i \text{ is the } i\text{-th output.} \quad (6)$$

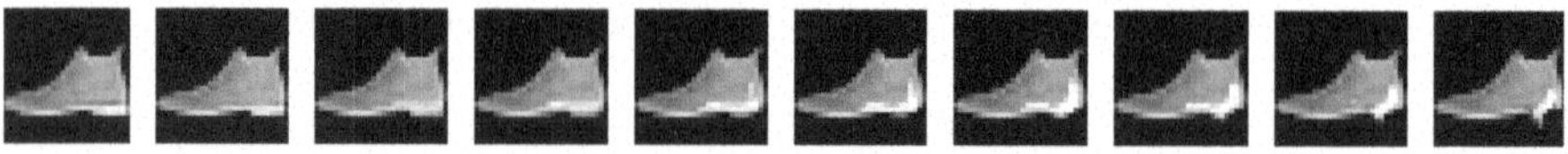

Fig. 5. Modification of the first variable from the first block of the latent vector associated to the shoe through MAX-AAE. This experiment shows an interpolation between a flat sole and a heel while leaving the other features unchanged.

Figure 5 shows that, by modifying one unique value of a dimension in a given block, only one feature in the image space is modified. In this specific example, from the first image (associated to the minimum value of dimension 0 from block 0) to the last image (associated to the maximum value) the only modified feature is the orientation lower part of the sole.

4 Application to Image Super Resolution

Image super-resolution consists in estimating a high resolution image from a low resolution image. To solve this ill-posed problem, a low-dimensional prior is necessary. In this Section, we study the performance of our new MAX-AAE architecture in this context. We define

$$y = \mathbf{A}\hat{x}, \quad (7)$$

where $y \in \mathbb{R}^m$ is the measurement vector representing a subsampled image, $\hat{x} \in \mathbb{R}^n$ the original high resolution image. The linear operator $\mathbf{A} \in \mathbb{R}^{m \times n}$ represents a low pass filtering followed by a subsampling, i.e. $\mathbf{A} = \mathbf{SF}$ where $\mathbf{S}$ is the sub sampling matrix (by a factor 2 in the experiments). The matrix $\mathbf{F}$ models the convolution by a Gaussian blur.

Given a prior low dimensional model Σ on x, we want to solve:

$$\mathbf{x}^* = \underset{x \in \Sigma}{\operatorname{argmin}} \|\mathbf{A}x - y\|_2^2. \quad (8)$$

Here, Σ is induced by a trained autoencoder f and is, then, non-convex. Many methods can be used to solve this problem. To compare the different autoencoders capacities, we chose the projected gradient descent (PGD) described by the following iterations, given an initialization x_0:

$$\mathbf{x}_{k+1} = P_\Sigma(x_k - \gamma \mathbf{A}^T(\mathbf{A}x_k - \mathbf{y})) \tag{9}$$

where γ is the step size and the projection step P_Σ is performed with the trained autoencoder (i.e. $P_\Sigma = f$, see e.g. [12]). For general low-dimensional models, it has been shown that this algorithm linearly converges to fixed points of the projection under a restricted isometry property of $\mathbf{A}$ and a *restricted* Lipschitz property of the projection [18].

We compare the results of Projected Gradient Descent with projection performed using the three autoencoders from the previous section over the two different datasets. We call PGD-AE, PGD-SUM-AAE and PGD-MAX-AAE the three variations of PGD. Table 2 compares them quantitatively over 600 images from the considered test set. On both MNIST and Fashion MNIST, we observe that PGD-MAX-AAE significantly outperforms PGD-SUM-AAE, and has comparable performance to PGD-AE.

Table 2. PSNR of estimation. Even MAX-AAE is more constrained, PGD-MAX-AAE gives equal results than the classical PGD-AE. It also outperforms PGD-SUM-AAE.

	MNIST	Fashion MNIST
PGD-AE	26.90 ± 2.63 dB	21.95 ± 2.92 dB
PGD-SUM-AAE	25.25 ± 2.22 dB	20.91 ± 2.58 dB
PGD-MAX-AAE	$\mathbf{27.02} \pm \mathbf{2.72}$ dB	$\mathbf{22.01} \pm \mathbf{3.45}$ dB

Visually, we see crucial differences between the reconstructions of PGD-MAX-AAE and PGD-AE (Fig. 6). While on simple data like MNIST, the visual results are similar, on more complex examples, there are significant differences. For example in Fig. 6, on the shirt – fifth column, the reconstruction is very different, with texture artefacts being hallucinated by both PGD-AE and PGD-SUM-AAE. We attribute this to the sparser representation induced by the max-sparsity model 3.1, which supports this model as a good image model. Note that for both MAX-AAE and SUM-AAE we can access the decomposition of the recovered image as in Fig. 3, i.e. an "explanation" of the estimation is provided.

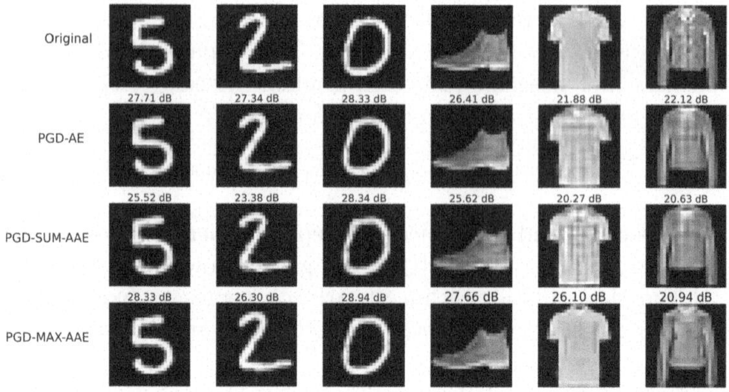

Fig. 6. Recovery of MNIST and Fashion MNIST images through PGD with different projection steps. Visually, the different PGDs perform equally on MNIST. For Fashion MNIST, PGD-MAX-AAE is more robust as it creates less artifacts.

5 Conclusion

We have introduced a new atomic autoencoder based on a max-sparsity model. This autoencoder provides sparser and sharper atomic decompositions of the input. It also achieves an atomic disentanglement of its latent space.

We observed that using this new low dimensional representation for the super-resolution inverse problem provides robust reconstruction of images compared to the classical autoencoder and the original proposition of atomic autoencoder.

Many future directions arise from these results. On the theoretical side, it is still an open question to give a mathematical proof of the atomic disentanglement achieved by atomic autoencoders. In terms of further applications, the next step for atomic autoencoders is to be extended to photorealistic textured images.

References

1. Aharon, M., Elad, M., Bruckstein, A.: K-SVD: an algorithm for designing overcomplete dictionaries for sparse representation. IEEE Trans. Signal Process. **54**(11), 4311–4322 (2006)
2. Chen, X., Duan, Y., Houthooft, R., Schulman, J., Sutskever, I., Abbeel,P.: InfoGAN: interpretable representation learning by information maximizing generative adversarial nets. In: Advances in Neural Information Processing Systems, vol. 29 (2016)
3. Cheung, B., Livezey, J., Bansal, A., Olshausen, B.: Discovering hidden factors of variation in deep networks. arXiv preprint arXiv:1412.6583 (2014)
4. Gousseau, Y., Roueff, F.: Modeling occlusion and scaling in natural images. Multiscale Model. Simul. **6**(1), 105–134 (2007)
5. Higgins, I., et al.: beta-VAE: learning basic visual concepts with a constrained variational framework. In: ICLR (Poster), vol. 3 (2017)

6. Hu, J., Tan, Y.: Nonlinear dictionary learning with application to image classification. Pattern Recogn. **75**, 282–291 (2018)
7. Joundi, A.: Code to reproduce experiments (2024). https://plmlab.math.cnrs.fr/ajoundi/max-aae-and-inverse-problems
8. Joundi, A., Traonmilin, Y., Newson, A.: Max-sparsity atomic autoencoders with application to inverse problems. Working paper or preprint (2025)
9. Lee, D., Seung, S.: Learning the parts of objects by non-negative matrix factorization. Nature **401**(6755), 788–791 (1999)
10. Mairal, J., Bach, F., Ponce, J.: Sparse modeling for image and vision processing. Found. Trends Comput. Graph. Vis. **8**(2–3), 85–283 (2014)
11. Newson, A., Traonmilin, Y.: Disentangled latent representations of images with atomic autoencoders. In: Fourteenth International Conference on Sampling Theory and Applications (2023)
12. Peng, P., Jalali, S., Yuan, X.: Solving inverse problems via auto-encoders. IEEE J. Sel. Areas Inf. Theory **1**(1), 312–323 (2020)
13. Scarlett, J., Heckel, R., Rodrigues, M., Hand, P., Eldar, Y.: Theoretical perspectives on deep learning methods in inverse problems. IEEE J. Sel. Areas Inf. Theory **3**(3), 433–453 (2022)
14. Shah, V., Hegde, C.: Solving linear inverse problems using GAN priors: an algorithm with provable guarantees. In: 2018 IEEE ICASSP, pp. 4609–4613. IEEE (2018)
15. Tariyal, S., Majumdar, A., Singh, R., Vatsa, M.: Deep dictionary learning. IEEE Access **4**, 10096–10109 (2016)
16. Traonmilin, Y., Gribonval, R.: Stable recovery of low-dimensional cones in hilbert spaces: one rip to rule them all. Appl. Comput. Harmon. Anal. **45**(1), 170–205 (2018)
17. Traonmilin, Y., Gribonval, R., Vaiter, S.: A theory of optimal convex regularization for low-dimensional recovery. Inf. Infer. J. IMA **13**(2), iaae013 (2024)
18. Traonmilin, Y., Aujol, J.F., Guennec, A.: Towards optimal algorithms for the recovery of low-dimensional models with linear rates. arXiv preprint arXiv:2410.06607 (2024)

Equivariant Bootstrap for Uncertainty Quantification in Image Classification

Andreas Decker[1(✉)], Alexander Effland[2], and Erich Kobler[3]

[1] German Center for Neurodegenerative Diseases (DZNE), Bonn, Germany
Andreas.Decker@dzne.de

[2] Institute of Applied Mathematics, University of Bonn, Bonn, Germany
effland@iam.uni-bonn.de

[3] Institute for Machine Learning, LIT AI Lab, Department of Virtual Morphology, Johannes Kepler University Linz, Linz, Austria
erich.kobler@jku.at

Abstract. Image classification has gained significant attention in recent years. However, the quantification of uncertainty in classification outcomes has only been more intensively studied in recent times. In this paper, we propose an equivariant bootstrap method, combined with an autoencoder, to quantify the inherent uncertainty of classification predictions. Specifically, the exact method involves applying group actions, such as rotations or translations, to the image. This process includes two applications of encoding, a single decoding step, and the application of the inverse of the group action, which generates a single bootstrap sample. Additionally, we introduce an inexact method that requires only a single pass through the encoder, trading off computational efficiency for slightly lower accuracy. We validate our methods through several numerical experiments, evaluating their performance using various metrics and benchmarking against other recent uncertainty quantification approaches. In particular, we focus on ablation studies that examine the impact of different group actions on the results.

Keywords: Image Classification · Autoencoder · Bootstrap

1 Introduction

Image classification is one of the most common tasks in imaging, involving the assignment of a label to an image from a predefined set of possible classes. In contrast to object detection or segmentation, where each pixel is assigned a label, *image classification* typically considers the image as a whole. The primary goal of image classification is to minimize the number of incorrect classifications. However, it is also valuable to assess the uncertainty in the algorithm's predictions, as this can provide deeper insights into its performance and reliability.

In this manuscript, we introduce two novel methods termed exact and inexact for quantifying uncertainty in image classification, combining autoencoders

T. A. Bubba et al. (Eds.): SSVM 2025, LNCS 15667, pp. 214–226, 2025.
https://doi.org/10.1007/978-3-031-92366-1_17

with (equivariant) bootstrap techniques. Our primary focus is on quantifying uncertainty for datasets of limited size.

The key insight is that applying group actions, such as rotations or translations, to the images enables the generation of new bootstrap samples. The distribution of these bootstrap samples can then be used to estimate the inherent uncertainty. In both the exact and inexact methods, all group actions are applied in image space, while the final classifications are performed in the latent space of the corresponding autoencoders. However, the inverse of the group action is applied only in the exact method.

Through several numerical experiments, we demonstrate the state-of-the-art performance of our relatively simple method. The quality of the method is primarily assessed using the Expected Calibration Error (ECE) and the Adaptive Calibration Error (ACE). For completeness, we also report the values of the negative log-likelihood loss, the Brier score, and the accuracy. We also study the influence of the latent space dimension and the choice of group action on the final output.

2 Related Work

In recent years, numerous methods have been developed to address the challenge of uncertainty quantification in image classification.

A significant fraction of these methods are designed for settings where a large amount of training data is available. One popular approach for uncertainty quantification in deep neural networks is DropConnect [14], which applies a Bernoulli distribution directly to the model weights. This enables the estimation of the posterior distribution for the respective weight matrices. An alternative method proposed in [7] leverages a multi-input, multi-output configuration, utilizing an ensemble of predictions generated by subnetworks. While these methods perform well in large-scale settings, they are less efficient when applied to relatively small networks [14], such as those used in our study.

A more recent approach, known as dual-focal loss, was introduced in [17]. This method proposes a novel loss function based on dual logits, aimed at balancing over-confidence and under-confidence in model predictions. A simple method for uncertainty quantification based on test-time data augmentation has been proposed in [1]. In [6], various calibration techniques were systematically compared, with temperature scaling emerging as the superior method. Furthermore, it has been speculated that both an increase in model size and a lack of regularization are strongly linked to model miscalibration, which leads to inferior uncertainty quantification.

Bootstrap methods [2,5] have gained considerable popularity in recent decades as reliable and versatile tools for uncertainty quantification. In [12], a novel adaptation of bootstrap techniques for classification tasks is presented. This method transforms the classification problem into a regression task, allowing the direct application of variational bootstrap techniques, which enables faster convergence and improved accuracy in uncertainty estimation. Our approach builds upon the recent work of Pereyra and Tachella [16], who introduced an equivariant formulation of a (parametric and equivariant) bootstrap algorithm specifically tailored for image reconstruction problems. Additionally, our method shares similarities with the work of [4], where both parametric and non-parametric bootstrap methods are applied in a lower-dimensional latent space. In [9], probabilistic and non-probabilistic methods are discussed in medical contexts as a promising field of application.

3 Methods

In this section, we begin with a brief overview of the image classification problem, emphasizing the role of autoencoders. We then introduce a novel equivariant bootstrap method designed for uncertainty quantification. Finally, we present various evaluation metrics used to assess the performance of the numerical experiments.

3.1 Problem Formulation

Throughout the paper, we consider a classical supervised image classification problem on the complete probability space $(\Omega = \mathcal{X} \times \mathcal{C}, \mathfrak{F}, \mathbb{P})$. Here, the input image $X \in \mathcal{X} \subseteq \mathbb{R}^{d_x \times d_y}$, with given dimensions $d_x, d_y \in \mathbb{N}$, and target label $Y \in \mathcal{C} := \{1, \ldots, C\}$ are random variables following a joint distribution $\mathbb{P} = p(X, Y)$.

We intend to learn multi-class classifiers $m : \mathcal{X} \to \triangle_C$ that predict a discrete probability over the possible class labels $\hat{P} = m(X) \in \triangle_C$ for any input $X \in \mathcal{X}$. The classifier's output space is the unit simplex $\triangle^C = \{y \in [0,1]^C : \|y\|_1 = 1\}$. For each model, the prediction is given by selecting the label with the highest predicted probability, i.e.,

$$\hat{Y} = \arg\max_{c=1,\ldots,C} (\hat{P} = m(X))_c.$$

In this paper, we focus primarily on the confidence of the predictions made by the image classifier. The overarching goal is to retrieve a *perfectly* calibrated classifier, which is according to Guo et al. [6] defined as

$$\mathbb{P}(\hat{Y} = Y | \hat{P}_{\hat{Y}} = p) = p$$

for all $p \in [0, 1]$. In other words, the predicted probabilities of a model are calibrated if they truly represent a probability in the frequentist sense. For example, assume we observe 100 realizations $\hat{y}$ of the prediction $\hat{Y}$, all with predicted probability $\hat{p}_{\hat{y}} = 0.7$, then we expect that 70 of theses samples are correctly classified. Thus, the predicted probability aligns with the confidence that the predicted label is the true one. Clearly, perfect calibration is not achievable in most practical applications.

3.2 Autoencoder-Based Classification

In this manuscript, we explore the calibration problem from the perspective of autoencoders (AEs) in combination with equivariant bootstrap methods. Recall that an AE consists of an encoder $\mathrm{Enc} : \mathcal{X} \to \mathcal{Z} \subseteq \mathbb{R}^{d_l}$ and a decoder $\mathrm{Dec} : \mathcal{Z} \to \mathcal{X}$. The encoder maps the data into the latent space $\mathcal{Z} \subseteq \mathbb{R}^{d_l}$, which has a significantly smaller dimension than the original space (i.e., $d_l \ll d_x d_y$). The decoder then reconstructs the data back into a subspace of $\mathcal{X}$ such that the reconstructed data should closely resemble the original data if the model is well-trained. The associated training problem amounts to

$$\min_{\xi} \sum_{i=1}^{n} \|(\mathrm{Dec}_{\xi} \circ \mathrm{Enc}_{\xi} - \mathrm{Id})(x_i)\|_2^2,$$

where ξ subsumes all learnable parameters of the autoencoder. Alternatively, variational autoencoders (VAEs) [8,11], which incorporate a Bayesian framework by introducing a prior distribution on the latent space, can also be considered. However, our numerical evaluation revealed no significant difference between AEs and VAEs. Therefore, following Occam's razor, we opt to use AEs for the remainder of the paper.

The starting point of the image classification algorithm is the heuristic observation that instances of the same class are expected to form clusters in the lower-dimensional latent space, facilitating the construction of classification functions to separate them. This observation allows us to directly train a neural network-based prediction algorithm $\mathrm{Pred}_{\zeta} : \mathcal{Z} \to [0,1]^C$ on the latent space. Combined with the fixed encoder Enc_{ξ}, we obtain a simple classification model for the original image by the composition

$$m = \mathrm{Pred}_{\zeta} \circ \mathrm{Enc}_{\xi} : \mathbb{R}^{d_x \times d_y} \to \triangle_C,$$

where ζ denotes the learnable parameters of the latent code-based classifier.

3.3 Equivariant Bootstrap for Uncertainty Quantification

Motivated by the equivariant bootstrap of Pereyra and Tachella [16] to quantify the uncertainty in image reconstruction problems, we present two similar, yet novel methods for optimizing the calibration metrics in image classification problems. Both of which connect autoencoders and the (equivariant) bootstrap. The *exact* method incorporates the equivariant bootstrap, which, however, requires two applications of the encoder and one of the decoder for each sample, as illustrated in Fig. 1. In contrast, the *inexact* method requires only a single application of the encoder, at the cost of often yielding inferior results.

Both methods are based on group actions applied to images, most notably rotations and translations (although numerous other group actions are possible, including shearing, scaling, or linear intensity transformations). However, most group actions, such as rotations, are difficult to realize directly in the latent

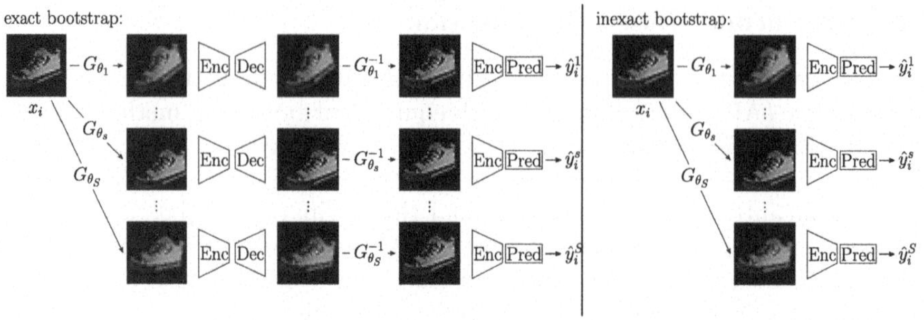

Fig. 1. Comparison of exact (left) and inexact (right) sampling schemes. Note that the exact scheme incorporates two encoder steps to facilitate the inverse group transform, while the inexact scheme only uses a single encoder step.

space, which is why all group actions must be performed in image space. Henceforth, we denote the group actions by G_θ, where θ denotes a particular choice of bootstrap parameters.

We emphasize that the exact method is equivariant in contrast to the inexact method. Equivariance in the exact method refers to the fact that the group action approximately commutes with the combined encoding and decoding of the autoencoder. A more formal definition can be found in [18]. However, the equivariance is not strict as the whole autoencoder is not precisely equal to the identity function and furthermore the actions of the group elements require the resampling of the input image.

Next, we present two methods for generating S-fold ($S \in \mathbb{N}$) bootstrap samples based on group actions G_θ, which we term *exact* and *inexact*. To create a single bootstrap sample, the *exact* method encompasses the subsequent steps:

S.1 Sample group actions G_θ according to probability distribution p_θ for θ.
S.2 Act with G_θ on the original images.
***S.3** Encode and decode the result with the pre-trained autoencoder.
***S.4** Act with the inverse G_θ^{-1} of G_θ on the output.
S.5 Encode and apply the prediction to obtain the newly sampled classification result of the bootstrap.

Here, p_θ denotes a probability distribution on the space of possible bootstrap parameters θ, which is most commonly uniform. The steps of the algorithm can be summarized as

$$m_\theta^{\text{exact}} = \text{Pred}_\zeta \circ \text{Enc}_\xi \circ G_\theta^{-1} \circ \text{Dec}_\xi \circ \text{Enc}_\xi \circ G_\theta \colon \mathbb{R}^{d_x \times d_y} \to \triangle_C.$$

The intuition behind this approach is as follows. We focus on datasets of limited size, where generating additional data samples is not feasible. As a result, we rely on the validation set to estimate the model's uncertainty. By constructing additional samples in the image space using the action of G_θ, we can substantially enlarge the validation set with inputs that remain representative of the original data distribution. To obtain a latent space vector that is more closely related to the original, we can decode the vector back into image space, invert the group action, and then encode the result again to return to latent space. Similar to [16], we can view the equivariant bootstrap procedure as a data augmentation strategy for latent space vectors, so this approach enables greater variation in the generated samples. Note that the samples are different from those generated by directly applying the group action on the image space.

We leverage the distinctive property of the autoencoder, where the composition of the functions Dec $\circ$ Enc is close to the identity function. Provided that the autoencoder performs well, the resulting images should be suitable for our purposes. However, in practice, this means that the autoencoder must have encountered similar images during training. Therefore, the parameter set θ must be constrained to ensure that the generated images are of high quality and relevant to the task.

If we directly apply the prediction to the output of the first encoding step (omitting the steps **S.3** and **S.4**), we refer to the method as *inexact*. In other words, the group action is not inverted, meaning the group transformation should not perturb the input too much such that the generated samples remain useful for model evaluation. Note that the inexact method can be interpreted as an ordinary data augmentation strategy advocated by [1] on the training set, but still utilizing the additional information generated by the bootstrap samples on the validation set.

In summary, the inexact method gives rise to the representation

$$m_\theta^{\text{inexact}} = \text{Pred}_\zeta \circ \text{Enc}_\xi \circ G_\theta \colon \mathbb{R}^{d_x \times d_y} \to \Delta_C.$$

In this paper, we consider only these group actions G_θ and their combinations:

- *Rotation*: For a given angle ϕ in the interval $[-\underline{\phi}, \underline{\phi}]$, let $R_\phi : \mathbb{R}^{d_x \times d_y} \to \mathbb{R}^{d_x \times d_y}$ be a rotation around the center by ϕ, with inverse $R_{-\phi}$. This inversion is only an approximation, as the rotation involves resampling data within the grid, which, in our case, is obtained via bilinear interpolation.
- *Translation*: Let $T_{(x,y)}$ be the translation of the image with an offset $(x, y) \in [-\underline{T}, \underline{T}]^2$ for a fixed $\underline{T} > 0$, with inverse $T_{(-x,-y)}$. Note that the inverse translation in the case of off-grid values is realized by bilinear interpolation.

For a more detailed theoretical justification of the overall method we refer the reader to [2,16].

3.4 Evaluation Metrics

In this section, we follow [15] and briefly review several metrics to quantify the calibration of a model m. To define these metrics, we consider a set of samples $(x_i)_{i=1}^N$. Since we are primarily interested in properly calibrated classifiers, we focus on the ECE and ACE metrics, the remaining evaluation metrics are reported for completeness.

The *negative log-likelihood* NLL is defined as

$$\mathrm{NLL}(m) = \frac{1}{N}\sum_{i=1}^{N} - \log(m(x_i)_{y_i}),$$

which depends solely on the predicted probability of the correct label y_i. The *Brier score* BS quantifies the mean squared difference between the predicted probabilities and the actual outcomes, i.e.,

$$\mathrm{BS}(m) = \frac{1}{N}\sum_{i=1}^{N}\sum_{c=1}^{C}(m(x_i)_c - \mathbf{1}_{c=y_i})^2.$$

Calibration errors quantify the mismatch between the confidence and the accuracy of our predictions. Since accuracy cannot be computed for individual predictions, we group the predictions into M bins B_j based on their predicted confidence scores $\hat{p}$. The confidence conf of a model m on a bin B_j is the mean confidence of its elements, and the accuracy acc is the fraction of correct predictions:

$$\mathrm{conf}(m; B_j) = \frac{1}{|B_j|}\sum_{i\in B_j} m(x_i)_{y_i}, \quad \mathrm{acc}(m; B_j) = \frac{1}{|B_j|}\sum_{i\in B_j} \mathbf{1}_{\hat{y}_i=y_i}, \tag{1}$$

where $\hat{y}_i$ and y_i are the predicted and true label of the ith sample, respectively. For the *Expected Calibration Error* ECE, the bins B_j are of equal size, i.e., B_j is the set of predictions with confidence probabilities in the interval $\left(\frac{j-1}{M}, \frac{j}{M}\right]$. The ECE is then defined as

$$\mathrm{ECE}(m) = \frac{1}{N}\sum_{j=1}^{M} |B_j| \left|\mathrm{acc}(m; B_j) - \mathrm{conf}(m; B_j)\right|. \tag{2}$$

Similarly, we can define the *Adaptive Calibration Error* ACE by choosing the bins in (2) such that they contain an equal number of elements. Throughout the manuscript, we solely use 15 bins in accordance with [6].

4 Numerical Experiments

In this section, we numerically validate the effectiveness of our proposed method for uncertainty quantification. To this end, we consider three widely used image datasets: MNIST [3], FashionMNIST [19], and CIFAR-10 [13].

4.1 Architecture and Training Details

In all experiments, we employ a symmetric autoencoder architecture comprising three encoding and decoding layers. Each encoding layer consists of a 2D-convolutional map, a SoftPlus activation function, and a MaxPool operation for downsampling. Following the encoding layers, a linear layer maps the features into the latent space.

Decoding is performed in reverse: it starts with a linear layer that maps back from the latent space, followed by three decoding layers. Each decoding layer begins with an upsampling operation, followed by a SoftPlus activation, and ends with a transposed 2D-convolutional map. The number of layers is fixed to three, with hidden layer dimensions of 4, 16, and 64. The latent space dimension is varied to account for the diverse characteristics of the datasets and to create AEs of differing capacities.

To classify the latent codes extracted by the encoder, we use a single linear layer followed by the SoftMax activation for the classifier Pred_ζ. Instead of AEs, we have additionally used variational AEs of the same structure with hardly any difference in the output, the results are not reported in this manuscript.

We train the autoencoders on various dataset variations related to the group actions that will later be used in bootstrap uncertainty quantification. First, we pad the input by 2 in both x- and y-directions to provide additional space, ensuring that after transformations, most of the image remains within the frame. Next, we upscale the input to 64×64 (a factor of two) to minimize information loss during the bilinear resampling. Data augmentation is performed through rotations up to $\underline{\phi} \in \{0°, 15°\}$ and translations up to $\underline{T} \in \{0, 4\}$. In addition, we also consider group actions combining translations and rotations, where $\underline{\phi}$ and $\underline{T}$ are chosen in the same way.

For $\underline{\phi} = 0°$ and $\underline{T} = 0$, the original dataset is used without augmentation. This implies that rotations and translations influence the dataset in two ways: first during data augmentation, and later during the actual bootstrap group action. Naturally, the loss for bootstrapped images depends significantly on the rotations applied during the autoencoder training process. For all group action combinations, the autoencoder parameters ξ are trained to minimize the mean squared error (MSE) loss on the train split of each dataset. Optimization is performed for 500 epochs using mini-batches of size 500 and Adam optimizer [10] with $\beta_1 = 0.9$, $\beta_2 = 0.999$, and an initial learning rate of 0.001. The learning rate is gradually reduced to 0.0001 using cosine annealing without restarts.

Given the latent encodings of the non-augmented training set, the simple prediction model Pred_ζ is learned to minimize the cross-entropy loss. As before, we use Adam with a mini-batch size of 500 with $\beta_1 = 0.9$, $\beta_2 = 0.999$, and a learning rate of 0.001. Training is terminated after 10000 epochs, where the cross-entropy losses stabilize.

4.2 Numerical Results

Next, we evaluate our approach and compare it to recent calibration approaches that are applicable in the context of (relatively) small-scale latent-code classification. We once again highlight that our objective is the minimization of the ECE and ACE scores, all other evaluation metrics are only reported for completeness. For all datasets, we use the same hyperparameters.

Table 1. Numerical results for MNIST dataset (8-dimensional latent space).

Cross-validation split ratio	Method	ECE ↓	ACE ↓	Brier score ↓	1+NLLLoss ↓	Accuracy ↑
10%	Baseline	0.0107 ± 0.0027	0.0100 ± 0.0038	0.0458 ± 0.0207	**0.0478 ± 0.0190**	0.9706 ± 0.0134
10%	Inexact BS (10-fold)	0.0069 ± 0.0019	0.0083 ± 0.0029	0.0572 ± 0.0232	0.0585 ± 0.0206	0.9638 ± 0.0150
10%	Inexact BS (100-fold)	0.0061 ± 0.0020	**0.0071 ± 0.0028**	0.0570 ± 0.0231	0.0583 ± 0.0205	0.9639 ± 0.0149
10%	Inexact BS (1000-fold)	**0.0060 ± 0.0020**	**0.0071 ± 0.0028**	0.0569 ± 0.0230	0.0581 ± 0.0205	0.9639 ± 0.0149
10%	Exact BS (10-fold)	0.0087 ± 0.0030	0.0088 ± 0.0040	0.0551 ± 0.0203	0.0518 ± 0.0185	0.9652 ± 0.0134
10%	Exact BS (100-fold)	0.0080 ± 0.0032	0.0083 ± 0.0041	0.0552 ± 0.0203	0.0519 ± 0.0185	0.9654 ± 0.0133
10%	Exact BS (1000-fold)	0.0080 ± 0.0033	0.0082 ± 0.0038	0.0553 ± 0.0204	0.0520 ± 0.0186	0.9653 ± 0.0133
10%	Dual-Focal Loss	0.0108 ± 0.0027	0.0101 ± 0.0037	0.0458 ± 0.0207	**0.0478 ± 0.0190**	**0.9707 ± 0.0135**
10%	Temperature Scaling	0.0109 ± 0.0029	0.0104 ± 0.0037	**0.0457 ± 0.0207**	**0.0478 ± 0.0190**	0.9706 ± 0.0134
100%	Baseline	0.0059	0.0060			
100%	Inexact BS (10-fold)	0.0033	0.0058			
100%	Inexact BS (100-fold)	0.0028	0.0052			
100%	Inexact BS (1000-fold)	**0.0025**	**0.0050**			
100%	Exact BS (10-fold)	0.0053	0.0052		same values as above	
100%	Exact BS (100-fold)	0.0052	0.0051			
100%	Exact BS (1000-fold)	0.0051	0.0052			
100%	Dual-Focal Loss	0.0057	0.0061			
100%	Temperature Scaling	0.0056	0.0062			

(a) Results for rotation with $\underline{\phi} = 15°$.

Cross-validation split ratio	Method	ECE ↓	ACE ↓	Brier score ↓	1+NLLLoss ↓	Accuracy ↑
10%	Baseline	0.0105 ± 0.0028	0.0096 ± 0.0044	**0.0351 ± 0.0192**	0.0378 ± 0.0181	0.9780 ± 0.0124
10%	Inexact BS (100-fold)	**0.0100 ± 0.0028**	0.0095 ± 0.0041	0.0352 ± 0.0192	0.0378 ± 0.0181	**0.9781 ± 0.0123**
10%	Exact BS (100-fold)	0.0106 ± 0.0031	**0.0085 ± 0.0029**	0.0396 ± 0.0203	0.0399 ± 0.0183	0.9741 ± 0.0126
10%	Dual-Focal Loss	0.0109 ± 0.0028	0.0097 ± 0.0043	0.0353 ± 0.0193	**0.0377 ± 0.0181**	0.9779 ± 0.0125
10%	Temperature Scaling	0.0105 ± 0.0028	0.0096 ± 0.0044	**0.0351 ± 0.0192**	0.0378 ± 0.0181	0.9780 ± 0.0124
100%	Baseline	0.0050	0.0058			
100%	Inexact BS (100-fold)	0.0050	0.0058			
100%	Exact BS (100-fold)	0.0049	**0.0046**		same values as above	
100%	Dual-Focal Loss	**0.0048**	0.0058			
100%	Temperature Scaling	0.0050	0.0058			

(b) Results for translation with $\underline{T} = 4$.

Cross-validation split ratio	Method	ECE ↓	ACE ↓	Brier score ↓	1+NLLLoss ↓	Accuracy ↑
10%	Baseline	0.0095 ± 0.0026	0.0084 ± 0.0034	**0.0325 ± 0.0174**	**0.0352 ± 0.0163**	**0.9800 ± 0.0115**
10%	Inexact BS (100-fold)	**0.0059 ± 0.0026**	**0.0055 ± 0.0020**	0.0428 ± 0.0209	0.0444 ± 0.0188	0.9732 ± 0.0134
10%	Exact BS (100-fold)	0.0064 ± 0.0025	**0.0055 ± 0.0026**	0.0415 ± 0.0209	0.0400 ± 0.0180	0.9739 ± 0.0132
10%	Dual-Focal Loss	0.0095 ± 0.0026	0.0084 ± 0.0034	0.0328 ± 0.0176	**0.0352 ± 0.0163**	0.9798 ± 0.0118
10%	Temperature Scaling	0.0095 ± 0.0026	0.0084 ± 0.0034	**0.0325 ± 0.0174**	**0.0352 ± 0.0163**	**0.9800 ± 0.0115**
100%	Baseline	0.0045	0.0055			
100%	Inexact BS (100-fold)	0.0034	0.0042			
100%	Exact BS (100-fold)	**0.0022**	**0.0027**		same values as above	
100%	Dual-Focal Loss	0.0040	0.0053			
100%	Temperature Scaling	0.0045	0.0055			

(c) Results for translation/rotation with $\underline{T} = 4$ and $\underline{\phi} = 15°$.

To quantify the calibration of each method, since ECE and ACE depend on the dataset and its distribution across bins, we utilize the validation set of each dataset and apply a cross-validation scheme. Each of the three evaluated datasets provides a total of 10000 images for validation.

In detail, we split the validation set consecutively into disjoint sets of relative size in $\{1\%, 10\%, 100\%\}$, resulting in 100, 10, or 1 different splits. On each of these three sets, we compute average statistics (mean and standard deviation) to the metrics presented in Sect. 3.4. For the 100% split, there is no standard deviation. For both the exact and inexact bootstrap methods, we extend each of the disjoint validation sets by computing S-fold bootstrap samples using $S \in \{10, 100, 1000\}$ according to the sampling schemes depicted in Fig. 1.

Table 1 presents the evaluation metrics for an autoencoder trained on the MNIST dataset with an 8-dimensional latent space, considering the group actions of rotation (Table 1a), translation (Table 1b), and their combination (Table 1c). We compare the baseline methods (i.e., without applying any calibration method) and our bootstrap approaches to the dual-focal loss method proposed in [17] and temperature scaling described in [6]. For the dual-focal loss comparison, we modify the baseline method by replacing the standard NLL-Loss in the final predictive linear layer with the dual-focal loss. During training, our bootstrap method incorporates rotations up to 15° wherever rotations are involved.

In terms of calibration metrics, specifically ECE and ACE, our proposed method consistently outperforms all competing approaches, where the absolute values decrease when increasing the cross-validation split ratio. Interestingly, in certain scenarios, the inexact bootstrap method surpasses the exact bootstrap method, despite lacking rigorous mathematical justification. For translations, however, the baseline and bootstrap methods show negligible differences in performance. Finally, the combined group actions of rotation and translation yield the best overall results, highlighting the advantage of leveraging multiple group actions simultaneously. The accuracy only exhibits small variations among the different settings, which, however, is slightly inferior in all bootstrap cases.

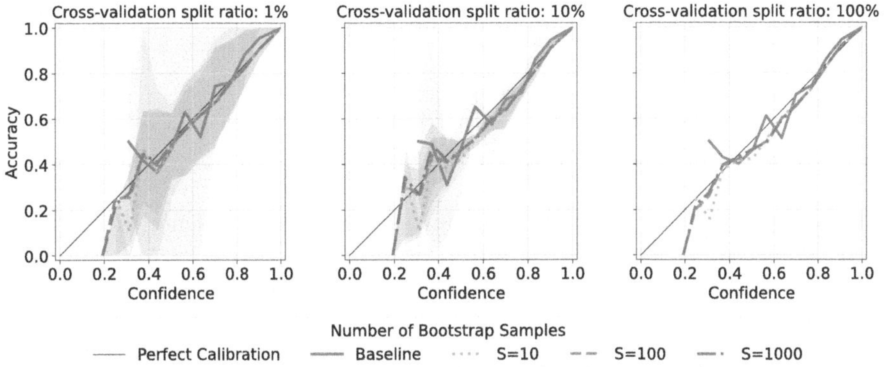

Fig. 2. Calibration curves for different sizes of validation set for MNIST dataset.

In Fig. 2 the calibration curves are visualized for the baseline method as well as the exact bootstrap method with various cross-validation split ratios and different S. The shaded areas represent the respective 25%- and 75%-quantiles

Table 2. Numerical results for rotations with $\underline{\phi} = 15°$ on different datasets.

Cross-validation split ratio	Method	ECE ↓	ACE ↓	Brier score ↓	1+NLLLoss ↓	Accuracy ↑
10%	Baseline	0.0262 ± 0.0057	0.0240 ± 0.0045	0.2376 ± 0.0108	0.2359 ± 0.0052	0.8303 ± 0.0087
	Inexact BS (100-fold)	0.0475 ± 0.0065	0.0471 ± 0.0065	0.3301 ± 0.0096	0.2970 ± 0.0054	0.7638 ± 0.0064
	Exact BS (100-fold)	**0.0181 ± 0.0048**	**0.0175 ± 0.0049**	0.2443 ± 0.0103	0.2432 ± 0.0051	0.8243 ± 0.0075
	Dual-Focal Loss	0.0256 ± 0.0059	0.0253 ± 0.0047	**0.2375 ± 0.0108**	**0.2358 ± 0.0052**	**0.8305 ± 0.0083**
	Temperature Scaling	0.0262 ± 0.0055	0.0240 ± 0.0045	0.2376 ± 0.0108	0.2359 ± 0.0052	0.8303 ± 0.0087
100%	Baseline	0.0088	0.0060	same values as above		
	Inexact BS (100-fold)	0.0468	0.0467			
	Exact BS (100-fold)	**0.0039**	**0.0034**			
	Dual-Focal Loss	0.0083	0.0068			
	Temperature Scaling	0.0087	0.0060			

(a) Results for FashionMNIST dataset (32-dimensional latent space).

Cross-validation split ratio	Method	ECE ↓	ACE ↓	Brier score ↓	1+NLLLoss ↓	Accuracy ↑
10%	Baseline	0.0257 ± 0.0067	0.0236 ± 0.0050	**0.2173 ± 0.0140**	0.2155 ± 0.0069	**0.8464 ± 0.0117**
	Inexact BS (100-fold)	0.0360 ± 0.0083	0.0357 ± 0.0084	0.3167 ± 0.0102	0.2917 ± 0.0053	0.7713 ± 0.0070
	Exact BS (100-fold)	**0.0200 ± 0.0068**	**0.0210 ± 0.0061**	0.2241 ± 0.0139	0.2243 ± 0.0067	0.8417 ± 0.0117
	Dual-Focal Loss	0.0258 ± 0.0045	0.0226 ± 0.0055	0.2174 ± 0.0140	0.2155 ± 0.0069	0.8459 ± 0.0118
	Temperature Scaling	0.0264 ± 0.0063	0.0235 ± 0.0052	**0.2173 ± 0.0141**	**0.2148 ± 0.0069**	**0.8464 ± 0.0117**
100%	Baseline	0.0088	0.0091	same values as above		
	Inexact BS (100-fold)	0.0357	0.0357			
	Exact BS (100-fold)	**0.0064**	**0.0049**			
	Dual-Focal Loss	0.0083	0.0093			
	Temperature Scaling	0.0092	0.0091			

(b) Results for FashionMNIST dataset (64-dimensional latent space).

Cross-validation split ratio	Method	ECE ↓	ACE ↓	Brier score ↓	1+NLLLoss ↓	Accuracy ↑
10%	Baseline	0.0466 ± 0.0072	0.0506 ± 0.0110	0.7379 ± 0.0082	0.7431 ± 0.0050	**0.4028 ± 0.0098**
	Inexact BS (100-fold)	0.1730 ± 0.0079	0.1730 ± 0.0079	0.8861 ± 0.0067	0.7762 ± 0.0037	0.2913 ± 0.0071
	Exact BS (100-fold)	**0.0375 ± 0.0115**	**0.0367 ± 0.0115**	0.7423 ± 0.0072	0.7488 ± 0.0042	0.3966 ± 0.0127
	Dual-Focal Loss	0.0450 ± 0.0110	0.0502 ± 0.0139	0.7379 ± 0.0081	0.7432 ± 0.0049	0.4024 ± 0.0115
	Temperature Scaling	0.0459 ± 0.0077	0.0493 ± 0.0103	**0.7377 ± 0.0082**	**0.7420 ± 0.0050**	**0.4028 ± 0.0098**
100%	Baseline	0.0323	0.0315	same values as above		
	Inexact BS (100-fold)	0.1730	0.1730			
	Exact BS (100-fold)	**0.0299**	**0.0286**			
	Dual-Focal Loss	0.0327	0.0311			
	Temperature Scaling	0.0303	0.0304			

(c) Results for CIFAR-10 dataset (128-dimensional latent space).

of the dataset splits with non-empty bins at the given confidence, hence there are no shaded areas for the 100% split ratio.

As a result, all methods show weak performance if the cross-validation split ratio is only 1%, which, however, significantly improves for larger split ratios, which is not unexpected due to the sparsity in the bins. In the latter cases, the exact method outperforms the baseline method.

Table 2 presents the results for more complex datasets: FashionMNIST (see Tables 2a and 2b) and CIFAR-10 (Table 2c). The exact bootstrap method, which uses rotations as a single group action with $\underline{\phi} = 15°$, demonstrates superior performance across the relevant calibration metrics, ECE and ACE, while maintaining accuracy comparable to the best-performing methods. In contrast, the inexact bootstrap method fails to produce reasonable results for these datasets.

The results for FashionMNIST indicate that, although the latent space dimension influences the calibration metrics, the impact of different methods on the outcomes remains relatively similar.

5 Conclusion

We propose two methods for quantifying uncertainty in image classification tasks by integrating autoencoders with (equivariant) bootstrap techniques. Among these, the exact method consistently demonstrates superior performance across most evaluated tasks in terms of relevant calibration metrics. Our findings reveal that the choice of group action and latent space dimensionality plays a critical role in determining performance. In future work, we plan to extend these methods to related imaging tasks and inverse problems, further exploring their applicability and robustness.

Acknowledgments. A. Effland was funded by the German Research Foundation under Germany's Excellence Strategy - EXC-2047/1 - 390685813 and - EXC2151 - 390873048. E. Kobler acknowledges funding from the German Research Foundation (DFG) KO 7162/1-1 project number 543939932, and Austrian Science Fund (FWF) [10.55776/COE12].

Disclosure of Interests. The authors have no competing interests to declare that are relevant to the content of this article.

References

1. Ayhan, M.S., Berens, P.: Test-time data augmentation for estimation of heteroscedastic aleatoric uncertainty in deep neural networks. In: Medical Imaging with Deep Learning (2018)
2. Davison, A.C., Hinkley, D.V.: Bootstrap Methods and Their Application, Cambridge Series in Statistical and Probabilistic Mathematics, vol. 1. Cambridge University Press, Cambridge (1997). https://doi.org/10.1017/CBO9780511802843
3. Deng, L.: The MNIST database of handwritten digit images for machine learning research. IEEE Signal Process. Mag. **29**(6), 141–142 (2012). https://doi.org/10.1109/MSP.2012.2211477
4. Dias, J.G., Vermunt, J.K.: Bootstrap methods for measuring classification uncertainty in latent class analysis. In: Rizzi, A., Vichi, M. (eds.) Compstat 2006 - Proceedings in Computational Statistics, pp. 31–41. Physica-Verlag HD (2006)
5. Efron, B., Tibshirani, R.J.: An Introduction to the Bootstrap, Monographs on Statistics and Applied Probability, vol. 57. Chapman and Hall, New York (1993). https://doi.org/10.1007/978-1-4899-4541-9
6. Guo, C., Pleiss, G., Sun, Y., Weinberger, K.Q.: On calibration of modern neural networks (2017). https://proceedings.mlr.press/v70/guo17a/guo17a.pdf
7. Havasi, M., et al.: Training independent subnetworks for robust prediction. In: International Conference on Learning Representations (2021). https://openreview.net/forum?id=OGg9XnKxFAH

8. Higgins, I., et al.: beta-VAE: learning basic visual concepts with a constrained variational framework. In: International Conference on Learning Representations (2017)
9. Huang, L., Ruan, S., Xing, Y., Feng, M.: A review of uncertainty quantification in medical image analysis: probabilistic and non-probabilistic methods. Med. Image Anal. **97**, 103223 (2024). https://doi.org/10.1016/j.media.2024.103223
10. Kingma, D.P., Ba, J.: Adam: a method for stochastic optimization. arXiv:1412.6980 (2014)
11. Kingma, D.P., Welling, M.: Auto-encoding variational bayes. In: International Conference on Learning Representations (2014)
12. Kozyrskiy, B., Milios, D., Filippone, M.: Variational bootstrap for classification. Procedia Comput. Sci. **207**, 1222–1231 (2022). https://doi.org/10.1016/j.procs.2022.09.178
13. Krizhevsky, A.: Cifar-10 (Canadian institute for advanced research) (2009). https://www.cs.toronto.edu/~kriz/cifar.html
14. Mobiny, A., Yuan, P., Moulik, S.K., Garg, N., Wu, C.C., Van Nguyen, H.: Dropconnect is effective in modeling uncertainty of bayesian deep networks. Sci. Rep. **11**(1), 5458 (2021). https://doi.org/10.1038/s41598-021-84854-x
15. Nixon, J., Dusenberry, M.W., Zhang, L., Jerfel, G., Tran, D.: Measuring calibration in deep learning. In: Computer Vision and Pattern Recognition Workshops (2019)
16. Pereyra, M., Tachella, J.: Equivariant bootstrapping for uncertainty quantification in imaging inverse problems. In: Dasgupta, S., Mandt, S., Li, Y. (eds.) Proceedings of The 27th International Conference on Artificial Intelligence and Statistics. Proceedings of Machine Learning Research, vol. 238, pp. 4141–4149. PMLR (2024). https://proceedings.mlr.press/v238/pereyra24a.html
17. Tao, L., Dong, M., Xu, C.: Dual focal loss for calibration. In: Proceedings of the 40th International Conference on Machine Learning, pp. 33833–33849 (2023). https://proceedings.mlr.press/v202/tao23a/tao23a.pdf
18. Weiler, M., Hamprecht, F.A., Storath, M.: Learning steerable filters for rotation equivariant CNNs. In: 2018 IEEE/CVF Conference on Computer Vision and Pattern Recognition, pp. 849–858 (2018). https://doi.org/10.1109/CVPR.2018.00095
19. Xiao, H., Rasul, K., Vollgraf, R.: Fashion-MNIST: a novel image dataset for benchmarking machine learning algorithms (2017). https://github.com/zalandoresearch/fashion-mnist

Equivariant Denoisers for Image Restoration

Marien Renaud[1,2(✉)], Arthur Leclaire[1], and Nicolas Papadakis[2]

[1] LTCI, Télécom Paris, IP Paris, Paris, France
marien.renaud@math.u-bordeaux.fr

[2] Univ. Bordeaux, CNRS, INRIA, Bordeaux INP, IMB, UMR 5251, 33400 Talence, France

Abstract. One key ingredient of image restoration is to define a realistic prior on clean images to complete the missing information in the observation. State-of-the-art restoration methods rely on a neural network to encode this prior. Moreover, typical image distributions are invariant to some set of transformations, such as rotations or flips. However, most deep architectures are not designed to represent an invariant image distribution. Recent works have proposed to overcome this difficulty by including equivariance properties within a Plug-and-Play paradigm. In this work, we propose a unified framework named Equivariant Regularization by Denoising (ERED) based on equivariant denoisers and stochastic optimization. We analyze the convergence of this algorithm and discuss its practical benefit.

Keywords: Image restoration · optimization · equivariance · plug-and-play

1 Introduction

Image restoration aims at recovering a proper image $x \in \mathbb{R}^d$ from a degraded observation $y \in \mathbb{R}^m$. A model to obtain y from x can be defined as $y \sim \mathcal{N}(\mathcal{A}(x))$, where $\mathcal{A} : \mathbb{R}^d \to \mathbb{R}^m$ is a deterministic operation on x and $\mathcal{N}$ is a noise distribution that corresponds to a known physical degradation. Typically, linear degradations with Gaussian noise can be written as $y = Ax + n$, with $A \in \mathbb{R}^{m \times d}$ and $n \sim \mathcal{N}(0, \sigma_y^2 I_m)$.

The Regularization Choice. In this context, the restoration task can be reformulated as a variational problem

$$\arg\min_{x \in \mathbb{R}^d} \mathcal{F}(x) = f(x) + \lambda r(x), \tag{1}$$

Supplementary Information The online version contains supplementary material available at https://doi.org/10.1007/978-3-031-92366-1_18.

T. A. Bubba et al. (Eds.): SSVM 2025, LNCS 15667, pp. 227–240, 2025.
https://doi.org/10.1007/978-3-031-92366-1_18

where the data-fidelity term $f = -\log p(y|\cdot)$ is the log-likelihood representing the degradation, while $r = -\log p$ is the regularization term encoding the prior distribution p, i.e. the model for clean images. The choice of the regularization r is crucial to complete the missing information in the observation y, e.g. by enforcing piecewise constant images [25] or wavelet sparsity [17]. To regularize the problem, recent generic methods rely on the use of an external denoiser within the Plug-and-Play (PnP) framework [31]. This denoiser D_σ, parametrized by a noise level σ, is generally a pre-trained deep neural network [33].

Image Invariances. Existing trained denoisers do not encode invariances on the underlying prior distribution. In fact, image distributions can be expected to be invariant to some transformations such as rotations or flips. Recent works develop Plug-and-Play restoration methods that try to enforce invariances on the underlying prior in order to improve restoration [1,4,9,19,28,30].

Stochastic Version of PnP. Another way to increase the performance of PnP focuses on the development of stochastic versions of PnP [15]. Stochasticity helps to reduce the computational cost [27,29]. Moreover, stochastic algorithms might compute better approximate solutions of problem (1) [12,23].

Contributions. (1) In this paper, we propose the unified π-equivariance formalism for the PnP framework, which generalizes so-called equivariant PnP [30] and other stochastic versions of PnP [12,23]. **(2)** We give theoretical insights on the convergence guarantees of the proposed Equivariant Regularization by Denoising (ERED, Algorithm 1). **(3)** We study the convergence of the ERED critical points (Problem 14) when the denoiser parameter σ goes to zero, in the case of a π-equivariant prior p. **(4)** We provide numerical experiments and comparisons for image restoration tasks, and discuss the practical benefits of equivariant approaches.

2 Background on Regularization by Denoising

In order to solve problem (1), when f and r are differentiable, one can use a gradient descent algorithm. However, the gradient of r, i.e. the score of the prior distribution p of clean images $s := \nabla \log p = -\nabla r$ is unknown. The authors of [24] proposed to make the following approximation

$$\nabla r = -\nabla \log p \approx \nabla r_\sigma := -\nabla \log p_\sigma, \tag{2}$$

where $p_\sigma = \mathcal{N}_\sigma * p$ is the convolution of p with the Gaussian $\mathcal{N}_\sigma$ with 0-mean and $\sigma^2 I_d$-covariance matrix. This is motivated by the Tweedie formula [8], which makes $\nabla \log p_\sigma$ tractable

$$-\nabla \log p_\sigma(x) = \frac{1}{\sigma^2}\left(x - D^*_\sigma(x)\right), \tag{3}$$

where $D^\star_\sigma$ is the Minimum Mean Square Error (MMSE) denoiser defined by $D^\star_\sigma(\tilde{x}) := \mathbb{E}[x|\tilde{x}] = \int_{\mathbb{R}^d} x p(x|\tilde{x}) dx$, for $\tilde{x} = x + \epsilon$ with $x \sim p(x), \epsilon \sim \mathcal{N}(0, \sigma^2 I_d)$.

This approximation leads to the Regularization By Denoising (RED) iterations:

$$x_{k+1} = x_k - \delta \nabla f(x_k) - \delta \frac{\lambda}{\sigma^2} (x_k - D_\sigma(x_k)), \tag{4}$$

where D_σ is a denoiser that is designed to approximate $D_\sigma^\star$.

The performance of RED can be improved by slightly modifying the algorithm with stochastic schemes [23,30], which incorporate invariance properties and enhance details in the restoration. In the next section, we propose a unified framework to generalize these approaches.

3 π-Equivariant Regularization by Denoising

We now introduce an extension of RED that is deduced from a notion of invariance, named π-equivariance, on the underlying prior.

Notations. We denote by $g : \mathbb{R}^d \to \mathbb{R}^d$ a differentiable transformation of $\mathbb{R}^d$; by $\mathcal{G}$ a measurable set of transformations of $\mathbb{R}^d$, and by $G \sim \pi$ a random variable of law π on $\mathcal{G}$.

3.1 π-Equivariant Image Distributions

Definition 1 (Invariance). *A density p on $\mathbb{R}^d$ is said to be invariant to a set of transformations $\mathcal{G}$ if $\forall g \in \mathcal{G}$, $p = p \circ g$ a.e.*

As shown in [16], natural images densities tend to be invariant to some set of transformations such as rotations or flips have been studied in.

Definition 2 (π-equivariance). *A density p on $\mathbb{R}^d$ is said to be π-equivariant if $\mathbb{E}_{G\sim\pi}[|\log(p \circ G)|] < \infty$ and $\log p = \mathbb{E}_{G\sim\pi}[\log(p \circ G)]$.*

Definition 2 relaxes the notion of invariance for a density in the following sense. If a density p is invariant to each $g \in \mathcal{G}$, p is π-equivariant, whatever the distribution π on $\mathcal{G}$.

Remark 1 (Key identity). For $p \in \mathcal{C}^1(\mathbb{R}^d, \mathbb{R}_+^*)$ and $g \in \mathcal{C}^1(\mathbb{R}^d, \mathbb{R}^d)$, we have

$$\nabla \log(p \circ g)(x) = \frac{\nabla(p \circ g)(x)}{(p \circ g)(x)} = \frac{J_g^T(x) \nabla p(g(x))}{(p \circ g)(x)} = J_g^T(x)(\nabla \log p)(g(x)), \tag{5}$$

with $x \in \mathbb{R}^d$. Thus, if p is π-equivariant, then $s = -\nabla \log p$, the score of p, verifies the identity $s = \mathbb{E}_{G\sim\pi}\left(J_G^T(s \circ G)\right)$.

In the context of equivariant transforms, Remark 1 suggests to apply the Jacobian of the transformation g instead of the inverse of the transformation g^{-1} as it is done in existing works [11,19,28,30]. Here the score (5) can be computed for any general differentiable transformation g, even if g^{-1} does not exist.

3.2 Equivariant Regularization

In order to encode the desired equivariance property in the regularization r, we introduce the **Equivariant REgularization by Denoising (ERED)** r_σ^π and the associated score s_σ^π respectively defined by

$$r_\sigma^\pi(x) := -\mathbb{E}_{G\sim\pi}\left(\log(p_\sigma \circ G)(x)\right) \tag{6}$$

$$s_\sigma^\pi(x) := -\mathbb{E}_{G\sim\pi}\left(J_G^T(x)(\nabla \log p_\sigma)(G(x))\right). \tag{7}$$

Note that under regularity assumptions on π and $p_\sigma \circ G$ (e.g. $\mathcal{G}$ finite and $p_\sigma \circ G$ differentiable), we get $s_\sigma^\pi = \nabla r_\sigma^\pi$. Thanks to the Tweedie formula (3), s_σ^π can be computed with an MMSE denoiser

$$s_\sigma^\pi(x) = \mathbb{E}_{G\sim\pi}\left(\frac{1}{\sigma^2} J_G^T(x)\left(G(x) - D_\sigma^*(G(x))\right)\right). \tag{8}$$

For a given denoiser D_σ, e.g. a supervised neural network, we can thus introduce the *equivariant denoiser* $\tilde{D}_\sigma$ defined by

$$\tilde{D}_\sigma(x) = \mathbb{E}_{G\sim\pi}\left[J_G^T(x) D_\sigma\left(G(x)\right)\right]. \tag{9}$$

Since the exact MMSE denoiser D_σ^* is not tractable, we make the following approximation:

$$s_\sigma^\pi(x) = \frac{1}{\sigma^2}\left(\mathbb{E}_\pi\left[J_G^T(x)G(x)\right] - \tilde{D}_\sigma^*(x)\right) \approx \frac{1}{\sigma^2}\left(\mathbb{E}_\pi\left[J_G^T(x)G(x)\right] - \tilde{D}_\sigma(x)\right).$$

No special structure is required on $\mathcal{G}$ or π, thus making the ERED framework a generic construction. However, in order to ensure that the ERED r_σ^π is indeed π-equivariant (Definition 2), more structure is required on $\mathcal{G}$ and π. In Proposition 1, a sufficient condition on $\mathcal{G}$ and π is provided for this property to hold. Before stating this result, let us recall that a compact Hausdorff topological group admits a unique right-invariant probability measure π, called Haar measure, that satisfies, for any integrable function $\varphi : \mathcal{G} \to \mathbb{R}$ and for any $g \in \mathcal{G}$, $\int_\mathcal{G} \varphi(g(x))d\pi(x) = \int_\mathcal{G} \varphi(x)d\pi(x)$ [10,20]. For a finite group $\mathcal{G}$, the Haar measure is the counting measure.

Proposition 1. *If $\mathcal{G}$ is a compact Hausdorff topological group and π the associated right-invariant Haar measure, then r_σ^π is π-equivariant.*

Proof. For $x \in \mathbb{R}^d$, by the right-invariance of π, we get

$$\mathbb{E}_{G'\sim\pi}\left(\log(r_\sigma^\pi(G'(x)))\right) = \int_\mathcal{G} \log(r_\sigma^\pi(G'(x))d\pi(G')$$

$$= \int_\mathcal{G} \log\left(-\int_\mathcal{G} \log(p_\sigma(G \circ G'(x))d\pi(G)\right) d\pi(G')$$

$$= \int_\mathcal{G} \log\left(-\int_\mathcal{G} \log(p_\sigma(G(x))d\pi(G)\right) d\pi(G') = \log\left(r_\sigma^\pi(x)\right).$$

□

Let us discuss the hypothesis on the set of transformations $\mathcal{G}$. First, $\mathcal{G}$ needs to be a group, i.e. $\forall g, g' \in \mathcal{G}, g^{-1} \circ g' \in \mathcal{G}$, to ensure that the composition $G \circ G'$ is still in $\mathcal{G}$ and that any transformation is invertible. Moreover, $\mathcal{G}$ needs to be a Hausdorff space (i.e. $\forall g, g' \in G$, there exist two neighbourhoods U and V of x and y, such that $U \cap V = \emptyset$) and is required to be compact to ensure that π is a probability measure. These hypotheses on $\mathcal{G}$ are general and cover in particular any finite discrete $\mathcal{G}$.

3.3 Examples of Equivariant Scores

The equivariant formulation generalizes recent works in the literature that we recall here. Moreover, we present new examples of equivariant scores that are included in our general framework.

Finite Set of Isometries. The authors of [30] proposed an equivariant version of PnP for a finite set of linear isometric transformations $\mathcal{G}$ with the uniform distribution on $\mathcal{G}$, i.e. $\forall g \in \mathcal{G}, \pi(g) = \frac{1}{|\mathcal{G}|}$. Since g is a linear isometry, $J_g^T(x) = g^{-1}$. In this case, the π-equivariant denoiser (9) and regularization (8) are respectively defined by

$$\tilde{D}_\sigma(x) = \frac{1}{|\mathcal{G}|} \sum_{g \in \mathcal{G}} g^{-1} \left[D_\sigma \left(g(x) \right) \right], \quad s_\sigma^\pi(x) \approx \frac{1}{\sigma^2} \left(x - \tilde{D}_\sigma(x) \right). \tag{10}$$

Infinite Set of Isometries. To generalize the previous example, we can propose an infinite set $\mathcal{G}$ of isometries, e.g. sub-pixel rotations for which π can be seen as the angular distribution. In this case, $\forall g \in \mathcal{G}, \forall x \in \mathbb{R}^d,\ J_g^T(x) = g^{-1}$ and then

$$\tilde{D}_\sigma(x) = \mathbb{E}_{G \sim \pi} \left(G^{-1} \left[D_\sigma \left(G(x) \right) \right] \right), \quad s_\sigma^\pi(x) \approx \frac{1}{\sigma^2} \left(x - \tilde{D}_\sigma(x) \right). \tag{11}$$

Noising-Denoising. The stochastic denoising regularization (SNORE) proposed in [23] consists in noising the image before denoising it. It can be interpreted as a π-equivariant denoiser for the set of translations $g_z(x) = x + \sigma z$, for $x, z \in \mathbb{R}^d$ and $\sigma > 0$ the noise level of D_σ, with the multivariable distribution, i.e. $\forall z \in \mathbb{R}^d$, $\pi(g_z) = \mathcal{N}(z; 0, \sigma^2 I_d)$. The Jacobian of the translation is $J_{g_z}(x) = I_d$. With this set of transformations, the π-equivariant denoiser (9) and regularization (8) can be expressed as

$$\tilde{D}_\sigma(x) = \mathbb{E}_{z \sim \mathbb{N}(0, \sigma I_d)} \left[D_\sigma \left(x + \sigma z \right) \right] \tag{12}$$

$$s_\sigma^\pi(x) \approx \frac{1}{\sigma^2} \left(\mathbb{E}_\pi \left[x + \sigma z \right] - \tilde{D}_\sigma(x) \right) = \frac{1}{\sigma^2} \left(x - \tilde{D}_\sigma(x) \right). \tag{13}$$

4 ERED Algorithm: Definition and Analysis

In this section, we define a generic π-equivariant PnP algorithm and demonstrate its convergence. Instead of solving Problem (1), we will tackle

$$\arg\min_{x \in \mathbb{R}^d} \mathcal{F}_\sigma^\pi(x) := f(x) + \lambda r_\sigma^\pi(x), \tag{14}$$

with the equivariant regularization r_σ^π defined in relation (6). We now introduce the equivariant Regularization by Denoising (ERED) algorithm (Algorithm 1), which is a biased stochastic gradient descent to solve Problem (14).

Algorithm 1. ERED

Parameters: $x_0 \in \mathbb{R}^d$, $\sigma > 0$, $\lambda > 0$, $\delta > 0$, $N \in \mathbb{N}$
Input: degraded image y
Output: restored image x_N
for $k = 0, 1, \ldots, N-1$ **do**
  Sample $G \sim \pi$
  $x_{k+1} = x_k - \delta \nabla f(x_k) - \frac{\delta\lambda}{\sigma^2} J_G^T(x_k)\left(G(x_k) - D_\sigma(G(x_k))\right)$
end for

Remark 2. If the denoiser D_σ is L-Lipschitz then the equivariant denoiser $\tilde{D}_\sigma$ defined in relation (9) is L-Lipschitz. Previous works have shown the link between the Lipschitz constant of the denoiser and the convergence of deterministic PnP algorithms [14,26,32]. However, no clue indicates that this property still holds for stochastic PnP algorithms, such as Algorithm 1. Thus, in this paper, we provide a new analysis to demonstrate the convergence of Algorithm 1.

4.1 Unbiased Convergence Analysis

In this section, we prove the convergence of the ERED (Algorithm 1) run with the exact MMSE denoiser $D_\sigma^\star$. With this denoiser, thanks to Tweedie formula, the iterations are computed by

$$x_{k+1} = x_k - \delta_k \nabla f(x_k) - \lambda \delta_k J_G^T(x_k) \nabla \log p_\sigma(G(x_k)), \tag{15}$$

with $G \sim \pi$ and $(\delta_k)_{k\in\mathbb{N}} \in \left(\mathbb{R}^+\right)^{\mathbb{N}}$ a non-increasing sequence of step-sizes.

Assumption 1. ***(a)*** *The step-size decreases to zero but not too fast:* $\sum_{k=0}^{+\infty} \delta_k = +\infty$ *and* $\sum_{k=0}^{+\infty} \delta_k^2 < +\infty$.

(b) *The data-fidelity term* $f : x \in \mathbb{R}^d \mapsto f(x) \in \mathbb{R}$ *is* $\mathcal{C}^\infty$.

(c) *The noisy prior score is sub-polynomial, i.e. there exist* $B \in \mathbb{R}^+$, $\beta \in \mathbb{R}$ *and* $n_1 \in \mathbb{N}$ *such that* $\forall \sigma > 0$, $\forall x \in \mathbb{R}^d$, $\|\nabla \log p_\sigma(x)\| \leq B\sigma^\beta(1 + \|x\|^{n_1})$.

Assumption 1(a) is standard in stochastic gradient descent analysis. It suggests a choice of the step-size rule to ensure convergence, for instance $\delta_k = \frac{\delta}{k^\alpha}$ with $\alpha \in]\frac{1}{2}, 1]$. Assumption 1(b) is typically verified for a linear degradation with additive Gaussian noise, i.e. $f(x) = \frac{1}{\sigma_y^2}\|y - Ax\|^2$. It ensures that the objective function of Problem (14) is $\mathcal{C}^\infty$. Under the so-called manifold hypothesis, i.e. p is supported on a compact, it is shown in [6] that Assumption 1(c) is verified with $n_1 = 1$ and $\beta = -2$.

Assumption 2. *(a) The random variable J_G has a uniform finite moment, i.e. $\exists \epsilon > 0, M_{2+\epsilon} \geq 0$ such that $\forall x \in \mathbb{R}^d, \mathbb{E}_{G\sim\pi}(\|\|J_G(x)\|\|^{2+\epsilon}) \leq M_{2+\epsilon} < +\infty$, with $\|\|\cdot\|\|$ the operator norm defined for $A \in \mathbb{R}^{d\times d}$ by $\|\|A\|\| = \sup_{\|x\|=1} \frac{\|Ax\|}{\|x\|}$.*

(b) The transformation has bounded moments on any compact, i.e. $\forall \mathbf{K} \subset \mathbb{R}^d$ compact, $\forall m \in \mathbb{N}$, $\exists C_{\mathbf{K},m} < +\infty$ such that $\forall x \in \mathbf{K}, \mathbb{E}_{G\sim\pi}(\|G(x)\|^m) \leq C_{\mathbf{K},m}$.

With Assumption 2, the behavior of the algorithm is controlled on each compact set. This assumption is verified for all examples presented in Sect. 3.3.

We now define $\mathbf{S}_\sigma = \{x \in \mathbb{R}^d | \nabla \mathcal{F}_\sigma^\pi(x) = 0\}$, the set of critical points of $\mathcal{F}_\sigma^\pi$ in $\mathbb{R}^d$, and $\Lambda_{\mathbf{K}}$, the set of random seeds for which the iterates of the algorithm are bounded in the compact K, by

$$\Lambda_{\mathbf{K}} = \bigcap_{k\in\mathbb{N}} \{x_k \in \mathbf{K}\}.$$

We finally denote the distance of a point to a set by $d(x, \mathbf{S}) = \inf_{y\in\mathbf{S}} \|x - y\|$, with $x \in \mathbb{R}^d$ and $\mathbf{S} \subset \mathbb{R}^d$. The restriction to realizations of the algorithm bounded in Λ_K will be referred to as the *boundedness assumption.*

Proposition 2. *Let $(x_k)_{k\in\mathbb{N}}$ be the iterates generated by Algorithm 1 with the exact MMSE Denoiser D_σ^*. Then, under Assumptions 1–2, we have almost surely on $\Lambda_{\mathbf{K}}$*

$$\lim_{k\to+\infty} d(x_k, \mathbf{S}_\sigma) = 0, \tag{16}$$

$$\lim_{k\to+\infty} \|\nabla \mathcal{F}_\sigma^\pi(x_k)\| = 0, \tag{17}$$

and $(\mathcal{F}_\sigma^\pi(x_k))_{k\in\mathbb{N}}$ converges to a value of $\mathcal{F}_\sigma^\pi(\mathbf{S}_\sigma)$.

Proof. The proof is obtained by applying [7, Theorem 2.1, (ii)], which is recalled in Appendix B.1 (Theorem 1) for completeness. To do so, we have to verify the different assumptions (Assumption 7 in Appendix B.1) of this theorem. First, Assumption 7(a) is verified by Assumption 1(a). Next p_σ is $\mathcal{C}^\infty$ by convolution with a Gaussian. Then $\log p_\sigma$ is also $\mathcal{C}^\infty$ [15] and so is r_σ^π defined in relation (6). By Assumption 7(b), $\mathcal{F}_\sigma^\pi = f + \lambda r_\sigma^\pi$ is $\mathcal{C}^\infty$. So Assumption 7(c) is verified.

Now, we are going to prove that Assumption 7(b) is verified, i.e. the noise fluctuation can be controlled. To that end, we define

$$\begin{aligned}\xi_k &= \nabla f(x_k) + \lambda J_G^T(x_k) \nabla \log p_\sigma(G(x_k)) - \nabla \mathcal{F}_\sigma^\pi(x_k) \\ &= \lambda \left(J_G^T(x_k) \nabla \log p_\sigma(G(x_k)) - \mathbb{E}_{G\sim\pi} \left(J_G^T(x_k) \nabla \log p_\sigma(G(x_k)) \right) \right),\end{aligned}$$

with $\lambda > 0, G \sim \pi$. By definition we have $\mathbb{E}(\xi_k) = 0$ and, from (15), we get

$$x_{k+1} = x_k - \delta_k(\nabla \mathcal{F}_\sigma^\pi(x_k) + \xi_k).$$

Lemma 1. *Under Assumptions 1–2, almost surely on $\Lambda_{\mathbf{K}}$, there exists $C > 0$, such that $\forall k \in \mathbb{N}$, $\mathbb{E}(\|\xi_k\|^2) \leq C$.*

Lemma 1 is demonstrated in Appendix B.2. By using Lemma 1, we get $\sum_{k\in\mathbb{N}} \delta_k^2 \mathbb{E}(\|\xi_k\|^2) \le C \sum_{k\in\mathbb{N}} \delta_k^2 < +\infty$, by Assumption 1. Then, we deduce from the Doob inequality (which holds because $\sum_{k=n}^{l} \xi_k$ is a martingale) that

$$\mathbb{E}\left(\sup_{n\le l\le m} \|\sum_{k=n}^{l} \delta_k \xi_k\|^2\right) \le 4 \sum_{k=n}^{m} \delta_k^2 \mathbb{E}\left(\|\xi_k\|^2\right). \tag{18}$$

By the monotone convergence theorem, this implies

$$\mathbb{E}\left(\sup_{n\le l} \|\sum_{k=n}^{l} \delta_k \xi_k\|^2\right) \le 4 \sum_{k=n}^{\infty} \delta_k^2 \mathbb{E}\left(\|\xi_k\|^2\right) \le 4C \sum_{k=n}^{\infty} \delta_k^2.$$

Thus the sequence $(\sup_{n\le l} \|\sum_{k=n}^{l} \delta_k \xi_k\|^2)_n$ tends to zero in L^1 and also almost surely (because it is non-increasing). The square function being non-decreasing, it implies that $(\sup_{n\le l} \|\sum_{k=n}^{l} \delta_k \xi_k\|)_n$ tends to zero almost surely. The process (15) thus verifies Assumption 7(b) almost surely. We can apply Theorem 1, which concludes the proof. □

4.2 Biased Convergence Analysis

In this section, we analyse the convergence of the ERED algorithm (Algorithm 1) run with a realistic denoiser $D_\sigma \neq D_\sigma^*$. In this case, ERED is a biased stochastic gradient descent for solving Problem (14). At each iteration, the algorithm writes

$$x_{k+1} = x_k - \delta_k \nabla f(x_k) - \frac{\delta_k \lambda}{\sigma^2} J_G^T(x_k)\left(G(x_k) - D_\sigma(G(x_k))\right), \tag{19}$$

with $G \sim \pi$. Defining the gradient estimator

$$\xi_k = \nabla f(x_k) + \frac{\lambda}{\sigma^2} J_G^T(x_k)\left(G(x_k) - D_\sigma(G(x_k))\right) - \nabla \mathcal{F}_\sigma^\pi(x_k),$$

the algorithm (19) can be reformulated as

$$x_{k+1} = x_k - \delta_k \left(\nabla \mathcal{F}_\sigma^\pi(x_k) + \xi_k\right). \tag{20}$$

Assumption 3. *The realistic denoiser D_σ is sub-polynomial, i.e. $\exists C > 0$ and $n_2 \in N$ such that $\forall x \in \mathbb{R}^d$, $\|D_\sigma(x)\| \le C(1 + \|x\|^{n_2})$.*

Assumption 3 is similar to Assumption 1(c) but it applies on the denoiser instead of the score of the underlying prior distribution. Notice that in this case the constant C might depend on the noise level σ. As an example, a bounded denoiser [3] verifies Assumption 3 with $n_2 = 1$.

Assumption 4. *For every compact $\mathbf{K}$, there exists $C_{\mathbf{K}}$, such that $\forall x \in \mathbf{K}$, $\forall g \in \mathcal{G}$, $\|g(x)\| \le C_{\mathbf{K}}$.*

Assumption 4 is verified as soon as $(g, x) \in \mathcal{G} \times \mathbb{R}^d \to g(x) \in \mathbb{R}^d$ is continuous and $\mathcal{G}$ compact. It is verified in particular for $\mathcal{G}$ being a finite set of isometries.

Proposition 3. *Let $(x_k)_{k \in \mathbb{N}}$ be the sequence provided by ERED (Algorithm 1) with an inexact denoiser D_σ. Then, under Assumptions 1–3, there exists $M_\mathbf{K}$ such that, almost surely on $\Lambda_\mathbf{K}$:*

$$\limsup_{k \to \infty} \|\nabla \mathcal{F}_\sigma^\pi(x_k)\| \le M_\mathbf{K} \eta^{\frac{1}{2}} \tag{21}$$

$$\limsup_{k \to \infty} \mathcal{F}_\sigma^\pi(x_k) - \liminf_{k \to \infty} \mathcal{F}_\sigma^\pi(x_k) \le M_\mathbf{K} \eta, \tag{22}$$

with the asymptotic bias $\eta = \limsup_{k \to \infty} \|\mathbb{E}(\xi_k)\|$.

Moreover, under Assumption 4, we have

$$\eta \le \frac{\lambda}{\sigma^2} \sup_{x \in \mathbf{K}} \mathbb{E}\left(|||J_G(x)|||\right) \|D_\sigma - D_\sigma^*\|_{\infty, \mathbf{L}}, \tag{23}$$

with $\mathbf{L} = \mathcal{B}(0, C_\mathbf{K})$, where $C_\mathbf{K}$ is introduced in Assumption 4.

Proof. First the bias is denoted by $\eta_k = \mathbb{E}(\xi_k)$ and the noise by $\gamma_k = \xi_k - \mathbb{E}(\xi_k)$. So we have $\xi_k = \gamma_k + \eta_k$ and $\mathbb{E}(\gamma_k) = 0$. We apply again Theorem 1. Assumptions 7(a)-7(c) are verified thanks to Assumptions 1(a) and 1(b).

Lemma 2. *Under Assumptions 2–3, almost surely on $\Lambda_\mathbf{K}$, there exists $C_2 > 0$, such that $\forall k \in \mathbb{N}$, $\mathbb{E}(\|\xi_k\|^2) \le C_2$.*

Lemma 2 is proved in Section B.3. Then, by using Lemma 2 and the Doob inequality as in relation (18), we demonstrated Assumption 7(b) of [7] (i.e. the noise fluctuation is controlled) and we can apply Theorem 1 to obtain equations (21)–(22). Under Assumption 4, we study the asymptotic behavior of η_k,

$$\begin{aligned} \|\eta_k\| = \|\mathbb{E}(\xi_k)\| &= \frac{\lambda}{\sigma^2} \|\mathbb{E}\left(J_G^T(x_k)\left(D_\sigma - D_\sigma^*\right)\left(G(x_k)\right)\right)\| \\ &\le \frac{\lambda}{\sigma^2} \mathbb{E}\left(|||J_G(x_k)|||\, \|\left(D_\sigma - D_\sigma^*\right)\left(G(x_k)\right)\|\right). \end{aligned}$$

By Assumption 4, because $x_k \in \mathbf{K}$, we know that $G(x_k) \in \mathbf{L} = \mathcal{B}(0, C_\mathbf{K})$. So we have $\|\left(D_\sigma - D_\sigma^*\right)\left(G(x_k)\right)\| \le \|D_\sigma - D_\sigma^*\|_{\infty, \mathbf{L}}$ and the desired inequality (23). □

4.3 Critical Points Analysis Geometrical Invariant Case

A critical point analysis for the noising-denoising case, presented in Eq. (12), is provided in [23]. However, to the best of our knowledge, no critical point analysis has been provided so far even in the case of geometrical invariance, i.e. $p \circ g = p$ for $g \in \mathcal{G}$. Here, we fill this gap by studying critical points under the relaxed π-equivariance property. First, we study the approximation $-\nabla \log p \approx s_\sigma^\pi$ when $\sigma \to 0$. Next, we deduce that critical points of Problem (14) converge to critical points of Problem (1) when $\sigma \to 0$.

Assumption 5. *(a) The prior distribution* $p \in C^1(\mathbb{R}^d,]0, +\infty[)$ *with* $\|p\|_\infty + \|\nabla p\|_\infty < +\infty$.

(b) J_G *has finite first moment, i.e.* $\sup_{x\in\mathbb{R}^d} \mathbb{E}_{G\sim\pi}(\||J_G(x)\||) < +\infty$.

Assumption 5(a) is needed to ensure that $\nabla \log p$ is well defined. Assumption 5(b) is verified in particular for a finite set of transformations, for a set of linear isometries or for the noising-denoising regularization.

Proposition 4. *Under Assumptions 4–5, for every compact* $\mathbf{K} \subset \mathbb{R}^d$, *if the prior* p *is* π*-equivariant, we have, when* $\sigma \to 0$,

$$\|s - s_\sigma^\pi\|_{\infty,\mathbf{K}} \to 0. \tag{24}$$

Proof. Due to the π-equivariance of p, we have $s = \mathbb{E}_{G\sim\pi}\left(J_G^T(s \circ G)\right)$ with a random variable $G \sim \pi$. With the definition of s_σ^π (8), we get for $x \in \mathbb{R}^d$

$$s_\sigma^\pi(x) - \nabla \log p(x) = \mathbb{E}_{G\sim\pi}\left(J_G^T(x)\left(\nabla \log p_\sigma - \nabla \log p\right)(G(x))\right). \tag{25}$$

By Assumption 4, we get that $\forall x \in \mathbf{K}, G(x) \in \mathcal{B}(0, C_\mathbf{K})$, with $\mathbf{L} = \mathcal{B}(0, C_\mathbf{K})$ the closed ball of center 0 and radius $C_\mathbf{K}$. By Assumption 5(a) and Proposition 1 in [15], we know that $\|\nabla \log p_\sigma - \nabla \log p\|_{\infty,\mathbf{L}} \to 0$ when $\sigma \to 0$.

Then, when $\sigma \to 0$, using Assumption 4, we obtain

$$\begin{aligned}\|s - s_\sigma^\pi\|_{\infty,\mathbf{K}} &\le \mathbb{E}_{G\sim\pi}\left(\||J_G\|| \|\nabla \log p_\sigma - \nabla \log p\|_{\infty,\mathbf{L}}\right)\\ &\le \|\nabla \log p_\sigma - \nabla \log p\|_{\infty,\mathbf{L}} \mathbb{E}_{G\sim\pi}\left(\||J_G\||\right) \to 0.\end{aligned}$$

□

From Proposition 4, we now deduce a critical point convergence when $\sigma \to 0$ in the sense of Kuratowski [2].

Assumption 6. *The data-fidelity term in* (1) *is continuously differentiable, i.e.* $f \in \mathcal{C}^1(\mathbb{R}^d, \mathbb{R})$.

Assumption 6 is needed to define the critical points of Problem (1). It is verified for a large set of inverse problems, including linear inverse problems with Gaussian noise, phase retrieval or despeckling.

We denote by $\mathbf{S}^*$ the set of critical point of $\mathcal{F}$, $\mathbf{S}_\sigma$ the set of critical points of $\mathcal{F}_\sigma^\pi$, and $\mathbf{S}$ the limit point of $\mathbf{S}_\sigma$ when $\sigma \to 0$, more precisely

$$\mathbf{S} = \{x \in \mathbb{R}^d | \exists \sigma_n > 0 \text{ decreasing to } 0, x_n \in \mathbf{S}_{\sigma_n} \text{ such that } x_n \xrightarrow[n\to\infty]{} x\}. \tag{26}$$

Proposition 5. *Under Assumptions 4–6, if the prior* p *is* π*-equivariant, we have*

$$\mathbf{S} \subset \mathbf{S}^*. \tag{27}$$

Proof. For $x \in \mathbf{S}$, we have $\sigma_n > 0$ decreasing to 0 and $x_n \in \mathbf{S}_{\sigma_n}$ such that $x_n \to x$. Because x_n is a converging sequence, there exists a compact $\mathbf{K}$ such that $\forall n \in \mathbb{N}, x_n \in \mathbf{K}$. Moreover, thanks to Proposition 4, $\|s - s_{\sigma_n}^{\mathcal{G}}\|_{\infty,\mathbf{K}} \xrightarrow[n\to\infty]{} 0$ and then $\|\mathcal{F} - \mathcal{F}_{\sigma_n}^\pi\|_{\infty,\mathbf{K}} \to 0$. So, $\|\mathcal{F}(x_n) - \mathcal{F}_{\sigma_n}^\pi(x_n)\| \to 0$ and by definition $\mathcal{F}_{\sigma_n}^\pi(x_n) = 0$ which gives $\|\mathcal{F}(x_n)\| \to 0$. It implies that $\|\nabla\mathcal{F}(x)\| = 0$. □

5 Experiments

In this section, we evaluate the practical gain of Algorithm 1 for image restoration. We focus on image deblurring with various blur kernels including fixed and motion kernels as proposed by [13]. The denoiser used in these experiments is the Gradient-Step DRUNet denoiser proposed by [13] with the provided weights, obtained with supervised training on natural color images. This denoiser reaches state-of-the-art denoising performance.

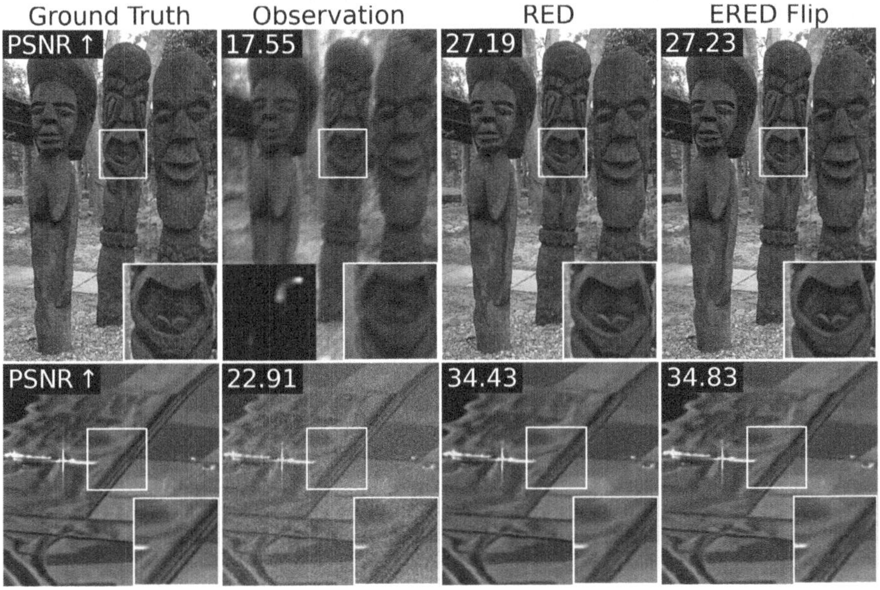

Fig. 1. Deblurring (a motion blur kernel with input noise level $\sigma_y = 5/255$) and despeckling (number of looks 50) with RED and ERED with a GS-denoiser trained on natural images or SAR images (respectively). The set of transformations for ERED is *random flip*. ERED produces a better qualitative result than RED. (Color figure online)

On Fig. 1, we observe that ERED outperforms the standard RED algorithm. The set of transformations for ERED is *random flip* of the two image axes. Especially, it succeeds to reduce artifacts generated by RED. In fact, ERED might impose more geometrical properties on the restored image due to its intrinsic equivariance.

Method	PSNR↑	SSIM↑	N↓
RED [24]	32.25	0.84	400
ERED rotation [30]	<u>32.53</u>	0.85	400
ERED translation	32.44	0.85	400
ERED flip	32.51	0.85	400
ERED subpixel rotation	32.32	0.85	400
ERED all transformations	31.94	0.83	400
SNORE [23]	32.45	<u>0.86</u>	1000
Annealed SNORE [23]	**32.89**	**0.87**	1500

Table 1. Quantitative comparison of image deblurring methods on 10 images from CBSD68 dataset with 10 different blur kernels (fixed and motion kernel of blur) and a noise level $\sigma_y = 5/255$. Best and second best results are respectively displayed in bold and underlined.

On Table 1, we present Algorithm 1 performance with various sets of transformations including random rotation of angle $\theta \in \{0, 1, 2, 3\} \times \pi/2$ named *rotation*, random subpixel rotation of angle $\theta \in [-\pi, \pi]$ named *subpixel rotation* using raster rotation [21], random flip of the two axis named *flip*, random translation along both axis named *translation*, random Gaussian noising named *SNORE*, and a random transformation taken randomly in between all the previous sets of transformations (including SNORE) named *all transformations*. We also present performances of RED and Annealed SNORE, another version of SNORE [23] where the denoiser parameter σ decreases through iterations. We notice that ERED obtains better quantitative performance than RED (+0.3dB) with the same computational cost. Moreover, the choice of the transformation impacts the restoration quality and our experiments suggest that flips and rotations are beneficial.

Additional experiments are provided in Appendix A on denoising, super-resolution, despeckling and deblurring with other denoisers.

6 Conclusion

In this paper, we propose ERED, an equivariant version of RED. We provide an interpretation of the ERED algorithm as an equivariant property of the underlying prior. We give theoretical convergence results (Propositions 2–3) and a critical point convergence with an equivariant prior p (Proposition 5). Experimental results illustrate the modest improvement brought by such methods.

Acknowledgements. This study has been carried out with financial support from the French Direction Générale de l'Armement. Experiments presented in this paper were carried out using the PlaFRIM experimental testbed, supported by Inria, CNRS (LABRI and IMB), Universite de Bordeaux, Bordeaux INP and Conseil Regional d'Aquitaine (see https://www.plafrim.fr).

References

1. Celledoni, E., Ehrhardt, M.J., Etmann, C., Owren, B., Schönlieb, C.B., Sherry, F.: Equivariant neural networks for inverse problems. IP **37**(8), 085006 (2021)
2. Chambolle, A., Duval, V., Machado, J.M.: 1D approximation of measures in wasserstein spaces. arXiv preprint arXiv:2304.14781 (2023)
3. Chan, S.H., Wang, X., Elgendy, O.A.: Plug-and-play ADMM for image restoration: fixed-point convergence and applications. IEEE TCI **3**(1), 84–98 (2016)

4. Chen, D., Tachella, J., Davies, M.E.: Equivariant imaging: learning beyond the range space. In: ICCV, pp. 4379–4388 (2021)
5. Dalsasso, E., Denis, L., Tupin, F.: As if by magic: self-supervised training of deep despeckling networks with merlin. IEEE TGRS **60**, 1–13 (2021)
6. De Bortoli, V.: Convergence of denoising diffusion models under the manifold hypothesis. Trans. Mach. Learn. Res. (2022). Expert Certification
7. Doucet, A., Tadic, V.: Asymptotic bias of stochastic gradient search. Ann. Appl. Probab. **27**(6) (2017)
8. Efron, B.: Tweedie's formula and selection bias. J. Am. Stat. Assoc. **106**(496), 1602–1614 (2011)
9. Fu, J., Xie, Q., Meng, D., Xu, Z.: Rotation equivariant proximal operator for deep unfolding methods in image restoration. IEEE TPAMI (2023)
10. Haar, A.: Der massbegriff in der theorie der kontinuierlichen gruppen. Ann. Math. **34**(1), 147–169 (1933)
11. Herbreteau, S., Moebel, E., Kervrann, C.: Normalization-equivariant neural networks with application to image denoising. In: NeurIPS, vol. 36 (2024)
12. Hu, Y., Peng, A., Gan, W., Milanfar, P., Delbracio, M., Kamilov, U.S.: Stochastic deep restoration priors for imaging inverse problems (2024)
13. Hurault, S., Leclaire, A., Papadakis, N.: Gradient step denoiser for convergent plug-and-play. In: International Conference on Learning Representations (2022)
14. Hurault, S., Leclaire, A., Papadakis, N.: Proximal denoiser for convergent plug-and-play optimization with nonconvex regularization. In: ICML. PMLR (2022)
15. Laumont, R., De Bortoli, V., Almansa, A., Delon, J., Durmus, A., Pereyra, M.: On map estimation with PNP priors and stochastic gradient descent. JMIV (2023)
16. Lenc, K., Vedaldi, A.: Understanding image representations by measuring their equivariance and equivalence. In: Proceedings of the IEEE Conference on Computer Vision and Pattern Recognition, pp. 991–999 (2015)
17. Mallat, S.: A wavelet tour of signal processing. Elsevier (1999)
18. Martin, D., Fowlkes, C., Tal, D., Malik, J.: A database of human segmented natural images and its application to evaluating segmentation algorithms and measuring ecological statistics. In: Proceedings Eighth IEEE International. ICCV 2001 (2001)
19. Mbakam, C.K., Giovannelli, J.-F., Pereyra, M.: Empirical Bayesian image restoration by Langevin sampling with a denoising diffusion implicit prior. arXiv preprint arXiv:2409.04384 (2024)
20. Neumann, J.: Zum haarschen mass in topologischen gruppen. Compos. Math. **1**, 106–114 (1935)
21. Paeth, A.W.: A fast algorithm for general raster rotation. In: Graphics Gems, pp. 179–195 (1990)
22. Pesquet, J.-C., Repetti, A., Terris, M., Wiaux, Y.: Learning maximally monotone operators for image recovery. SIAM J. Imaging Sci. **14**(3), 1206–1237 (2021)
23. Renaud, M., Prost, J., Leclaire, A., Papadakis, N.: Plug-and-play image restoration with stochastic denoising regularization. In: ICML (2024)
24. Romano, Y., Elad, M., Milanfar, P.: The little engine that could: regularization by denoising (red). SIAM J. Imag. Sci. **10**(4), 1804–1844 (2017)
25. Rudin, L.I., Osher, S., Fatemi, E.: Nonlinear total variation based noise removal algorithms. Physica D **60**(1–4), 259–268 (1992)
26. Ryu, E., Liu, J., Wang, S., Chen, X., Wang, Z., Yin, W.: Plug-and-play methods provably converge with properly trained denoisers. In: ICML. PMLR (2019)
27. Sun, Y., Liu, J., Kamilov, U.: Block coordinate regularization by denoising. In: Advances in Neural Information Processing Systems, vol. 32 (2019)

28. Tachella, J., Pereyra, M.: Equivariant bootstrapping for uncertainty quantification in imaging inverse problems. arXiv (2023)
29. Tang, J., Davies, M.: A fast stochastic plug-and-play ADMM for imaging inverse problems. arXiv preprint arXiv:2006.11630 (2020)
30. Terris, M., Moreau, T., Pustelnik, N., Tachella, J.: Equivariant plug-and-play image reconstruction. arXiv, pp. 25255–25264 (2024)
31. Venkatakrishnan, S.V., Bouman, C.A., Wohlberg, B.: Plug-and-play priors for model based reconstruction. In: IEEE GlobalSIP, pp. 945–948. IEEE (2013)
32. Wei, D., Chen, P., Li, F.: Learning pseudo-contractive denoisers for inverse problems. In: ICML (2024)
33. Zhang, K., Li, Y., Zuo, W., Zhang, L., Van Gool, L., Timofte, R.: Plug-and-play image restoration with deep denoiser prior. IEEE TPAMI **44**(10), 6360–6376 (2021)

Max-Normalized Radon Cumulative Distribution Transform for Limited Data Classification

Matthias Beckmann[1], Robert Beinert[2(✉)], and Jonas Bresch[2]

[1] Center for Industrial Mathematics, University of Bremen, Bibliothekstraße 5, 28359 Bremen, Germany
research@mbeckmann.de

[2] Institut für Mathematik, Technische Universität Berlin, Straße des 17. Juni 136, 10623 Berlin, Germany
{beinert,bresch}@math.tu-berlin.de

Abstract. The Radon cumulative distribution transform (R-CDT) exploits one-dimensional Wasserstein transport and the Radon transform to represent prominent features in images. It is closely related to the sliced Wasserstein distance and facilitates classification tasks, especially in the small data regime, like the recognition of watermarks in filigranology. Here, a typical issue is that the given data may be subject to affine transformations caused by the measuring process. The aim of this paper is to make the R-CDT and the related sliced Wasserstein distance invariant under affine transformations. For this, we propose a two-step normalization of the R-CDT and prove that our novel transform allows linear separation of affinely transformed image classes. The theoretical results are supported by numerical experiments showing a significant increase of the classification accuracy compared to the original R-CDT.

Keywords: Radon-CDT · sliced Wasserstein distance · feature representation · image classification · pattern recognition · small data regime

1 Introduction

Automated pattern recognition and classification play a central role in numerous applications and disciplines, be it in medical imaging, biometrics, or document analysis. Nowadays, in the big data regime, end-to-end deep neural networks provide the latest state of the art. In the small data regime, however, hand-crafted feature extractors and classifiers still stand their ground. Ideally, the feature extractor is designed to transform different classes to linear separable subsets. This may, for instance, be achieved by the so-called Radon cumulative distribution transform (R-CDT) introduced in [8], which is based on one-dimensional optimal transport maps that are generalized to two-dimensional data by applying the Radon transform, known from tomography [11,14]. This approach shows

T. A. Bubba et al. (Eds.): SSVM 2025, LNCS 15667, pp. 241–254, 2025.
https://doi.org/10.1007/978-3-031-92366-1_19

great potential in many applications [5,9,15] and is closely related to the sliced Wasserstein distance [3,16]. A similar approach for data on the sphere is studied in [12,13], for multi-dimensional optimal transport maps in [10], and for optimal Gromov–Wasserstein transport maps in [2].

A central inspiration for this paper is the application of pattern recognition techniques in filigranology—the study of watermarks. These play a central role in dating historical manuscripts as well as identifying scribes and papermills. For automatic classification, the main issue is the enormous number of classes with only few members per class, see WZIS[1]. An end-to-end processing pipeline for thermograms of watermarks including an R-CDT-based classification is proposed in [7], where the authors report classification invariance with respect to translation and dilation of the watermark. Other affine transformations caused, e.g., by unstandardized recording methods are, however, not included yet.

Contribution. The aim of this paper is to incorporate invariance with respect to affine transformations into the R-CDT, i.e., to make the sliced Wasserstein distance invariant to these transforms. In contrast to [15], where the dataset is augmented to encode invariances, we propose a two-step normalization of the R-CDT for probability measures on $\mathbb{R}^2$. To this end, we first generalize the classical Radon transform to measures in Sect. 2 and, thereon, introduce the novel *max-normalized R-CDT* ($_{\mathrm{m}}$NR-CDT) in Sect. 3. Our main theoretical contribution is Theorem 1 ensuring the linear separability of affinely transformed measure classes by $_{\mathrm{m}}$NR-CDT. The theoretical findings are supported by proof-of-concept experiments in Sect. 4 showing a significant improvement of the classification accuracy by the proposed normalization, especially in the small data regime.

2 Radon Transform

The main idea behind the classical Radon transform [11] is to integrate a given bivariate function along all parallel lines pointing in a certain direction. This integral transform can also be interpreted as projection of the given function onto the line with orthogonal orientation. In the following, we briefly review the classical Radon transform for functions and generalize the concept to measures. Finally, we study the effect of affine transformations on the Radon transform, which is crucial to solve the classification task at hand.

2.1 Radon Transform of Functions

Depending on $\boldsymbol{\theta} \in \mathbb{S}_1 := \{\mathbf{x} \in \mathbb{R}^2 \mid \|\mathbf{x}\| = 1\}$, we introduce the *slicing operator* $S_{\boldsymbol{\theta}} \colon \mathbb{R}^2 \to \mathbb{R}$ by

$$S_{\boldsymbol{\theta}}(\mathbf{x}) := \langle \mathbf{x}, \boldsymbol{\theta} \rangle, \quad \mathbf{x} \in \mathbb{R}^2.$$

Its preimages $S_{\boldsymbol{\theta}}^{-1}(t)$, $t \in \mathbb{R}$, are the lines $\ell_{t,\boldsymbol{\theta}}$ in direction $\boldsymbol{\theta}^\perp := (\theta_2, -\theta_1)^\top \in \mathbb{S}_1$ with distance t to the origin. More precisely, we have

$$\ell_{t,\boldsymbol{\theta}} := S_{\boldsymbol{\theta}}^{-1}(t) = \{t\boldsymbol{\theta} + \tau\boldsymbol{\theta}^\perp \mid \tau \in \mathbb{R}\} \subset \mathbb{R}^2.$$

[1] Wasserzeichen-Informationssystem: www.wasserzeichen-online.de.

Using the bijection $\varphi_{\boldsymbol{\theta}}\colon \mathbb{R}^2 \to \mathbb{R}^2$ defined as $\varphi_{\boldsymbol{\theta}}(t,\tau) := t\,\boldsymbol{\theta} + \tau\,\boldsymbol{\theta}^\perp$, whose inverse is given by $\varphi_{\boldsymbol{\theta}}(\mathbf{x})^{-1} = (\langle \mathbf{x}, \boldsymbol{\theta}\rangle, \langle \mathbf{x}, \boldsymbol{\theta}^\perp\rangle)$, we parameterize $\ell_{t,\boldsymbol{\theta}}$ via $\tau \mapsto \varphi_{\boldsymbol{\theta}}(t,\tau)$.

For $f \in L^1(\mathbb{R}^2)$, we define its *Radon transform* $\mathcal{R}[f]\colon \mathbb{R}\times\mathbb{S}_1 \to \mathbb{R}$ as the line integral

$$\mathcal{R}[f](t,\boldsymbol{\theta}) := \int_{\ell_{t,\boldsymbol{\theta}}} f(s)\mathrm{d}s, \quad (t,\boldsymbol{\theta}) \in \mathbb{R}\times\mathbb{S}_1,$$

where $\mathrm{d}s$ denotes the arc length element of $\ell_{t,\boldsymbol{\theta}}$. This defines the *Radon operator* $\mathcal{R}\colon L^1(\mathbb{R}^2) \to L^1(\mathbb{R}\times\mathbb{S}_1)$. For fixed $\boldsymbol{\theta} \in \mathbb{S}_1$, we set $\mathcal{R}_{\boldsymbol{\theta}} := \mathcal{R}(\cdot,\boldsymbol{\theta})$, which is referred to as the *restricted Radon operator* $\mathcal{R}_{\boldsymbol{\theta}}\colon L^1(\mathbb{R}^2) \to L^1(\mathbb{R})$. The action of the Radon operator is illustrated in Fig. 1. The Radon transform is also well-defined for all $f \in L^p(\mathbb{R}^2)$ with $p \geq 1$ and $\mathrm{supp}(f) \subseteq \mathbb{B}_2 := \{\mathbf{x} \in \mathbb{R}^2 \mid \|\mathbf{x}\| \leq 1\}$, in which case $\mathcal{R}[f] \in L^p(\mathbb{R}\times\mathbb{S}_1)$ with $\mathrm{supp}(\mathcal{R}[f]) \subseteq \mathbb{I}\times\mathbb{S}_1$, where $\mathbb{I} := [-1,1]$.

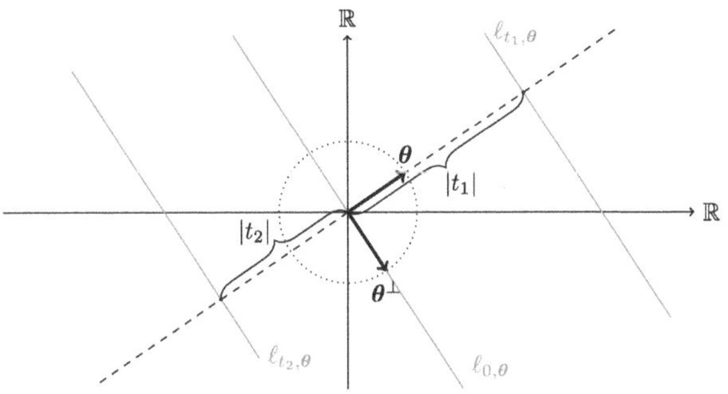

Fig. 1. Illustration of the bivariate Radon transform with distance $t \in \mathbb{R}$ and normal direction $\boldsymbol{\theta} \in \mathbb{S}_1$. The given function is integrated along the lines $\ell_{t,\boldsymbol{\theta}}$.

According to [11], the adjoint $\mathcal{R}^*\colon L^\infty(\mathbb{R}\times\mathbb{S}_1) \to L^\infty(\mathbb{R}^2)$ of the Radon operator $\mathcal{R}\colon L^1(\mathbb{R}^2) \to L^1(\mathbb{R}\times\mathbb{S}_1)$ is given by the *back projection*

$$\mathcal{R}^*[g](\mathbf{x}) := \int_{\mathbb{S}_1} g(S_{\boldsymbol{\theta}}(\mathbf{x}),\boldsymbol{\theta})\mathrm{d}\sigma_{\mathbb{S}_1}(\boldsymbol{\theta}), \quad \mathbf{x} \in \mathbb{R}^2,$$

where $\sigma_{\mathbb{S}_1}$ denotes the surface measure on $\mathbb{S}_1$.

2.2 Radon Transform of Measures

The concept of the Radon transform is now translated to signed, regular, finite measures $\mu \in \mathcal{M}(\mathbb{R}^2)$. For a fixed direction $\boldsymbol{\theta} \in \mathbb{S}_1$, we generalize the *restricted Radon transform* $\mathcal{R}_{\boldsymbol{\theta}}$ to measures by setting

$$\mathcal{R}_{\boldsymbol{\theta}}\colon \mathcal{M}(\mathbb{R}^2) \to \mathcal{M}(\mathbb{R}), \quad \mu \mapsto (S_{\boldsymbol{\theta}})_{\#}\mu = \mu \circ S_{\boldsymbol{\theta}}^{-1},$$

which corresponds to the integration along $\ell_{t,\boldsymbol{\theta}}$. Note that $\mathcal{R}_{\boldsymbol{\theta}}[\mu](\mathbb{R}) = \mu(\mathbb{R}^2)$ for all $\boldsymbol{\theta} \in \mathbb{S}_1$ and, thus, the mass of μ is preserved by $\mathcal{R}_{\boldsymbol{\theta}}$. In measure theory,

$\mathcal{R}_{\boldsymbol{\theta}}$ can be considered as a disintegration family. Heuristically, we may generalize the Radon transform by integrating $\mathcal{R}_{\boldsymbol{\theta}}$ along $\boldsymbol{\theta} \in \mathbb{S}_1$. Therefore, we define the *Radon transform* $\mathcal{R}\colon \mathcal{M}(\mathbb{R}^2) \to \mathcal{M}(\mathbb{R} \times \mathbb{S}_1)$ via

$$\mathcal{R}[\mu] := \mathcal{I}_{\#}[\mu \times u_{\mathbb{S}_1}] \tag{1}$$

with $\mathcal{I}(\mathbf{x}, \boldsymbol{\theta}) := (S_{\boldsymbol{\theta}}(\mathbf{x}), \boldsymbol{\theta})$ for $(\mathbf{x}, \boldsymbol{\theta}) \in \mathbb{R}^2 \times \mathbb{S}_1$. Here $\mu \times u_{\mathbb{S}_1}$ denotes the product measure between the given μ and the uniform measure $u_{\mathbb{S}_1} := \sigma_{\mathbb{S}_1}/2\pi$ on $\mathbb{S}_1$.

Proposition 1. *Let $\mu \in \mathcal{M}(\mathbb{R}^2)$. Then, $\mathcal{R}[\mu]$ can be disintegrated into the family $\mathcal{R}_{\boldsymbol{\theta}}[\mu]$ with respect to $u_{\mathbb{S}_1}$, i.e., for all continuous $g \in C_0(\mathbb{R} \times \mathbb{S}_1)$ vanishing at infinity, we have*

$$\langle \mathcal{R}[\mu], g \rangle = \int_{\mathbb{S}_1} \langle \mathcal{R}_{\boldsymbol{\theta}}[\mu], g(\cdot, \boldsymbol{\theta}) \rangle \mathrm{d}u_{\mathbb{S}_1}(\boldsymbol{\theta}).$$

Proof. By definition in (1), we obtain

$$\begin{aligned}\langle \mathcal{R}[\mu], g \rangle &= \int_{\mathbb{R}\times\mathbb{S}_1} g(t, \boldsymbol{\theta}) \mathrm{d}\mathcal{I}_{\#}[\mu \times u_{\mathbb{S}_1}](t, \boldsymbol{\theta}) = \int_{\mathbb{S}_1} \int_{\mathbb{R}^2} g(S_{\boldsymbol{\theta}}(\mathbf{x}), \boldsymbol{\theta}) \mathrm{d}\mu(\mathbf{x}) \mathrm{d}u_{\mathbb{S}_1}(\boldsymbol{\theta}) \\ &= \int_{\mathbb{S}_1} \int_{\mathbb{R}^2} g(t, \boldsymbol{\theta}) \mathrm{d}[(S_{\boldsymbol{\theta}})_{\#}\mu](t) \mathrm{d}u_{\mathbb{S}_1}(\boldsymbol{\theta}) = \int_{\mathbb{S}_1} \langle \mathcal{R}_{\boldsymbol{\theta}}[\mu], g(\cdot, \boldsymbol{\theta}) \rangle \mathrm{d}u_{\mathbb{S}_1}(\boldsymbol{\theta})\end{aligned}$$

using Fubini's theorem. □

One can find the measure-valued Radon transform $\mathcal{R}\colon \mathcal{M}(\mathbb{R}^2) \to \mathcal{M}(\mathbb{R}\times\mathbb{S}_1)$ as the adjoint of the function-valued adjoint $\mathcal{R}^*: L^\infty(\mathbb{R}\times\mathbb{S}_1) \to L^\infty(\mathbb{R}^2)$, similar to the case of distributions with compact support, cf. [14].

Proposition 2. *The Radon transform of $\mu \in \mathcal{M}(\mathbb{R}^2)$ satisfies*

$$\langle \mathcal{R}[\mu], g \rangle = \langle \mu, \mathcal{R}^*[g] \rangle \quad \forall\, g \in L^\infty(\mathbb{R} \times \mathbb{S}_1).$$

Proof. For all $\mu \in \mathcal{M}(\mathbb{R}^2)$ and $g \in L^\infty(\mathbb{R} \times \mathbb{S}_1)$, applying Fubini's theorem gives

$$\langle \mathcal{R}[\mu], g \rangle = \int_{\mathbb{R}\times\mathbb{S}_1} g(t, \boldsymbol{\theta}) \mathrm{d}\mathcal{I}_{\#}[\mu \times u_{\mathbb{S}_1}](t, \boldsymbol{\theta}) = \int_{\mathbb{R}^2} \int_{\mathbb{S}_1} g(S_{\boldsymbol{\theta}}(\mathbf{x}), \boldsymbol{\theta}) \mathrm{d}u_{\mathbb{S}_1}(\boldsymbol{\theta}) \mathrm{d}\mu(\mathbf{x}).$$

□

Note that, for $f \in L^1(\mathbb{R}^2)$ and the Lebesgue measure $\lambda_{\mathbb{R}^2}$ on $\mathbb{R}^2$, the Radon transform satisfies

$$\mathcal{R}[f\, \lambda_{\mathbb{R}^2}] = \mathcal{R}[f]\, \sigma_{\mathbb{R}\times\mathbb{S}_1},$$

where $\sigma_{\mathbb{R}\times\mathbb{S}_1}$ denotes the surface measure on $\mathbb{R} \times \mathbb{S}_1$. In particular, the Radon transform of an absolutely continuous measure is again absolutely continuous.

2.3 Radon Transform of Affine Transformations

We now consider the Radon transform of an affinely transformed finite measure $\mu \in \mathcal{M}(\mathbb{R}^2)$. To this end, let $\mathbf{A} \in \mathrm{GL}(2)$ and $\mathbf{y} \in \mathbb{R}^2$, i.e., $\mathbf{A}$ is contained in the general linear group GL of regular matrices. We define $\mu_{\mathbf{A},\mathbf{y}} \in \mathcal{M}(\mathbb{R}^2)$ via

$$\mu_{\mathbf{A},\mathbf{y}} := (\mathbf{A} \cdot + \mathbf{y})_{\#}\mu = \mu \circ (\mathbf{A}^{-1}(\cdot - \mathbf{y})). \tag{2}$$

Proposition 3. *For any $\boldsymbol{\theta} \in \mathbb{S}_1$, the restricted Radon transform satisfies*

$$\mathcal{R}_{\boldsymbol{\theta}}[\mu_{\mathbf{A},\mathbf{y}}] = (\|\mathbf{A}^\top\boldsymbol{\theta}\| \cdot + \langle \mathbf{y}, \boldsymbol{\theta}\rangle)_{\#}\mathcal{R}_{\frac{\mathbf{A}^\top\boldsymbol{\theta}}{\|\mathbf{A}^\top\boldsymbol{\theta}\|}}[\mu] = \mathcal{R}_{\frac{\mathbf{A}^\top\boldsymbol{\theta}}{\|\mathbf{A}^\top\boldsymbol{\theta}\|}}[\mu] \circ \left(\tfrac{\cdot - \langle \mathbf{y}, \boldsymbol{\theta}\rangle}{\|\mathbf{A}^\top\boldsymbol{\theta}\|}\right).$$

Proof. Direct calculations yield

$$\begin{aligned}\mathcal{R}_{\boldsymbol{\theta}}[\mu_{\mathbf{A},\mathbf{y}}] &= (S_{\boldsymbol{\theta}})_{\#}[(\mathbf{A} \cdot + \mathbf{y})_{\#}\mu] = (\langle \mathbf{A} \cdot + \mathbf{y}, \boldsymbol{\theta}\rangle)_{\#}\mu = (\langle \cdot, \mathbf{A}^\top\boldsymbol{\theta}\rangle + \langle \mathbf{y}, \boldsymbol{\theta}\rangle)_{\#}\mu \\ &= (\|\mathbf{A}^\top\boldsymbol{\theta}\| \langle \cdot, \tfrac{\mathbf{A}^\top\boldsymbol{\theta}}{\|\mathbf{A}^\top\boldsymbol{\theta}\|}\rangle + \langle \mathbf{y}, \boldsymbol{\theta}\rangle)_{\#}\mu = (\|\mathbf{A}^\top\boldsymbol{\theta}\| \cdot + \langle \mathbf{y}, \boldsymbol{\theta}\rangle)_{\#}\mathcal{R}_{\frac{\mathbf{A}^\top\boldsymbol{\theta}}{\|\mathbf{A}^\top\boldsymbol{\theta}\|}}[\mu],\end{aligned}$$

and the proof is complete. □

Table 1. Summary of common transformations for $\mu \in \mathcal{M}(\mathbb{R}^2)$ with $a, b > 0$ and $c, \varphi \in \mathbb{R}$. The unit circle is parameterized by $\boldsymbol{\theta}(\vartheta) := (\cos(\vartheta), \sin(\vartheta))^\top$. The Radon transform for the left half of $\mathbb{S}_1$ follows by symmetry.

transformation	$\mathbf{A}$	$\mathbf{y}$	$\mathcal{R}_{\boldsymbol{\theta}(\vartheta)}[\mu_{\mathbf{A},\mathbf{y}}]$, $\vartheta \in (-\frac{\pi}{2}, \frac{\pi}{2})$
translation	$\mathbf{I}$	$\mathbb{R}^2$	$\mathcal{R}_{\boldsymbol{\theta}(\vartheta)}[\mu] \circ (\cdot - \langle \mathbf{y}, \boldsymbol{\theta}(\vartheta)\rangle)$
rotation	$\begin{pmatrix} \cos(\varphi) & -\sin(\varphi) \\ \sin(\varphi) & \cos(\varphi) \end{pmatrix}$	$\mathbf{0}$	$\mathcal{R}_{\boldsymbol{\theta}(\vartheta-\varphi)}[\mu]$
reflection	$\begin{pmatrix} \cos(\varphi) & \sin(\varphi) \\ \sin(\varphi) & -\cos(\varphi) \end{pmatrix}$	$\mathbf{0}$	$\mathcal{R}_{\boldsymbol{\theta}(\varphi-\vartheta)}[\mu]$
anisotropic scaling	$\begin{pmatrix} a & 0 \\ 0 & b \end{pmatrix}$	$\mathbf{0}$	$\mathcal{R}_{\boldsymbol{\theta}(\arctan(\frac{b}{a}\tan(\vartheta)))}[\mu] \circ ([a^2\cos^2(\vartheta) + b^2\sin^2(\vartheta)]^{-1/2} \cdot)$
vertical shear	$\begin{pmatrix} 1 & 0 \\ c & 1 \end{pmatrix}$	$\mathbf{0}$	$\mathcal{R}_{\boldsymbol{\theta}(\arctan(c+\tan(\vartheta)))}[\mu] \circ ([1 + c^2\cos^2(\vartheta) + c\sin(2\vartheta)]^{-1/2} \cdot)$

The effect of common affine transformations on the Radon transform is given in Table 1. In order to describe the deformation with respect to $\boldsymbol{\theta}$, we over-parameterize the unit circle $\mathbb{S}_1$ via $\boldsymbol{\theta}(\vartheta) := (\cos(\vartheta), \sin(\vartheta))^\top$, $\vartheta \in \mathbb{R}$. As by Proposition 3, an affine transformation essentially causes a translation and dilation of the transformed measure together with a non-affine remapping in $\boldsymbol{\theta}$.

3 Optimal Transport-Based Transforms

The aim of the following is to introduce an image distance that is unaware of affine transformations. Methodologically, we rely on the *Radon cumulative distribution transform* (R-CDT) introduced in [8], which allows to utilize the fast-to-compute, one-dimensional Wasserstein distance in the context of image processing due to a Radon-based slicing technique. As the R-CDT is not invariant under affine transformation by itself, we propose a two-step normalization scheme, which is essentially grounded on our observations regarding the Radon transform under affine transformations in Sect. 2.3. Finally, we study the linear separability of affinely transformed image classes by our novel normalized R-CDT.

3.1 R-CDT for Measures

The R-CDT traces back to Kolouri et al. [8] and transforms smooth, bivariate density functions. In contrast to [8], we introduce the concept for arbitrary probability measures, similar to [5]. In a first step, we consider probability measures $\mathcal{P}(\mathbb{R}) \subset \mathcal{M}(\mathbb{R})$ defined on the real line. For $\mu \in \mathcal{P}(\mathbb{R})$, the *cumulative distribution function* $F_\mu \colon \mathbb{R} \to [0,1]$ is given by $F_\mu(t) := \mu((-\infty, t])$, $t \in \mathbb{R}$. Its generalized inverse, known as *quantile function*, reads as

$$F_\mu^{[-1]}(t) := \inf\{s \in \mathbb{R} \mid F_\mu(s) > t\}, \quad t \in \mathbb{R}.$$

Based on a reference measure $\rho \in \mathcal{P}(\mathbb{R})$ that does not give mass to atoms, e.g., the uniform distribution $u_{[0,1]}$ on $[0,1]$, we define the *cumulative distribution transform* $\widehat{\mu} \colon \mathbb{R} \to \mathbb{R}$, in short CDT, via

$$\widehat{\mu} := F_\mu^{[-1]} \circ F_\rho.$$

For any convex cost function $c \colon \mathbb{R} \to [0, \infty)$, the CDT (with respect to ρ) solves the Monge–Kantorovich transportation problem [17], i.e.,

$$\widehat{\mu} = \operatorname*{arg\,min}_{T_\# \rho = \mu} \int_{\mathbb{R}} c(s - T(s)) \mathrm{d}\rho(s),$$

where the minimum is taken over all measurable functions $T \colon \mathbb{R} \to \mathbb{R}$. In other words, $\widehat{\mu} \colon \mathbb{R} \to \mathbb{R}$ is an optimal Monge map transporting ρ to μ while minimizing the cost. If $\mu \in \mathcal{P}_2(\mathbb{R})$, i.e., μ has finite 2nd moment, then $\widehat{\mu}$ is square integrable with respect to ρ, i.e., $\widehat{\mu} \in L^2_\rho(\mathbb{R})$. Moreover, for $\mu, \nu \in \mathcal{P}_2(\mathbb{R})$, the norm distance

$$\|\widehat{\mu} - \widehat{\nu}\|_\rho := \left(\int_{\mathbb{R}} |\widehat{\mu}(t) - \widehat{\nu}(t)|^2 \mathrm{d}\rho(t) \right)^{\frac{1}{2}}$$

equals the well-established Wasserstein-2 distance [17].

To deal with a probability measure $\mu \in \mathcal{P}(\mathbb{R}^2)$ defined on the plane, we first determine the Radon transform $\mathcal{R}[\mu] \in \mathcal{M}(\mathbb{R} \times \mathbb{S}_1)$ with its disintegration family

$\{\mathcal{R}_{\boldsymbol{\theta}}[\mu] \in \mathcal{P}(\mathbb{R}) \mid \boldsymbol{\theta} \in \mathbb{S}_1\}$. Then, for each fixed $\boldsymbol{\theta} \in \mathbb{S}_1$, we consider the CDT $\widehat{\mathcal{R}}_{\boldsymbol{\theta}}[\mu]$ (with respect to the same reference measure $\rho \in \mathcal{P}(\mathbb{R})$ for all $\boldsymbol{\theta} \in \mathbb{S}_1$) of the Radon projection $\mathcal{R}_{\boldsymbol{\theta}}[\mu]$, yielding the *R-CDT* $\widehat{\mathcal{R}}[\mu] \colon \mathbb{R} \times \mathbb{S}_1 \to \mathbb{R}$ of μ via

$$\widehat{\mathcal{R}}[\mu](t, \boldsymbol{\theta}) := \widehat{\mathcal{R}}_{\boldsymbol{\theta}}[\mu](t), \quad (t, \boldsymbol{\theta}) \in \mathbb{R} \times \mathbb{S}_1.$$

If $\mu \in \mathcal{P}_2(\mathbb{R}^2)$, then the Radon projection $\mathcal{R}_{\boldsymbol{\theta}}[\mu] \in \mathcal{P}_2(\mathbb{R})$ has finite 2nd moment as well. Consequently, $\widehat{\mathcal{R}}[\mu] \in L^2_{\rho \times u_{\mathbb{S}_1}}(\mathbb{R} \times \mathbb{S}_1)$. For $\mu, \nu \in \mathcal{P}_2(\mathbb{R}^2)$, the norm distance

$$\|\widehat{\mathcal{R}}[\mu] - \widehat{\mathcal{R}}[\nu]\|_{\rho \times u_{\mathbb{S}_1}} := \Big(\int_{\mathbb{S}_1} \int_{\mathbb{R}} |\widehat{\mathcal{R}}[\mu](t, \boldsymbol{\theta}) - \widehat{\mathcal{R}}[\nu](t, \boldsymbol{\theta})|^2 \mathrm{d}\rho(t) \mathrm{d}u_{\mathbb{S}_1}(\boldsymbol{\theta}) \Big)^{\frac{1}{2}}$$

is also called sliced Wasserstein-2 distance in the literature [3].

3.2 Normalized R-CDT

The R-CDT is by itself not invariant under affine transformations, which emerge in various applications. More precisely, the R-CDT inherits the behavior of the Radon transform observed in Sect. 2.3. Notice that the translation and dilation of $\mathcal{R}_{\boldsymbol{\theta}}[\mu]$ causes a horizontal shift (addition of a constant) and a scaling (multiplication with a constant) of $\widehat{\mathcal{R}}_{\boldsymbol{\theta}}[\mu]$, respectively. In the first normalization step, we revert these effects by ensuring zero mean and unit standard deviation of the R-CDT projection. More precisely, we define the *normalized R-CDT* (NR-CDT) $\mathcal{N}[\mu] \colon \mathbb{R} \times \mathbb{S}_1 \to \mathbb{R}$ of $\mu \in \mathcal{P}_2(\mathbb{R}^2)$ via

$$\mathcal{N}[\mu](t, \boldsymbol{\theta}) := \mathcal{N}_{\boldsymbol{\theta}}[\mu](t) := \frac{\widehat{\mathcal{R}}_{\boldsymbol{\theta}}[\mu](t) - \operatorname{mean}(\widehat{\mathcal{R}}_{\boldsymbol{\theta}}[\mu])}{\operatorname{std}(\widehat{\mathcal{R}}_{\boldsymbol{\theta}}[\mu])}, \quad (t, \boldsymbol{\theta}) \in \mathbb{R} \times \mathbb{S}_1,$$

where, for $g \in \mathrm{L}^2_\rho(\mathbb{R})$,

$$\operatorname{mean}(g) := \int_{\mathbb{R}} g(s) \mathrm{d}\rho(s), \qquad \operatorname{std}(g) := \Big(\int_{\mathbb{R}} |g(s) - \operatorname{mean}(g)|^2 \mathrm{d}\rho(s) \Big)^{\frac{1}{2}}.$$

To ensure that the NR-CDT is well defined, we have to guarantee that the standard deviation of the R-CDT projection does not vanish. For this, we restrict ourselves to measures whose supports are not contained in a straight line. More precisely, we consider the class

$$\mathcal{P}^*_c(\mathbb{R}^2) := \{\mu \in \mathcal{P}(\mathbb{R}^2) \mid \operatorname{supp}(\mu) \subset\subset \mathbb{R}^2 \wedge \dim(\operatorname{supp}(\mu)) > 1\} \subset \mathcal{P}_2(\mathbb{R}^2).$$

Here, $\subset\subset$ denotes a compact subset, and dim the dimension of the affine hull. For these, the standard deviation of the restricted Radon transform is bounded away from zero and cannot vanish.

Proposition 4. *Let $\mu \in \mathcal{P}^*_c(\mathbb{R}^2)$. Then, there exists a constant $c > 0$ such that*

$$\operatorname{std}(\widehat{\mathcal{R}}_{\boldsymbol{\theta}}[\mu]) \geq c \quad \forall \boldsymbol{\theta} \in \mathbb{S}_1.$$

For the proof, we first show the following continuity.

Lemma 1. *For fixed* $\mu \in \mathcal{P}_c^*(\mathbb{R}^2)$*, the functions* $\boldsymbol{\theta} \in \mathbb{S}_1 \mapsto \operatorname{mean}(\widehat{\mathcal{R}}_{\boldsymbol{\theta}}[\mu]) \in \mathbb{R}$ *and* $\boldsymbol{\theta} \in \mathbb{S}_1 \mapsto \operatorname{std}(\widehat{\mathcal{R}}_{\boldsymbol{\theta}}[\mu]) \in \mathbb{R}_{\geq 0}$ *are continuous.*

Proof. We rewrite the mean as

$$\operatorname{mean}(\widehat{\mathcal{R}}_{\boldsymbol{\theta}}[\mu]) = \int_{\mathbb{R}} \widehat{\mathcal{R}}_{\boldsymbol{\theta}}[\mu](t)\mathrm{d}\rho(t) = \int_{\mathbb{R}} t\mathrm{d}\mathcal{R}_{\boldsymbol{\theta}}[\mu](t) = \int_{\mathbb{R}^2} \langle \mathbf{x}, \boldsymbol{\theta}\rangle \mathrm{d}\mu(\mathbf{x}).$$

Since the integrand is continuous in $\boldsymbol{\theta}$ and uniformly bounded by $|\langle\cdot,\boldsymbol{\theta}\rangle| \leq \|\cdot\|$, the dominated convergence yields the assertion. Analogously, we have

$$\begin{aligned}\operatorname{std}(\widehat{\mathcal{R}}_{\boldsymbol{\theta}}[\mu]) &= \Big(\int_{\mathbb{R}} |\widehat{\mathcal{R}}_{\boldsymbol{\theta}}[\mu](t) - \operatorname{mean}(\widehat{\mathcal{R}}_{\boldsymbol{\theta}}[\mu])|^2 \mathrm{d}\rho(t)\Big)^{\frac{1}{2}} \\ &= \Big(\int_{\mathbb{R}^2} |\langle \mathbf{x}, \boldsymbol{\theta}\rangle - \operatorname{mean}(\widehat{\mathcal{R}}_{\boldsymbol{\theta}}[\mu])|^2 \mathrm{d}\mu(\mathbf{x})\Big)^{\frac{1}{2}}.\end{aligned}$$

The integrand is again continuous in $\boldsymbol{\theta}$ and uniformly bounded by

$$|\langle\cdot,\boldsymbol{\theta}\rangle - \operatorname{mean}(\widehat{\mathcal{R}}_{\boldsymbol{\theta}}[\mu])|^2 \leq 2\|\cdot\|^2 + 2\max_{\boldsymbol{\theta}\in\mathbb{S}_1} (\operatorname{mean}(\widehat{\mathcal{R}}_{\boldsymbol{\theta}}[\mu]))^2;$$

thus, the standard deviation is continuous by dominated convergence. □

Proof (Proposition 4). Assume the contrary, i.e., $c = 0$. Then, due to the continuity of $\boldsymbol{\theta} \mapsto \operatorname{std}(\widehat{\mathcal{R}}_{\boldsymbol{\theta}}[\mu])$, there exists a minimizing and convergent sequence in $\mathbb{S}_1$ whose limit $\boldsymbol{\theta}$ is attained and satisfies $\operatorname{std}(\widehat{\mathcal{R}}_{\boldsymbol{\theta}}[\mu]) = 0$, i.e.,

$$\int_{\mathbb{R}^2} |\langle \mathbf{x}, \boldsymbol{\theta}\rangle - \operatorname{mean}(\widehat{\mathcal{R}}_{\boldsymbol{\theta}}[\mu])|^2 \mathrm{d}\mu(\mathbf{x}) = 0.$$

Hence, the support of μ is contained in the line $\{\mathbf{x} \in \mathbb{R}^2 \mid \langle \mathbf{x}, \boldsymbol{\theta}\rangle = \operatorname{mean}(\widehat{\mathcal{R}}_{\boldsymbol{\theta}}[\mu])\}$ in contradiction to $\mu \in \mathcal{P}_c^*(\mathbb{R}^2)$. □

The NR-CDT is nearly invariant under affine transformations up to bijective remappings of the directions, i.e., up to a resorting of the family $\{\mathcal{N}_{\boldsymbol{\theta}}[\mu] \mid \boldsymbol{\theta} \in \mathbb{S}_1\}$.

Proposition 5. *Let* $\mu \in \mathcal{P}_c^*(\mathbb{R}^2)$, $\mathbf{A} \in \mathrm{GL}(2)$, $\mathbf{y} \in \mathbb{R}^2$, *and* $\mu_{\mathbf{A},\mathbf{y}}$ *as in* (2). *Then, for any* $\boldsymbol{\theta} \in \mathbb{S}_1$*, the NR-CDT satisfies*

$$\mathcal{N}_{\boldsymbol{\theta}}[\mu_{\mathbf{A},\mathbf{y}}] = \mathcal{N}_{\frac{\mathbf{A}^\top\boldsymbol{\theta}}{|\mathbf{A}^\top\boldsymbol{\theta}|}}[\mu].$$

Proof. Transferring Proposition 3 to the CDT space, we have

$$\widehat{\mathcal{R}}_{\boldsymbol{\theta}}[\mu_{\mathbf{A},\mathbf{y}}](t) = \|\mathbf{A}^\top\boldsymbol{\theta}\|\, \widehat{\mathcal{R}}_{h_{\mathbf{A}}(\boldsymbol{\theta})}[\mu](t) + \langle \mathbf{y}, \boldsymbol{\theta}\rangle$$

with the bijection $h_{\mathbf{A}}(\boldsymbol{\theta}) := (\mathbf{A}^\top\boldsymbol{\theta})/\|\mathbf{A}^\top\boldsymbol{\theta}\|$, $\boldsymbol{\theta} \in \mathbb{S}_1$; so that

$$\operatorname{mean}(\widehat{\mathcal{R}}_{\boldsymbol{\theta}}[\mu_{\mathbf{A},\mathbf{y}}]) = \|\mathbf{A}^\top\boldsymbol{\theta}\| \operatorname{mean}(\widehat{\mathcal{R}}_{h_{\mathbf{A}}(\boldsymbol{\theta})}[\mu]) + \langle \mathbf{y}, \boldsymbol{\theta}\rangle$$

and

$$\mathrm{std}(\widehat{\mathcal{R}}_{\boldsymbol{\theta}}[\mu_{\mathbf{A},\mathbf{y}}]) = \|\mathbf{A}^\top \boldsymbol{\theta}\| \, \mathrm{std}(\widehat{\mathcal{R}}_{h_{\mathbf{A}}(\boldsymbol{\theta})}[\mu]).$$

Consequently,

$$\mathcal{N}_{\boldsymbol{\theta}}[\mu_{\mathbf{A},\mathbf{y}}](t) = \frac{\widehat{\mathcal{R}}_{h_{\mathbf{A}}(\boldsymbol{\theta})}[\mu](t) - \mathrm{mean}(\widehat{\mathcal{R}}_{h_{\mathbf{A}}(\boldsymbol{\theta})}[\mu])}{\mathrm{std}(\widehat{\mathcal{R}}_{h_{\mathbf{A}}(\boldsymbol{\theta})}[\mu])} = \mathcal{N}_{h_{\mathbf{A}}(\boldsymbol{\theta})}[\mu](t).$$

□

3.3 Max-Normalized R-CDT

In the final normalization step, we treat the resorting of $\{\mathcal{N}_{\boldsymbol{\theta}}[\mu] \mid \boldsymbol{\theta} \in \mathbb{S}_1\}$. Since the underlying mapping is unknown in general, we propose to take the supremum over all directions. More precisely, for $\mu \in \mathcal{P}_c^*(\mathbb{R}^2)$, we define its *max-normalized R-CDT* ($_{\mathrm{m}}$NR-CDT) $\mathcal{N}_{\mathrm{m}}[\mu] \colon \mathbb{R} \to \mathbb{R}$ via

$$\mathcal{N}_{\mathrm{m}}[\mu](t) := \sup_{\boldsymbol{\theta} \in \mathbb{S}_1} \mathcal{N}_{\boldsymbol{\theta}}[\mu](t), \quad t \in \mathbb{R}.$$

We show that $\mathcal{N}_{\mathrm{m}}$ maps a given measure to a bounded function so that the $_{\mathrm{m}}$NR-CDT space $\mathcal{N}_{\mathrm{m}}[\mathcal{P}_c^*(\mathbb{R}^2)]$ is contained in $L^\infty_\rho(\mathbb{R})$ for the underlying reference measure $\rho \in \mathcal{P}(\mathbb{R})$.

Proposition 6. *Let $\mu \in \mathcal{P}_c^*(\mathbb{R}^2)$. Then, $\mathcal{N}_{\mathrm{m}}[\mu] \in L^\infty_\rho(\mathbb{R})$.*

Proof. The restricted Radon operator cannot enlarge the size of the support $\mathrm{diam}(\mu) := \sup_{\mathbf{x},\mathbf{y} \in \mathrm{supp}(\mu)} \|\mathbf{x} - \mathbf{y}\|$, i.e., $\mathrm{diam}(\mathcal{R}_{\boldsymbol{\theta}}[\mu]) \leq \mathrm{diam}(\mu)$. Moreover, the range of $\widehat{\mathcal{R}}_{\boldsymbol{\theta}}[\mu]$ coincides with the support of $\mathcal{R}_{\boldsymbol{\theta}}[\mu]$. Using that the mean lies in the convex hull of the support, we thus have

$$|\widehat{\mathcal{R}}_{\boldsymbol{\theta}}[\mu](t) - \mathrm{mean}(\widehat{\mathcal{R}}_{\boldsymbol{\theta}}[\mu])| \leq \mathrm{diam}(\mu) \quad \forall \boldsymbol{\theta} \in \mathbb{S}_1.$$

Since $\mu \in \mathcal{P}_c^*(\mathbb{R}^2)$, Proposition 4 gives $c := \min_{\boldsymbol{\theta} \in \mathbb{S}_1} \mathrm{std}(\widehat{\mathcal{R}}_{\boldsymbol{\theta}}[\mu]) > 0$. Thus, the $_{\mathrm{m}}$NR-CDT is bounded by $|\mathcal{N}_{\mathrm{m}}[\mu](t)| \leq \mathrm{diam}(\mu)/c$ for all $t \in \mathbb{R}$. □

With the $_{\mathrm{m}}$NR-CDT, we accomplish our objective to define a transport-based transform that is invariant under affine transformations.

Proposition 7. *Let $\mu \in \mathcal{P}_c^*(\mathbb{R}^2)$, $\mathbf{A} \in \mathrm{GL}(2)$, $\mathbf{y} \in \mathbb{R}^2$, and $\mu_{\mathbf{A},\mathbf{y}}$ as in (2). Then, the $_m$NR-CDT satisfies $\mathcal{N}_{\mathrm{m}}[\mu_{\mathbf{A},\mathbf{y}}] = \mathcal{N}_{\mathrm{m}}[\mu]$.*

Proof. Since the mapping $h_{\mathbf{A}}(\boldsymbol{\theta}) := (\mathbf{A}^\top \boldsymbol{\theta})/\|\mathbf{A}^\top \boldsymbol{\theta}\|$ is a bijection on $\mathbb{S}_1$, we obtain

$$\mathcal{N}_{\mathrm{m}}[\mu_{\mathbf{A},\mathbf{y}}](t) = \sup_{\boldsymbol{\theta} \in \mathbb{S}_1} \mathcal{N}_{\boldsymbol{\theta}}[\mu_{\mathbf{A},\mathbf{y}}](t) = \sup_{\boldsymbol{\theta} \in \mathbb{S}_1} \mathcal{N}_{h_{\mathbf{A}}(\boldsymbol{\theta})}[\mu](t) = \mathcal{N}_{\mathrm{m}}[\mu](t).$$

□

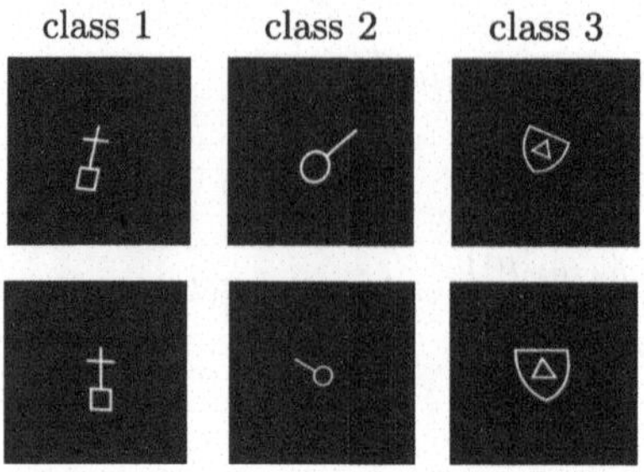

Fig. 2. Samples of the academic dataset consisting of randomly affine-transformed synthetic template images.

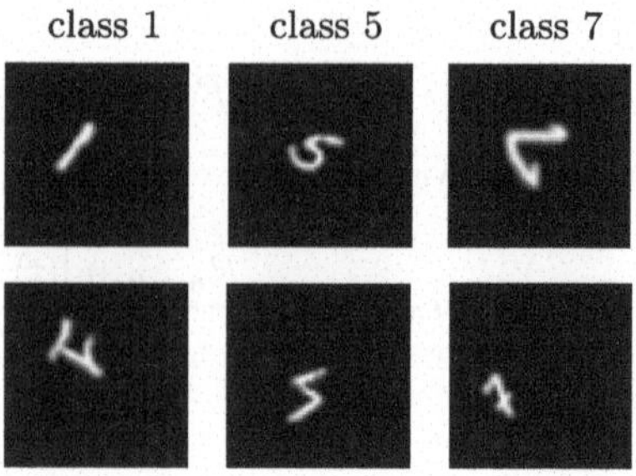

Fig. 3. Samples of the LinMNIST dataset (random choices of ones, fives and sevens) based on affine-transformed MNIST digits.

Table 2. Accuracy of nearest neighbour classification for the academic dataset with 10 images per class and the LinMNIST dataset with 50 images per class, in both cases with equispaced angles in $[0, \pi)$.

num. angles	academic		LinMNIST	
	$\|\cdot\|_\infty$	$\|\cdot\|_2$	$\|\cdot\|_\infty$	$\|\cdot\|_2$
2	0.5666	0.6333	0.5814 ± 0.0890	0.5970 ± 0.0467
4	0.9333	1.0000	0.6444 ± 0.0613	0.7155 ± 0.0492
8	0.9333	1.0000	0.7533 ± 0.0500	0.8200 ± 0.0454
16	0.9667	1.0000	0.7844 ± 0.0333	0.8362 ± 0.0401
32	0.9000	1.0000	0.8014 ± 0.0380	0.8429 ± 0.0409
64	0.8333	1.0000	0.8059 ± 0.0468	0.8503 ± 0.0377
128	0.8333	1.0000	0.8022 ± 0.0392	0.8570 ± 0.0300

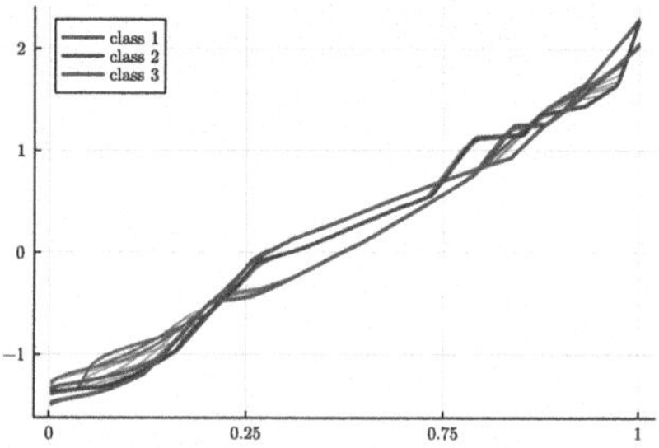

Fig. 4. Visualization of ${}_{\mathrm{m}}$NR-CDT for the academic dataset and 120 angles in $[0, \pi)$.

The invariance under affine transformations immediately yields the linear separability of affine measure classes, which originate from a single template.

Theorem 1. *For template measures $\mu_0, \nu_0 \in \mathcal{P}_c^*(\mathbb{R}^2)$ with*

$$\mathcal{N}_{\mathrm{m}}[\mu_0] \neq \mathcal{N}_{\mathrm{m}}[\nu_0]$$

consider the classes

$$\mathbb{F} = \{\mu \in \mathcal{P}(\mathbb{R}^2) \mid \exists\, \mathbf{A} \in \mathrm{GL}(2),\, \mathbf{y} \in \mathbb{R}^2 \colon \mu = (\mathbf{A} \cdot + \mathbf{y})_\# \mu_0\}, \tag{3a}$$

$$\mathbb{G} = \{\nu \in \mathcal{P}(\mathbb{R}^2) \mid \exists\, \mathbf{A} \in \mathrm{GL}(2),\, \mathbf{y} \in \mathbb{R}^2 \colon \nu = (\mathbf{A} \cdot + \mathbf{y})_\# \nu_0\}. \tag{3b}$$

Then, $\mathbb{F}$ and $\mathbb{G}$ are linearly separable in ${}_{\mathrm{m}}$NR-CDT space.

Proof. Due to the affine construction of $\mathbb{F}$ and $\mathbb{G}$, Proposition 7 yields $\mathcal{N}_{\mathrm{m}}[\mathbb{F}] = \{\mathcal{N}_{\mathrm{m}}[\mu_0]\}$ and $\mathcal{N}_{\mathrm{m}}[\mathbb{G}] = \{\mathcal{N}_{\mathrm{m}}[\nu_0]\}$. Hence, the assumption $\mathcal{N}_{\mathrm{m}}[\mu_0] \neq \mathcal{N}_{\mathrm{m}}[\nu_0]$ implies the linear separability of $\mathcal{N}_{\mathrm{m}}[\mathbb{F}]$ and $\mathcal{N}_{\mathrm{m}}[\mathbb{G}]$ in $L^\infty_\rho(\mathbb{R})$. □

The proof of Theorem 1 reveals that the classes $\mathbb{F}$ and $\mathbb{G}$ collapse to single points in ${}_{\mathrm{m}}$NR-CDT space, directly allowing for linear separability. This is in contrast to the R-CDT results in [8], which can only tackle a subset of affine transforms, and where the considered classes form convex sets in R-CDT space.

4 Numerical Experiments

By the following proof-of-concept experiments, we support our linear separability result in Theorem 1 with numerical evidence. For this, the proposed ${}_{\mathrm{m}}$NR-CDT is implemented in Julia[2]. All experiments[3] are performed on an off-the-shelf MacBookPro 2020 with Intel Core i5 Chip (4-Core CPU, 1.4 GHz) and 8 GB RAM.

Datasets. For our simulations, we rely on two datasets. For academic purposes, the first dataset is based on (up to) three synthetic template symbols, which are randomly translated, rotated, dilated, and sheared, cf. Fig. 2. In this manner, we construct perfect affine classes as needed for our theory, see (3). For a more realistic scenario, we also consider the LinMNIST dataset [1] consisting of affinely transformed MNIST digits [4], cf. Fig. 3. In contrast to the first dataset, this data does not originate from a common ground truth. Therefore, the second dataset can be considered as a collection of imperfect affine classes.

4.1 Nearest Neighbour Classification

In the first experiment, we aim to validate the theoretical result from Theorem 1. Looking at the proof, we recall that ${}_{\mathrm{m}}$NR-CDT maps each entire affine class to a single point. The easiest way for classification is the nearest neighbour method, which can be immediately generalized to an arbitrary number of classes. For the first dataset, we use the template symbols as references and classify all class members in ${}_{\mathrm{m}}$NR-CDT space based on the nearest neighbour rule using the Chebychev norm $\|\cdot\|_\infty$ or Euclidean norm $\|\cdot\|_2$, cf. Table 2 (columns 2 and 3). For illustration, the ${}_{\mathrm{m}}$NR-CDTs of the entire dataset are depicted in Fig. 4, where the thick lines correspond to the templates and the thin lines to the transformed images. In theory, the classes should yield three curves. However, due to approximation errors, we observe slight perturbations.

For the second dataset, since we have no templates, we iteratively select five instances per class as references and classify the remaining class members again based on the nearest neighbour rule. Thereon, we compute the mean and standard deviation of the achieved accuracy, see Table 2 (columns 4 and 5). For the discretization of the ${}_{\mathrm{m}}$NR-CDT, we use 2 to 128 angles in $[0, \pi)$, reported in column 1 of Table 2. As expected, due to Theorem 1, the classification of the first dataset is (nearly) perfect; remarkable, already for a very small number of chosen angles. For the LinMNIST dataset, the achieved accuracy ranges from 60% to 86%, which is still significantly better than random guessing, achieving

[2] The Julia Programming Language – Version 1.9.2 (https://docs.julialang.org).
[3] The code will be available at https://github.com/DrBeckmann/NR-CDT.

an accuracy of 33% as we deal with a three class problem. Let us stress that perfect classification is not to be expected since LinMNIST does not satisfy our theoretical assumptions.

Table 3. Classification accuracy (mean±std based on 10-fold cross validation) for the academic dataset. The first two classes in Fig. 2 are used with different class sizes and numbers of equispaced angles in $[0, \pi)$.

class size	Euclidean	R-CDT			$_{\text{m}}$NR-CDT		
		2	4	8	2	4	8
10	0.5277 ± 0.1118	0.5555 ± 0.0868	0.5555 ± 0.0785	0.5500 ± 0.0714	0.6666 ± 0.1458	0.8777 ± 0.0819	$\mathbf{0.9944 \pm 0.0175}$
30	0.4777 ± 0.0697	0.5314 ± 0.0771	0.6111 ± 0.0740	0.6111 ± 0.0872	0.8129 ± 0.0512	0.9537 ± 0.0463	$\mathbf{1.0000 \pm 0.0000}$
90	0.5283 ± 0.0422	0.8074 ± 0.0673	0.9024 ± 0.0765	0.9259 ± 0.0770	0.9111 ± 0.0578	0.9728 ± 0.0184	$\mathbf{1.0000 \pm 0.0000}$
270	0.5413 ± 0.0286	0.9806 ± 0.0144	0.9983 ± 0.0045	0.9979 ± 0.0065	0.9938 ± 0.0073	0.9952 ± 0.0061	$\mathbf{1.0000 \pm 0.0000}$

4.2 Support Vector Machine Classification

In this second set of numerical experiments, we compare three different ansätze in combination with linear support vector machines (SVMs) [6]. The naïve approach uses the Euclidean representation of the images as basis for the SVM. Inspired by [8], the second approach makes use of the plain R-CDT projections using a fixed number of angles. Finally, the third approach utilizes our $_{\text{m}}$NR-CDT projections over the same set of angles. For all these methods, a 10-fold cross validation is performed. This means that the dataset is partitioned into ten subsets, of which one is successively used for training, whereas the remaining nine are reserved for testing. The results for different class sizes and numbers of angels are summarized for the academic dataset in Table 3 and for the LinMNIST dataset in Table 4. We observe that our approach outperforms all others, especially in the small data regime and for few angles. For large data sizes, R-CDT and $_{\text{m}}$NR-CDT perform at nearly the same accuracy.

Table 4. Classification accuracy (mean±std based on 10-fold cross validation) for the LinMNIST dataset. The first and last class in Fig. 3 are used with different class sizes and numbers of equispaced angles in $[0, \pi)$.

class size	Euclidean	R-CDT			$_{\text{m}}$NR-CDT		
		2	4	8	2	4	8
10	0.5055 ± 0.1841	0.5277 ± 0.1315	0.5333 ± 0.1233	0.5277 ± 0.1178	0.7000 ± 0.1340	0.7944 ± 0.1229	$\mathbf{0.8500 \pm 0.0457}$
20	0.4472 ± 0.0606	0.5166 ± 0.0644	0.5194 ± 0.0435	0.5222 ± 0.0520	0.8111 ± 0.0597	0.8250 ± 0.1134	$\mathbf{0.8666 \pm 0.0772}$
50	0.4911 ± 0.0214	0.5744 ± 0.6200	0.6200 ± 0.0755	0.6400 ± 0.0944	0.8088 ± 0.0536	0.9055 ± 0.0328	$\mathbf{0.9311 \pm 0.0533}$
250	0.5291 ± 0.0318	0.8346 ± 0.0252	0.9057 ± 0.0274	0.9206 ± 0.0191	0.8997 ± 0.0128	0.9493 ± 0.0109	$\mathbf{0.9551 \pm 0.0105}$
500	0.5408 ± 0.0181	0.8596 ± 0.0130	0.9296 ± 0.0093	0.9433 ± 0.0095	0.9031 ± 0.0101	0.9479 ± 0.0119	$\mathbf{0.9598 \pm 0.0122}$
1.000	0.5348 ± 0.0213	0.8675 ± 0.0090	0.9437 ± 0.0053	0.9538 ± 0.0067	0.9016 ± 0.0124	0.9602 ± 0.0042	$\mathbf{0.9675 \pm 0.0055}$
5.000	0.5421 ± 0.0068	0.8843 ± 0.0055	0.9544 ± 0.0024	0.9587 ± 0.0020	0.9110 ± 0.0042	0.9650 ± 0.0015	$\mathbf{0.9739 \pm 0.0012}$

5 Conclusion

In this work, we proposed the novel max-normalized R-CDT for feature representation and proved linear separability of classes generated by affine transforms of given templates. This was validated by numerical experiments showing a significant increase in classification accuracy over original R-CDT. Potential future directions include the control of perturbations either in the templates or in the transforms as well as a more in-depth numerical study in various applications.

References

1. Beckmann, M., Heilenkötter, N.: Equivariant neural networks for indirect measurements. SIAM J. Math. Data Sci. **6**(3), 579–601 (2024). https://doi.org/10.1137/23M1582862
2. Beier, F., Beinert, R., Steidl, G.: On a linear Gromov–Wasserstein distance. IEEE Trans. Image Process. **31**, 7292–7305 (2022). https://doi.org/10.1109/TIP.2022.3221286
3. Bonneel, N., Rabin, J., Peyré, G., Pfister, H.: Sliced and radon Wasserstein Barycenters of measures. J. Math. Imaging Vis. **51**(1), 22–45 (2015). https://doi.org/10.1007/s10851-014-0506-3
4. Deng, L.: The MNIST database of handwritten digit images for machine learning research. IEEE Signal Process. Mag. **29**(6), 141–142 (2012). https://doi.org/10.1109/MSP.2012.2211477
5. Diaz Martin, R., Medri, I.V., Rohde, G.K.: Data representation with optimal transport (2024). https://doi.org/10.48550/arXiv.2406.15503. arXiv:2406.15503
6. Fan, R.E., Chang, K.W., Hsieh, C.J., Wang, X.R., Lin, C.J.: LIBLINEAR: a library for large linear classification. J. Mach. Learn. Res. **9**, 1871–1874 (2008). https://www.csie.ntu.edu.tw/~cjlin/liblinear/
7. Hauser, D., Beckmann, M., Koliander, G., Stiehl, H.S.: On image processing and pattern recognition for thermograms of watermarks in manuscripts – a first proof-of-concept. In: International Conference on Document Analysis and Recognition (ICDAR), pp. 91–107 (2024). https://doi.org/10.1007/978-3-031-70543-4_6
8. Kolouri, S., Park, S.R., Rohde, G.K.: The Radon cumulative distribution transform and its application to image classification. IEEE Trans. Image Process. **25**(2), 920–934 (2016). https://doi.org/10.1109/TIP.2015.2509419
9. Kolouri, S., Park, S.R., Thorpe, M., Slepcev, D., Rohde, G.K.: Optimal mass transport. IEEE Signal Process. Mag. **34**(4), 43–59 (2017). https://doi.org/10.1109/MSP.2017.2695801
10. Moosmüller, C., Cloninger, A.: Linear optimal transport embedding: provable Wasserstein classification for certain rigid transformations and perturbations. Inf. Inference J. IMA **12**(1), 363–389 (2023). https://doi.org/10.1093/imaiai/iaac023
11. Natterer, F.: The Mathematics of Computerized Tomography. SIAM, Philadelphia (2001). https://doi.org/10.1137/1.9780898719284
12. Quellmalz, M., Beinert, R., Steidl, G.: Sliced optimal transport on the sphere. Inverse Prob. **39**(10), 105005 (2023). https://doi.org/10.1088/1361-6420/acf156
13. Quellmalz, M., Buecher, L., Steidl, G.: Parallelly sliced optimal transport on spheres and on the rotation group. J. Math. Imaging Vis. **66**(6), 951–976 (2024). https://doi.org/10.1007/s10851-024-01206-w

14. Ramm, A.G., Katsevich, A.I.: The Radon Transform and Local Tomography. CRC Press (1996). https://doi.org/10.1201/9781003069331
15. Shifat-E-Rabbi, M., et al.: Radon cumulative distribution transform subspace modeling for image classification. J. Math. Imaging Vis. **63**, 1185–1203 (2021). https://doi.org/10.1007/s10851-021-01052-0
16. Shifat-E-Rabbi, M., Zhuang, Y., Li, S., Rubaiyat, A.H.M., Yin, X., Rohde, G.K.: Invariance encoding in sliced-Wasserstein space for image classification with limited training data. Pattern Recognit. **137**, 109268 (2023). https://doi.org/10.1016/j.patcog.2022.109268
17. Villani, C.: Topics in Optimal Transportation. American Mathematical Society (2003). https://doi.org/10.1090/gsm/058

Prediction of Parametric Surfaces for Multi-object Segmentation in 3D Biological Imaging

Quentin Rapilly[1(✉)], Anaïs Badoual[1], Pierre Maindron[2], Guenaelle Bouet[2], and Charles Kervrann[1]

[1] Centre Inria de l'Université de Rennes, Equipe SAIRPICO, Unité Chimie et Biologie de la Cellule, U1143 INSERM, UMR3666 CNRS, Institut Curie, Université PSL, Rennes, France
quentin.rapilly@inria.fr
[2] Mines Saint Etienne, Université Jean Monnet Saint-Étienne, INSERM, SAINBIOSE U1059, 42023 Saint-Étienne, France

Abstract. Multi-object segmentation algorithms are of great interest in a very large range of fields. Deep learning brought major improvements in terms of processing speed or prediction accuracy. Nevertheless, some traditional methods such as active surfaces have features that conventional deep learning methods cannot provide, especially representing the object in a continuous geometrical way and encoding prior information on the shapes to segment. Those features are of particular interest in biology to efficiently segment noisy and poorly resolved data, and then understand the interactions between segmented cells. We introduce NAGINI-3D (N-Active shapes for seGmentINg 3D biological Images), a new hybrid segmentation method dedicated to multi-object segmentation of 3D images that combines the efficiency of deep learning and the powerful representation of active surfaces. We evaluate our method on real and synthetic 3D datasets of fluorescence microscopy.

Keywords: Multi-object segmentation · Parametric surfaces · Convolutional Neural Networks

1 Introduction

The study of biological phenomena requires to capture a lot of observations and produces many microscopy images. That big amount of data can be tedious for experts to analyse. It has led to the development of automatic methods for various tasks, such as counting, classifying or segmenting objects.

More than three decades ago, active contours in 2D, also called "snakes", then generalised to active surfaces in 3D, were widely used for segmentation tasks in biomedical imaging. In recent years, deep learning brought new powerful segmentation methods able to overcome difficulties with snakes, in particular the management of multiple objects. They are also faster and more accurate but

T. A. Bubba et al. (Eds.): SSVM 2025, LNCS 15667, pp. 255–268, 2025.
https://doi.org/10.1007/978-3-031-92366-1_20

cannot represent objects in a continuous way. In this paper, we propose a method that combines the benefits of both paradigms by using a Convolutional Neural Network (CNN) to predict, for each object in an image, the optimal parameters of a surface representing its boundaries.

Snakes are known to be very popular for various segmentation tasks in medical imaging [1–3]. Initially proposed for 2D image segmentation, parametric snakes consist in a contour parameterized by control points that are roughly initialized to approximate the boundaries of the object of interest. Their positions are then iteratively optimized by minimizing an energy functional, to delineate the object. Their main advantage is that they can represent the object to segment with few parameters, in a continuous and smooth way, that can be used afterwards for morphological and geometrical analysis. This method has been generalized to 3D [4,5] with active surfaces. However, the need of an initialization and the time consuming optimization made them quite obsolete for many use-case despite their interesting properties. Moreover, most of them are dedicated to single object segmentation and are limited for multi-object problems.

More recently, the revolution of deep learning led to powerful generic methods dedicated to segmentation tasks that produce pixel-wise segmentation masks and deal with multi-object easily [6,7]. Recent foundation models [8] can segment objects unseen in their training dataset, as they learn the concept of boundaries between objects. However, these models often require a huge quantity of annotated data to be trained. If microscopy imaging produces big amount of data, few labels are available. It makes generic models less efficient than deep learning methods dedicated to biological images that incorporate geometrical features: StarDist represents objects with star-convex polygon [9] while Cellpose and more recently Omnipose [10,11] predict gradients used to construct the cells by solving the heat diffusion equation. These methods, dedicated to 2D images, are widely used by biologists because of their accurate multi-object detection and segmentation results. Concerning the segmentation of 3D images, there is no consensus within the community. Some methods are direct 3D extensions of 2D counteparts such as StarDist [12] and Omnipose which provides a reconstruction algorithm with 3D gradient flows [11]. These methods are often criticized because 3D annotated datasets are rare and the trained algorithms often lack of generalization. Other approaches overcome this limitation by applying the 2D algorithm on each slice of a 3D volume and combine the 2D results in the final step [10,13,14]. These methods can suffer from spatial incoherence along the third dimension and require a lot of hyper-parameters tuning. Moreover, they are often less accurate than the 3D methods if 3D annotations are available.

Among hybrid methods, SplineDist [15] proposed an approach for 2D images using a CNN to predict parametric contours. Inspired from StarDist 2D, SplineDist offers a way to segment objects in a continuous manner and is not restricted to star-convex shapes. A variant of StarDist 3D, using spherical harmonics instead of star-convex polyhedron to represent object continuously, is presented in [16]. Nevertheless, spherical harmonics still limit the approach to star-convex objects. To our knowledge, the idea of SplineDist has never been extended to 3D before because of several technical barriers explained below. We present a first approach to overcome them.

In this paper, we propose a segmentation method that is not limited to star-convex shapes, inspired from both SplineDist 2D and StarDist 3D, that 1) performs multi-object segmentation on 3D images in a fast manner, taking benefits of the CNN architectures; 2) represents objects in a smooth and continuous way using parametric surfaces; 3) approximates any object homeomorphic to a sphere. In our experiments, we show that our method achieves competitive results with the state-of-the-art algorithms. We validate our approach on a synthetic dataset depicting non star-convex objects and compare our method on a dataset depicting nuclei in fluorescence microscopy images. Our code is available on GitHub at https://github.com/QuentinRapilly/NAGINI-3D/.

2 Parametric Spline Surfaces for Object Representation

We represent each object with a parametric surface:

$$\forall(u,v) \in [0,1]^2, \quad \boldsymbol{s}(u,v) = \sum_{k=-\infty}^{\infty} \sum_{l=-\infty}^{\infty} \boldsymbol{c}_{k,l}\varphi_1(M_1 u - k)\varphi_2(M_2 v - l), \tag{1}$$

where $\{\boldsymbol{c}_{k,l}\}_{\substack{k\in\mathbb{Z}\\ l\in\mathbb{Z}}} \in (\mathbb{R}^3)^{\mathbb{Z}\times\mathbb{Z}}$ are the 3D control points describing the shape. In (1), M_1 and M_2 are two parameters determining the complexity of the surface. The higher they are, the more accurately the surface can represent local variations. Moreover, φ_1 and φ_2 are the basis functions, whose choice allows to enforce various properties to the surface (smoothness, reproduction properties, ...). Here, we use the parameterization as described in [5] for surfaces with a spherical topology. The functions $\varphi_1 = \phi_{M_1,per}$ and $\varphi_2 = \phi_{2M_2}$ allows to reproduce ellipsoids using (1), where the function $\phi_{M,per}(t) = \sum_{n=-\infty}^{\infty} \phi_M(t - Mn)$ is the M periodic version of ϕ_M, an exponential B-Spline defined as:

$$\phi_M : t \rightarrow \frac{1}{1-\cos\frac{2\pi}{M}} \Big[\mathbb{1}_{[0,\frac{1}{2}[}(|t|)(\cos\frac{2\pi|t|}{M}\cos\frac{\pi}{M} - \cos\frac{2\pi}{M}) + \mathbb{1}_{[\frac{1}{2},\frac{3}{2}[}(|t|)(\sin\frac{\pi(\frac{3}{2}-|t|)}{M})^2\Big].$$

Using φ_1 periodicity and φ_2 compact support, we rewrite (1) in the form:

$$\forall(u,v) \in [0,1]^2, \quad \boldsymbol{s}(u,v) = \sum_{k=0}^{M_1-1} \sum_{l=-1}^{M_2+1} \boldsymbol{c}_{k,l}\varphi_1(M_1 u - k)\varphi_2(M_2 v - l). \tag{2}$$

Note that, for surfaces homeomorphic to a sphere, the variables u and v can be viewed as the longitude and the latitude parameters of the surface.

In (2), to guarantee that the surface is closed and smooth, the control points have to satisfy the following constraints [5]:

$$\boldsymbol{c}_{k,-1} = \boldsymbol{c}_{k,1} + \frac{c_{M_1}[k]\boldsymbol{T}_{1,N} + s_{M_1}[k]\boldsymbol{T}_{2,N}}{M_2\varphi_2'(1)}, \tag{3a}$$

$$\boldsymbol{c}_{k,0} = \frac{\boldsymbol{c}_N}{\varphi_2(0)} - \frac{\varphi_2(1)(\boldsymbol{c}_{k,-1} + \boldsymbol{c}_{k,1})}{\varphi_2(0)}, \tag{3b}$$

$$\boldsymbol{c}_{k,M_2+1} = \boldsymbol{c}_{k,M_2-1} - \frac{c_{M_1}[k]\boldsymbol{T}_{1,S} + s_{M_1}[k]\boldsymbol{T}_{2,S}}{M_2\varphi_2'(1)}, \tag{3c}$$

$$\boldsymbol{c}_{k,M_2} = \frac{\boldsymbol{c}_S}{\varphi_2(0)} - \frac{\varphi_2(1)(\boldsymbol{c}_{k,M_2-1} + \boldsymbol{c}_{k,M_2+1})}{\varphi_2(0)}, \tag{3d}$$

with $c_{M_1}[n] = \Gamma_{M_1} \cos \frac{2\pi n}{M_1}$, $s_{M_1}[n] = \Gamma_{M_1} \sin \frac{2\pi n}{M_1}$ and $\Gamma_{M_1} = \frac{2(1-\cos\frac{2\pi}{M_1})}{\cos\frac{\pi}{M_1} - \cos\frac{3\pi}{M_1}}$. Here, $\boldsymbol{c}_S$, $\boldsymbol{c}_N$ denote the south and north poles of the surface respectively, and the vectors $\boldsymbol{T}_{1,N}, \boldsymbol{T}_{2,N} \in \mathbb{R}^3$, ($\boldsymbol{T}_{1,S}, \boldsymbol{T}_{2,S} \in \mathbb{R}^3$, respectively) characterize the tangent plane to the north pole (south, respectively).

Finally, the set of $I = M_1(M_2-1)+6$ parameters $\{\boldsymbol{f}_i\}_{1\leq i\leq I} \in (\mathbb{R}^3)^I$ provides a way to represent any object homeomorphic to a sphere by a closed and smooth parametric surface:

$$\{\boldsymbol{f}_i\}_{1\leq i\leq I} = \{\boldsymbol{c}_N, \boldsymbol{c}_S, \boldsymbol{T}_{1,N}, \boldsymbol{T}_{2,N}, \boldsymbol{T}_{1,S}, \boldsymbol{T}_{2,S}, \boldsymbol{c}_{0,1}, ..., \boldsymbol{c}_{M_1-1,M_2-1}\}. \tag{4}$$

It contains a subpart of the control points $\{\boldsymbol{c}_{k,l}\}_{\substack{0\leq k\leq M_1-1 \\ 1\leq l\leq M_2-1}}$. The missing ones are constructed using (3). An example of parameterized surface is shown on Fig. 1a.

3 A CNN for the Prediction of the Surface Parameters

Our idea is to train a CNN to predict for each object in an image, an optimal set of parameters $\{\boldsymbol{f}_i\}_{1\leq i\leq I}$ that describes its boundaries in a single forward pass.

The CNN architecture we consider is a U-Net [7]. The input of the network is a 3D image of shape (C_{in}, H, W, D), where H, W and D denote the height, width and depth of the image, respectively, and C_{in} is the number of channels. A channel in microscopy often represents the intensity of the signal representing a dedicated molecule. The network predicts a volume of shape (C_{out}, H, W, D). For each voxel of the input image $\boldsymbol{x} = (x, y, z) \in \Omega$ with $\Omega = \{1, ..., H\} \times \{1, ..., W\} \times \{1, ..., D\}$, the network outputs $C_{out} = 3I + 1$ channels. The $3I$ first channels correspond to the 3 spherical coordinates of the I estimated parameters $\{\hat{\boldsymbol{f}}_{\boldsymbol{x},i}\}$ of a surface delineating the object to which $\boldsymbol{x}$ belongs. The spherical coordinates are predicted in a frame centered on $\boldsymbol{x}$. The last channel is the probability $\hat{p}(\boldsymbol{x})$ of $\boldsymbol{x} \in \Omega$ to be the barycenter of an object. This channel is used in the object detection step during the inference. It allows to extract a single location in each object. At each location, we use the predicted parameters to construct the surface representing the corresponding object. More details are given in Sect. 3.2.

3.1 Training Description

The training pipeline (see Fig. 2) consists in comparing the predicted channels to the ground-truth (GT). To compare the predicted probability map with a GT one, we use the Binary Cross Entropy (BCE). However, evaluating how well a predicted surface describes the object of interest is not a trivial task. One cannot just compare the predicted control points with "optimal ones", as very different sets of parameters can describe nearly identical surfaces. Instead of comparing the parameters, we compare the surface generated with (2) to the GT surface computed from the boundaries of a binary mask.

The idea, inspired from [15] (in 2D), is to sample the two surfaces and compare the resulting points clouds. However, the translation from 2D contours to surfaces is still difficult for the two following reasons: 1) Sampling the predicted surface cannot be performed by evenly sampling the variables u and v definition domains, as it would result in a concentration of points in the vicinity of the poles (Fig. 1b); 2) The points sampled on a contour are ordered (by following it) which leads to a very simple formulation for the training loss [17]. On surfaces, the sampled points are unordered.

Surface Sampling Procedure. We aim at sampling points, evenly distributed, on a surface described by (2), using parameters predicted by the CNN. Evenly partitioning the definition domains of variables u and v: $(u_p)_{0 \leq p \leq P-1}$ and $(v_q)_{0 \leq q \leq Q-1}$ with $u_p = \frac{p}{P}$ and $v_q = \frac{q}{Q}$ would result in oversampled regions around the poles and undersampled regions around the equator of the surface [18]. A solution consists in generating a Fibonacci lattice containing $P = 2N+1$ points ($N \in \mathbb{N}$), giving the latitudes and longitudes $(u_i, v_i)_{-N \leq i \leq N}$ to use to sample the surface: $\forall i \in \{-N, ..., N\}$, $u_i = \text{frac}(i \cdot G)$ and $v_i = \frac{1}{\pi} \cdot \arcsin(\frac{2 \cdot i}{P}) + 0.5$, where $\text{frac} : x \rightarrow x - \lfloor x \rfloor$ denotes the fractional part function and $G = \frac{\sqrt{5}-1}{2}$. The two samplings methods are shown on Fig. 1b and 1c. By substituting these values in (2) we obtain P points $\hat{S} = (\hat{\boldsymbol{S}}_i)_{-N \leq i \leq N}$ evenly sampling the surface:

$$\forall i \in \{-N, ..., N\}, \quad \hat{\boldsymbol{S}}_i = \sum_{k=0}^{M_1-1} \sum_{l=-1}^{M_2+1} \boldsymbol{c}_{k,l} \varphi_1(M_1 u_i - k) \varphi_2(M_2 v_i - l). \tag{5}$$

For fixed values of P, M_1 and M_2, we pre-compute and store the terms $w_{i,k,l}^{P,M_1,M_2} = \varphi_1(M_1 u_i - k)\varphi_2(M_2 v_i - l)$, $\forall (i,k,l) \in \{-N, ..., N\} \times \{0, ..., M_1 - 1\} \times \{-1, ..., M_2 + 1\}$, which are identical for all surfaces. This leads to an efficient way to sample any surface by just computing a weighted sum of the control points:

$$\forall i \in \{-N, ..., N\}, \quad \hat{\boldsymbol{S}}_i = \sum_{k=0}^{M_1-1} \sum_{l=-1}^{M_2+1} \boldsymbol{c}_{k,l} w_{i,k,l}^{P,M_1,M_2}. \tag{6}$$

This formulation makes the computation of multiple sampling in the mean time very fast, as it can be parallelized on GPU architectures.

To sample a GT surface, we first compute the boundary of the segmentation mask (using a dilation deformation). Then we apply a "Farthest Point Sampling"

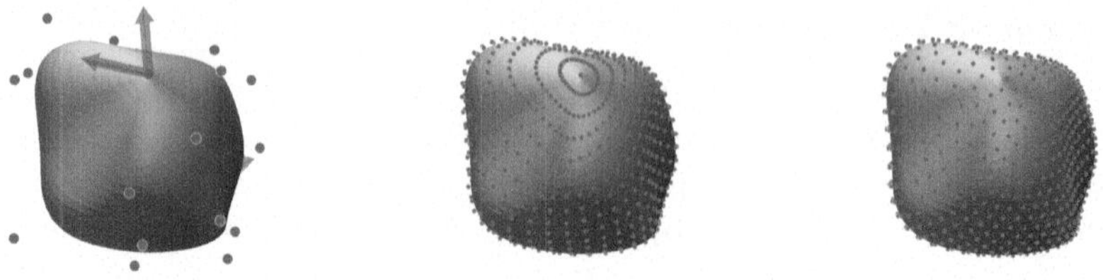

(a) Parametric surface (b) Even sampling of u, v (c) Fibonacci lattice

Fig. 1. (a) Parametric surface (green arrows: normal at poles $\boldsymbol{T}_{1,N}, \boldsymbol{T}_{2,N}, \boldsymbol{T}_{1,S}, \boldsymbol{T}_{2,S}$; green dots: control points $\{\boldsymbol{c}_{k,l}\}$, see (4)). (b) Sampling method evenly dividing u and v domains. (c) Sampling method using the Fibonacci lattice. (Color figure online)

(FPS) algorithm [19], that samples a surface evenly. It consists in choosing one initial point on the surface then iteratively adding to the sampling the farthest point on the surface (using geodesic distance) to all the others already added. The process is stopped when the desired number of points is reached. That leads to a set of P points $S = (\boldsymbol{S}_i)_{1 \leq i \leq P}$ evenly distributed on the GT surface. The sampling of each GT surface is computed once and stored.

Surface Matching. The next issue to deal with is to perform matching between the samplings of the GT and predicted surfaces. According to [17], the ordering of the points on sampled contours and cyclic permutations allows to recover the best matching. Nevertheless, in 3D, the points are unordered on the surface and there is no such trick to speed up the computation of a distance to our knowledge. To overcome this difficulty, the Sliced Wasserstein Distance (SW) [20], has been investigated to compare points clouds [21]. SW is more representative than the chamfer distance and faster to compute than the Wasserstein Distance (WD). In our case, SW of order p between $\hat{S}$ and S is defined as:

$$SW_p(\hat{S}, S) = \int_{\theta \in \mathbb{U}} W_p(\pi_{\boldsymbol{\theta}}(\hat{S}), \pi_{\boldsymbol{\theta}}(S)) d\theta, \tag{7}$$

where $\mathbb{U}$ is the unit sphere, $\pi_{\theta}(S) = \{\boldsymbol{\theta}^T \boldsymbol{S_i},\ \boldsymbol{S}_i \in S\}$ is the set of projections of each element of S on the 1D subspace of $\mathbb{R}^3$ directed by $\boldsymbol{\theta}$. In our framework, $W_p(X, Y)$ is the 1D WD between two discrete one-dimensional distributions. It consists in sorting the two sets in increasing order and comparing the elements two by two: $W_p(X, Y) = \left(\frac{1}{P} \sum_{i=1}^{P} |x_{\sigma_X(i)} - y_{\sigma_Y(i)}|^p\right)^{1/p}$, where σ_X and σ_Y are the permutations operators sorting X and Y in increasing order.

According to [21], the integral in (7) can be approximated by a Monte-Carlo procedure that randomly picks $K \geq 100$ directions $\boldsymbol{\theta}$. Finally, we can rewrite the distance:

$$SW_p(\hat{S}, S) = \frac{1}{K} \sum_{\theta \in \mathbb{U}_K} W_p(\pi_{\boldsymbol{\theta}}(\hat{S}), \pi_{\boldsymbol{\theta}}(S)), \tag{8}$$

where $\mathbb{U}_K$ contains K random elements, uniformly sampled in $\mathbb{U}$. In the following, we use $p = 1$ and note SW the SW_1 of order 1.

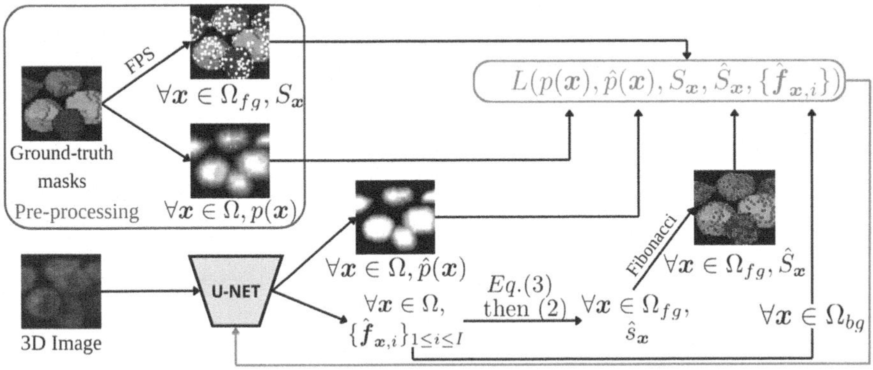

Fig. 2. Training loop of the network.

Computation of the GT Probability Map. For each voxel $x \in \Omega$, we define the GT probability $p(x)$ as follow: $p(x) = 0$ if x belongs to the background. If x is inside an object of index j ($x \in \Omega_j$), we define $d_j(x) = ||b_j - x||_2$ the distance to the barycenter b_j of the object Ω_j. We note $D(x) = \max_{y \in \Omega_j} d_j(y) - d_j(x)$ and then $p(x) = \frac{D(x)}{\max_{y \in \Omega_j} D(y)}$ its normalized version. With this definition, $p(x)$ is positive, increases as x get closer to the barycenter of an object of interest and is maximum equal to 1 if $\exists j, x = b_j$. For particular shapes, when the barycenter does not belong to the interior of the object, it is maximum for the point in this object that is the closest to the barycenter.

Training Loss. We define the training loss for each voxel x of an image Ω as:

$$L(p(x), \hat{p}(x), S_x, \hat{S}_x, \{\hat{f}_{x,i}\}) = BCE(p(x), \hat{p}(x)) + \lambda_s L_s(p(x), S_x, \hat{S}_x, \{\hat{f}_{x,i}\}), \tag{9}$$

where the parameter λ_s weights the importance given to the surface loss L_s. The surface loss compares the two samplings $\hat{S}_x$ and S_x using SW, if x belongs to an object ($x \in \Omega_{fg}$). For the background voxels ($x \in \Omega_{bg}$) it minimizes the norm of the predicted parameters. This acts as a regularization term to penalize the values of the parameters to be too large (leading to unrealistic shapes). The surface loss is defined as:

$$L_s(p(x), S_x, \hat{S}_x, \{\hat{f}_{x,i}\}) = p(x) \cdot \mathbb{1}_{\Omega_{fg}}(x) \cdot SW(\hat{S}_x, S_x) + \lambda_{reg} \cdot \mathbb{1}_{\Omega_{bg}}(x) \cdot \sum_i ||\hat{f}_{x,i}||_2, \tag{10}$$

where $\mathbb{1}$ denotes the indicator function. The SW is weighted by the GT probability to give more importance to the voxels that are more likely to be a voxel of interest. The parameter λ_{reg} controls the importance of the regularization term.

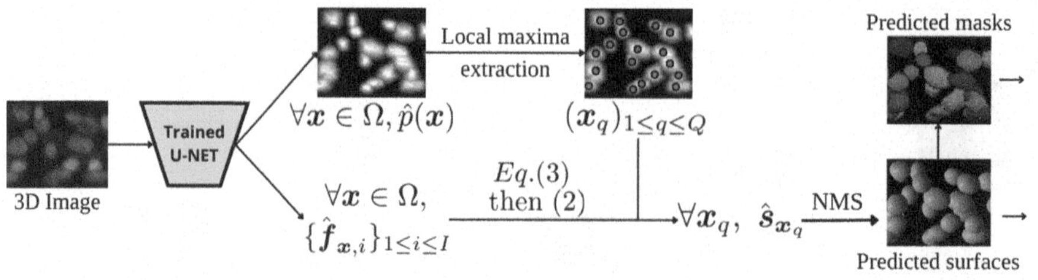

Fig. 3. Inference pipeline applied to generate multi-object segmentation.

3.2 Inference Procedure for Segmentation

Surface Prediction. During the inference step, the goal is to predict one surface for each object on a given input image Ω. To determine at which positions we need to construct parametric surfaces we extract the local maxima $(\boldsymbol{x}_q)_{1 \leq q \leq Q}$ of $\hat{p}$ having a probability greater than a chosen threshold $\gamma > 0$. These voxels are supposed to represent the barycenters (or the closest voxel to them) in the objects of interest. We generate the surface parameterized by $\{\hat{\boldsymbol{f}}_{\boldsymbol{x}_q,i}\}_{1 \leq i \leq I}$ for each selected location $\boldsymbol{x}_q$. Finally, we process to a Non Maximum Suppression (NMS) that consists in evaluating the Intersection-over-Union (IoU) between each pair of surface and removing potential duplicated surfaces representing the same object. Our inference pipeline is illustrated in Fig. 3.

Refinement of Predicted Surfaces. We observed experimentally that the network is much easier to train with small values of M_1 and M_2 in (2), mainly because it avoids unrealistic self-intersecting shapes. In a refinement step, we decided to add an optional post-processing algorithm: we up-sample the number of control points of the predicted shapes and then apply a classical snake optimization process with an energy based on image gradients just before the NMS step. We can then train a network for small values of M_1 and M_2, and still segment more complex shapes (with local variations or non-convex parts) that require a higher number of control points to be represented. Even with this step, segmenting an image that contains hundreds of objects takes only a few seconds.

3.3 Settings of Model Hyper Parameters

The U-Net implementation we used is the very versatile one "General U-Net" described in [22]. It has 3 layers. Each layer is a double 3×3 convolution, with batch-normalisation and ReLU activation function. The number of channels of the first layer is 32. At each down-sampling step, the number of channels is multiplied by a factor 2 and each dimension of the image is divided by a factor 2. The parameters used are $\lambda_s = 1$ and $\lambda_{reg} = 10^{-3}$. We trained our model over 6000 epochs, with a batch size of 1, using the AdamW optimizer with a learning rate of 10^{-4}. We applied data augmentation on the images using random flips

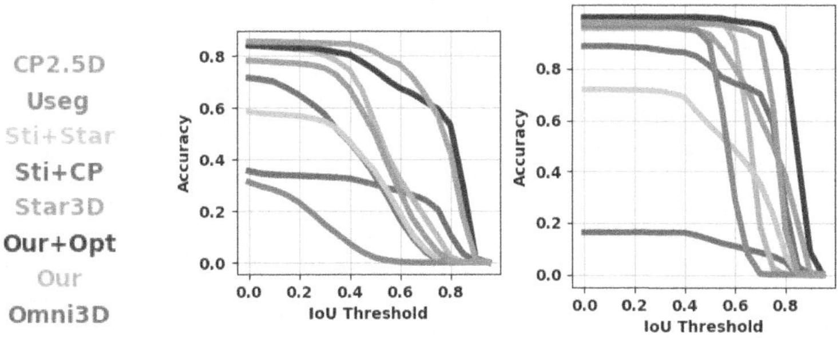

(a) BONES dataset (b) CAPS dataset

Fig. 4. Accuracy curves $Acc(\sigma)$ depending on the IoU threshold σ for each method.

along each axis. The number of sampled points is set to $P = 151$. We chose small values of the surface parameters $M_1 = 4$ and $M_2 = 2$ ($I = 10$) as explained in Sect. 3.2. For the experiments, we present the results with the up-sampling of control points $M_1 \leftarrow 2M_1$ and $M_2 \leftarrow 2M_2$ ($I = 30$) and the snake refinement step, denoted as **Our+Opt**. The results without refinement are denoted as **Our**. The algorithms are trained on a Nvidia A100 GPU.

The probability threshold γ used during the local maxima extraction and the IoU threshold used for the NMS step are selected automatically to maximize $\int_0^1 Acc(\sigma)d\sigma$ (see in Sect. 4), on the validation set, at the end of the training.

4 Experiments

In this section, we present the results obtained on a dataset of real microscopy images depicting nuclei and a dataset of synthetic images: **BONES** is a dataset of 28 images of fluorescence microscopy representing nuclei of bone cells cultivated on furrows of various angles to evaluate the influence of support on cell development. We split the set into 18 for training, 3 for validation and 7 for testing. Each image ($80 - 320 \times 180 \times 310$ voxels) (the first dimension depends of the furrow angle) contains approximately 240 objects. The GT masks were segmented automatically and curated by an expert. **CAPS** is a synthetic dataset we created, representing spherical caps (portion of a sphere cut off by a plane). The geometry of these objects could be assimilated to those of erythrocytes (red blood cells). We generated caps of various size and thickness randomly cut off to represent star-convex and non star-convex objects. The proportion of non-star-convex objects is driven by the ranges of possible positions of the cut-off plane (see Fig. 6a) and of possible thicknesses. The thicker the sphere is and the smaller part of the sphere the cut-off plane keeps, the more likely the corresponding cap will be star-convex. We added blur and Poisson-Gaussian noise to simulate fluorescence microscopy images. We generated 10 training images, 3 for validation, 5 for testing. Each image ($200 \times 200 \times 200$ voxels) contains 220 objects.

Table 1. Mean inference time per image evaluated on the CAPS test set. The experiments were conducted on an Nvidia A100 80GB.

	Star3D	Omni3D	Sti+CP	Sti+Star	Useg	CP2.5D	Our	Our+Opt
Time (s)	65.59	15.14	64.56	67.69	70.23	29.27	**7.11**	9.06

Table 2. Detection and segmentation scores of tested methods. The recall and precision are both computed with $TP(0)$, $FP(0)$ and $FN(0)$ (i.e. $\sigma = 0$) to assess the quality of the detection. The mean IoU is given among the TP to assess the quality of the segmentation for the objects correctly detected.

	Star3D	Omni3D	Sti+CP	Sti+Star	Useg	CP2.5D	Our	Our+Opt
BONES								
Recall	0.965	**0.984**	0.930	0.976	0.888	0.955	0.947	0.939
Precision	0.882	0.356	0.755	0.593	0.325	0.809	**0.884**	**0.884**
Mean IoU	0.793	0.717	0.511	0.555	0.313	0.584	0.606	**0.796**
CAPS								
Recall	**1.0**	**1.0**	0.979	0.983	0.974	0.986	**1.0**	0.999
Precision	0.958	0.162	0.905	0.729	0.996	0.994	**1.0**	**1.0**
Mean IoU	0.779	0.688	0.764	0.692	0.592	0.776	0.678	**0.858**

We compared our method with **direct 3D methods**: StarDist 3D denoted **Star3D** with 96 radii as in [12], and Omnipose 3D [11] denoted **Omni3D**. We trained each method on the dataset considered prior to evaluation. We also compared with **aggregation algorithms** using pre-trained 2D models: U-Segment3D [14] denoted **Useg** provides an implementation using pre-trained 2D Cellpose models, Cellstitch [13] which can easily be linked after any 2D segmentation models, we applied it with StarDist 2D (**Sti+Star**) and Cellpose 2D (**Sti+CP**). For StarDist we used the pre-trained model dedicated to fluorescence. For Cellpose we tried the nuclei and cytoplasm models and kept the best for each use case. We also evaluated the aggregation method built into Cellpose, that have been improved in version 3.1.0, denoted **CP2.5D**. For each method we tuned the hyper-parameters to get the best segmentation results as possible.

We assessed the tested algorithms using the method depicted in [12]. We find matching between the predicted and GT cells by solving an assignment problem maximizing the sum of the IoU of all matches. A GT cell with no corresponding prediction is considered a false-negative (FN) and a predicted cell with no corresponding GT is considered a false-positive (FP). A predicted cell having a corresponding GT is considered a true-positive (TP) prediction and the corresponding IoU is computed. For each IoU threshold $\sigma \in [0, 1]$, $TP(\sigma)$ is the number of TP with an IoU greater or equal than σ. $FN(\sigma)$ ($FP(\sigma)$, respectively) is the number of GTs (predicted cells, respectively) with no matching prediction (GT, respectively) or in a matching with an IoU smaller than σ. The evaluation metric that we consider is the accuracy for a given σ: $Acc(\sigma) = \frac{TP(\sigma)}{TP(\sigma)+FN(\sigma)+FP(\sigma)}$.

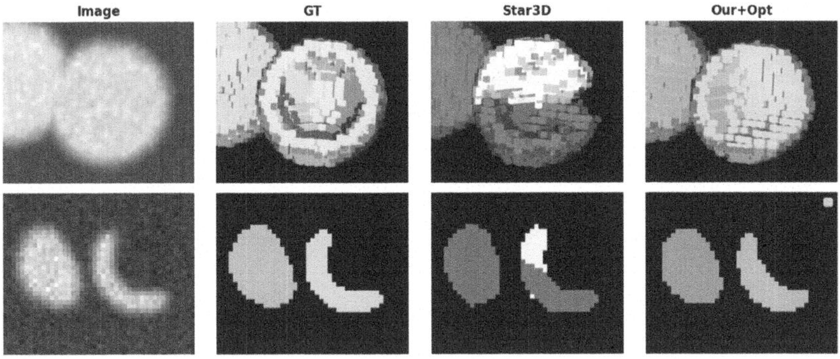

Fig. 5. Example of **Star3D** splitting non star-convex objects in two. **Our+Opt** predicts a single object. The first row shows the 3D volumes and the second row 2D slices.

Figure 4 shows the accuracy curve for each method and dataset. In Table 2, we reported several scores assessing the quality of detection and segmentation. In Table 1 we report the inference time of each method on CAPS test dataset.

On BONES, we note that the 3D methods outperform the aggregation algorithms in term of both detection and segmentation. The only exception is **Omni3D**, that predicts very accurate masks but tends to predict residual groups of pixels not merged with other labels or nuclei split into multiple labels, resulting in many FP as shown by its precision score. The best methods for this dataset are **Star3D** and **Our+Opt** that have very similar performances in detection and segmentation. On CAPS, the difference between 3D methods and aggregation algorithms is less noticeable: **Useg** gets detection scores similar to the best 3D methods; **Sti+CP** perform really well in term of segmentation accuracy; **CP2.5D** achieves detection and segmentation scores comparable to 3D methods. Unlike **Star3D**, whose performances drop to that of the best aggregation algorithms, **Our+Opt** still outperforms all the other methods in terms of detection and segmentation. This difference can be explained by the presence of non-star-convex objects in this dataset: **Star3D** cannot represents such objects and split them into several parts (Fig. 5), leading to FP predictions and under-segmented objects. In our framework, surfaces are not limited to star-convex shapes and can represent all the objects of the set, allowing **Our+Opt** to get almost perfect detection scores and the best mean IoU by a noticeable margin. In addition, we see that our method is the fastest compared to all other tested methods (Table 1).

To demonstrate that our method performs equivalently on star-convex and non-star-convex objects, we evaluated it on images with a higher proportion of non-star-convex objects, driven by the range of possible positions of the cut-off plane when creating the caps (see Fig. 6a). We thus, generated two other test sets of 5 images, one with significantly more non-star-convex objects but still an important part of star-convex objects (denoted "medium prop. set") and

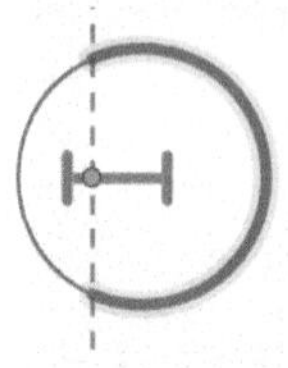

(a) Cap generation

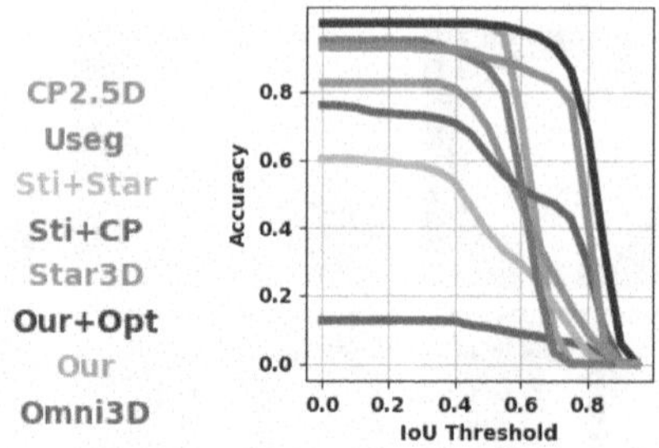

(b) Medium proportion

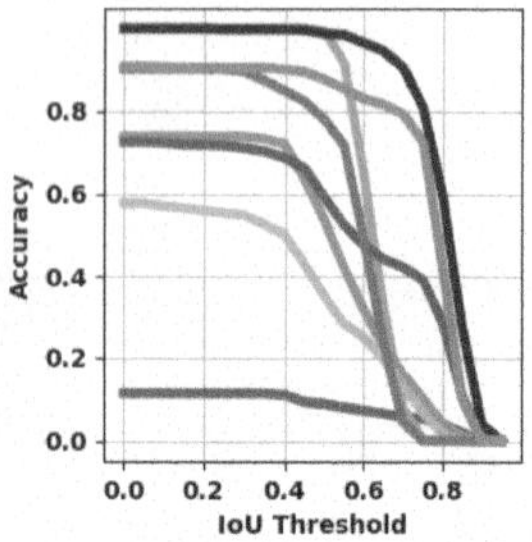

(c) Strong proportion

Fig. 6. (a) Cap (in red) generated by the cut-off plane (dashed line) randomly positioned in the chosen blue range; (b) and (c) Evolution of accuracy according to the proportion of non-star-convex objects in testing images. (Color figure online)

one with almost no star-convex object (denoted "strong prop. set"). On Fig. 6, we see that there is a drop in performance for almost every methods as the proportion of non-star-convex objects increases, especially for **Star3D**. However, **Our**, **CP2.5D** and **Our+Opt** show very little loss of performance.

5 Conclusion

We presented an approach to train a CNN to predict parametric surfaces allowing to perform multi-object segmentation on 3D images in a fast way. Our modeling represents any closed shape continuously. We demonstrated that it outperforms aggregation algorithms based on 2D models and is competitive against other direct 3D algorithms, especially on images depicting complicated shapes such as non star-convex objects. The main limitation of the method is the need for a GPU with dozens of gigabytes of memory when training with large sampling sizes (several hundred points). However, it is not required to use too many points if one only wants a good approximation that will be refined by applying the snake optimization step. For future work, we plan to adapt some features so that it can be used on highly anisotropic images. We also investigate the potential of the method on images depicting objects with various scales.

Acknowledgments. Many thanks to Cambridge Advanced Imaging Centre for letting us use their computing resources at Cambridge University to train the models.

Author contributions. Q.R. designed the method, implemented the code and conducted the experiments. P.M. and G.B. provided the fluorescence images and the corresponding labels. Q.R. generated the synthetic images and labels. Q.R., A.B and C.K. wrote the article. A.B. and C.K. supervised the research.

References

1. Kass, M., Witkin, A., Terzopoulos, D.: Snakes: active contour models. Int. J. Comput. Vis. **1**(4), 321–331 (1988)
2. Blake, A., Isard, M.: Active Contours: The Application of Techniques From Graphics, Vision, Control Theory and Statistics to Visual Tracking of Shapes in Motion. Springer (2012)
3. Badoual, A., Schmitter, D., Uhlmann, V., Unser, M.: Multiresolution subdivision snakes. IEEE Trans. Image Process. **26**(3), 1188–1201 (2016)
4. Dufour, A., Thibeaux, R., Labruyere, E., Guillen, N., Olivo-Marin, J.-C.: 3-D active meshes: fast discrete deformable models for cell tracking in 3-D time-lapse microscopy. IEEE Trans. Image Process. **20**(7), 1925–1937 (2010)
5. Delgado-Gonzalo, R., Chenouard, N., Unser, M.: Spline-based deforming ellipsoids for interactive 3D bioimage segmentation. IEEE Trans. Image Process. **22**(10), 3926–3940 (2013)
6. He, K., Gkioxari, G., Dollár, P., Girshick, R.: Mask R-CNN. In: IEEE International Conference on Computer Vision (ICCV) (2017)
7. Ronneberger, O., Fischer, P., Brox, T.: U-net: convolutional networks for biomedical image segmentation. In: Navab, N., Hornegger, J., Wells, W.M., Frangi, A.F. (eds.) MICCAI 2015. LNCS, vol. 9351, pp. 234–241. Springer, Cham (2015). https://doi.org/10.1007/978-3-319-24574-4_28
8. Kirillov, A., et al.: Segment anything. In: IEEE International Conference on Computer Vision (ICCV), pp. 4015–4026 (2023)
9. Schmidt, U., Weigert, M., Broaddus, C., Myers, G.: Cell detection with star-convex polygons. In: Frangi, A.F., Schnabel, J.A., Davatzikos, C., Alberola-López, C., Fichtinger, G. (eds.) MICCAI 2018. LNCS, vol. 11071, pp. 265–273. Springer, Cham (2018). https://doi.org/10.1007/978-3-030-00934-2_30
10. Stringer, C., Wang, T., Michaelos, M., Pachitariu, M.: Cellpose: a generalist algorithm for cellular segmentation. Nat. Meth. **18**(1), 100–106 (2021)
11. Cutler, K.J., et al.: Omnipose: a high-precision morphology-independent solution for bacterial cell segmentation. Nat. Meth. **19**(11), 1438–1448 (2022)
12. Weigert, M., Schmidt, U., Haase, R., Sugawara, K., Myers, G.: Star-convex polyhedra for 3D object detection and segmentation in microscopy. In: IEEE Winter Conference on Applications of Computer Vision (WACV), pp. 3655–3662 (2020)
13. Liu, Y., Jin, Y., Azizi, E., Blumberg, A.J.: Cellstitch: 3D cellular anisotropic image segmentation via optimal transport. BMC Bioinform. **24**(1), 480 (2023)
14. Zhou, F.Y., et al.: A general algorithm for consensus 3D cell segmentation from 2D segmented stacks. BioRxiv (2024)
15. Mandal, S., Uhlmann, V.: Splinedist: automated cell segmentation with spline curves. In: IEEE International Symposium Biomedical Imaging (ISBI), pp. 1082–1086 (2021)
16. Eschweiler, D., Rethwisch, M., Koppers, S., Stegmaier, J.: Spherical harmonics for shape-constrained 3D cell segmentation. In: IEEE International Symposium on Biomedical Imaging (ISBI), pp. 792–796 (2021)
17. Mandal, S., Uhlmann, V.: A learning-based formulation of parametric curve fitting for bioimage analysis. In: Vermolen, F.J., Vuik, C. (eds.) Numerical Mathematics and Advanced Applications ENUMATH 2019. LNCSE, vol. 139, pp. 1031–1038. Springer, Cham (2021). https://doi.org/10.1007/978-3-030-55874-1_102
18. González, Á.: Measurement of areas on a sphere using fibonacci and latitude-longitude lattices. Math. Geosci. **42**, 49–64 (2010)

19. Peyré, G., Cohen, L.D.: Geodesic methods for shape and surface processing, pp. 29–56. Springer, Dordrecht (2009)
20. Bonneel, N., Rabin, J., Peyré, G., Pfister, H.: Sliced and radon wasserstein barycenters of measures. J. Math. Imaging Vision **51**, 22–45 (2015)
21. Nguyen, T., Pham, Q.H., Le, T., Pham, T., Ho, N., Hua, B.S.: Point-set distances for learning representations of 3D point clouds. In: IEEE International Conference on Computer Vision (ICCV), pp. 10478–10487 (2021)
22. Meunier, E., Bouthemy, P.: Unsupervised space-time network for temporally-consistent segmentation of multiple motions. In: IEEE/CVF Conference on Computer Vision and Pattern Recognition, pp. 22139–22148 (2023)

Plug-and-Play Half-Quadratic Splitting for Ptychography

Alexander Denker(✉), Johannes Hertrich, Zeljko Kereta, Silvia Cipiccia, Ecem Erin, and Simon Arridge

University College London, London, UK
{a.denker,j.hertrich,z.kereta,s.cipiccia,ecem.erin.23}@ucl.ac.uk, s.arridge@cs.ucl.ac.uk

Abstract. Ptychography is a coherent diffraction imaging method that uses phase retrieval techniques to reconstruct complex-valued images. It achieves this by sequentially illuminating overlapping regions of a sample with a coherent beam and recording the diffraction pattern. Although this addresses traditional imaging system challenges, it is computationally intensive and highly sensitive to noise, especially with reduced illumination overlap. Data-driven regularisation techniques have been applied in phase retrieval to improve reconstruction quality. In particular, plug-and-play (PnP) offers flexibility by integrating data-driven denoisers as implicit priors. In this work, we propose a half-quadratic splitting framework for using PnP and other data-driven priors for ptychography. We evaluate our method both on natural images and real test objects to validate its effectiveness for ptychographic image reconstruction.

Keywords: Ptychography · Phase Retrieval · Plug-and-Play

1 Introduction

Phase retrieval aims to recover a complex-valued signal $\mathbf{x} \in \mathbb{C}^n$ from intensity-only measurements

$$\mathbf{y} = |\mathbf{A}\mathbf{x}|,$$

for a linear forward operator $\mathbf{A}$. The reconstruction of $\mathbf{x}$ from $\mathbf{y}$ is a non-linear and ill-posed inverse problem, due to the lack of phase information.

Phase retrieval is ubiquitous in imaging applications, e.g., X-ray crystallography [31] or holography [40], see also the review [7]. Moreover, phase retrieval is a crucial component in many computational microscopy pipelines. Standard microscopy techniques use a lens between the object and the detector to form an image. Optical imperfections cause aberrations that limit the resolution of the imaging system. This is particularly severe in the X-ray regime where manufacturing high-quality optics is both challenging and extremely costly. Coherent diffraction imaging (CDI) offers an alternative by replacing the lens with a computational algorithm, thereby eliminating optical aberrations [4]. CDI makes use

T. A. Bubba et al. (Eds.): SSVM 2025, LNCS 15667, pp. 269–281, 2025.
https://doi.org/10.1007/978-3-031-92366-1_21

of diffraction patterns to reconstruct the object of interest. The far-field diffraction patterns can be approximated by the Fourier transform.

Ptychography is a form of CDI that enables imaging of extended objects by shifting coherent illumination, smaller than the object, across the sample at overlapping positions, see Fig. 1 for a visualisation of the measurement procedure. For each position, a coherent beam illuminates a small region of the sample, and the intensity of the diffraction pattern from the illuminated region is recorded [27]. The overlap between illuminations introduces redundancy, which mitigates the ill-posed nature of the problem and ensures a stable reconstruction. However, increased overlap also increases the number of scanning positions and consequently, the total scanning time. Thus, minimising overlap is desirable.

Ptychographic reconstruction methods mostly use iterative phase retrieval algorithms [8,14,20,37]. The performance of these algorithms often deteriorates in the presence of high noise and reduced illumination overlap. To address these challenges and stabilise the reconstruction process, regularisation techniques can be employed [2]. Regularisation introduces prior knowledge to improve reconstruction stability and guide it towards a more plausible and stable solution. This can take the form of sparsity-enforcing priors, smoothness constraints, or more sophisticated models that reflect expected properties of the object of interest. In particular, data-driven priors such as (weakly) convex ridge regularisers [11,12] and patch-priors [1] have seen increased attention, see [15] for a recent overview.

A particularly promising approach to leverage data-driven priors is the use of plug-and-play (PnP) algorithms [36]. The PnP framework provides a flexible way to incorporate image denoising methods as implicit priors without explicitly defining a regularisation term. Building on the success of modern denoisers in phase retrieval [24,32,34,38], we explore the application of PnP to ptychography.

Plug-and-Play. The core concept of PnP algorithms is to implicitly define the regularisation term using a denoiser. PnP algorithms build on classical splitting methods, which separate the data fidelity and regularisation components of the optimisation problem. In these methods, the regularisation term often manifests through a proximal operator, typically corresponding to a Gaussian denoising task. Hence, the main idea of PnP [36] is to replace the proximal operator in variational reconstruction algorithms by a more general denoiser. While the authors of [5,36] used classical denoisers, modern approaches predominantly use neural network-based denoisers [3,23,41,42]. Performance of PnP algorithms can be further improved by using dataset-specific denoisers derived from diffusion or flow-matching generative models [13,22]. However, this approach reduces the generality of the method. Regularisation by denoising (RED) [28] is a concept related to PnP, which does not replace a proximal operator. Instead, it employs gradient descent with a Laplacian-type, differentiable regulariser. While PnP methods work well in practice, they usually do not converge, see e.g. [33] for numerical examples. As a remedy, most implementations use only finitely many steps of the underlying reconstruction algorithm. For linear inverse problems, convergence can be achieved by restricting either the Lipschitz constant or the architecture of the denoiser [16–18,26,30].

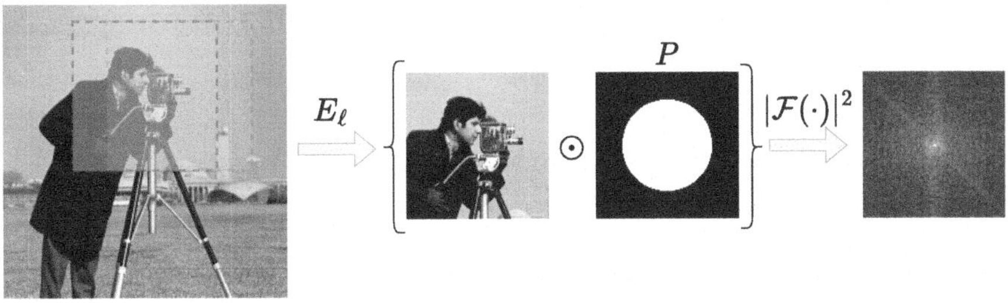

Fig. 1. The measurement model for ptychography. We first extract a patch from the image. We take the element-wise product of the extracted patch with the probe. Finally, we obtain the magnitude of the Fourier transform. This process is repeated for all probe positions in the image. The blue squares in the left image show two consecutive overlapping regions. (Color figure online)

Plug-and-Play for Phase Retrieval. Most existing PnP algorithms were studied for linear inverse problems. However, some algorithms, including PnP with half-quadratic splitting (HQS) and forward-backward splitting (FBS), can also be applied for non-linear inverse problems provided data-fidelity steps can be solved. In particular, [34] propose PnP-FISTA for Fourier ptychography and [24] propose a PnP-FASTA for phase retrieval. Studies show that PnP-HQS usually works better than PnP-FBS and is less sensitive to the step-size parameter [41]. Following this line of research, we propose using PnP-HQS for phase retrieval and ptychography. In Sect. 2, we show that the corresponding data-fidelity steps can be explicitly solved. Further, most PnP algorithms for phase retrieval only deal with real-valued images, see [24,32,34,38], and exploit pre-trained denoisers for natural grey-scale images. However, objects of interest in ptychography are inherently complex-valued with the amplitude and phase encoding important physical information. Thus, we require denoisers for complex-valued images.

Mathematical Model of the Forward Operator

The measurement process in ptychography, as visualised in Fig. 1, can be formally described by defining a window operator (coded illumination) E_ℓ, for $\ell \in \{1, ..., L\}$, which extracts an $N \times N$ window around a probe position p_ℓ. Measurements $\mathbf{y}_\ell$ are generated by first applying the window extraction operator E_ℓ followed by an element-wise multiplication with P, representing the illumination function or a mask. Finally, the wave is transported to the far-field via the Fourier transform and its magnitude is recorded. This process can be summarised by

$$\mathbf{y}_\ell^2 = |\mathcal{F}(A_\ell \mathbf{x})|^2 \quad \ell = 1, \ldots, L, \tag{1}$$

where $A_\ell = P \circ E_\ell$ is the concatenation of the probe with the window extraction operator, and $\mathcal{F}$ is the 2D Fourier transform. For notational simplicity, in the following we identify all the images, such as $\mathbf{x}$ or $\mathbf{y}$, through their vectorisations. This allows representing A_ℓ as diagonal matrices (since they encode element-wise operations), with adjoints denoted A_ℓ^*. Thus, $A_\ell^* A_\ell$ are diagonal matrices.

Noise Model and Data-Fidelity Term. We use a shot noise model as in [24]

$$\mathbf{y}_\ell^2 := |\mathcal{F}(A_\ell \mathbf{x})|^2 + \alpha \operatorname{diag}(|\mathcal{F}(A_\ell \mathbf{x})|^2)\eta, \text{ for } \eta \sim \mathcal{N}(0, \mathbf{I}).$$

This approximates a Poisson noise model with $\mathbf{y}_\ell^2/\alpha^2 \sim \text{Poisson}(|\mathcal{F}(A_\ell \mathbf{x})|^2/\alpha^2)$, where α controls the signal-to-noise ratio. Even though we do not deal with Gaussian noise, we follow [39] and use $\frac{1}{2}\|\mathbf{y} - G(\mathbf{z})\|^2$ as the data-fidelity term, where G is the forward operator and $\mathbf{y}$ the measurements obtained by stacking the forward operators $|(\mathcal{F} \circ A_\ell)(\cdot)|$ and amplitudes $|\mathcal{F}(A_\ell \mathbf{x})|$ of each probe. This simplification is widely used in ptychographic applications [24,39] in hard X-ray regimes where measurements are collected using photon counting detectors.

2 Half-Quadratic Splitting

In this section we derive a half-quadratic splitting (HQS) algorithm [10] for ptychography. To this end, we consider the variational regularisation problem

$$\arg\min_{\mathbf{x}} \frac{1}{2}\|\mathbf{y} - G(\mathbf{x})\|^2 + \lambda R(\mathbf{x}), \tag{2}$$

where R is a convex regularisation term, $\lambda > 0$ a regularisation parameter, and G is either the forward operator (1), in case of ptychography, or $G(\mathbf{x}) = |\mathcal{F}(\mathbf{x})|$, in case of phase retrieval. We first focus on phase retrieval, where we show that data-consistency steps can be solved analytically. We then turn our focus to ptychography, where we introduce multiple auxiliary variables and reduce the resulting subproblems to the phase retrieval case.

2.1 Half-Quadratic Splitting for Fourier Phase Retrieval

For $G(\mathbf{x}) = |\mathcal{F}(\mathbf{x})|$, the variational problem (2) can be reformulated as

$$\arg\min_{\mathbf{x},\mathbf{z}} \frac{1}{2}\|\mathbf{y} - |\mathcal{F}(\mathbf{z})|\|^2 + \lambda R(\mathbf{x}), \quad \text{s.t.} \quad \mathbf{x} = \mathbf{z}.$$

Relaxing the constraint $\mathbf{x} = \mathbf{z}$ by a quadratic penalisation, we obtain the problem

$$\arg\min_{\mathbf{x},\mathbf{z}} \mathcal{L}_\mu(\mathbf{x}, \mathbf{z}) := \frac{1}{2}\|\mathbf{y} - |\mathcal{F}(\mathbf{z})|\|^2 + \frac{\mu}{2}\|\mathbf{z} - \mathbf{x}\|^2 + \lambda R(\mathbf{x}).$$

The original problem is recovered by letting $\mu \to \infty$. In the following, for its solution we consider the alternating minimisation method

$$\begin{aligned} \mathbf{z}_k &= \arg\min_{\mathbf{z}} \|\mathbf{y} - |\mathcal{F}(\mathbf{z})|\|^2 + \mu\|\mathbf{z} - \mathbf{x}_{k-1}\|^2, \\ \mathbf{x}_k &= \arg\min_{\mathbf{x}} \frac{\mu}{2\lambda}\|\mathbf{x} - \mathbf{z}_k\|^2 + R(\mathbf{x}) = \operatorname{prox}_{\frac{\lambda}{\mu}R}(\mathbf{z}_k). \end{aligned} \tag{3}$$

Whereas for linear inverse problems subproblem (3) has an efficient closed form solution, for non-linear inverse problems this can be challenging to solve. However, in the case of Fourier phase retrieval, we can exploit the structure of the forward operator to derive an explicit solution. Substituting $\mathbf{z} = \mathcal{F}^{-1}(\hat{\mathbf{z}})$, subproblem (3) is equivalent to

$$\begin{aligned} \mathbf{z}_k = \mathcal{F}^{-1}(\hat{\mathbf{z}}_k), \quad \hat{\mathbf{z}}_k &= \arg\min_{\hat{\mathbf{z}}} \|\mathbf{y} - |\mathcal{F}(\mathcal{F}^{-1}(\hat{\mathbf{z}}))|\|^2 + \mu\|\mathcal{F}^{-1}(\hat{\mathbf{z}}) - \mathbf{x}_{k-1}\|^2 \\ &= \arg\min_{\hat{\mathbf{z}}} \|\mathbf{y} - |\hat{\mathbf{z}}|\|^2 + \frac{\mu}{n}\|\hat{\mathbf{z}} - \mathcal{F}(\mathbf{x}_{k-1})\|^2, \end{aligned} \tag{4}$$

where the equality comes from the fact that $\frac{1}{\sqrt{n}}\mathcal{F}$ is an isometry, with n representing the number of components (pixels) of $\mathbf{x}$. In the following, we denote $\hat{\mathbf{x}}_{k-1} = \mathcal{F}(\mathbf{x}_{k-1})$. The objective in (4) is fully separable with respect to the argument. In particular, the minimiser $\hat{\mathbf{z}}_k = (\hat{z}_{k,i})_{i=1}^n$ is given by

$$\hat{z}_{k,i} = \arg\min_{\hat{z}_i} (y_i - |\hat{z}_i|)^2 + \frac{\mu}{n}|\hat{z}_i - \hat{x}_{k-1,i}|^2.$$

Since the first term is independent from the phase of $\hat{z}_i$ and the second term is minimised for $\text{phase}(\hat{z}_i) = \text{phase}(\hat{x}_{k-1,i})$ for any fixed amplitude $|\hat{z}_i|$, we have that $\text{phase}(\hat{z}_{k,i}) = \text{phase}(\hat{x}_{k-1,i})$. It remains to minimise with respect to the amplitude of $\hat{z}_{k,i}$. Inserting the solution with respect to the phase into the above minimisation problem, the amplitude $|\hat{z}_{k,i}|$ is given by

$$|\hat{z}_{k,i}| = \arg\min_r (y_i - r)^2 + \frac{\mu}{n}(r - |\hat{x}_{k-1,i}|)^2 = c y_i + (1-c)|\hat{x}_{k-1,i}|,$$

where $c = \frac{n}{n+\mu}$. In summary, the solution is given by

$$\mathbf{z}_k = \mathcal{F}^{-1}(\hat{\mathbf{z}}_k), \text{ for } \hat{\mathbf{z}}_k = (c\mathbf{y} + (1-c)|\hat{\mathbf{x}}_{k-1}|)\exp(i\,\text{phase}(\hat{\mathbf{x}}_{k-1})), \tag{5}$$

where $c = \frac{n}{n+\mu\sigma^2}$, $\hat{\mathbf{x}}_{k-1} = \mathcal{F}(\mathbf{x}_{k-1})$ and all operations are applied pointwise.

2.2 Multiple Auxiliary Variables in HQS for Ptychography

For ptychography, we reformulate the variational problem (2) as

$$\arg\min_{\mathbf{x},\mathbf{z}} \sum_{\ell=1}^{L} \frac{1}{2}\|\mathbf{y}_\ell - |\mathcal{F}(\mathbf{z}_\ell)|\|^2 + \lambda R(\mathbf{x}), \quad \text{s.t.} \quad A_\ell \mathbf{x} = \mathbf{z}_\ell.$$

Relaxing the constraints, we aim to minimise

$$\mathcal{J}_\mu(\mathbf{x},\mathbf{z}) := \sum_{\ell=1}^{L} \frac{1}{2}\|\mathbf{y}_\ell - |\mathcal{F}(\mathbf{z}_l)|\|^2 + \sum_{\ell=1}^{L} \frac{\mu}{2}\|\mathbf{z}_\ell - A_\ell \mathbf{x}\|^2 + \lambda R(\mathbf{x}),$$

Algorithm 1. Half-quadratic splitting algorithm for ptychography

Input: Measurements $(\mathbf{y}_\ell)_{\ell=1}^L$, denoising parameters $(\tau_k)_{k\in\mathbb{N}}$, regularisation strength λ, initialisation $\mathbf{x}_0$
for $k = 1, 2, ...$ **do**
  $\mu_k = \lambda/\tau_k^2$
  $c_k = \frac{n}{n+\mu_k}$
  for $\ell = 1, ..., L$ **do**
    $\hat{\mathbf{x}}_{k-1,\ell} = \mathcal{F}(A_\ell \mathbf{x}_{k-1})$
    $\mathbf{z}_{k,\ell} = \mathcal{F}^{-1}(\hat{\mathbf{z}}_{k,\ell})$ with $\hat{\mathbf{z}}_{k,\ell} = (c_k \mathbf{y}_\ell + (1-c_k)|\hat{\mathbf{x}}_{k-1,\ell}|)\exp(\mathrm{i}\,\mathrm{phase}(\hat{\mathbf{x}}_{k-1,\ell}))$
  end for
  $\tilde{\mathbf{z}}_k = \frac{\sum_{\ell=1}^L A_\ell^* \mathbf{z}_{k,\ell}}{\sum_{\ell=1}^L |A_\ell|^2}$
  $\mathbf{x}_k = D^{-1}\mathrm{prox}_{\tau_k R\circ D^{-1}}(D\tilde{\mathbf{z}}_k)$
end for

where the original problem formulation is recovered in the limit $\mu \to \infty$. Applying an alternating minimisation scheme leads to iterations

$$\mathbf{z}_{k,\ell} = \arg\min_{\mathbf{z}} \|\mathbf{y}_\ell - |\mathcal{F}(\mathbf{z}_\ell)|\|^2 + \mu\|\mathbf{z}_\ell - A_\ell \mathbf{x}_{k-1}\|^2, \quad \ell = 1, ..., L, \tag{6}$$

$$\mathbf{x}_k = \arg\min_{\mathbf{x}} \frac{\mu}{2\lambda}\sum_{\ell=1}^L \|A_\ell \mathbf{x} - \mathbf{z}_{k,\ell}\|^2 + R(\mathbf{x}). \tag{7}$$

The first step (6) is the same as the first step (3) for phase retrieval and admits an explicit solution given by (5) for each ℓ. To solve the second step (7), we first rewrite the sum of the squared norms as

$$\begin{aligned}\sum_{\ell=1}^L \|A_\ell \mathbf{x} - \mathbf{z}_{k,\ell}\|^2 &\propto \mathbf{x}^*\left(\sum_{\ell=1}^L |A_\ell|^2\right)\mathbf{x} - 2\left(\sum_{\ell=1}^L \mathbf{z}_{k,\ell}^* A_\ell\right)\mathbf{x},\\ &\propto \left\|\left(\sum_{\ell=1}^L |A_\ell|^2\right)^{\frac{1}{2}}\mathbf{x} - \left(\sum_{\ell=1}^L |A_\ell|^2\right)^{-\frac{1}{2}}\left(\sum_{\ell=1}^L A_\ell^* \mathbf{z}_{k,\ell}\right)\right\|^2,\end{aligned} \tag{8}$$

where $\propto$ indicates equality up to additive constants independent of $\mathbf{x}$, and we use the diagonality of A_ℓ to establish $A_\ell^* A_\ell = |A_\ell|^2$. Here, $\sum_{\ell=1}^L |A_\ell|^2$ corresponds to the accumulated squared intensity of a probe hitting a pixel, and $\left(\sum_{\ell=1}^L |A_\ell|^2\right)^{-\frac{1}{2}}\left(\sum_{\ell=1}^L A_\ell^* \mathbf{z}_{k,\ell}\right)$ is the average of vectors $\mathbf{z}_{k,\ell}$ weighted with respect to probe matrices A_ℓ. Inserting (8) and denoting the diagonal matrix $D = \left(\sum_{\ell=1}^L |A_\ell|^2\right)^{\frac{1}{2}}$, and $\tilde{\mathbf{z}}_k = \frac{\sum_{\ell=1}^L A_\ell^* \mathbf{z}_{k,\ell}}{\sum_{\ell=1}^L |A_\ell|^2}$, we can reformulate (7) as

$$\mathbf{x}_k = \arg\min_{\mathbf{x}} \frac{\mu}{2\lambda}\|D\mathbf{x} - D\tilde{\mathbf{z}}_k\|^2 + R(\mathbf{x}) = D^{-1}\mathrm{prox}_{\frac{\lambda}{\mu}R\circ D^{-1}}(D\tilde{\mathbf{z}}_k). \tag{9}$$

This corresponds to a denoising problem with spatially varying noise level, which has been used in PnP algorithms [25], albeit in a different context.

Remark 1. When the entries of the probe and the probe positions are sufficiently regular we can assume $D^2 \approx \gamma I$ for some $\gamma > 0$ such that

$$\sum_{\ell=1}^{L} \|A_\ell \mathbf{x} - \mathbf{z}_{k,\ell}\|^2 \approx \gamma \|\mathbf{x} - \tilde{\mathbf{z}}_k\|^2 \quad \text{with} \quad \tilde{\mathbf{z}}_k = \frac{\sum_{\ell=1}^{L} A_\ell^* \mathbf{z}_{k,\ell}}{\sum_{\ell=1}^{L} |A_\ell|^2},$$

where division is understood element-wise. Thus, problem (7) can be written as

$$\mathbf{x}_k = \arg\min_{\mathbf{x}} \frac{\mu\gamma}{2\lambda} \|\mathbf{x} - \tilde{\mathbf{z}}_k\|^2 + R(\mathbf{x}) = \operatorname{prox}_{\frac{1}{\mu\gamma} R}(\tilde{\mathbf{z}}_k).$$

In particular, when $R(\mathbf{x}) = \iota_{[0,\infty)^n}(\mathbf{x})$ is the indicator function of the non-negative orthant, we recover the commonly used simultaneous PIE algorithm [19] for ptychography and the error reduction algorithm [9] for phase retrieval.

In practice, we choose an increasing sequence $(\mu_k)_{k \in \mathbb{N}}$ and for each k perform only one step of HQS. To determine μ_k we follow a similar principle to that in [41]. For this, note that if we replace the proximal mapping with a learned denoiser, the denoising strength is given by $\tau_k = \sqrt{\lambda/\mu_k}$. As the denoisers are trained for a specific range of noise levels, it is easier to choose τ_k due to a clear interpretation. Choosing τ_k defines a sequence $\mu_k = \lambda/\tau_k^2$ with λ as the remaining hyper-parameter. We choose $\tilde{\lambda}$ by $\lambda = \tilde{\lambda}\hat{\sigma}^2$, where $\hat{\sigma}^2$ is an estimate of the noise level. This results in the algorithm summarised in Algorithm 1.

2.3 Complex Denoiser

To achieve a complex-valued reconstruction, we have several ways for applying the proximal mapping, depending on the choice of the regulariser. Decomposing into real and imaginary parts, or into magnitude and phase, we can apply a denoiser pre-trained on real-valued greyscale images. For instance, for a regulariser of the form $R(\mathbf{z}) = R_1(\Re(\mathbf{z})) + R_2(\Im(\mathbf{z}))$ we can write the proximal mapping as

$$\operatorname{prox}_R(\mathbf{z}) = \operatorname{prox}_{R_1}(\Re(\mathbf{z})) + i \operatorname{prox}_{R_2}(\Im(\mathbf{z})),$$

due to the separability of the proximal mapping. This motivates to apply a pre-trained denoiser to the real and imaginary parts independently. To ensure non-negativity of the real and imaginary parts, before denoising we add a constant equal to the maximum amplitude. This pre-processing step is required as most pre-trained denoisers are designed to work only for non-negative images.

3 Experiments

We compare PnP approaches against two commonly used classical algorithms: Sequential PIE (SeqPie) and Simultaneous PIE (SimPie) [19]. All PnP algorithms are initialised with 100 iterations of SimPIE. Moreover, we compare different choices of denoisers for PnP algorithms. First, we use two regularisers for

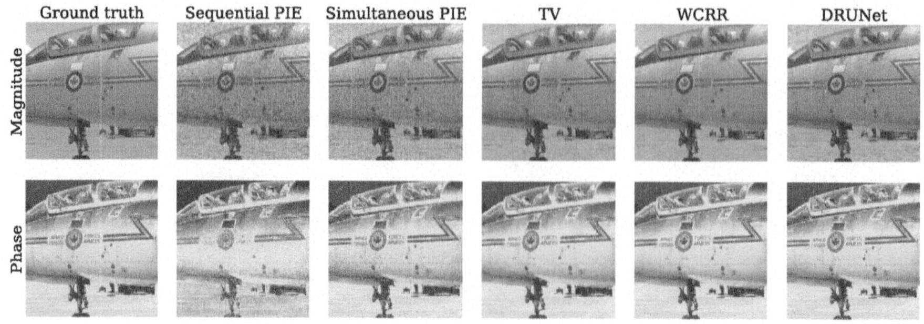

Fig. 2. Reconstructions for the 7×7 probe setting with $\alpha = 20.0$. We only show the centre of the image where PSNR is calculated.

which we solve the spatially varying proximal mapping as in Eq. (9), applied independently to the real and imaginary components of the image. In particular, we use Total Variation (TV) [29] and the weakly convex ridge regulariser (WCRR) [12]. Finally, we use a pre-trained DRUNet [41][1] and use the equivariant evaluation from [35]. For the pre-trained DRUNet, we use the approximation from Remark 1. We compute the PSNR on the amplitude and on the phase. For the phase, we correct for a global phase shift and calculate the phase PSNR as

$$\mathrm{PSNR}_\phi(\phi_{\mathrm{reco}}, \phi_{\mathrm{gt}}) = 10 \log_{10} \left(\frac{(2\pi)^2}{\mathrm{MSE}_\phi(\phi_{\mathrm{reco}}, \phi_{\mathrm{gt}})} \right),$$

where MSE_ϕ is the mean squared error on the sphere defined as

$$\mathrm{MSE}_\phi(\phi_{\mathrm{reco}}, \phi_{\mathrm{gt}}) = \frac{1}{N} \sum_{i=1}^{N} ((\phi_{\mathrm{reco},i} - \phi_{\mathrm{gt},i} + \pi) \bmod 2\pi - \pi)^2,$$

and ϕ_{reco} and ϕ_{gt} are the phases of the reconstruction and ground truth image.

3.1 Natural Images

We evaluate our choices of denoisers on the BSD500 dataset [21] of natural RGB colour images. We first crop the central 256×256px region, and create a complex valued image by setting the amplitude to be the mean of the red and green channels. For the phase we scale the blue colour channel to $(-\pi, \pi]$ and add a random global phase shift. We use patches of size 96×96px and a binary circular probe of radius 37px, cf. the experimental setup in Fig. 1. Probe positions are equidistantly distributed in both coordinate directions. Since measurements do not cover the entire image and are sparse near the image borders, some regions cannot be reconstructed. Therefore, we exclude 20px regions from each border and focus on the central part of the image when calculating the quality

[1] Implementation from the `deepinv` package https://deepinv.github.io.

Table 1. PSNR (mean ± standard deviation) for amplitude and phase for the complex-valued BSD500 images with a noise level $\alpha = 20.0$ for different probe setups with a circular probe. Best results are highlighted.

Probes	7×7 (38% overlap)		9×9 (48% overlap)		11×11 (56% overlap)		15×15 (68% overlap)	
	$PSNR_a$	$PSNR_\phi$	$PSNR_a$	$PSNR_\phi$	$PSNR_a$	$PSNR_\phi$	$PSNR_a$	$PSNR_\phi$
SeqPIE	21.36±1.30	23.03±5.68	21.93±1.22	23.85±5.53	21.99±1.18	23.50±5.70	21.91±1.17	23.78±5.64
SimPIE	23.29±1.25	25.35±5.95	26.25±0.99	28.08±6.50	28.39±0.92	30.10±6.86	31.40±0.99	33.27±7.16
TV	26.13±2.85	28.72±6.11	28.90±2.03	31.33±6.28	30.56±1.63	32.81±6.70	32.70±1.63	35.19±6.88
DRUNet	27.73±2.35	29.76±6.65	30.24±2.07	32.11±6.63	31.07±2.63	32.85±6.87	32.00±2.10	34.30±5.70
WCRR	27.03±2.90	30.01±5.93	29.56±2.10	32.38±6.46	31.13±1.84	33.86±6.20	33.12±1.84	35.81±6.99

Table 2. PSNR (mean ± standard deviation) for amplitude and phase for the complex-valued BSD500 images for 7×7 circular probes (38% overlap) for different noise levels. Best results are highlighted.

Noise level	$\alpha = 10.0$		$\alpha = 20.0$		$\alpha = 30.0$		$\alpha = 40.0$	
	$PSNR_a$	$PSNR_\phi$	$PSNR_a$	$PSNR_\phi$	$PSNR_a$	$PSNR_\phi$	$PSNR_a$	$PSNR_\phi$
SimPIE	28.79±0.79	30.89±6.45	23.29±1.25	25.35±5.95	20.39±1.61	22.43±5.33	18.61±1.85	20.76±4.87
TV	28.09±2.32	30.16±6.21	26.13±2.85	28.72±6.11	23.39±3.29	26.17±6.08	21.18±3.35	24.23±5.67
DRUNet	32.35±2.13	34.89±5.98	27.73±2.35	29.76±6.65	24.01±2.43	25.79±6.42	21.47±2.57	23.35±5.72
WCRR	31.20±2.42	33.92±5.87	27.03±2.90	30.01±5.93	24.28±2.98	27.38±5.47	22.17±3.04	25.35±5.18

metrics. For all methods we use $K = 600$ iterations and choose hyperparameters to maximise the amplitude PSNR on a small validation set for 38% overlap and $\alpha = 20$. For all methods we choose a logarithmic schedule for the denoising strength τ_k. For TV we choose $\tau_1 = 30$, $\tau_K = 6$ and $\lambda = 0.0001$. For the WCRR we choose $\tau_1 = 1500, \tau_K = 200$ and $\lambda = 0.01$. Finally, for the DRUNet we use $\tau_1 = 30, \tau_K = 5$ and $\lambda = 0.0001$. For SimPIE and SeqPIE, we used 2000 iterations, as additional iterations yielded only small improvements.

Influence of the Overlap. We evaluate how the overlap percentage, determined by the number of probe positions, affects the performance. We test with overlap percentages from 38% (49 probe locations) to 68% (225 probe locations). With classical algorithms (e.g. PIE) an overlap of 60–70% is required to produce good results [20]. For binary probes, overlap is computed as the intersection over union of the support. Results are presented in Table 1 and an example reconstruction is shown in Fig. 2. For all probe setting we consistently see that HQS with different denoisers outperforms the (unregularised) PIE. In particular, the performance gap is bigger for a smaller overlap. The neural network based regularisers outperform the TV regulariser for all settings.

Influence of Noise. For a fixed probe setting, we evaluate the performance of different algorithms under varying noise levels. Here, we use the 7×7 probe setting, which results in a overlap of 38%. We test noise levels $\alpha = 10, 20, 30, 40$.

The results are in Table 1. Similar to previous results, PnP approaches outperform SimPIE. Among the data-driven denoisers, the WCRR has a slightly better performance compared to the DRUNet for higher noise levels. Whereas the DRUNet and TV result in a similar performance at this level. For the unregularised SimPIE we observe a bigger drop in performance with increasing noise level (Table 2).

3.2 Brain Phantom

Finally, we employ the PnP algorithms to a brain phantom with a resolution of 800×2200px [6]. We simulate measurements using a circular probe, noise level $\alpha = 40$ and a probe setting of 57×23, resulting in an overlap of 36.3% in the x-axis and 44.7% in the y-axis. The phantom has a low constant amplitude, so we only show the phase in Fig. 3. For this phantom, PnP regularisers again outperform SimPIE. The TV reconstruction exhibits the typical stair-casing effect,

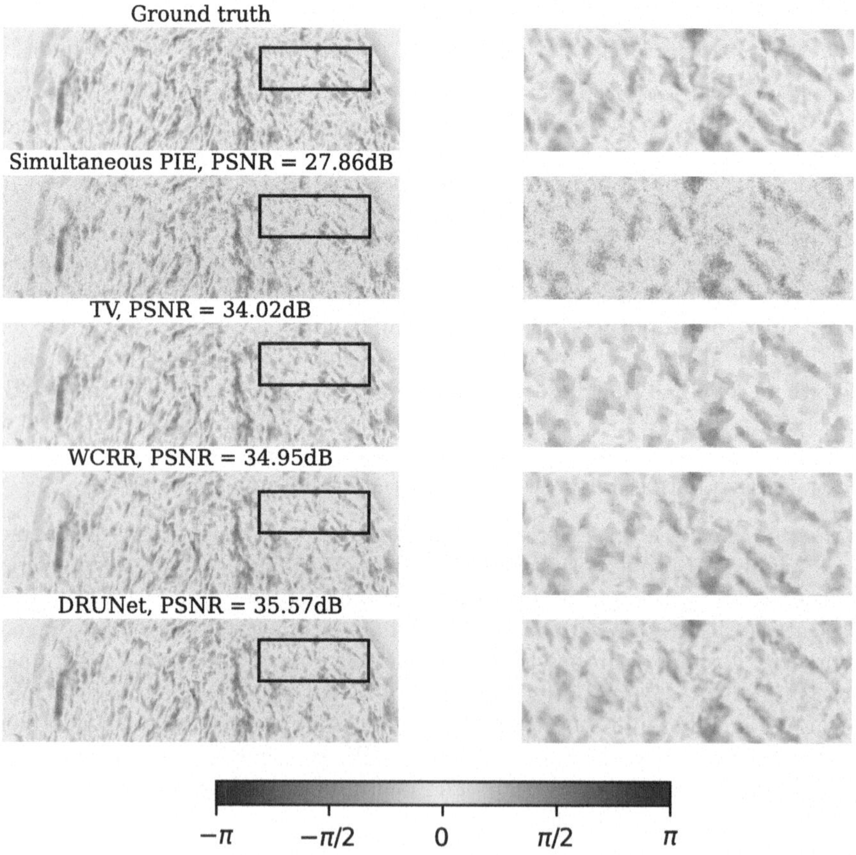

Fig. 3. Reconstruction of the phase of the brain phantom. The ground truth is given as the converged ePIE [20] solution from oversampled measured data.

whereas both the WCRR and the DRUNet reconstruction appear more realistic, despite being trained on natural images.

4 Conclusion and Further Work

In this work we extended the PnP-HQS algorithm to ptychographic image reconstruction. We show that the data-fidelity step can be solved explicitly, reducing the computational cost. In our initial experiments, the PnP-HQS algorithm was able to produce good reconstructions while using lower probe overlap than classical methods. Thus, the developed framework could be a first step to reducing the acquisition time. The next important step in applying this methodology to realistic ptychography systems is to reconstruct both the object and the probe (which is in this work assumed to be known) simultaneously [20].

Acknowledgements. Alexander Denker, Simon Arridge and Johannes Hertrich are supported by the EPSRC programme grant EP/V026259/1. Johannes Hertrich also acknowledges funding by the German Research Foundation (DFG) within the Walter Benjamin Programme with project number 530824055. Zeljko Kereta is supported by EPSRC programme grant EP/X010740/1. Silvia Cipiccia is supported by EPSRC New Investigator Award EP/X020657/1.

References

1. Altekrüger, F., Denker, A., Hagemann, P., Hertrich, J., Maass, P., Steidl, G.: PatchNR: learning from very few images by patch normalizing flow regularization. Inverse Prob. **39**(6), 064006 (2023)
2. Benning, M., Burger, M.: Modern regularization methods for inverse problems. Acta Numer. **27**, 1–111 (2018)
3. Buzzard, G.T., Chan, S.H., Sreehari, S., Bouman, C.A.: Plug-and-play unplugged: optimization-free reconstruction using consensus equilibrium. SIAM J. Imag. Sci. **11**(3), 2001–2020 (2018)
4. Candes, E.J., Li, X., Soltanolkotabi, M.: Phase retrieval from coded diffraction patterns. Appl. Comput. Harmon A. **39**(2), 277–299 (2015)
5. Chan, S.H., Wang, X., Elgendy, O.A.: Plug-and-play ADMM for image restoration: fixed-point convergence and applications. IEEE Trans. Comput. Imaging **3**(1), 84–98 (2016)
6. Cipiccia, S., et al.: Fast x-ray ptychography: towards nanoscale imaging of large volume of brain. Eur. Phys. J. Plus **139**(5), 434 (2024)
7. Dong, J., Valzania, L., Maillard, A., Pham, T., Gigan, S., Unser, M.: Phase retrieval: from computational imaging to machine learning. IEEE Sig. Process. Mag. **40**(1), 45–57 (2023)
8. Enders, B., Thibault, P.: A computational framework for ptychographic reconstructions. Proc. Royal Soc. A: Math. Phys. Eng. Sci. **472**(2196), 20160640 (2016)
9. Fienup, J.R.: Phase retrieval algorithms: a comparison. Appl. Opt. **21**(15), 2758–2769 (1982)
10. Geman, D., Yang, C.: Nonlinear image recovery with half-quadratic regularization. IEEE Trans. Image Process. **4**(7), 932–946 (1995)

11. Goujon, A., Neumayer, S., Bohra, P., Ducotterd, S., Unser, M.: A neural-network-based convex regularizer for inverse problems. IEEE Trans. Comput. Imag. **9**, 781–795 (2023)
12. Goujon, A., Neumayer, S., Unser, M.: Learning weakly convex regularizers for convergent image-reconstruction algorithms. SIAM J. Imag. Sci. **17**(1), 91–115 (2024)
13. Graikos, A., Malkin, N., Jojic, N., Samaras, D.: Diffusion models as plug-and-play priors. NeurIPS **35**, 14715–14728 (2022)
14. Guizar-Sicairos, M., Fienup, J.: Phase retrieval with transverse translation diversity: a nonlinear optimization approach. Optics express **16**, 7264–7278 (2008)
15. Habring, A., Holler, M.: Neural-network-based regularization methods for inverse problems in imaging. In: GAMM-Mitteilungen, p. e202470004 (2024)
16. Hertrich, J., Neumayer, S., Steidl, G.: Convolutional proximal neural networks and plug-and-play algorithms. Linear Algebra Appl. **631**, 203–234 (2021)
17. Hurault, S., Leclaire, A., Papadakis, N.: Gradient step denoiser for convergent plug-and-play. In: ICLR (2022)
18. Hurault, S., Leclaire, A., Papadakis, N.: Proximal denoiser for convergent plug-and-play optimization with nonconvex regularization. In: ICML, pp. 9483–9505 (2022)
19. Konijnenberg, A.P., Coene, W., Pereira, S.F., Urbach, H.P.: Combining ptychographical algorithms with the hybrid input-output (HIO) algorithm. Ultramicroscopy **171**, 43–54 (2016)
20. Maiden, A., Rodenburg, J.M.: An improved ptychographical phase retrieval algorithm for diffractive imaging. Ultramicroscopy **109**(10), 1256–62 (2009)
21. Martin, D., Fowlkes, C., Tal, D., Malik, J.: A database of human segmented natural images and its application to evaluating segmentation algorithms and measuring ecological statistics. In: IEEE ICCV, vol. 2, pp. 416–423 (2001)
22. Martin, S., Gagneux, A., Hagemann, P., Steidl, G.: PnP-flow: plug-and-play image restoration with flow matching (2024). arXiv:2410.02423
23. Meinhardt, T., Moller, M., Hazirbas, C., Cremers, D.: Learning proximal operators: using denoising networks for regularizing inverse imaging problems. In: ICCV, pp. 1781–1790 (2017)
24. Metzler, C., Schniter, P., Veeraraghavan, A., Baraniuk, R.: prDeep: robust phase retrieval with a flexible deep network. In: ICML, pp. 3501–3510. PMLR (2018)
25. Pendu, M.L., Guillemot, C.: Preconditioned plug-and-play ADMM with locally adjustable denoiser for image restoration. SIAM J. Imag. Sci. **16**(1), 393–422 (2023)
26. Pesquet, J.C., Repetti, A., Terris, M., Wiaux, Y.: Learning maximally monotone operators for image recovery. SIAM J. Imag. Sci. **14**(3), 1206–1237 (2021)
27. Rodenburg, J., Maiden, A.: Ptychography. In: Hawkes, P.W., Spence, J.C.H. (eds.) Springer Handbook of Microscopy. SH, pp. 819–904. Springer, Cham (2019). https://doi.org/10.1007/978-3-030-00069-1_17
28. Romano, Y., Elad, M., Milanfar, P.: The little engine that could: regularization by denoising (RED). SIAM J. Imag. Sci. **10**(4), 1804–1844 (2017)
29. Rudin, L.I., Osher, S., Fatemi, E.: Nonlinear total variation based noise removal algorithms. Physica D **60**(1–4), 259–268 (1992)
30. Ryu, E., Liu, J., Wang, S., Chen, X., Wang, Z., Yin, W.: Plug-and-play methods provably converge with properly trained denoisers. In: ICML, pp. 5546–5557 (2019)
31. Sayre, D.: Some implications of a theorem due to Shannon. Acta Crystallogr. A **5**(6), 843–843 (1952)
32. Shastri, S.K., Schniter, P.: Fast and robust phase retrieval via deep expectation-consistent approximation (2024). arxiv:2407.09687

33. Sommerhoff, H., Kolb, A., Moeller, M.: Energy dissipation with plug-and-play priors. In: NeurIPS Workshop on Solving Inverse Problems with Deep Networks (2019)
34. Sun, Y., Xu, S., Li, Y., Tian, L., Wohlberg, B., Kamilov, U.S.: Regularized Fourier ptychography using an online plug-and-play algorithm. In: IEEE ICASSP, pp. 7665–7669. IEEE (2019)
35. Terris, M., Moreau, T., Pustelnik, N., Tachella, J.: Equivariant plug-and-play image reconstruction. In: CVPR, pp. 25255–25264 (2024)
36. Venkatakrishnan, S.V., Bouman, C.A., Wohlberg, B.: Plug-and-play priors for model based reconstruction. In: IEEE GlobalSIP, pp. 945–948 (2013)
37. Wakonig, K., et al.: It ptychoshelves, a versatile high-level framework for high-performance analysis of ptychographic data. J. Appl. Crystallogr. **53**(2), 574–586 (2020)
38. Wang, Y., Sun, X., Fleischer, J.: When deep denoising meets iterative phase retrieval. In: ICML, pp. 10007–10017. PMLR (2020)
39. Yeh, L.H., et al.: Experimental robustness of Fourier ptychography phase retrieval algorithms. Opt. Express **23**(26), 33214–33240 (2015)
40. Zhang, J., Pégard, N., Zhong, J., Adesnik, H., Waller, L.: 3D computer-generated holography by non-convex optimization. Optica **4**(10), 1306–1313 (2017)
41. Zhang, K., Li, Y., Zuo, W., Zhang, L., Van Gool, L., Timofte, R.: Plug-and-play image restoration with deep denoiser prior. IEEE TPAMI **44**(10), 6360–6376 (2021)
42. Zhang, K., Zuo, W., Gu, S., Zhang, L.: Learning deep CNN denoiser prior for image restoration. In: CVPR, pp. 3929–3938 (2017)

Deep Unrolling for Learning Optimal Spatially Varying Regularisation Parameters for Total Generalised Variation

Thanh Trung Vu[1,2], Andreas Kofler[3], and Kostas Papafitsoros[2](✉)

[1] University of Cambridge, Cambridge, UK
ttv22@cam.ac.uk
[2] School of Mathematical Sciences, Queen Mary University of London, London, UK
k.papafitsoros@qmul.ac.uk
[3] Physikalisch-Technische Bundesanstalt (PTB), Braunschweig and Berlin, Germany
andreas.kofler@ptb.de

Abstract. We extend a recently introduced deep unrolling framework for learning spatially varying regularisation parameters in inverse imaging problems to the case of Total Generalised Variation (TGV). The framework combines a deep convolutional neural network (CNN) inferring the two spatially varying TGV parameters with an unrolled algorithmic scheme that solves the corresponding variational problem. The two subnetworks are jointly trained end-to-end in a supervised fashion and as such the CNN learns to compute those parameters that drive the reconstructed images as close as possible to the ground truth. Numerical results in image denoising and MRI reconstruction show a significant qualitative and quantitative improvement compared to the best TGV scalar parameter case as well as to other approaches employing spatially varying parameters computed by unsupervised methods. We also observe that the inferred spatially varying parameter maps have a consistent structure near the image edges, asking for further theoretical investigations. In particular, the parameter that weighs the first-order TGV term has a triple-edge structure with alternating high-low-high values whereas the one that weighs the second-order term attains small values in a large neighborhood around the edges.

Keywords: Spatially Varying Regularisation Parameters · Inverse Problems · Total Generalised Variation · Denoising · Magnetic Resonance Imaging · Neural networks · Unrolling

1 Introduction

In inverse imaging problems, variational regularisation problems of the type

$$\min_{u \in X} \mathcal{D}(Au, f) + \mathcal{R}(u; \Lambda) \tag{1}$$

T. A. Bubba et al. (Eds.): SSVM 2025, LNCS 15667, pp. 282–294, 2025.
https://doi.org/10.1007/978-3-031-92366-1_22

are widely used to compute an estimation $u \in X$ of some ground truth imaging data u_{true}, X being a Banach space, given data f that satisfy the equation

$$f = Au_{\text{true}} + \eta. \tag{2}$$

Here, A denotes the forward operator of the typically ill-posed inverse problem and η is a random noise component. Solving (1) using spatially varying regularisation parameters Λ, instead of scalar ones, has been the subject of many works, see [6,12,20] and the references therein. The goal is to compute and subsequently use a regularisation parameter $\Lambda : \Omega \to (\mathbb{R}_+)^\ell$ that balances the data fidelity $\mathcal{D}$ and the – in general ℓ components – of the regularisation term $\mathcal{R}$ with a different strength at every point $x \in \Omega \subset \mathbb{R}^d$ (every pixel in the pixel-domain Ω, in the discrete setting). In the case where $\mathcal{R}(u) = \text{TV}(u)$, the Total Variation of the (grayscale) function $u : \Omega \to \mathbb{R}$ and assuming Gaussian noise, problem (1) amounts to

$$\min_{u \in X} \frac{1}{2}\|Au - f\|^2_{L^2(\Omega)} + \int_\Omega \Lambda(x) d|Du|. \tag{3}$$

Small values of $\Lambda : \Omega \to \mathbb{R}_+$ impose little local regularity and are thus suitable for preserving detailed parts of the image and edges. On the other hand, high values of Λ impose large regularity and are preferable for smooth, homogeneous areas.

For higher-order extensions of TV, especially those defined in an infimal convolution manner, the role and the interplay of spatially varying regularisation parameters on the resulting image quality and structure are not as straightforward. A prominent example is the Total Generalised Variation (TGV) [4]

$$\text{TGV}_{\Lambda_0,\Lambda_1}(u) := \min_{w \in \text{BD}(\Omega)} \int_\Omega \Lambda_1(x) d|Du - w| + \int_\Omega \Lambda_0(x) d|\mathcal{E}w|, \tag{4}$$

where $\Lambda : \Omega \to (\mathbb{R}_+)^2$, with $\Lambda = (\Lambda_0, \Lambda_1)$. Here, $\mathcal{E}$ denotes the measure that represents the distributional symmetrised gradient of $w \in \text{BD}(\Omega)$, the space of functions of bounded deformations. The combined values of Λ_0 and Λ_1 not only regulate the regularisation intensity but also control the staircasing effect, which is a characteristic limitation of TV, with suitable values of these parameters promoting piecewise affine structures.

In contrast to TV, where multiple works have considered computing spatially varying Λ in (3), see aforementioned reviews, there are limited works that focus on the computation of such spatially varying Λ_0, Λ_1 for TGV. In [13], a bilevel unsupervised scheme was used, employing a statistics-based upper level energy. The approach produced satisfactory results, albeit with a high computational cost and a need to impose continuity assumptions to Λ_0, Λ_1 for robustness reasons. The latter regularity for the parameter-maps limits their adaptability to the image structure, thus not fully exploiting the potential of spatial adaptivity.

Recently, there has been a series of works focusing on learning regularisation parameters for TV using neural networks [2,9,18]. There, a network is initially suitably trained in a supervised manner to infer regularisation parameters, which in a second phase are used in the corresponding variational problem that is solved as usual. Thus, one exploits the versatility of neural networks to inform a model-based variational regularisation scheme whose solution is interpretable since the regulariser remains handcrafted.

Contribution

Here, we adapt the approach introduced in [16] to compute spatially varying regularisation maps for TGV. It involves training a network that consists of two subnetworks in a supervised fashion. The first subnetwork is a deep convolutional neural network (CNN), that takes as an input the data f and outputs the maps Λ_0, Λ_1. To tie these maps to the variational problem, they are fed into a second appended subnetwork, an unrolled PDHG algorithm [8], that solves the TGV minimisation problem considering the regularisation maps to be fixed. The entire network is trained end-to-end with pairs (f^i, u^i_{true}) and thus, the CNN is implicitly trained to output those maps that drive the approximate solution u^i of the variational problem as close to as possible to u^i_{true}. Since the CNN is expressive enough, given some new data f^{test}, it can produce meaningful parameter-maps adapted to f^{test}. We show that this approach produces far better results than similar approaches in image denoising and MR image reconstruction, significantly boosting the performance of TGV. We also show that in contrast to the TV, the structure of the resulting TGV parameter maps is non-trivial and asks for further theoretical investigation.

Outline: In Sect. 2, we set some notation and recall some preliminaries about TV and TGV regularisation with scalar and spatially varying parameter maps. We proceed to describe the deep unrolled PDHG network for learning these maps. In Sects. 3 and 4, we present our numerical results in denoising and MRI reconstruction, respectively, and discuss the structure of the resulting maps near the image edges. We conclude in Sect. 5.

2 The Framework

2.1 Preliminaries

We initially define the several notions in function space to recall a few properties that stem from their analysis. As usual, $\Omega \subset \mathbb{R}^d$ is a bounded, Lipschitz domain and we consider grayscale images, i.e. $\mathbb{R}$-valued. For positive functions $\Lambda, \Lambda_0, \Lambda_1 : \Omega \to \mathbb{R}$ we define the spatially varying TV and TGV functionals as

$$\mathrm{TV}_{\Lambda}(u) = \int_{\Omega} \Lambda(x) d|Du|, \tag{5}$$

$$\mathrm{TGV}_{\Lambda_0,\Lambda_1}(u) = \min_{w \in \mathrm{BD}(\Omega)} \int_{\Omega} \Lambda_1(x) d|Du - w| + \int_{\Omega} \Lambda_0(x) d|\mathcal{E}w|. \tag{6}$$

We also denote by TV_{λ} and $\mathrm{TGV}_{\lambda_0,\lambda_1}$ the corresponding functionals with scalar parameters. It holds that when the weights $\Lambda, \Lambda_0, \Lambda_1$ are bounded, lower semi-continuous functions, bounded away from zero, the functionals (5) and (6) are well-defined for $u \in \mathrm{BV}(\Omega)$ [10]. Moreover, the following proposition suggests that continuity of the weights is not essential for well-posedness.

Proposition 1. *Let* $(\mathcal{H}, \|\cdot\|_{\mathcal{H}})$ *be a Hilbert space,* $f \in \mathcal{H}$ *and* $\Lambda, \Lambda_0, \Lambda_1 \in L^\infty(\Omega)$ *be lower semicontinuous, bounded away from zero. Let also* $A \in \mathcal{L}(L^p(\Omega), \mathcal{H})$ *with* $p \in (1, d^*]$, $d^* = d/(d-1)$ ($d^* = \infty$ *if* $d = 1$). *Then both problems*

$$\min_{u \in \mathrm{BV}(\Omega)} \frac{1}{2}\|Au - f\|_{\mathcal{H}}^2 + \mathrm{TV}_\Lambda(u), \tag{7}$$

$$\min_{u \in \mathrm{BV}(\Omega)} \frac{1}{2}\|Au - f\|_{\mathcal{H}}^2 + \mathrm{TGV}_{\Lambda_0,\Lambda_1}(u), \tag{8}$$

admit a solution in $\mathrm{BV}(\Omega)$.

Proof. The proof readily follows by combining [10, Thm 3.2, Prop. 5.9], [11, Prop. 5.1] and [5, Thm. 2.11 & Prop. 5.17] and we omit it here.

We also recall the following proposition from [19], which dictates that when the ratio λ_0/λ_1 is large enough, $\mathrm{TGV}_{\lambda_0,\lambda_1}$ essentially behaves like TV_{λ_1}.

Proposition 2 (from [19]). *There exists a constant* $C > 0$ *depending only on the domain* Ω *such that if* $\lambda_0, \lambda_1 > 0$ *satisfy* $\lambda_0/\lambda_1 > C$ *then*

$$\mathrm{TGV}_{\lambda_0,\lambda_1}(u) = \lambda_1|Du - m_{\mathcal{E}}(\nabla u)|(\Omega), \quad \textit{for all } u \in \mathrm{BV}(\Omega), \tag{9}$$

where for $v \in L^1(\Omega, \mathbb{R}^d)$ *we define* $m_{\mathcal{E}} := \operatorname{argmin}\{\|v - w\|_{L^1(\Omega,\mathbb{R}^d)} : w \in Ker\mathcal{E}\}$. *Recall that* $Ker\mathcal{E} = \{r(x) = Bx + c : \ c \in \mathbb{R}^d,\ B \in \mathbb{R}^{d\times d}$ *skew symmetric*$\}$.

2.2 U-Net Plus Unrolled PDHG for TGV

The combined neural network-based algorithm unrolling technique for the estimation of spatially varying Λ_0, Λ_1 that we adopt here follows [16]. Let $u_n = S^n(\tilde{u}_0, f, (\Lambda_0, \Lambda_1), A)$, $n \in \mathbb{N}$ denote the iterates of some algorithm that solves (8). That is, $u_n \to u^*$ as $n \to \infty$, where u^* is a solution of (8). Here $\tilde{u}_0$ denotes a suitable initialisation in the image space e.g. $\tilde{u}_0 = A^* f$, with A^* denoting the adjoint of A. Next we denote with $\mathrm{NET}_\theta : A^* f \mapsto (\Lambda_0, \Lambda_1)$ a deep convolutional neural network with learnable parameters θ (a U-Net [21] in our implementations). Then, for a *fixed* $N \in \mathbb{N}$, we define the overall network

$$\mathcal{N}_\theta^N(f) = S^N(A^*(f), f, \mathrm{NET}_\theta(A^* f), A). \tag{10}$$

The unrolled network (10) can then be trained end-to-end in a supervised fashion using a dataset of data-ground truth pairs $(f^i, u^i_{\mathrm{true}})_{i=1}^M$, and an appropriate pairwise distance function l,

$$\min_\theta \mathrm{Loss}(\theta) := \frac{1}{M}\sum_{i=1}^M l(\mathcal{N}_\theta^N(f^i), u^i_{\mathrm{true}}), \tag{11}$$

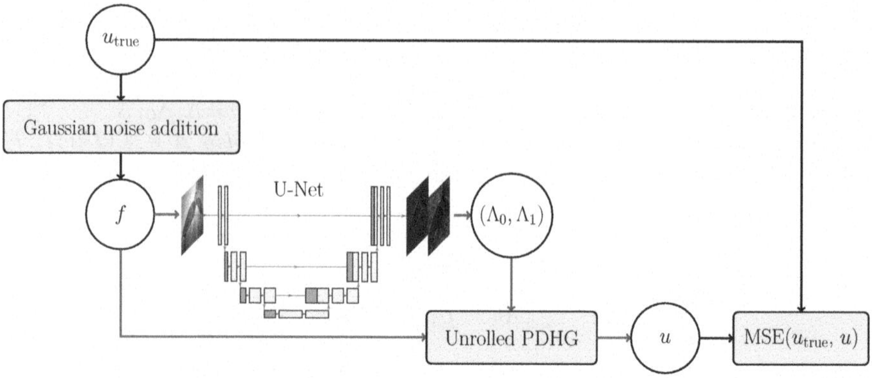

Fig. 1. Visualisation of the network $\mathcal{N}_\theta^N : f \mapsto (\Lambda_0, \Lambda_1) \mapsto u$ of (10) (red arrows) and its training procedure (11) (black arrows) for the denoising case. (Color figure online)

see also Fig. 1 for an illustration of the denoising case. Hence, the network NET_θ implicitly learns to output spatially varying regularisation parameters that force an approximate solution of (8) to be close to the ground truth. Because NET_θ will be overparameterised and thus expressive enough, it is expected that when the trained $\mathcal{N}_\theta^N$ is applied to new unseen data f^{test} it will still produce suitable data-adaptive parameters even though it does not have access to the ground truth. Henceforth, we denote the corresponding networks that follow the above approach by U-TV and U-TGV. We stress that, even though U-TV was introduced in [16], that work did not include 2D denoising and 2D MRI results, which we provide here for comparison with U-TGV. We finally note that here we used the anisotropic versions of TV and TGV with their standard discretisations.

3 Numerical Experiments in Image Denoising

3.1 Set-Up

The Training Dataset: To train the U-TV and U-TGV models for denoising we used the SeaTurtleID2022 dataset [1], which is originally designed for re-identification tasks. The rationale was to test among others the model performance when trained on a specific image distribution, here underwater images of sea turtles, and subsequently tested to images outside this distribution. We randomly selected 500 and 50 images for the training and the validation set respectively. All images were cropped and rescaled to 512×512, converted to grayscale and normalised to $[0, 1]$. During training, the noisy input images were generated on the fly by corrupting the target images by zero-mean Gaussian noise with standard deviation randomly chosen in $[0, 0.2]$.

The Architecture of NET_θ*:* We used a U-Net architecture [21] consisting of three encoding and three decoding blocks, and two final convolutional layers, resulting in 1-128-256-512-1024-512-256-128-2 structure (-1 for U-TV). All convolutional layers have a kernel size of 3×3 and a stride of 1. We used the Leaky ReLU as activation function with a negative slope of 0.01. The map $0.1 \times \text{softplus}(\cdot)$ was applied to the final layer to guarantee positiveness for the parameters $\Lambda, \Lambda_0, \Lambda_1$. Overall, the number of trainable parameters θ was 28,712,577 and 28,712,706 for U-TV and U-TGV respectively.

The Choice of Solution Algorithm: As an algorithm for TV minimisation, we used the standard PDHG algorithm [8], see also [16, Alg. 5.1]. Regarding step sizes, we chose $\sigma = \tau = \text{sigmoid}(10)/\sqrt{13}$. We also used the PDHG algorithm for U-TGV as outlined in [3], setting $\sigma = \tau = 0.29$. In both algorithms, we set the value of extrapolation parameter to sigmoid(10), where $\text{sigmoid}(y) = 1/(1 + \exp(-y))$. We note that these values guarantee the convergence of the algorithms.

The Choice of N: The number of unrolled PDHG iterations is crucial since it should be large enough to approximate the solution of the variational problem. However, setting it too large potentially unnecessarily increases the computational cost (GPU-memory and time) during training. We set $N = 256$ which, according to our observations, achieved a good balance.

The Training Procedure: We used the Adam optimiser [14] with a learning rate of 10^{-4}, a batch size of 1, and the MSE loss. The number of training epochs for U-TV and U-TGV was 200 and 100 respectively, beyond which, the performance of the models when applied to the validation data no longer improved. Training took approximately 10 and 20 h for U-TV and U-TGV respectively, on an RTX 4090 (24 GB VRAM).

The full code for all the experiments can be found in https://github.com/trung-vt/LearningRegularizationParametersForTGV.

3.2 Results

Figures 2 and 3 show extensive results for the "turtle" and "parrot" images. The "turtle" image belongs to the same distribution of images (the SeaTurtleID2022 dataset) the models were trained with, but was not part of the training set. Apart from U-TV and U-TGV, we also show the standard TV and TGV results (best scalar parameter with respect to SSIM, found by a grid-search), as well the spatially varying bilevel TGV result of [13] (unsupervised). We also report the spatially varying TV results (denoted by WTV) obtained by the algorithm proposed in [7] using a maximum-likelihood-type procedure (unsupervised). We see that for both images U-TGV gives the best reconstruction, slightly better than U-TV, with both approaches significantly outperforming the other methods, both visually and quantitatively. We also performed experiments with various noise levels in 50 natural images from [22]. The results, shown in Table 1, are aligned with the ones of the previous figures indicating that the models perform well in a wider distribution of images than the one trained on.

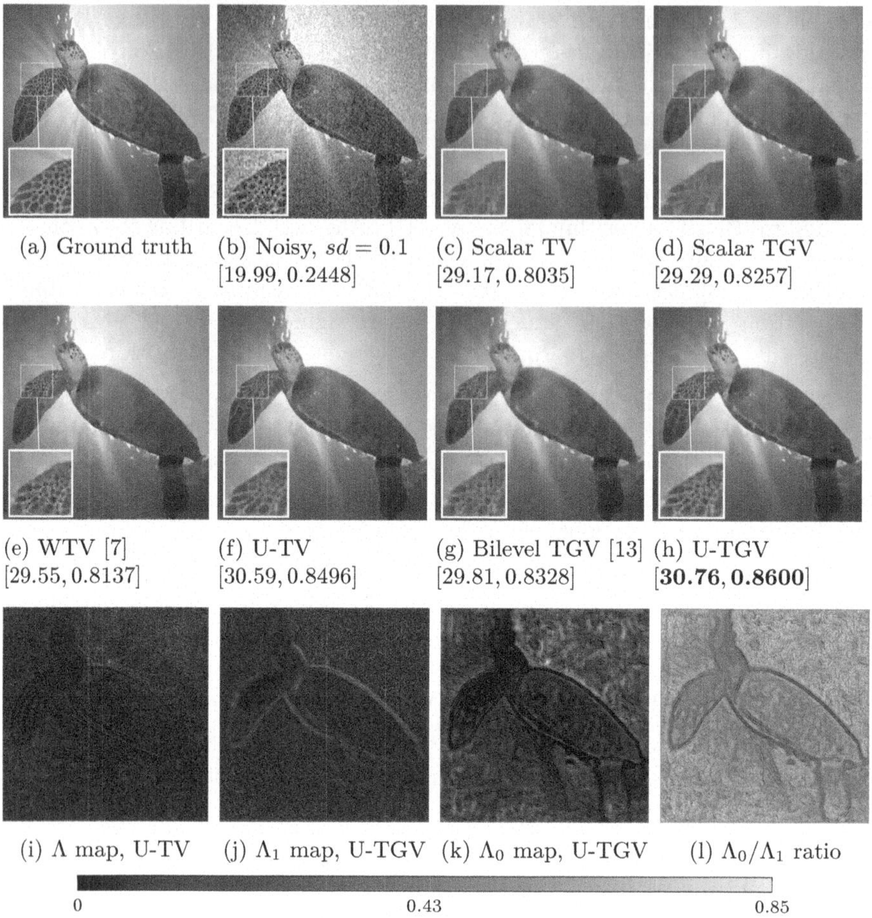

Fig. 2. Denoising results for the "turtle" image. The numbers is brackets show the [PSRN, SSIM] values. The colorbar is common among the visualisations of the parameter maps, except the Λ_0/Λ_1 ratio shown in logarithmic scale.

3.3 Structure of the Parameter-Maps

Further insights are gained by examining the parameter maps in Figs. 2 and 3. By inspecting the ratio Λ_0/Λ_1 we can see the areas of the image where TGV locally behaves like TV, c.f. Proposition 2. This ratio is indeed higher in flatter, almost constant, areas. A surprising observation is that the structure of Λ for U-TV is significantly different than the one of Λ_1 for U-TGV. While Λ takes small values at edges and detailed areas of the image, the values of Λ_1 alternate between being high-low-high at the edges. We also observe that the values of Λ_0 are small at a wider neighbourhood around the edges. To obtain a better visualisation of this *triple-edge* phenomenon, in Fig. 4, we show the denoising results of the "square" image and the corresponding Λ_0, Λ_1 maps. We observe

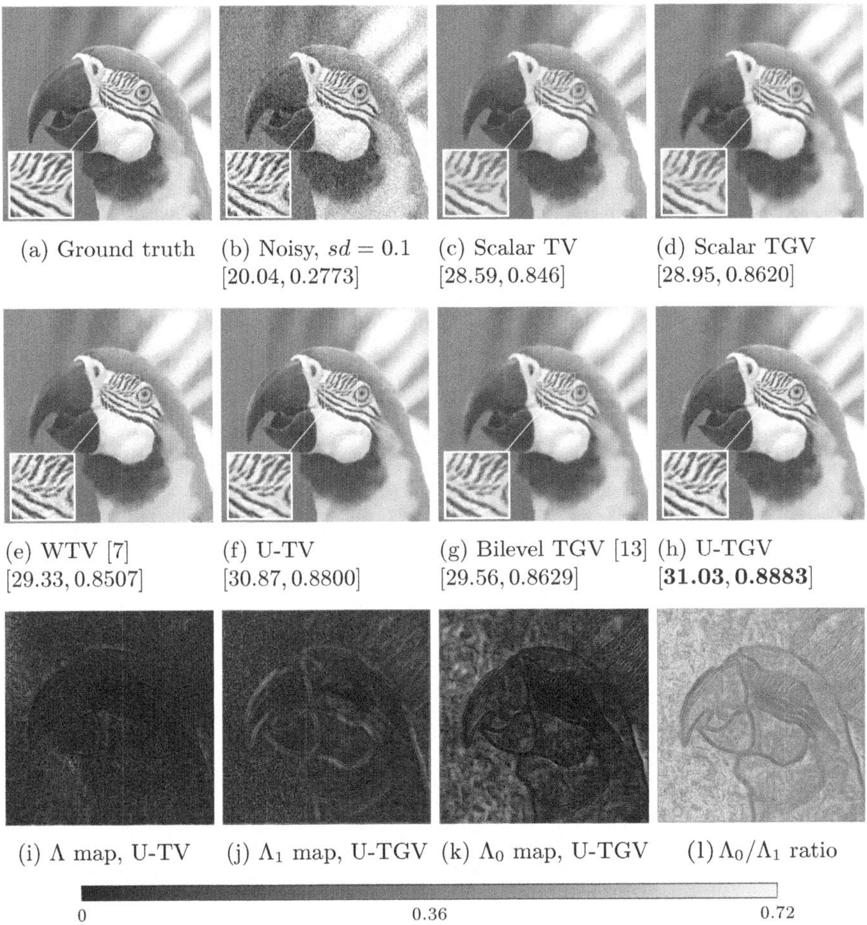

Fig. 3. Denoising results for the "parrot" image. The numbers in brackets show the [PSRN, SSIM] values. The colorbar is common among the visualisations of the parameter maps, except the Λ_0/Λ_1 ratio shown in logarithmic scale.

that Λ_1 takes small values exactly at the edges, resulting in small penalisation of the gradient there, while Λ_0 takes small values at a wider area. Both parameter maps take larger values further away from the edges.

4 Numerical Experiments in MRI Reconstruction

Here, we consider the case of accelerated MR image reconstruction, where the forward model in (2) is given by $A = PF$, where F denotes the 2D Fourier transform and P denotes a projection onto the set of the acquired measurements.

Table 1. Denoising results for 50 natural images and for various noise levels.

sd	metric	scalar TV	scalar TGV	U-TV	U-TGV
0.05	PSNR	30.58 ± 1.78	30.81 ± 1.89	30.86 ± 2.70	**31.03 ± 2.60**
	SSIM	0.8485 ± 0.033	0.8556 ± 0.033	0.8567 ± 0.051	**0.8590 ± 0.051**
0.1	PSNR	27.10 ± 2.19	27.36 ± 2.26	27.96 ± 2.67	**28.11 ± 2.60**
	SSIM	0.7399 ± 0.058	0.7533 ± 0.056	0.7746 ± 0.068	**0.7786 ± 0.069**
0.15	PSNR	25.24 ± 2.37	25.49 ± 2.42	26.41 ± 2.67	**26.49 ± 2.61**
	SSIM	0.6668 ± 0.078	0.6822 ± 0.074	0.7135 ± 0.085	**0.7174 ± 0.086**
0.2	PSNR	23.89 ± 2.42	24.14 ± 2.46	25.38 ± 2.65	**25.40 ± 2.58**
	SSIM	0.6133 ± 0.092	0.6286 ± 0.086	0.6686 ± 0.098	**0.6706 ± 0.098**

(a) Ground truth (b) noisy, $sd = 0.05$ (c) U-TGV (d) U-TGV, close-up

(e) Λ_1 map (f) Λ_0 map (g) Λ_1, close-up (h) Λ_0, close-up

Fig. 4. Visualisation of the structure of the parameters Λ_0 and Λ_1 at the image edges. Both maps alternate between high-low-high values at the edges but Λ_0 takes small values at a larger neighbourhood around the edges.

4.1 Set-Up

The Training Dataset: The dataset consists of $M = 3452$ pairs of retrospectively undersampled MR measurements and target images (f^i, u^i_{true}) that are generated according to (2), where the ground-truth images u^i_{true} are random samples extracted from 598 subjects from the fastMRI multi-coil brain dataset [24]. Since MR images are typically complex-valued and the fastMRI only provides root-sum-of-square (RSS) reconstructions as target images, we first estimated coil sensitivity maps from the fully sampled multi-coil measurements with `MRpro` [25] using the method described in [23]. Then, to obtain complex-valued images, we applied the adjoint of the multi-coil MR forward operator to the multi-coil

Table 2. Summary of MRI reconstruction results for all 302 test images.

R	sd	metric	zero-filled	Scalar TV	Scalar TGV	U-TV	U-TGV
4	0.05	PSNR	24.93 ± 1.56	27.21 ± 1.81	28.58 ± 2.14	**32.60** ± 2.30	32.43 ± 2.28
		SSIM	0.6361 ± 0.045	0.7134 ± 0.088	0.7619 ± 0.064	0.7129 ± 0.091	**0.8022** ± 0.079
	0.1	PSNR	24.83 ± 1.55	27.15 ± 1.82	28.39 ± 2.12	32.01 ± 2.20	**32.05** ± 2.22
		SSIM	0.6235 ± 0.046	0.7144 ± 0.090	0.7590 ± 0.064	0.7039 ± 0.093	**0.7969** ± 0.079
	0.2	PSNR	24.41 ± 1.53	26.91 ± 1.79	27.93 ± 2.09	30.94 ± 2.11	**31.23** ± 2.14
		SSIM	0.5607 ± 0.048	0.7096 ± 0.088	0.7515 ± 0.063	0.6976 ± 0.095	**0.7909** ± 0.077
8	0.05	PSNR	23.08 ± 1.55	23.92 ± 1.71	25.02 ± 1.99	**29.65** ± 2.25	29.62 ± 2.31
		SSIM	0.5723 ± 0.053	0.6262 ± 0.085	0.6736 ± 0.070	0.6681 ± 0.095	**0.7611** ± 0.081
	0.1	PSNR	23.02 ± 1.55	23.87 ± 1.71	24.93 ± 1.98	29.22 ± 2.19	**29.36** ± 2.28
		SSIM	0.5648 ± 0.056	0.6264 ± 0.085	0.6719 ± 0.069	0.6607 ± 0.095	**0.7578** ± 0.080
	0.2	PSNR	22.76 ± 1.54	23.77 ± 1.71	24.66 ± 1.94	28.38 ± 2.14	**28.73** ± 2.24
		SSIM	0.5065 ± 0.056	0.6249 ± 0.085	0.6676 ± 0.068	0.6608 ± 0.095	**0.7549** ± 0.077

measurement data and finally cropped the resulting images to a size of 320×320. We then retrospectively generated k-space data with random acceleration factor R from $4, 5, \ldots, 8$ and adding zero-mean Gaussian noise with a random standard deviation in $[0, 0.2]$. We used 3000, 150 and 302 images for training, validation and testing, respectively.

The Architecture of NET_θ*:* The employed U-Net has a similar structure as before.

The Choice of Solution Algorithm: As in the denoising case, we unrolled the PDHG algorithm as described in [15,16]. This time we let the step sizes σ, τ to be trainable starting from $\sigma = \tau = 1/\sqrt{3}$, after which they took the values $\sigma = 0.3414$, $\tau = 0.3255$ for U-TV and $\sigma = 0.1695$, $\tau = 0.6553$ for U-TGV. We set the extrapolation parameter $\theta = 1$ for both models.

The Choice of N*:* As before, we set $N = 256$ for both models.

The Training Procedure: We used the AdamW optimiser with a learning rate of 10^{-4}, a weight decay of 10^{-5} [17], a batch size of 1 and the MSE loss. Both models were trained for 100 epochs, and the epoch with the lowest validation error was chosen as the final model, 43rd and 72nd for U-TV and U-TGV respectively. Training took approximately 55 and 87 h for U-TV and U-TGV respectively, done on the same machine as in denoising.

4.2 Results

Figure 5 compares reconstructions for a test image under $R = 8$, and Gaussian noise of $sd = 0.05$. The best result is that of U-TGV, with a significant margin over U-TV, particularly with respect to SSIM, and more than 4dB higher PSNR than the scalar version. We observe again a triple-edge structure for the parameter maps, with Λ_0 taking small values in larger areas than Λ_1. In Table 2, we summarise the results for all test images for $R = 4, 8$ and various noise levels.

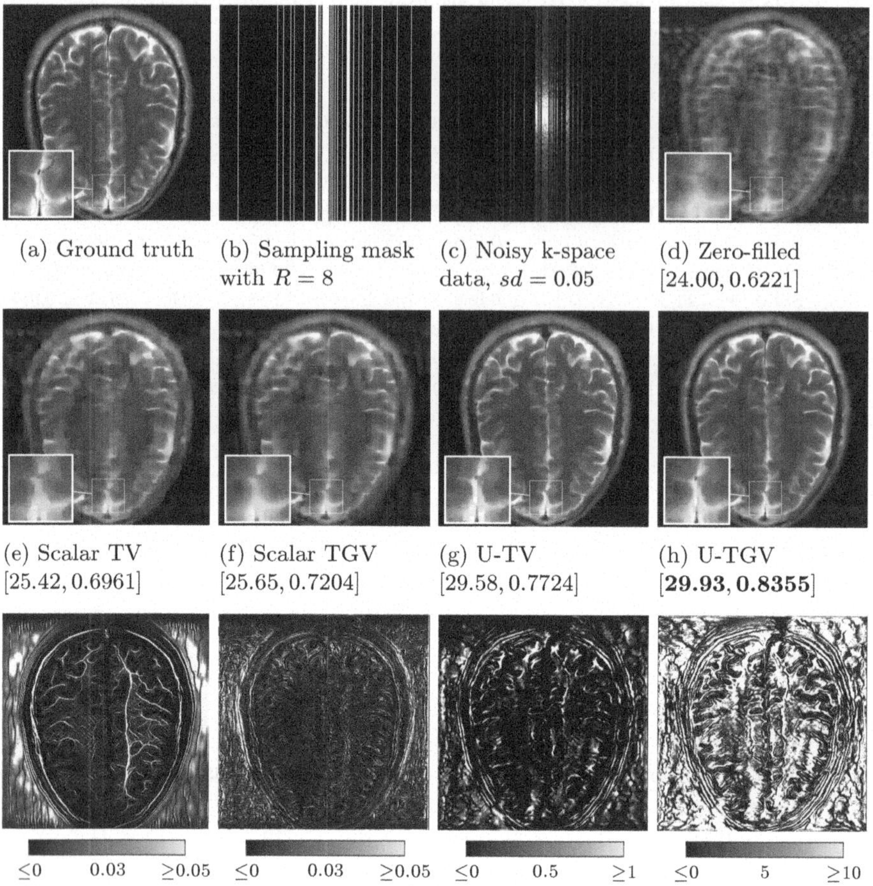

(a) Ground truth (b) Sampling mask with $R = 8$ (c) Noisy k-space data, $sd = 0.05$ (d) Zero-filled $[24.00, 0.6221]$

(e) Scalar TV $[25.42, 0.6961]$ (f) Scalar TGV $[25.65, 0.7204]$ (g) U-TV $[29.58, 0.7724]$ (h) U-TGV $[\mathbf{29.93}, \mathbf{0.8355}]$

(i) Λ map, U-TV (j) Λ_1 map, U-TGV (k) Λ_0 map, U-TGV (l) Λ_0/Λ_1 ratio

Fig. 5. MRI reconstruction results for a Brain image. The numbers in brackets show the [PSNR, SSIM] values.

Apart from the low noise regime, U-TGV performs slightly better than U-TV with respect to PSNR, but significantly better with respect to SSIM in all cases. Both models heavily outperform the scalar versions.

5 Conclusion

We showed that using a CNN to infer spatially varying TGV regularisation parameters that are highly adaptive to the data, significantly boosts its performance both in denoising and MRI reconstruction. The results enjoy high interpretability being solutions of variational problems with a handcrafted prior, as all the neural network-related "black-boxness" is transferred to the regularisation parameters rather than the reconstructed images themselves. Related to that, a further theoretical investigation of the structure of the parameter maps (triple-edge phenomenon) and their interplay in the regularisation process is of great interest.

Acknowledgments. We thank Luca Calatroni for producing the WTV results.

References

1. Adam, L., Čermák, V., Papafitsoros, K., Picek, L.: SeaTurtleID2022: a long-span dataset for reliable sea turtle re-identification. In: Proceedings of the IEEE/CVF Winter Conference on Applications of Computer Vision (WACV), pp. 7146–7156 (2024)
2. Afkham, B.M., Chung, J., Chung, M.: Learning regularization parameters of inverse problems via deep neural networks. Inverse Probl. **37**(10), 105017 (2021)
3. Bredies, K.: Recovering piecewise smooth multichannel images by minimization of convex functionals with total generalized variation penalty. In: Efficient Algorithms for Global Optimization Methods in Computer Vision, pp. 44–77 (2014)
4. Bredies, K., Kunisch, K., Pock, T.: Total generalized variation. SIAM J. Imaging Sci. **3**(3), 492–526 (2010)
5. Bredies, K., Holler, M.: Higher-order total variation approaches and generalisations. Inverse Probl. **36**(12), 123001 (2020)
6. Calatroni, L., Chung, C., De Los Reyes, J.C., Schönlieb, C.B., Valkonen, T.: Bilevel approaches for learning of variational imaging models. In: Radon Series on Computational and Applied Mathematics, vol. 18. De Gruyter, Berlin, Boston (2017)
7. Calatroni, L., Lanza, A., Pragliola, M., Sgallari, F.: A flexible space-variant anisotropic regularization for image restoration with automated parameter selection. SIAM J. Imaging Sci. **12**(2), 1001–1037 (2019)
8. Chambolle, A., Pock, T.: A first-order primal-dual algorithm for convex problems with applications to imaging. J. Math. Imaging Vision **40**(1), 120–145 (2011)
9. Cuomo, S., De Rosa, M., Izzo, S., Piccialli, F., Pragliola, M.: Speckle noise removal via learned variational models. Appl. Numer. Math. **200**, 162–178 (2024)
10. Davoli, E., Ferreira, R., Fonseca, I., Iglesias, J.A.: Dyadic partition-based training schemes for TV/TGV denoising. J. Math. Imaging Vis. 1–39 (2024)
11. Hintermüller, M., Holler, M., Papafitsoros, K.: A function space framework for structural total variation regularization with applications in inverse problems. Inverse Probl. **34**(6), 064002 (2018)
12. Hintermüller, M., Papafitsoros, K.: Generating structured nonsmooth priors and associated primal-dual methods. In: Kimmel, R., Tai, X.C. (eds.) Processing, Analyzing and Learning of Images, Shapes, and Forms: Part 2, Handbook of Numerical Analysis, vol. 20, pp. 437–502 (2019)
13. Hintermüller, M., Papafitsoros, K., Rautenberg, C.N., Sun, H.: Dualization and automatic distributed parameter selection of total generalized variation via bilevel optimization. Numer. Funct. Anal. Optim. **43**(8), 887–932 (2022)
14. Kingma, D.P., Ba, J.: Adam: a method for stochastic optimization. arXiv preprint arXiv:1412.6980 (2014)
15. Knoll, F., Bredies, K., Pock, T., Stollberger, R.: Second order total generalized variation (TGV) for MRI. Magn. Reson. Med. **65**(2), 480–491 (2011)

16. Kofler, A., et al.: Learning regularization parameter-maps for variational image reconstruction using deep neural networks and algorithm unrolling. SIAM J. Imaging Sci. **16**(4), 2202–2246 (2023)
17. Loshchilov, I., Hutter, F.: Decoupled weight decay regularization. In: ICLR (2019)
18. Nekhili, R.R.D., Descombes, X., Calatroni, L.: A hybrid approach combining CNNs and variational modelling for blind image denoising. HAL preprint (2022)
19. Papafitsoros, K., Valkonen, T.: Asymptotic behaviour of total generalised variation. In: Aujol, J.F., Nikolova, M., Papadakis, N. (eds.) Scale Space and Variational Methods in Computer Vision, pp. 702–714 (2015)
20. Pragliola, M., Calatroni, L., Lanza, A., Sgallari, F.: On and beyond total variation regularization in imaging: the role of space variance. SIAM Rev. **65**(3), 601–685 (2023)
21. Ronneberger, O., Fischer, P., Brox, T.: U-net: convolutional networks for biomedical image segmentation. In: MICCAI, pp. 234–241 (2015)
22. University of Granada, Computer Vision Group: Dataset of Standard 512 × 512 Grayscale Test Images (2003). https://ccia.ugr.es/cvg/CG/base.htm
23. Walsh, D.O., Gmitro, A.F., Marcellin, M.W.: Adaptive reconstruction of phased array MR imagery. Magn. Reson. Med. **43**(5), 682–690 (2000)
24. Zbontar, J., et al.: fastMRI: an open dataset and benchmarks for accelerated MRI. arXiv preprint arXiv:1811.08839 (2018)
25. Zimmermann, F.F., et al.: MRpro - pytorch-based MR image reconstruction and processing package (2025). https://doi.org/10.5281/zenodo.14886714

Identifying Memorization of Diffusion Models Through p-Laplace Analysis

Jonathan Brokman[1,2(✉)], Amit Giloni[2], Omer Hofman[2], Roman Vainshtein[2], Hisashi Kojima[3], and Guy Gilboa[1]

[1] Technion - Israel Institute of Technology, Haifa, Israel
jonathanbrok@gmail.com
[2] Fujitsu Research, Beijing, China
[3] Fujitsu Limited, Kawasaki, Japan

Abstract. Diffusion models, today's leading image generative models, estimate the score function, i.e. the gradient of the log probability of (perturbed) data samples, without direct access to the underlying probability distribution. This work investigates whether the estimated score function can be leveraged to compute higher-order differentials, namely p-Laplace operators. We show here these operators can be employed to identify memorized training data. We propose a numerical p-Laplace approximation based on the learned score functions, showing its effectiveness in identifying key features of the probability landscape. We analyze the structured case of Gaussian mixture models, and demonstrate the results carry-over to image generative models, where memorization identification based on the p-Laplace operator is performed for the first time.

Keywords: p-laplace · Diffusion Models · Score-function · Memorization

1 Introduction

Memorization in generative models occurs when the model produces outputs that closely replicate samples from its training data, rather than generating novel content. This issue is notably present even in contemporary popular diffusion models. This highlights important questions about generalization, and also raises privacy concerns in generative AI, as it can lead to the unintended leakage of sensitive training data [22,32].

A key aspect of memorization lies in its relationship with the probability density of data points. Research findings often link memorization to "bumps" or "delta regions" in the learned data distribution. These bumps may arise in sparsely populated areas of the training dataset due to limited data size. They also occur in inherently low-probability regions (of valid data samples), or result from unintended amplification of probability mass around replicated samples within the training set [10,17,23].

T. A. Bubba et al. (Eds.): SSVM 2025, LNCS 15667, pp. 295–307, 2025.
https://doi.org/10.1007/978-3-031-92366-1_23

The p-Laplace is a fundamental differential operator which can be used to quantify degrees of smoothness (also in high dimensions, as desired for our case). We outline below some of its essential properties. Recall the discrete Hilbert space $\mathbb{R}^d$ with standard discrete Euclidean inner product and norm definitions, i.e. $\langle u, v\rangle = \sum_i u_i \cdot v_i$ and $\|u\|^2 = \langle u, u\rangle$, respectively. The p-Dirichlet energy for a function $u : \mathbb{R}^d \rightarrow \mathbb{R}$ is defined by,

$$\begin{aligned} J_p(u) :=& \frac{1}{p}\langle |\nabla u|^p, 1\rangle = \frac{1}{p}\langle |\nabla u|^{p-2}\nabla u, \nabla u\rangle = \frac{1}{p}\langle -\text{div}\left(|\nabla u|^{p-2}\nabla u\right), u\rangle \\ =& \frac{1}{p}\langle -\Delta_p u, u\rangle, \end{aligned} \tag{1}$$

where $\Delta_p u$ is the p-Laplace operator defined as

$$\Delta_p u := \nabla \cdot \left(|\nabla u|^{p-2}\nabla u\right), \tag{2}$$

$\nabla\cdot$ and ∇ are the discrete divergence and gradient operators respectively, and $|\nabla u|^{p-2} = \left[\left(\frac{\partial u}{\partial x(1)}\right)^2 + \cdots + \left(\frac{\partial u}{\partial x(d)}\right)^2\right]^{\frac{p-2}{2}}$. We note that $-\Delta_p$ is a positive semi-definite operator and that $J_p(u) \geq 0$, $\forall u$. The p-Laplace is $p-1$ homogeneous i.e.

$$\Delta_p(a \cdot u) = a|a|^{p-2} \cdot \Delta_p(u), \quad \forall a \in \mathbb{R}. \tag{3}$$

We focus on function analysis via the p-Laplace and its ability to reliably characterize important phenomena.

We investigate the ability to use the p-Laplace to characterize memorized samples in diffusion models. In this setting, the probability distribution is that of natural image data, which is unknown. To address this challenge we propose to capture the "probability bump" around memorized samples using p-Laplace approximations, a technique not yet explored. This leads us to several fundamental questions:

- How can we estimate the p-Laplace operator in the settings of diffusion models?
- What estimation techniques and values of p are best suited for practical applications?
- Can the p-Laplace operator effectively characterize points of memorization?

Our findings outline both recommended and less desired approaches for this task, thereby advancing the research on the geometry of the implicitly learned probability function in diffusion models. Furthermore, we provide a novel mathematical perspective that supports the concept of a "bumpy probability" around memorized samples.[1]

[1] Official implementation: https://github.com/JonathanBrok/Identifying-Memorization-of-Diffusion-Models-through-p-Laplace-Analysis/.

2 Previous Work

Memorization is a well-researched phenomenon. [17] provided theory for long-tailed distributions, claiming that memorizing rare samples from low-probability regions enhances generalization. [18] further showed that neural networks allocate probability mass to these sparse instances. [23] observed that limited data for diffusion model training causes memorization, and analyzed the generalization that occurs as the data increases in size. An alternative research branch, such as [10, 32], focused on data replication in diffusion model training sets and its effect of memorization amplification.

Effective uses of the p-Laplace in image processing is demonstrated in [4, 11, 12, 20, 27] for $1 \leq p \leq 2$ with p dependent on the smoothness of the image. Properties of the p-Laplace gradient flow in the context of image filtering are discussed in [16, 24, 25, 34]. For $p \rightarrow 1$, the energy approaches total-variation and the gradient flow approximates the total variation (TV) flow [3]. Properties of the graph p-Laplace were analyzed in [6, 7, 15, 19, 31, 35], showing several data-processing applications, including for message passing in pre-trained image-language models [36]. Analysis of the operator as $p \rightarrow \infty$ was performed in [9, 14]. Nonlinear p-Laplace eigenvalue problems and the p-Laplace spectra were investigated in [5, 13].

Nonlinear operators and PDEs (Partial Differential Equations) have been studied in contexts related to generative modeling and diffusion processes. For instance, [2] introduced flow-based generative models for lattice field theory, using PDE frameworks. [26] extended score-based diffusion models to function spaces, showcasing the flexibility of diffusion models in handling complex data structures and function representations. Recently, [8] tackled generated versus real image detection using the cross-similarity between predicted noise and signal, which they re-interpreted as a combined criterion of curvature (1-Laplace), gradient, and bias of the noise predictions.

To the best of our knowledge, we are the first to estimate the p-Laplace with diffusion models and to use it to characterize the learned probability distribution. Namely, we are the first to leverage p-Laplace for memorization analysis.

3 Method

3.1 Hypothesis: Measuring Memorization with p-Laplace

Building upon prior observations that relate memorized samples to "bumps" in the training data distribution, we hypothesize that these memorized samples will manifest as local maxima in the learned (log) probability distribution. The p-Laplace, which quantifies the outward flux of $|\nabla u|^{p-2}\nabla u$ around a point (as elaborated below in Sect. 3.3), serves as a tool for this analysis. Typically, around local maxima, the gradient vectors point inward, indicating a negative flux. Therefore, we expect that memorized points will exhibit lower values of the p-Laplace for their learned probabilities.

3.2 Diffusion Model Settings

Let us describe basic principals of diffusion models which we will use for our method. Given a dataset $X \subset \mathbb{R}^d$ drawn from an unknown distribution, diffusion models learn to approximate the *score function* $s(x) := \nabla \log p(x)$, where x is a perturbed data sample with probability density function $p(\cdot)$.

Score Function Approximation via Noise Prediction: Diffusion models rely on the result that estimating $s(x)$ is similar to predicting the noise, when x is a noisy data-sample [29,33]. This connection can be traced back to the theoretical foundations laid out by Miyasawa (1961) [28], showing that the minimum mean square error (MMSE) estimator of s is obtained by denoising x under a suitable Gaussian perturbation. More concretely, consider a noising process governed by:

$$q(x_t \mid x_0) := \mathcal{N}(x_t; x_0, \sigma_t^2 I), \tag{4}$$

where σ_t increases with time t, x_t is a corresponding noised sample and $\mathcal{N}(a; b, c)$ denotes the probability density function of a Gaussian RV a with mean and covariance b, c^2. The MMSE predictor of the noise added at t aligns with $-s(x_t)$. Thus, by training a neural network to predict this noise, we effectively learn the score function for varying degrees of noise.

Forward and Backward Diffusion Processes: The *forward* diffusion process gradually corrupts data samples x_0 with increasing noise levels over discrete timesteps $t = 0, \ldots, T$ as follows:

$$x_t = \sqrt{1 - \alpha_t} x_0 + \sqrt{\alpha_t} \epsilon, \ \epsilon \sim N(0, I), \tag{5}$$

i.e. we have $q(x_t \mid x_0) = \mathcal{N}(x_t; \sqrt{1 - \alpha_t} x_0, \alpha_t I)$. The *backward* (reverse) diffusion process is constructed to invert this corruption, starting from pure Gaussian noise at $t = T$ and iteratively denoising, aiming to recover a sample from the clean (un-perturbed) data distribution at $t = 0$. To obtain this, as $t \to 0$, the diffusion model predictions of added noise become increasingly similar to the score function of an unperturbed sample $s(x_0)$.

Langevin Sampling and Stochastic Differential Equation (SDE): Once the score is known or approximated, fixing $\alpha_t \leftarrow \alpha$, $x \leftarrow \sqrt{1 - \alpha} x_0 + \sqrt{\alpha} \epsilon$, we can sample from $p(x)$ via a stochastic process following Langevin dynamics:

$$dx = \frac{\alpha}{2} s(x)\, dt + \sqrt{\alpha}\, dW_t,$$

where W_t is a Wiener process (Brownian motion), and $\alpha > 0$ controls step size or "temperature". This stochastic differential equation, known as "Ito's SDE" [21], ensures that, under suitable conditions, samples drawn from its stationary distribution approximate samples from $p(x)$. When α is not constant but scheduled to

[2] This $\mathcal{N}(a; b, c)$ is standard notation in the diffusion models literature.

reduce with time, or "annealed", we obtain *annealed Langevin dynamics*, which further improves mixing and coverage by gradually lowering noise levels as the chain proceeds. Formally, if we have a decreasing schedule $\{\alpha_t\}$, $\alpha_t \in \mathbb{R}^+ \, \forall t$, we write:

$$dx = \frac{\alpha_t}{2} s(x)\, dt + \sqrt{\alpha_t}\, dW_t,$$

thereby guiding samples from a broad noisy prior toward the target distribution as $t \to 0$.

Importantly, this entire process often takes place in latent space, as introduced by [30]. In their work, they proposed to run diffusion in a learned latent space rather than in pixel space. First, a Variational Autoencoder (VAE) maps data

$$x \in \mathbb{R}^d \quad \longmapsto \quad z = E(x) \in \mathbb{R}^m,$$

where E is the encoder. The forward noising in latent space is then

$$z_t = \sqrt{1 - \alpha_t}\, z_0 + \sqrt{\alpha_t}\, \epsilon, \quad \epsilon \sim \mathcal{N}(0, I), \tag{6}$$

and the model is trained to invert (6) via noise prediction. After denoising, the final latent sample z_0 is decoded to pixel space:

$$\widehat{x}_0 = D(z_0), \tag{7}$$

where D is the VAE decoder.

Working in latent space lowers compute costs and speeds up training, compared to direct diffusion on high-resolution images. Artifacts may appear due to imperfect VAE reconstruction, yet our p-Laplace analysis, introduced below, applies if we treat the learned latent distribution as $p(z)$ and measure memorized "bumps" within that domain.

3.3 Method: p-Laplace Estimation

Consider $u(x) = \log p(x)$, and consequently $s(x) = \nabla u(x)$. Denote $\hat{s}(x)$ as the approximation of s given by the diffusion model - i.e. the relationships are:

$$\nabla u(x) = s(x) \approx \hat{s}(x). \tag{8}$$

By Eq. (8), the p-Laplace $\Delta_p u$ can be written as:

$$\Delta_p u(x) = \nabla \cdot \left(|s(x)|^{p-2} s(x)\right). \tag{9}$$

Volume and Boundary Integral Formulations: To approximate $\Delta_p u(x_0)$ locally, we consider a d-dimensional ball $B_R(x_0)$ of radius R. The average p-Laplace in the ball is:

$$\overline{\Delta_p u}(x_0) = \frac{1}{|B_R(x_0)|} \int_{B_R(x_0)} \Delta_p u(x)\, dx = \frac{1}{|B_R(x_0)|} \int_{B_R(x_0)} \nabla \cdot \left(|s(x)|^{p-2} s(x)\right) dx,$$

where $|B_R(x_0)|$ is the volume of the ball. Alternatively, by the divergence theorem,

$$\overline{\Delta_p u}(x_0) = \frac{1}{|B_R(x_0)|} \int_{\partial B_R(x_0)} |s(x)|^{p-2} s(x) \cdot n\, ds.$$

with n the outward unit normal on the boundary $\partial B_R(x_0)$.

d-Dimensional Ball and Sphere Measures: The volume of a d-dimensional ball of radius R is:

$$|B_R(x_0)| = \frac{\pi^{d/2}}{\Gamma(\frac{d}{2}+1)} R^d,$$

where Γ denotes the gamma function. The surface measure of the d-sphere S^{d-1} of radius R is:

$$|\partial B_R(x_0)| = \frac{2\pi^{d/2}}{\Gamma(d/2)} R^{d-1}.$$

Discrete Monte Carlo Approximation: Since we do not have closed-form integrals, we approximate these integrals numerically:
1. **Volume Integral Approximation:** Uniformly sample N points $\{x_i\}$ inside $B_R(x_0)$. Compute $\Delta_p u(x_i)$ at each sampled point:

$$\overline{\Delta_p u}(x_0) \approx \frac{1}{N} \sum_{i=1}^{N} \nabla \cdot \left(|s(x)|^{p-2} s(x)\right), \tag{10}$$

since the factor $|B_R(x_0)|/N$ is the average volume element for uniform sampling. The normalization ensures that the final average approximates the integral.
2. **Boundary Integral Approximation:** Uniformly sample N points $\{y_i\}$ on the sphere $\partial B_R(x_0)$. Evaluate $|s(y_i)|^{p-2} s(y_i) \cdot n_i$ and sum:

$$\overline{\Delta_p u}(x_0) \approx \frac{|\partial B_R(x_0)|}{|B_R(x_0)|} \left(\frac{1}{N} \sum_{i=1}^{N} |s(y_i)|^{p-2} s(y_i) \cdot n_i \right), \tag{11}$$

where the $\frac{|\partial B_R(x_0)|}{|B_R(x_0)|}$ factor was introduced, since without it the Monte Carlo converges to the mean over the sphere $\partial B_R(x_0)$. This factor ensures that we do not normalize by the sphere's surface, and correctly normalize by the volume.

Regularity for Applying the Divergence Theorem. The perturbations in Eq. (5) induce the probability density

$$p_\alpha(x) = (\sqrt{1-\alpha} p_0 * \mathcal{N}(x; \sqrt{1-\alpha} x_0, \alpha I) = \sqrt{1-\alpha} \int_{\mathbb{R}^d} p_0(y)\, G_\alpha(x-y)\, dy, \tag{12}$$

where p_0 is the underlying (unperturbed) and $\mathcal{N}$ denotes the Gaussian probability density function similarly to Eq. (4). As is well known, convolving any locally integrable density (and assuming p_0 is such) with a Gaussian kernel yields a C^∞ function on $\mathbb{R}^d$ - (see for instance "properties of mollifications" in [1]). Hence, the resulting $p_\alpha(x)$ is smooth for all $\alpha > 0$, and strictly positive where p_0 is non-zero. Therefore, the log-likelihood inherits sufficient regularity for the divergence theorem to apply in our derivations.

Using Score from Diffusion Model: In practice, $s(x)$ may be unknown. If we have a trained diffusion model that estimates the score function, we plug in its approximation $\hat{s}(x)$ instead of $s(x)$. Note that this introduces an additional degree of approximation to Eqs. (10), (11).

By comparing p-Laplace estimates computed from the true score (if available) and the learned score, we assess the fidelity of the diffusion model p-Laplace approximation. We also compare between the volume-based and boundary-based formulations, and, finally, experiment with p-Laplace upon memorization of the diffusion model.

4 Experiments

In this section, we evaluate the proposed numerical approximations and test the efficacy of p-Laplace in measuring memorization. Following [33], we conduct two experiments using a two-dimensional Gaussian mixture model (GMM). Initially, we assess our numerical approximations, and subsequently, we explore the memorization phenomenon, identifying the best choice of approximation and order p of the p-Laplace for capturing memorized instances.

The final experiment extends these methods to a full-scale diffusion model, applying the insights gained from the GMM experiments. We demonstrate that our findings generalize effectively to the unknown, complex, high-dimensional probability landscape of natural images.

4.1 GMM Setup

In this setting analytic computations such as gradient and divergence expressions can be performed in order to obtain error statistics, considering the underlying scalar function $u(\mathbf{x})$ to be the log-pobability of the GMM. For all GMM experiments, we use 3 mixture components, each with covariance $\sigma^2 I$ and equal mixture weights, and 100 samples per integral estimate (same number for both boundary and volume formulations). Aggregated statistics (when applicable) are across 100 independent runs per neighborhood-circle and p value. Radius of integration spheres: $R = 1.0$. Finite difference step for numerical derivatives is set to $\delta = 10^{-3}$.

We train a diffusion model on the GMM data with 1000 samples. As architecture we choose a MLP with one hidden layer (128 neurons) and sinusoidal time embeddings discretized to $T = 100$ timesteps. Training is done for 500 epochs with learning rate of $1e^{-3}$.

4.2 Experiment 1: GMM Error Analysis and Field Approximations

In this experiment, we assess the correctness of our approach, where the error is estimated using access to true quantities: The true log probability density $u(x)$ and hence its p-Laplace $\Delta_p u(x)$, enabling precise analysis and visualization - see Fig. 1.

We compare various p-Laplace operators, namely $p = 1, 2, 3$, computed from:

- The **oracle** gradient field s (direct numerical differentiation of u).
- The **learned** gradient field $\hat{s}$, obtained by training a diffusion model on the GMM data.

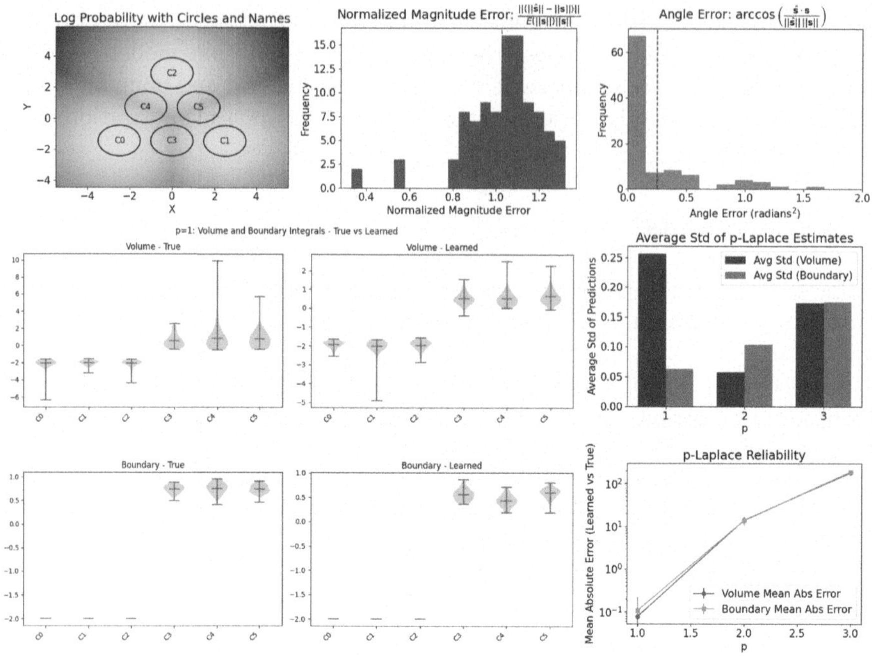

Fig. 1. GMM error analysis (take-away: 1-Laplace boundary formulation is the most reliable). Here we test the fidelity of our approach, to the values of the accessible true probability and its p-Laplace. We test learned, and oracle fields, using volume and boundary formulations. **Top left:** 6 neighborhoods are tested - 3 maxima and 3 non-maxima points. **Top middle-right:** $\hat{s}$ achieves low error rate in direction estimation, however it is not the case for magnitudes, with relative error of $\sim \times 1.2$ receiving non-negligible frequency. **Bottom-right:** 1-Laplace is approximated significantly better than in $p = 2, 3$, due to invariance to errors in magnitude. **Middle-right:** The volume integral formulation has very high variance rendering it less reliable than the boundary formulation, especially for $p = 1$. **Middle-bottom, left-middle:** we provide further per-neighborhood violin plots for $p = 1$ showing that, despite the high variance in volume formulation, maxima neighborhoods are still distinguishable from non-maxima.

We test both boundary and volume integral formulations, Eqs. (11), (10) of Sect. 3.3.

Figure 1 incorporates a thorough multi-faceted analysis, with a simple take-away: If we are interested in a reliable and robust p-Laplace, the best choice is the 1-Laplace using boundary formulation for its approximation. We further reveal an important insight: $\hat{s}$ approximates the direction of s far better than its magnitude. 1-Laplace only uses normalized gradients, canceling the imprecise magnitude estimations - which indeed leads to its superior performance.

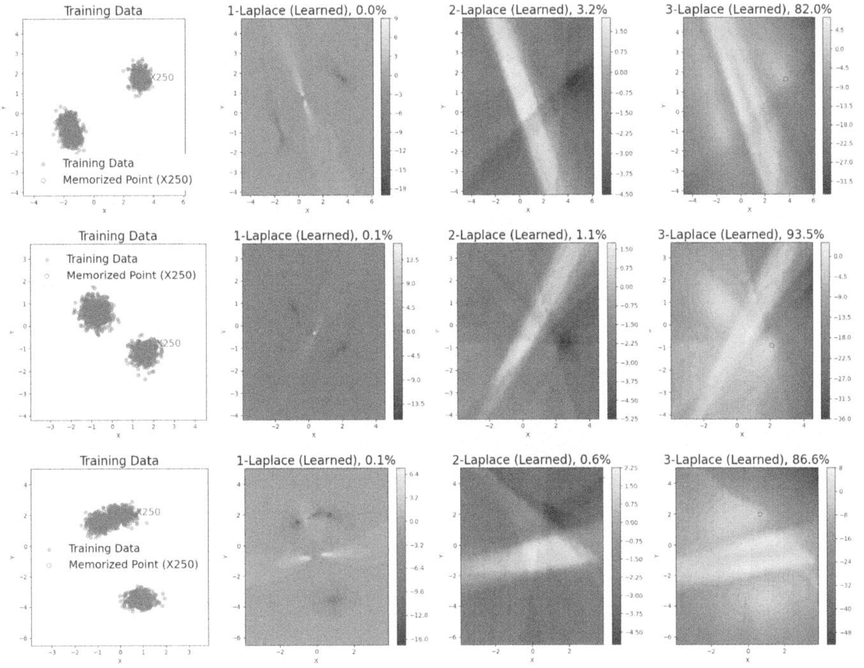

Fig. 2. We are interested in the ability of the p-Laplace of the implicitly learned log probability to reliably reflect memorization, and provide distinguishable values at memorized points. In each row, the scatter-plot shows the GMM training set with randomly drawn peak locations, including one red sample that was replicated 250 times to induce memorization. The colormaps portray the p-Laplace, where we see again how the values of the 1-Laplace pin-point better the memorized sample, assigning it the lowest percentile (percentages in title) compared to the other $p = 2, 3$-Laplace. (Color figure online)

4.3 Experiment 2: Memorization Detection in GMM Setups

Here we explore whether our approach for the p-Laplace approximations can distinguish memorized points from non-memorized ones. Since memorization is a synthetic phenomenon of the model, it is not captured by the true distribution - hence we will not experiment with the true distribution here.

We introduce a controlled memorization scenario within the GMM framework. By intentionally replicating a single training sample multiple times (adding 250 duplicates of one point to a 1000-sample training set), we induce a synthetic "spike" in the model's learned distribution. Such replications were indeed previously observed to cause memorization in diffusion models, e.g. [10].

The true distribution is naturally unaffected by the data sampling process, and consequently the true p-Laplace is not expected to exhibit any anomaly at the memorized point (since the true distribution remains unchanged), whereas

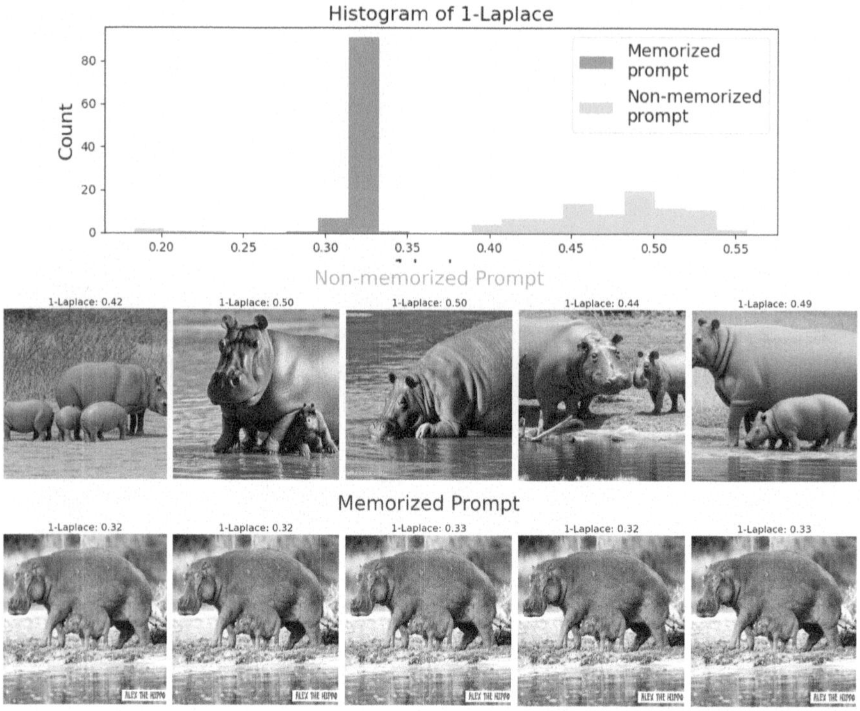

Fig. 3. Testing our approach, using a pre-trained stable diffusion v1.4 to predict the score-function $\hat{s}$. $\hat{s}$ is plugged to the 1-Laplace boundary integral formulation (Eq. (11)) to distinguish memorized from non-memorized samples. Memorized prompt: Mothers influence on her young hippo. Non-memorized prompt is the exact same prompt but with Mother's (with an apostrophe). We generated 100 images with each prompt, and aggregated the criterions to the histogram - which indeed shows good distinguishability. Qualitative examples of 5 generations per prompt are provided below, with their associated 1-Laplace estimation reported in the title.

the *learned* p-Laplace potentially reveals a distinct signature at the memorized location.

To quantify detectability, we:

- Train the diffusion model on 1000 GMM samples plus 250 replicates of a randomly chosen point, simulating memorization.
- Compute the learned p-Laplace across a grid points
- Assess the percentile rank of the memorized point's p-Laplace value among all grid points.

The results of this analysis, comparing p-Laplace of $p = 1, 2, 3$, reveals that $p = 1$ best highlights the memorized samples as outliers, as these samples occupy very low percentile values, see Fig. 2.

4.4 Experiment 3: Large-Scale Image Generative Model Application

Here we apply our approach to a large-scale, pre-trained image diffusion model trained on LAION-5B. We choose Stable Diffusion v1.4 [30] as a testbed due to its popularity as well as well-studied memorization issues for certain prompts. We test a known memorized prompt of this model, "Mothers influence on her young hippo" (see e.g. [32]), and a nearly identical non-memorized prompt, "Mother's influence on her young hippo," generating 100 samples from each, see Fig. 3. As recommended by our previous experiments, we employ the Boundary integral formulation with 64 samples and p = 1-Laplace. We use $T = 500$ generation steps. We leave out the $\frac{|\partial B_R(x_0)|}{|B_R(x_0)|}$ factor, since we are only interested in distinguishing memorized from non-memorized points, and this factor is uniform across samples (it only depends on the size of the neighborhood ball, which is fixed).

The memorized prompt yields 100% memorized generations, whereas the non-memorized prompt results in 96% varied outcomes (and 4% memorization still). By examining the approximated p-Laplace values at generated samples, we find that the approach can reliably identify and distinguish these memorized outputs.

5 Conclusion

We introduce a novel approach, using score-function analysis of diffusion models to approximate the p-Laplace operator, and use it to distinguish memorized from non-memorized generations. We proposed and tested two formulas to approximate the average of the p-Laplace over a neighborhood, approximating either a hyper-ball or a hyper-sphere integral. Our experiments indicate that the 1-Laplace with the boundary integral formula is most effective. We demonstrated this approach's efficacy on stable-diffusion 1.4, a practical and widely used image generative model. To summarize, this work advances the understanding of diffusion models by offering new insights into the approximation of the p-Laplace operator and demonstrating its utility in characterizing training data memorization in image generative models. This initiates a novel theoretical research direction for this timely task.

References

1. Adams, R.A., Fournier, J.J.: Sobolev Spaces, vol. 140. Elsevier (2003)
2. Albergo, M.S., Kanwar, G., Shanahan, P.E.: Flow-based generative models for Markov chain monte carlo in lattice field theory. Phys. Rev. D **100**(3), 034515 (2019)
3. Andreu, F., Ballester, C., Caselles, V., Mazón, J.M., et al.: Minimizing total variation flow. Diff. Integr. Eqn. **14**(3), 321–360 (2001)
4. Baravdish, G., Svensson, O., Åström, F.: On backward p (x)-parabolic equations for image enhancement. Num. Functional Anal. Opt. **36**(2), 147–168 (2015)
5. Bozorgnia, F., Bungert, L., Tenbrinck, D.: The infinity laplacian eigenvalue problem: reformulation and a numerical scheme. J. Sci. Comput. **98**(2), 40 (2024)

6. Brokman, J., Burger, M., Gilboa, G.: Spectral total-variation processing of shapes-theory and applications. ACM Trans. Graph. **43**(2), 1–20 (2024)
7. Brokman, J., Gilboa, G.: Nonlinear spectral processing of shapes via zero-homogeneous flows. In: Elmoataz, A., Fadili, J., Quéau, Y., Rabin, J., Simon, L. (eds.) SSVM 2021. LNCS, vol. 12679, pp. 40–51. Springer, Cham (2021). https://doi.org/10.1007/978-3-030-75549-2_4
8. Brokman, J., Giloni, A., Hofman, O., Vainshtein, R., Kojima, H., Gilboa, G.: Manifold induced biases for zero-shot and few-shot detection of generated images. In: The Thirteenth International Conference on Learning Representations (2025)
9. Bungert, L.: The inhomogeneous p-laplacian equation with Neumann boundary conditions in the limit p to infinity. Adv. Continuous Discr. Models **2023**(1), 8 (2023)
10. Carlini, N., et al.: Extracting training data from diffusion models. In: 32nd USENIX Security Symposium (USENIX Security 23), pp. 5253–5270 (2023)
11. Chen, J., Guo, J.: Image restoration based on adaptive p-laplace diffusion. In: 3rd International Congress on Image and Signal Processing, vol. 1, pp. 143–146. IEEE (2010)
12. Chen, Y., Levine, S., Rao, M.: Variable exponent, linear growth functionals in image restoration. SIAM J. Appl. Math. **66**(4), 1383–1406 (2006)
13. Cohen, I., Gilboa, G.: Introducing the p-laplacian spectra. Signal Process. **167**, 107281 (2020)
14. Deidda, P., Burger, M., Putti, M., Tudisco, F.: The graph ∞-laplacian eigenvalue problem. arXiv preprint arXiv:2410.19666 (2024)
15. Elmoataz, A., Toutain, M., Tenbrinck, D.: On the p-laplacian and infinity-laplacian on graphs with applications in image and data processing. SIAM J. Imag. Sci. **8**(4), 2412–2451 (2015)
16. Fazeny, A., Tenbrinck, D., Burger, M.: Hypergraph p-laplacians, scale spaces, and information flow in networks. In: Calatroni, L., Donatelli, M., Morigi, S., Prato, M., Santacesaria, M. (eds.)International Conference on Scale Space and Variational Methods in Computer Vision, vol. 14009, pp. 677–690. Springer, Cham (2023). https://doi.org/10.1007/978-3-031-31975-4_52
17. Feldman, V.: Does learning require memorization? A short tale about a long tail. In: Proceedings of the 52nd Annual ACM SIGACT Symposium on Theory of Computing, pp. 954–959 (2020)
18. Feldman, V., Zhang, C.: What neural networks memorize and why: discovering the long tail via influence estimation. Adv. Neural. Inf. Process. Syst. **33**, 2881–2891 (2020)
19. Hafiene, Y., Fadili, J., Elmoataz, A.: Nonlocal p-laplacian evolution problems on graphs. SIAM J. Numer. Anal. **56**(2), 1064–1090 (2018)
20. Huang, C., Zeng, L.: Level set evolution model for image segmentation based on variable exponent p-laplace equation. App. Math. Mod. **40**(17–18), 7739–7750 (2016)
21. Itô, K.: On Stochastic Differential Equations. no. 4, American Mathematical Soc. (1951)
22. Jahanian, A., Puig, X., Tian, Y., Isola, P.: Generative models as a data source for multiview representation learning. arXiv preprint arXiv:2106.05258 (2021)
23. Kadkhodaie, Z., Guth, F., Simoncelli, E.P., Mallat, S.: Generalization in diffusion models arises from geometry-adaptive harmonic representation. arXiv preprint arXiv:2310.02557 (2023)
24. Kuijper, A.: Image processing by minimising l^p norms. Pattern Recognit. Image Anal. **23**(2), 226–235 (2013)

25. Kuijper, A.: p-laplacian driven image processing. In: Image Processing, 2007. ICIP 2007. IEEE International Conference, vol. 5, pp. V–257. IEEE (2007)
26. Lim, J.H., et al.: Score-based diffusion models in function space. arXiv preprint arXiv:2302.07400 (2023)
27. Liu, Q., Guo, Z., Wang, C.: Renormalized solutions to a reaction-diffusion system applied to image denoising. DCDS-B **21**(6), 1839–1858 (2016)
28. Miyasawa, K., et al.: An empirical Bayes estimator of the mean of a normal population. Bull. Inst. Internat. Statist **38**(181–188), 1–2 (1961)
29. Nichol, A.Q., Dhariwal, P.: Improved denoising diffusion probabilistic models. In: International Conference on Machine Learning, pp. 8162–8171. PMLR (2021)
30. Rombach, R., Blattmann, A., Lorenz, D., Esser, P., Ommer, B.: High-resolution image synthesis with latent diffusion models. In: Proceedings of the IEEE/CVF Conference on Computer Vision and Pattern Recognition, pp. 10684–10695 (2022)
31. Slepcev, D., Thorpe, M.: Analysis of p-laplacian regularization in semisupervised learning. SIAM J. Math. Anal. **51**(3), 2085–2120 (2019)
32. Somepalli, G., Singla, V., Goldblum, M., Geiping, J., Goldstein, T.: Diffusion art or digital forgery? Investigating data replication in diffusion models. In: Proceedings of the IEEE/CVF Conference on Computer Vision and Pattern Recognition, pp. 6048–6058 (2023)
33. Song, Y., Ermon, S.: Generative modeling by estimating gradients of the data distribution. In: Advances in Neural Information Processing Systems, vol. 32 (2019)
34. Wei, W., Zhou, B.: A p-laplace equation model for image denoising. Inform. Technol. J **11**, 632–636 (2012)
35. Weihs, A., Fadili, J., Thorpe, M.: Discrete-to-continuum rates of convergence for nonlocal p-laplacian evolution problems. Inf. Inference J. IMA **13**(4), iaae031 (2024)
36. Wu, H., Zhang, X., Xu, P., Liao, P., Yao, X., Yu, B.: p-laplacian adaptation for generative pre-trained vision-language models. In: Proceedings of the AAAI Conference on Artificial Intelligence, vol. 38, pp. 6003–6011 (2024)

Learning Anisotropic Metrics for Geodesic Distances via the Heat Equation for Image Segmentation

Nicolas Makaroff, Théo Bertrand, and Laurent D. Cohen(✉)

CEREMADE, UMR CNRS 7534, University Paris Dauphine,
PSL Research University, 75775 Paris, France
{bertrand,makaroff,cohen}@ceremade.dauphine.fr

Abstract. In this paper, we investigate the computation of anisotropic metrics for geodesic distances using the heat equation and its application to image segmentation, particularly for tubular structures. Building upon the work of Bertrand et al. [2], we extend this approach to anisotropic media by incorporating spatially varying and direction-dependent diffusion tensors derived from the image's structure tensor field. Our contributions are twofold: first, we formulate anisotropic geodesic distance computation using the heat equation and integrate it within deep learning models for image segmentation; second, we propose two methods that learn the anisotropic metric directly from image data—one requiring explicit seed point selection and another eliminating the need for seed points by predicting a probability map that serves as the initial condition for heat diffusion. Experiments on synthetic and medical image datasets demonstrate the effectiveness of our methods in accurately segmenting vascular tree structures by leveraging the anisotropic properties inherent in the images without relying on manual seed point selection.

Keywords: Image Segmentation · Deep Learning · Geodesic Distance · Fast Marching · Heat Equation · Medical Image Analysis

1 Introduction

Segmenting regions in images based on geodesic distances is a fundamental problem in computer vision and image analysis, with applications ranging from medical imaging to robotics and geophysics. Traditionally, geodesic distances and curves are computed by explicitly constructing a Riemannian metric g from data, leveraging prior knowledge of the task. This metric is then used to compute geodesic distances, which are integral to understanding the underlying geometry of the data. However, this approach introduces bias through manual parameter tuning and subjective decisions in metric selection, potentially limiting the adaptability and generalisation of the method to diverse datasets.

To address these limitations, we propose a novel framework that learns the metric tensor directly from data using deep learning techniques. Specifically,

T. A. Bubba et al. (Eds.): SSVM 2025, LNCS 15667, pp. 308–321, 2025.
https://doi.org/10.1007/978-3-031-92366-1_24

we design convolutional neural networks (CNNs) that predict the anisotropic metric and potential fields required for geodesic distance computation. By optimising the network parameters through supervised learning on training data, we eliminate the need for manual metric selection and parameter tuning. This approach allows the network to learn the intrinsic geometry of the data adaptively, enhancing segmentation accuracy and robustness.

Traditional methods, like the Fast Marching Method (FMM) introduced by Sethian [8], compute geodesic distances by solving the Eikonal equation. Several prior works have leveraged geodesic distances for segmentation tasks. Malladi and Sethian [6] used minimal path distances and fast marching methods for 3D brain image segmentation. Cohen and Deschamps [3], segmented 3D vascular trees by propagating fronts based on minimal path distances, while Benmansour [1] introduced anisotropic metrics determined dynamically during fast marching computations to segment vascular structures. These methods rely on predefined metrics and do not directly integrate metric learning from data.

Recent efforts have explored learning metrics from data. Scarvelis et al. [7] and Heitz et al. [5] investigated fitting metric tensors to spatio-temporal data to capture velocity fields and underlying geometry. However, these approaches are not tailored for end-to-end segmentation tasks and often involve complex computations unsuitable for integration within deep learning frameworks.

Our work builds upon the work of Bertrand et al. [2]. In order to extend that approach to anisotropic metric, we use the heat method introduced by Crane et al. [4], and extended to anisotropic heat equation by Yang and Cohen [11], both based on Varadhan's asymptotic formula [10], which relate the behaviour of the heat equation to geodesic distances on a Riemannian manifold. By incorporating anisotropic diffusion tensors derived from the image's structure tensor field, we extend these methods to handle anisotropic media, capturing directional dependencies essential for accurately modelling complex pathways within tubular structures.

In contrast, our approach integrates metric learning and geodesic distance computation within a deep learning framework, enabling end-to-end training and inference. We propose two distinct methods: 1. Seed Point-Based Anisotropic Geodesic Distance Computation: Extending prior isotropic models, we predict both the anisotropic metric tensor and a Gaussian potential that initiates heat propagation from predefined seed points. This method leverages the strengths of deep learning for feature extraction and analytical computations of geodesic distances, enhancing precision in heterogeneous media. 2. Seedless Anisotropic Geodesic Distance Computation: To address challenges associated with seed point selection in non-convex regions, we introduce a seedless approach. Our neural network comprises two branches: one predicts the anisotropic diffusion tensor, and the other predicts a probability map serving as an initial condition for the heat equation. This design eliminates the need for explicit seed points, using an attention mechanism to guide heat propagation and refine segmentation.

Our contributions are as follows:

- We formulate anisotropic geodesic distance computation using the heat equation, integrating it within deep learning models for image segmentation.
- We propose two methods that learn the anisotropic metric directly from data, with and without the need for explicit seed point selection.
- We demonstrate the effectiveness of our methods on synthetic and real-world datasets, particularly in segmenting tubular structures in medical images.

The remainder of the paper is organised as follows. Section 2 provides background on geodesic distance computation, anisotropic heat diffusion, and the structure tensor field. Section 3 details our two proposed methodologies, distinguishing between the seed point-based and seedless approaches. Section 4 presents experimental results for both methods, including data descriptions, training procedures, and performance evaluations.

2 Background

2.1 Fast Marching Energy CNN for Image Segmentation

Bertrand et al. introduced the Fast Marching Energy CNN (FMECNN), integrating geodesic distance computations with convolutional neural networks for image segmentation, particularly in medical imaging. This framework learns isotropic metrics via a CNN to impose geometric and topological constraints on the segmentation process. By predicting both the potential function and the barycenter (seed point) of target regions, it generates segmentation masks corresponding to geodesic balls, ensuring path-connected and coherent outputs.

FMECNN extends traditional segmentation methods by incorporating a fast marching module within the network architecture, enabling end-to-end learning of geometrically meaningful segmentations. This addresses the limitations of conventional CNN-based methods, which often lack explicit structural priors and may produce irregular or disconnected segmentations. Applied to brain tumour segmentation using MRI data, FMECNN demonstrated competitive performance compared to architectures like U-Net and ResNet-U-Net, while enforcing geometric consistency.

However, FMECNN faces challenges that limit its applicability:

- Seed Point Prediction: The method relies on accurately predicting seed points from which geodesic distances are computed. Identifying optimal seed points in complex or non-convex regions is difficult, and inaccuracies can degrade segmentation performance.
- Isotropic Metric Assumption: Assuming isotropic diffusion restricts effectiveness in images with anisotropic properties, such as tubular structures (e.g., blood vessels), where diffusion varies with direction.
- Computational Overhead: Incorporating the fast marching algorithm introduces computational overhead, especially when dealing with multiple seed points or high-resolution images, hindering scalability and efficiency.

These challenges highlight the need for methods that handle anisotropic media, eliminate dependency on explicit seed point selection, and maintain computational efficiency.

Figure 1 illustrates FMECNN applied to vascular tree segmentation using predefined seed points. The ground truth shows an intricate network of vessels, including small branches. While the model successfully segments larger vessels, it misses many small vessels and fine branches. The learned potential captures primary vascular pathways but struggles with detailed branching patterns. This indicates that FMECNN, in its current form, is not well-suited for tasks requiring the segmentation of small and intricate vascular structures.

Relying on predefined seed points and isotropic metrics poses significant challenges in complex, anisotropic vascular networks. The isotropic assumption fails to account for the directional diffusion inherent in blood vessels, leading to inadequate segmentation of smaller vessels. Manually selecting seed points is impractical and may not cover all regions of interest. These limitations underscore the need for advanced methods that can learn anisotropic metrics directly from data and eliminate the need for explicit seed point selection. Our proposed methods aim to address these challenges by incorporating anisotropic diffusion tensors and seedless segmentation approaches.

2.2 Isotropic Heat Diffusion

The heat equation is a partial differential equation (PDE) that describes heat propagation in a medium over time. In the isotropic case, where the thermal diffusivity is constant and the same in all directions, the heat equation in $\mathbb{R}^2$ is given by:

$$\frac{\partial u}{\partial t} = \alpha \Delta u, \tag{1}$$

where $u(x,t)$ represents the temperature at position $x \in \mathbb{R}^2$ and time $t \geq 0, \alpha > 0$ is the thermal diffusivity coefficient, and Δ is the Laplace operator.

The Laplace operator in two dimensions is:

$$\Delta u = \frac{\partial^2 u}{\partial x^2} + \frac{\partial^2 u}{\partial y^2}. \tag{2}$$

In homogeneous media, the thermal diffusivity α is constant, implying uniform heat propagation in all directions.

2.3 Spatially Varying Thermal Diffusivity

In heterogeneous media, the thermal diffusivity α can vary spatially. Following Yang et al. [11], we define α as:

$$\alpha(x) = |1 - |p(x_0) - p(x)||^d + \varepsilon, \tag{3}$$

where $p(x)$ represents a potential function (e.g., image intensity), x_0 is the heat source location, $d > 0$ controls the contrast between features, and ε ensures positivity.

Varying α implies that the rate of heat diffusion varies spatially within the domain, allowing for modelling complex propagation scenarios, such as barriers or preferred diffusion paths.

2.4 Anisotropic Heat Diffusion

To model anisotropic diffusion, we introduce the diffusion tensor $D(x)$, leading to the anisotropic heat equation:

$$\frac{\partial u}{\partial t} = \nabla \cdot (D(x)\nabla u), \tag{4}$$

where $D(x)$ is a symmetric positive-definite matrix at each point x. This allows us to control the diffusion rate in different directions.

By constructing $D(x)$ using features derived from the image, we can guide the heat propagation along preferred directions, such as the principal orientations of tubular structures.

2.5 Structure Tensor Field

The structure tensor field $T(x)$ captures local image orientations and anisotropies. It is defined as:

$$T(x) = \begin{pmatrix} \left(\frac{\partial I}{\partial x}\right)^2 & -\frac{\partial I}{\partial x}\frac{\partial I}{\partial y} \\ -\frac{\partial I}{\partial y}\frac{\partial I}{\partial x} & \left(\frac{\partial I}{\partial y}\right)^2 \end{pmatrix} * G_\sigma, \tag{5}$$

where $I(x)$ is the image intensity function, G_σ is a Gaussian kernel with standard deviation σ, and $*$ denotes convolution.

The structure tensor is diagonalised to obtain its eigenvalues $\mu_1(x) \geq \mu_2(x)$ and corresponding eigenvectors $e_1(x)$ and $e_2(x)$, representing the principal orientations and magnitudes of local variations:

$$T(x) = \mu_1(x)e_1(x)e_1(x)^\top + \mu_2(x)e_2(x)e_2(x)^\top. \tag{6}$$

This decomposition allows us to construct the tensor $T(x)$ to model anisotropic diffusion aligned with the image features.

2.6 Varadhan's Asymptotic Formula

Varadhan [10] established a relationship between the behaviour of the heat kernel and the geodesic distance on a Riemannian manifold. Specifically, as $t \to 0$, the heat kernel satisfies:

$$\lim_{t \to 0} \left(-4t \log u(x, y, t)\right) = d_\phi^2(x, y), \tag{7}$$

where $u(x, y, t)$ is the heat kernel, and $d_\phi(x, y)$ is the geodesic distance between points x and y with respect to the metric defined by ϕ.

This formulation allows us to compute anisotropic geodesic distances by solving the heat equation and analysing the behaviour of the solution as $t \to 0$.

3 Methodology

This section presents two methods for computing anisotropic geodesic distances using the heat equation integrated within deep learning frameworks for image segmentation. The first method extends prior work by incorporating seed points and predicting the anisotropic metric and the initial potential field. The second method addresses the challenges of seed point selection by introducing a seedless approach that leverages an attention mechanism within the neural network.

3.1 First Method: Seed Point-Based Anisotropic Geodesic Distance Computation

Overview. Our first method builds upon the framework introduced in the isotropic case, merging deep learning with geodesic distance computation for segmentation tasks. We extend this approach to handle anisotropic geodesic distances by incorporating the heat method based on Varadhan's formulation [10,11]. The key idea is to predict the anisotropic metric tensor and a Gaussian potential that initiates heat propagation from predefined seed points.

Deep Learning Model. We employ a convolutional neural network (CNN) based on the U-Net architecture, renowned for its effectiveness in medical image segmentation. The network is designed to predict:

- Anisotropic Metric $\phi(x)$: Ensuring positivity and non-zero values by defining $\phi(x) = f_\theta(u)^2 + \varepsilon$, where $f_\theta(u)$ is the network output for input image u, θ represents the network parameters, and ε is a small constant to maintain numerical stability.
- Gaussian Potential Field: Serving as the source for heat propagation, the network predicts a Gaussian potential field that determines the initial conditions for the diffusion process.

The model comprises convolutional layers with decreasing filter sizes, followed by batch normalisation and ReLU activation functions. The final layer uses a softmax activation to produce a probability distribution, ensuring that the output values sum to one and can be interpreted as a potential field for initiating heat propagation.

Anisotropic Geodesic Distance Computation. Using the predicted anisotropic metric $\phi(x)$ and the Gaussian potential field, we construct a structure tensor $D(x)$ that models the directional heat flow properties of the medium. The structure tensor is crucial for accurately simulating heat diffusion in media where thermal properties, such as tubular structures, vary with direction. We use the formulation of (6) to construct the structure tensor. We solve the anisotropic heat equation:

$$\frac{\partial u}{\partial t} = \nabla \cdot (D(x)\nabla u), \tag{8}$$

With the initial condition provided by the Gaussian potential field. As $t \to 0$, Varadhan's formulation allows us (see [11]) to compute the geodesic distance map:

$$d_\phi(x_0, x) = \lim_{t \to 0} \sqrt{-4t \log u(x, t)}, \tag{9}$$

where x_0 are the seed points from which heat propagation begins.

Segmentation with Multiple Seed Points. We accommodate multiple seed points $S = \{x_0^i\}_{1 \leq i \leq q}$, q the number of seed points, for complex tubular networks with branching paths or intersections. The geodesic distance to the closest seed point is computed as:

$$d_\phi(S, x) = \min_{x_0^i \in S} d_\phi(x_0^i, x), \forall i \in \{1, \cdots, q\}. \tag{10}$$

This computation efficiently extends the model's applicability to tubular networks of varied complexity without imposing additional computational burdens on the heat method.

Mask Generation. To generate the segmentation mask, we use a differentiable approximation of the characteristic function of a geodesic ball:

$$\chi^\delta(d_\phi(S, x)) = 1 - \frac{1}{1 + \exp\left(-\frac{d_\phi(S,x)-1}{\delta}\right)}, \tag{11}$$

where δ is a small positive parameter related to the image resolution, controlling the smoothness of the transition. This approach allows for gradient-based optimisation during training, facilitating the integration of the geodesic distance computation within the neural network framework.

Advantages and Limitations. This method effectively leverages seed points to guide the segmentation process, capturing the complex pathways within tubular structures by accurately modelling the anisotropic properties of the medium. However, selecting appropriate seed points can be challenging in practice, especially in non-convex regions, and may introduce computational overhead when dealing with multiple seeds.

3.2 Second Method: Seedless Anisotropic Geodesic Distance Computation

Motivation. While the first method relies on predefined seed points, this requirement poses challenges in non-convex regions like vascular networks, where identifying a unique optimal seed point is difficult. Additionally, computing geodesic distances from multiple seed points can be computationally expensive. To address these issues, we propose a second method that avoids explicit seed point selection by modifying the initialisation of the heat flow.

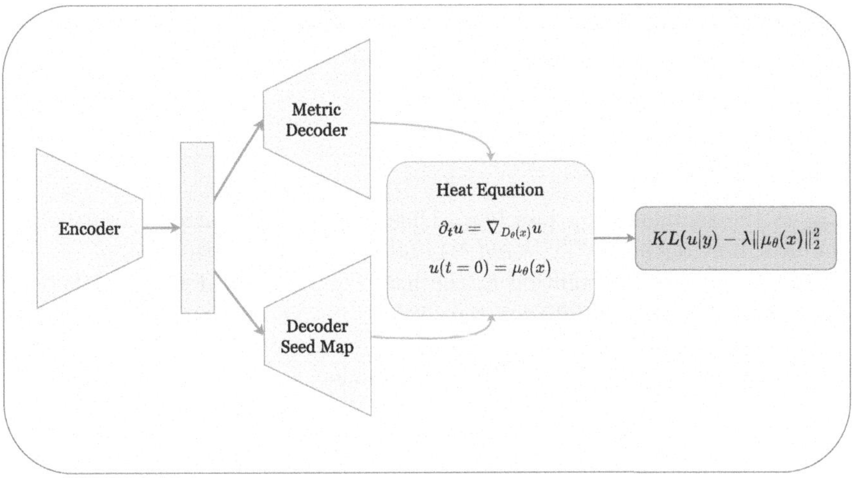

Fig. 1. Architecture of the alternative approach for anisotropic tubular structure segmentation. The network comprises two branches: predicting the diffusion tensor $D(x)$ and predicting the probability map $\mu(x)$.

Deep Learning Model. Our alternative approach employs a neural network composed of two branches, as illustrated in Fig. 1:

1. Metric Prediction Branch: Predicts the anisotropic diffusion tensor $D(x)$, which defines the heat flow and is related to the inverse of the classic metric tensor.
2. Probability Map Prediction Branch: Predicts a 2D probability map $\mu(x)$, indicating likely vascular landmarks or regions of interest that guide the heat flow and refine the segmentation.

This design allows the network to learn the medium's anisotropic properties and the areas from which heat should propagate without specifying exact seed points.

Anisotropic Heat Flow without Seed Points. We initialise the heat equation with the predicted probability map $\mu(x)$ instead of a Dirac delta function at seed points:

$$\frac{\partial u}{\partial t} = \nabla \cdot (D(x)\nabla u), \quad u(x,0) = \mu(x). \tag{12}$$

This formulation enables the heat to propagate from high-probability regions, effectively serving as an attention mechanism focusing on areas of interest within the image. The heat flow $u(x,t)$ captures the influence of the anisotropic diffusion and the initial probability distribution $\mu(x)$.

Loss Function and Training. Given input images x and their corresponding normalised ground truth masks y, we define the loss function as:

$$\mathcal{L} = \mathcal{L}_{\text{seg}}(\Phi_t^{D_{\theta(x)}}(\mu_\theta(x)), y) - \lambda\|\mu_\theta(x)\|_2^2, \tag{13}$$

where:

- $\mathcal{L}_{\text{seg}}$ is the segmentation loss (e.g., Dice loss or cross-entropy loss) between the heat flow output $\Phi^{D\theta(x)}t(\mu\theta(x))$ and the ground truth mask y.
- $\Phi_t^D(\mu)$ represents the solution to the heat equation (heat flow) applied to μ until time t, using the diffusion tensor D.
- λ is a regularisation parameter promoting sparsity in $\mu(x)$, encouraging the network to focus on the most relevant regions.

The sparsity-promoting penalty $\|\mu_\theta(x)\|_2^2$ prevents the probability map from spreading too diffusely across the image, which helps in refining the segmentation by concentrating the heat propagation in areas of interest.

Advantages and Limitations. This seedless approach simplifies the segmentation pipeline by eliminating the need for explicit seed point selection, reducing computational complexity. Using a probability map as an initial condition allows the network to learn where to focus the heat propagation, effectively functioning as an attention mechanism. However, this method relies on the network's ability to predict accurate probability maps, and the interpretation of the probability map may differ from explicit seed point methods.

3.3 Comparison of Methods

The two methods present distinct strategies for integrating anisotropic geodesic distance computation into Deep Learning models for image segmentation. The first method, a seed point-based approach, directly incorporates seed points, granting explicit control over the initiation of heat propagation. It efficiently handles multiple seed points for complex structures and leverages the predicted anisotropic metric and initial potential to capture the geometry of tubular formations. However, this method requires selecting or predicting appropriate seed points, which can be challenging in practice and may introduce computational overhead when numerous seed points are involved.

In contrast, the second method employs a seedless approach using an attention mechanism. It eliminates the need for explicit seed point selection, thereby simplifying the segmentation process. Using a probability map to guide heat propagation effectively focuses on regions of interest and reduces the computational complexity associated with handling multiple seed points. Nonetheless, this method relies on the network's ability to accurately predict the probability map $\mu(x)$. The initial probability distribution may not precisely represent discrete seed locations, which can potentially affect segmentation accuracy in some cases.

Both methods effectively integrate anisotropic diffusion within a deep learning framework to enhance the segmentation of complex structures like vascular networks. The choice between these approaches depends on the application's specific requirements and the data's nature. The first method offers explicit control over heat propagation when seed point selection is feasible and advantageous. Conversely, the second method provides a flexible and computationally efficient alternative in scenarios where seed point selection is impractical.

4 Experiments

4.1 Data

We conducted experiments on synthetic datasets and real-world medical images.

- Synthetic Tree Structures: Designed to test the model's capability in detecting arboreal structures. The dataset includes 20 images for training and 20 for testing, with key points represented by Gaussian heatmaps.
- DRIVE Dataset [9]: This dataset comprises 40 colour retinal images with corresponding manual vessel annotations provided by experts. We used only the green channel of the images, as it offers the highest contrast for blood vessels.

The model was trained using the Adam optimiser with a learning rate 0.01. Data augmentation techniques were applied, including horizontal and vertical flipping, random rotations, Geometric transformations (shifting and scaling).

These augmentations expose the model to various structural orientations and scales, improving generalisation.

4.2 Results

Our methods effectively segmented tubular structures, accurately capturing complex geometries and anisotropic properties.

Isotropic vs. Anisotropic Heat Propagation. Figure 2 illustrates the difference between isotropic and anisotropic heat propagation across a fingerprint image. The anisotropic diffusion tensors guide the heat along the fingerprint ridges, adapting to the local geometry.

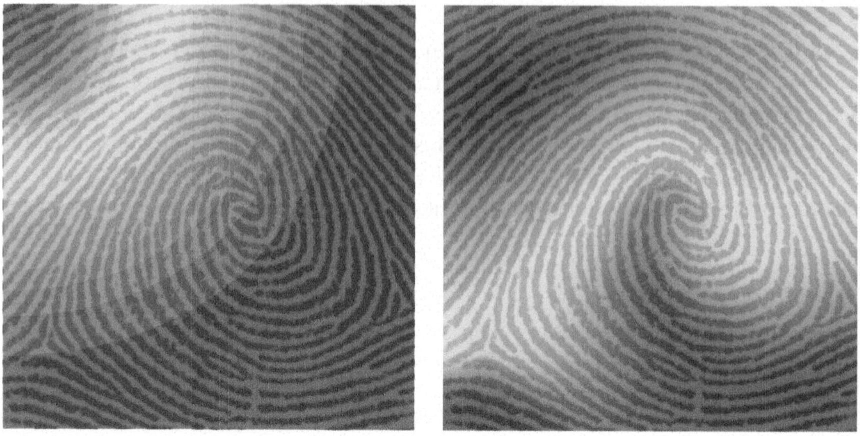

(a) Isotropic heat propagation. (b) Anisotropic heat propagation.

Fig. 2. Comparison of isotropic and anisotropic heat propagation across a fingerprint image.

Segmentation Results. Figure 3 demonstrates the results of anisotropic tree structure analysis. In the first image (a), the output of the anisotropic model on the tree structure is shown, illustrating the heat propagation pattern. In (b), the potential field generated from the anisotropic diffusion is visualised, highlighting how the heat flows through the tree branches. Image (c) shows the modulation of the geodesic distance with the coefficient α, the thermal diffusion, affecting the diffusion process based on local variations. Finally, in (d), the predicted segmentation or outline of the tree structure is displayed, highlighting the model's ability to capture the geometric features of the tree.

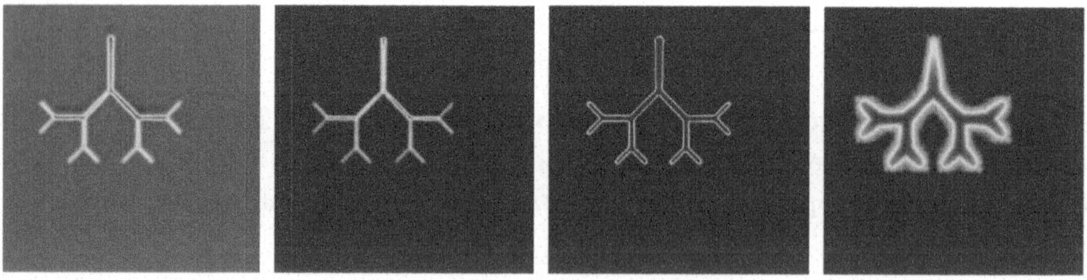

(a) Output of the proposed model. (b) Predicted potential field. (c) Modulated geodesic distance. (d) Predicted segmentation.

Fig. 3. Anisotropic segmentation results for a tree structure.

4.3 Second Model Results

In Fig. 4, the barycenter predicted by the Seed decoder branch of our model does not coincide with the barycenter maps produced by the original FMECNN model. Instead, it predicts segmentations of the vascular network. This divergence occurs because the second branch of our network is not explicitly trained to predict seed points, and the sparsity-promoting penalty in the loss function does not collapse the barycenter to a set of discrete points, even for larger values of λ. This behaviour can be interpreted as an attention mechanism, where the solution to the heat equation, computed from the two network outputs, refines the segmentation map.

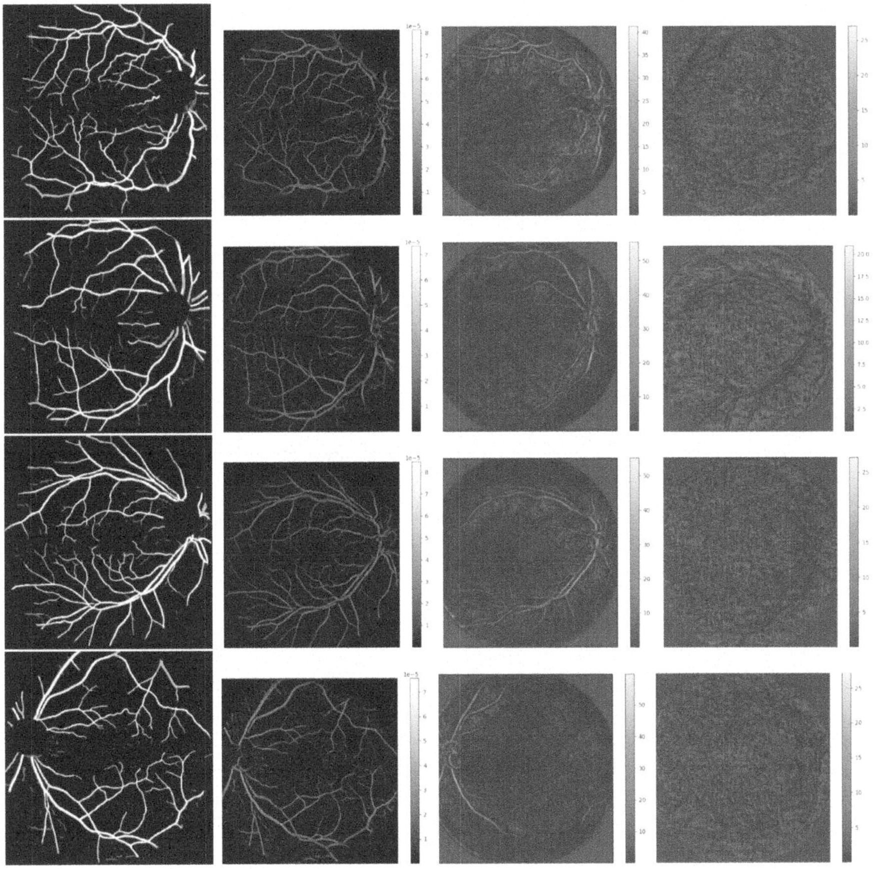

Fig. 4. Output of our method on a sample from the IOSTAR dataset. Left: Comparison of proposed segmentation versus Ground Truth. Center Left: Barycenter map output by the network. Center Right: Sum of the metric elements in both directions. Right: (log of) Anisotropy factor.

5 Conclusion

We introduced two methods for integrating anisotropic geodesic distance computation into deep learning models for image segmentation of complex structures like vascular networks. The first method employs seed points to control heat propagation, capturing intricate tubular geometries but requiring accurate seed placement. The second method is seedless, using a probability map to guide heat propagation, simplifying the segmentation process but relying on the network's ability to predict this map accurately. Both methods effectively segment complex, anisotropic structures without manual seed selection. Our experiments demonstrate that learning anisotropic metrics directly from data and integrating geodesic computation into deep learning models advance image segmentation techniques for complex, directionally dependent structures.

Acknowledgments. This work is in part supported by the French government under management of Agence Nationale de la Recherche as part of the "Investissements d'avenir" program, reference ANR-19-P3IA-0001 (PRAIRIE 3IA Institute).

Disclosure of Interests. The authors have no competing interests to declare that they are relevant to the content of this article.

References

1. Benmansour, F., Cohen, L.D.: Tubular structure segmentation based on minimal path method and anisotropic enhancement. Int. J. Comput. Vision **92**(2), 192–210 (2011)
2. Bertrand, T., Makaroff, N., Cohen, L.D.: Fast marching energy CNN. In: International Conference on Scale Space and Variational Methods in Computer Vision (SSVM) 2023, pp. 276–287. Springer, Cham (2023). https://doi.org/10.1007/978-3-031-31975-4
3. Cohen, L.D., Deschamps, T.: Segmentation of 3D tubular objects with adaptive front propagation and minimal tree extraction for 3D medical imaging. Comput. Methods Biomech. Biomed. Eng. **10**(4), 289–305 (2007)
4. Crane, K., Weischedel, C., Wardetzky, M.: Geodesics in heat: a new approach to computing distance based on heat flow. ACM Trans. Graph. (TOG) **32**(5), 1–11 (2013)
5. Heitz, M., Bonneel, N., Coeurjolly, D., Cuturi, M., Peyré, G.: Ground metric learning on graphs. J. Math. Imag. Vis. **63**, 89–107 (2021)
6. Malladi, R., Sethian, J.A.: A real-time algorithm for medical shape recovery. In: Sixth International Conference on Computer Vision (IEEE Cat. No. 98CH36271), pp. 304–310. IEEE (1998)
7. Scarvelis, C., Solomon, J.: Riemannian metric learning via optimal transport. arXiv Preprint arXiv:2205.09244 (2022)
8. Sethian, J.A.: A fast marching level set method for monotonically advancing fronts. Proc. Natl. Acad. Sci. **93**(4), 1591–1595 (1996)
9. Staal, J., Abràmoff, M.D., Niemeijer, M., Viergever, M.A., Van Ginneken, B.: Ridge-based vessel segmentation in color images of the retina. IEEE Trans. Med. Imag. **23**(4), 501–509 (2004)

10. Varadhan, S.: On the behavior of the fundamental solution of the heat equation with variable coefficients. Commun. Pure Appl. Math. **20**(2), 431–455 (1967)
11. Yang, F., Cohen, L.D.: Geodesic distance and curves through isotropic and anisotropic heat equations on images and surfaces. J. Math. Imag. Vis. **55**, 210–228 (2016)

Deceptive Diffusion: Generating Synthetic Adversarial Examples

Lucas Beerens[1,2], Catherine F. Higham[3], and Desmond J. Higham[1,2(✉)]

[1] School of Mathematics, University of Edinburgh, Edinburgh EH9 3FD, UK
d.j.higham@ed.ac.uk
[2] Maxwell Institute, University of Edinburgh, Edinburgh EH8 9BT, UK
[3] School of Computing Science, University of Glasgow, Glasgow G12 8QQ, UK

Abstract. We introduce the concept of deceptive diffusion—training a generative AI model to produce adversarial images. Whereas a traditional adversarial attack algorithm aims to perturb an existing image to induce a misclassificaton, the deceptive diffusion model can create an arbitrary number of new, misclassified images that are not directly associated with training or test images. Deceptive diffusion offers the possibility of strengthening defence algorithms by providing adversarial training data at scale, including types of misclassification that are otherwise difficult to find. In our experiments, we also investigate the effect of training on a partially attacked data set. This highlights a new type of vulnerability for generative diffusion models: if an attacker is able to stealthily poison a portion of the training data, then the resulting diffusion model will generate a similar proportion of misleading outputs.

Keywords: Image classification · Generative AI · Stability

1 Motivation

In this work, we combine two types of algorithm that have come to prominence in artificial intelligence (AI): adversarial and generative. Adversarial attack algorithms are designed to reveal vulnerabilities in classification systems; for example by perturbing a chosen image in a way that is imperceptible to the human eye, but causes a change in classification [13,31]. Generative models are designed to create outputs that are similar to, but not simply copies of, the examples on which they were trained [8,16]. Here, we show that by training on data that consists of adversarially perturbed images, a generative diffusion model can be made to create fresh examples of adversarial images that do not correspond directly to any underlying real images.

In Sect. 2 we give some background information on the two main ingredients of our work: adversarial attack algorithms and generative diffusion models. Section 3 describes the results of computational experiments where we investigate the idea of training a diffusion model on adversarially-perturbed data. We finish with a brief discussion in Sect. 4.

T. A. Bubba et al. (Eds.): SSVM 2025, LNCS 15667, pp. 322–333, 2025.
https://doi.org/10.1007/978-3-031-92366-1_25

1.1 Related Work

We refer to [5] for an overview of recent attempts to use generative AI tools to produce adversarial inputs. The AdvDiffuser algorithm of [5] appears to be the first and only approach to generating new, synthesized, examples of adversarial images using a diffusion model. In that work, the authors take an existing, trained diffusion model and adapt the denoising, or backward, process by adding adversarial perturbations at each time step. This change increases computational complexity, since an extra gradient step is required at each time point. Our approach differs by building a new diffusion model, which then generates images with a standard de-noising algorithm. In addition to lowering the computational cost, our *deceptive diffusion* method reveals a new type of security threat that arises when standard generative diffusion models are created on training data that has been attacked. In particular, we find that the drop in classification success is in direct proportion to the fraction of training data that is adversarially perturbed. Hence, if an attacker is able to poison some portion of the training data, the builders of a generative diffusion model may inadvertently create a tool that produces a corresponding proportion of adversarial images.

In the conceptually different, and more traditional, setting of computing an adversarial perturbation to an existing image, we mention that the DiffAttack algorithm [4] also makes use of a diffusion model. We also note that the earlier work [12], which is not concerned with generative AI, showed that adversarial examples can be effective for data poisoning.

2 Background

2.1 Adversarial Attack Algorithms

State of the art image classification tools are known to possess inherent vulnerabilites. In particular, they can be fooled by adversarial attacks, where an existing image undergoes a small perturbation that would not be noticeable to a human, but causes a change in the predicted class. Since this effect was first pointed out, [13,31], a wide range of attack and defence strategies have been put forward, [1,2,24,25], and bigger picture questions concerning the inevitability of attack success have been investigated, [7,11,28,29,32,33]. The susceptibility of AI systems to attack is a serious issue in many application areas and it is pertinent to the recent calls for AI regulation. For example, the amendment of June 2023 [10] to Article 15 - paragraph 4 - subparagraph 1 of the EU AI act [9] requires that: *"High-risk AI systems shall be resilient as regards to attempts by unauthorised third parties to alter their use, behaviour, outputs or performance by exploiting the system vulnerabilities."*

2.2 Generative Diffusion Models

A generative diffusion model for creating realistic, but synthetic, images can be built by first training a neural network to de-noise a collection of noisy images, and then asking the network to de-noise a new sample of pure noise [3].

In Algorithms 1 and 2 we summarize the basic unconditional diffusion model setting from [16]; see also [15,23] for detailed explanations of the steps involved. Here, the α_t are parameters taking values between zero and one. They have the form $\alpha_t = 1 - \beta_t$, where the predetermined sequence $\beta_1, \beta_2, \ldots, \beta_T$ is known as the *variance schedule*. In [16], linearly increasing values from $\beta_1 = 10^{-4}$ to $\beta_T = 0.02$ are used. We also let $\overline{\alpha}_t = \prod_{i=1}^{t} \alpha_i$, and $\sigma_q^2(t) = (1-\alpha_t)(1-\overline{\alpha}_{t-1})/(1-\overline{\alpha}_t)$.

In step 5 of Algorithm 1, $\boldsymbol{\epsilon}_\theta$ denotes the output from a neural network. Given a version of the noisy image, $\sqrt{\overline{\alpha}_t}\, \mathbf{x}_0 + \sqrt{1-\overline{\alpha}_t}\, \boldsymbol{\epsilon}$, corresponding to a time t, the job of the network is to predict the noise $\boldsymbol{\epsilon}$. Here, a simple least-squares loss function is used.

Algorithm 1. Training with the forward process [16]

```
1: repeat
2:     x_0 ~ q(x_0)                                  ▷ choose an image from training set
3:     t ~ Uniform({1, 2, ..., T})
4:     ε ~ N(0, I)                                   ▷ standard Gaussian sample
5:     Take gradient step w.r.t. θ on ||ε − ε_θ(√ᾱ_t x_0 + √(1 − ᾱ_t) ε, t)||_2^2
6: until converged
```

Algorithm 2 from [16] summarizes the sampling process. Here, a set of pure noise pixel values is de-noised from time T to time 0 in order to produce a new synthetic image.

Algorithm 2. Sampling with the backward process [16]

```
1: x_T ~ N(0, I)                                     ▷ standard Gaussian sample
2: for t = T, T − 1, ..., 1 do
3:     z ~ N(0, I)                                   ▷ standard Gaussian sample
4:     x_{t−1} = (1/√α_t) (x_t − ((1 − α_t)/√(1 − ᾱ_t)) ε_θ) + σ_q(t) z
5: end for
6: return x_0
```

3 Experimental Results

We now outline the key components in our computational experiments.

We use the MNIST data set [21], which contains 60,000 training images and 10,000 test images of handwritten digits, with labels indicating the categories: '0', '1','2',...,'9'.

As a classifier, we use a convolutional neural network (CNN) based on the architecture of LeNet [19,20]. The exact architecture can be found in our code. After training, this classifier achieves an accuracy of 99.02% on the test images.

For the adversarial attack algorithm we use PGDL2 [17], a PyTorch implementation of the projected gradient method from [24]. This attack algorithm uses a robust optimization approach to seek an optimal perturbation in an ℓ_2 sense, using gradients of the loss function. We use the default setting in PGDL2 where an attack is declared successful if it finds a sufficiently small class-changing perturbation within a specified number of iterations of a first order gradient method. The bound on the ℓ_2 norm of the attack was set to 2 (each of the 784 pixels takes values between 0 and 1). We chose a large bound of 1000 on the number of iterations in order to maximize the size of the attacked image dataset for training the diffusion model. We used PGDL2 in untargeted mode, so that any change of classification is acceptable.

In the diffusion model, we used a neural network with a UNet2DModel architecture from https://huggingface.co/docs/diffusers/en/api/models/unet2d which is motivated by the original version in [27].

3.1 Initial Sanity Check

Before moving on to adversarial images, we first report on an initial test which confirms that the diffusion model is capable of producing outputs that are acceptable to the classifier.

In this test, we train the diffusion model using the original MNIST training data. We supply the labels during the training process, so we use a conditional version of Algorithm 1, where in step 5 the network learns to remove noise and produce an image when given both a time t and a label. This is built in to the UNet2DModel. A trainable encoder maps the label into the same space as the timestep. These two quantities are then added and passed to the model in the same way that the time is usually passed [26].

We found that 99.5% of the outputs from the trained model were classified with the intended label.

3.2 Deceptive Diffusion Model

Our aim is now to build a *deceptive diffusion model* that takes a label i and generates a new image that looks like digit i but is misclassified.

Using PGDL2 for untargeted attacks on the 60,000 MNIST training images gave a success rate of 86.5%, thereby producing 51,918 perturbed images that are classified differently to their nearby original images. We trained the diffusion model on these adversarial images, using the original labels. Figure 1 illustrates the process. Here, the image of the three on the left is from the MNIST training set, and the image in the middle arises from a successful attack by PGDL2 (classified as an eight). After training the diffusion model on all 51,918 adversarial images, asking for an output from the '3' category produced the result shown (classified as a five).

After using the trained diffusion model to generate 100 new images from each of the ten categories and passing these through the CNN classifier, we found that

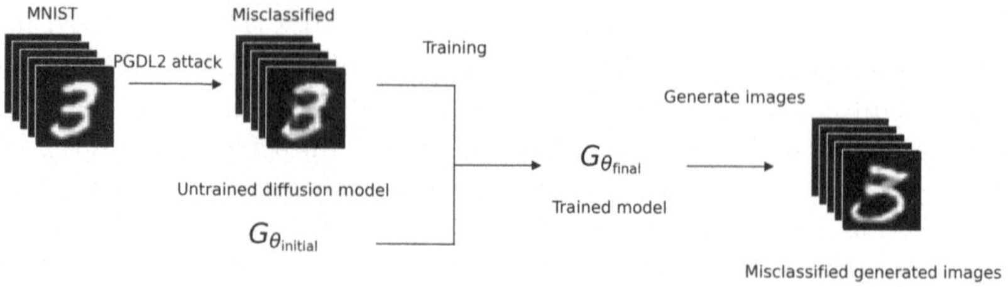

Fig. 1. Building the deceptive diffusion model. Images that were successfully attacked by PGDL2 are used as training data, with the original labels retained. The trained diffusion model, $G_{\theta_{\text{final}}}$, produces adversarial images associated with a given a label. (For the images in this diagram, the image from PGDL2 is classified as an '8' and the image from the deceptive diffusion model, which was supplied with the label '3', is classified as a '5'.)

93.6% of the outputs were classified differently to their requested labels. Figure 2 gives a confusion matrix showing the performance by category. For comparison, Fig. 3 shows a confusion matrix for the PGDL2 attacks on the 60,000 training images.

Table 1 shows the correlation between the rows of the confusion matrices in Figs. 2 and 3. The high correlation values indicate that the two confusion matrices are similar. We emphasize that PGDL2 was used in untargeted mode: an image from category i can be perturbed so that the classifier predicts any new category $j \neq i$. From Table 1 we see that although the deceptive diffusion model was not provided with a target class j, it tends to produce new $i \mapsto j$ misclassifications of the same type as PGDL2.

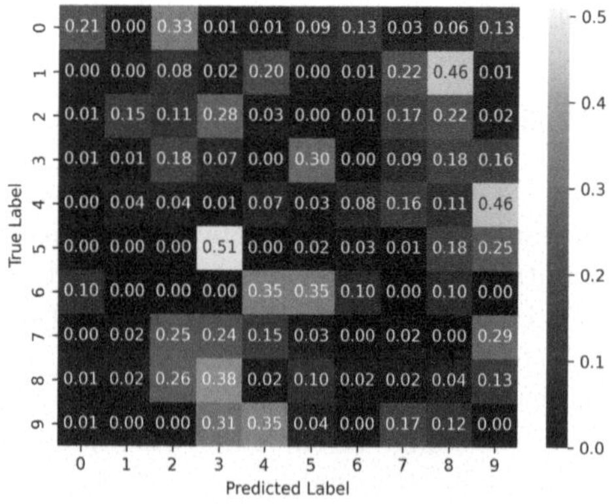

Fig. 2. Confusion matrix for the deceptive diffusion model. For a given label (row) we show the frequency with which the classifier assigned each label (column) to the output. Entries on the diagonal therefore correspond to unsuccessful attempts to create an adversarial image. Overall misclassification rate is 93.6%.

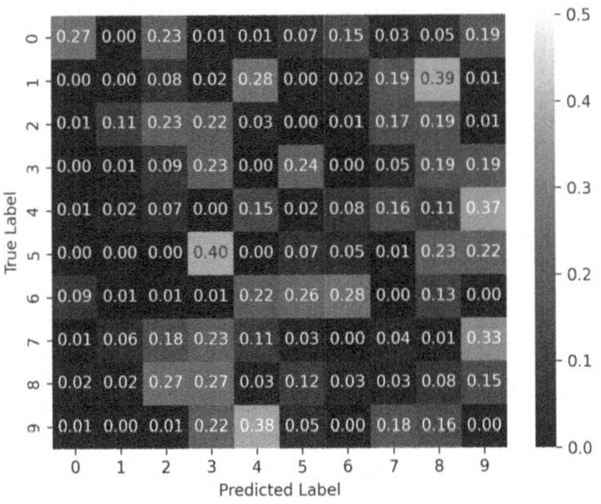

Fig. 3. Confusion matrix for PGDL2 attacks on the 60,000 MNIST training images. With training images corresponding to each label (row) we show the frequency with which the classifier assigned each label (column) after the attack. Entries on the diagonal therefore correspond to unsuccessful attacks. Overall success rate is 86.5%.

Table 1. Correlation of confusion matrix rows for PGDL2 attack and generated data.

Class	0	1	2	3	4	5	6	7	8	9
Correlation	0.90	0.97	0.88	0.79	0.96	0.98	0.82	0.96	0.96	0.96

To give a feel for the outputs from the deceptive diffusion model, Fig. 4 (upper) shows 100 independent outputs corresponding to the label '9'. We note from Fig. 2 that 0% of such outputs are classified as nines. Hence, we see that the model is capable of producing convincing adversarial images. For comparison, Fig. 4 (lower) shows the results of PGDL2 on images from the '9' category.

Partial Attacks. So far, we have looked at two options for the training data. Either all training data was attacked, or all training data was clean. Now we look at a third case: partially attacked training data. Again we choose the same MNIST images that were successfully attacked using PGDL2. Consider $p \in \{0, 20, 40, 60, 80, 100\}$. For each class, we replace $p\%$ of the clean images with their successfully attacked counterpart. Now using these six datasets, we train six models.

For each class, 100 images are generated using each of the trained models. In Fig. 5 we show the resulting accuracy of the classifier on these generated images for the models trained on varying levels of poisoned data. We see that the classification accuracy degrades roughly in proportion with the amount of poisoned training data. This result is intuitively reasonable, under the assumption that all training images carry equal weight when the diffusion model is created.

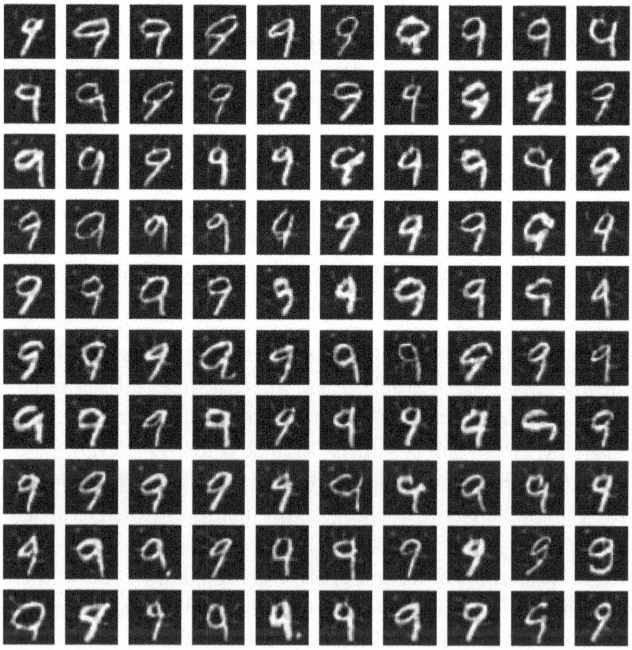

Fig. 4. Upper: example of 100 images arising when the deceptive diffusion model was given the label '9'. Lower: example of 100 images arising from successful PGDL2 attacks on images that had label '9'.

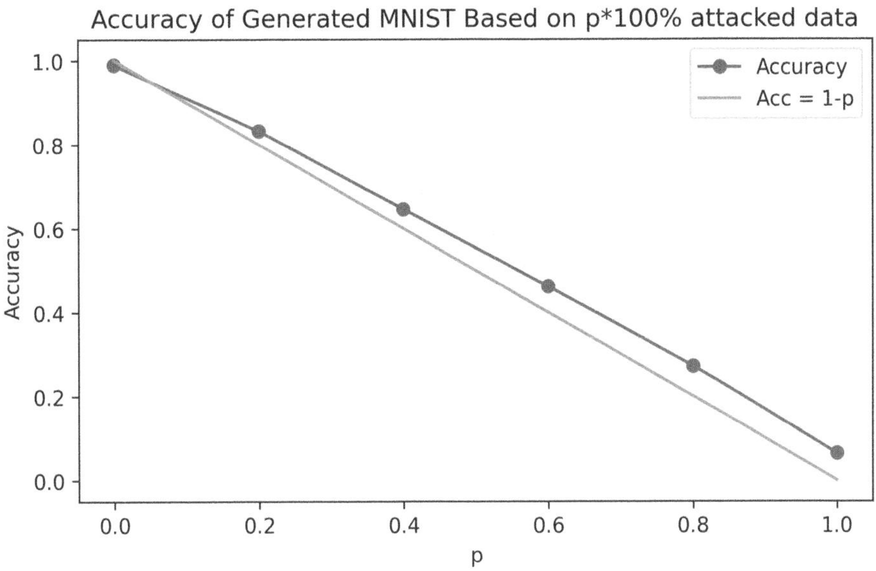

Fig. 5. Classification accuracy (vertical axis) for output from a deceptive diffusion model where a fraction of the training data (horizontal axis) is replaced by its adversarially attacked counterpart. The slope representing linear proportionality is also shown.

Fréchet Inception Distance. A widely used measure for generated image quality is the Fréchet Inception Distance (FID) [14], where lower is better. It compares a generated dataset to a ground truth dataset. First, a classifier is used to extract features. Then the Fréchet distance between these feature sets is computed. Typically the Inception v3 classifier [30] without its last layer is used. To take into account that the generator is conditioned on the class, we use the Class-Aware Fréchet Distance (CAFD), which computes the FID for every class and takes the average [22].

Since our dataset is of low resolution, instead of Inception v3 we use the classifier that we trained earlier, with its last layer removed. This way the output is in $\mathbb{R}^{128}$.

In Fig. 6, the CAFD is shown for the diffusion models trained with partially poisoned data. These values are compared with the CAFD for the test set and the PGDL2 attacked training set. These are displayed at $p = 0$ and $p = 1$ respectively, because they represent samples from the ground truths for the clean and attacked case respectively. To avoid bias, these two sets are limited to contain the same number of samples as the generated sets, [6].

The results in Fig. 6 show that the CAFD increases monotonically as the level of poisoning increases. This seems reasonable, because, as shown in Fig. 5, higher levels of poisoning lead to higher levels of misclassification. The CAFD relies on the feature extraction of an MNIST classifier. Since the attacks target the classifier, it makes sense that the extracted features are different. The key observation here is that the fully adversarial model ($p = 1$) corresponds to a CAFD that is similar to that of the PGDL2 attacked data set, indicating that deception diffusion can mimic adversarially attacked data successfully according to this metric.

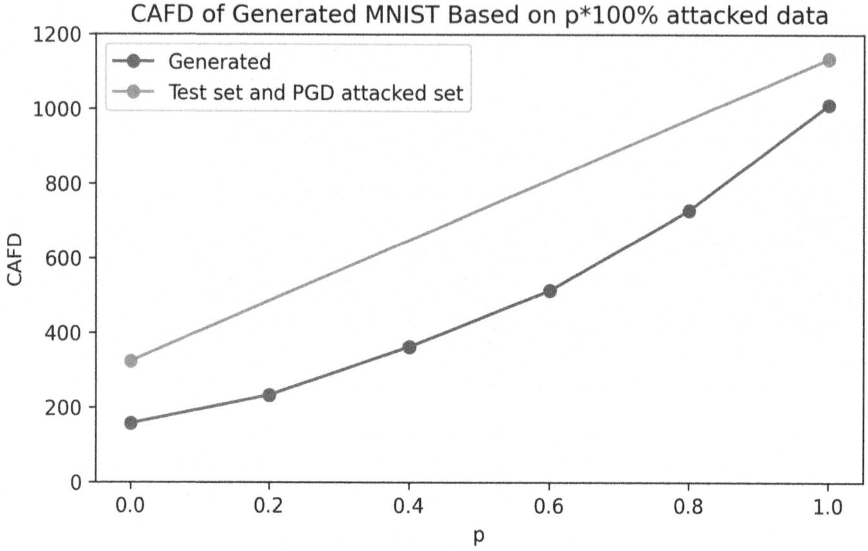

Fig. 6. Class-aware Fréchet Distance for a deceptive diffusion model where a percentage of the training data (horizontal axis) is replaced by its adversarially attacked counterpart. The ground truth dataset is MNIST. The straight line joins the CAFD for the test set at $p = 0$ and the PGDL2 attacked training set at $p = 1$. These two sets contain the same number of samples as the generated sets.

4 Conclusions

A traditional adversarial attack algorithm aims to perturb an existing image across a decision boundary. Instead, by training a generative diffusion model on adversarial data, we are able to create synthetic images that automatically lie on the wrong side of a decision boundary. This observation, which we believe to have been made for the first time in this work, reveals a new type of vulnerability for generative AI: if a diffusion model is inadvertently trained on fully or partially poisoned data then a tool may be produced that generates unlimited amounts of classifier-fooling examples.

In common with the AdvDiffuser algorithm in [5], when deliberately trained on adversarial data, a *deceptive diffusion* model has the potential to

- create effective adversarial images at scale, independently of the amount of training and test data available,

- create examples of misclassification that are difficult to obtain with a traditional adversarial attack; for example, in a healthcare setting when certain classes are underrepresented in the data [18].

This technique has applications for defence as well as attack, since it provides valuable new sources of data for adversarial training algorithms that aim to improve robustness.

There are many directions in which the deceptive diffusion idea could be pursued; notably, testing on other types of labeled image data, generating adversarial images that are successful across a range of independent classifiers, and finding computable signatures with which to identify this new type of threat.

Acknowledgments. LB is supported by the MAC-MIGS Centre for Doctoral Training under Engineering and Physical Sciences Research Council (EPSRC) grant EP/S023291/1. CFH received funding under EPSRC grants EP/T00097X/1, EP/R018634/1, and EP/T021020/1. DJH is supported by a fellowship from the Leverhulme Trust.

Disclosure of Interests. The authors declare no competing interests.

Data Statement. Code for these experiments is available from https://github.com/LucasBeerens/Deceptive_Diffusion.

References

1. Akhtar, N., Mian, A.: Threat of adversarial attacks on deep learning in computer vision: a survey. IEEE Access **6**, 14410–14430 (2018)
2. Beerens, L., Higham, D.J.: Adversarial ink: componentwise backward error attacks on deep learning. IMA J. Appl. Math. **89**, 175–196 (2024)
3. Cao, H., et al.: A survey on generative diffusion models. IEEE Trans. Knowl. Data Eng. **36**, 2814–2830 (2024)
4. Chen, J., Chen, H., Chen, K., Zhang, Y., Zou, Z., Shi, Z.: Diffusion models for imperceptible and transferable adversarial attack. IEEE Trans. Pattern Anal. Mach. Intell. **47**(2), 961–977 (2024)
5. Chen, X., Gao, X., Zhao, J., Ye, K., Xu, C.-Z.: AdvDiffuser: natural adversarial example synthesis with diffusion models. In: Proceedings of the IEEE/CVF International Conference on Computer Vision (ICCV), October 2023, pp. 4562–4572 (2023)
6. Chong, M.J., Forsyth, D.: Effectively unbiased FID and inception score and where to find them. In: Proceedings of the IEEE/CVF Conference on Computer Vision and Pattern Recognition, pp. 6070–6079 (2020)
7. Colbrook, M.J., Antun, V., Hansen, A.C.: The difficulty of computing stable and accurate neural networks: On the barriers of deep learning and Smales 18th problem. Proceedings of the National Academy of Sciences (2021)

8. Dhariwal, P., Nichol, A.: Diffusion models beat GANs on image synthesis. In: Ranzato, M., Beygelzimer, A., Dauphin, Y., Liang, P., Vaughan, J.W. (eds.) Advances in Neural Information Processing Systems, vol. 34, pp. 8780–8794. Curran Associates, Inc. (2021)
9. European Comission: Proposal for a regulation of the European parliament and of the council laying down harmonised rules on artificial intelligence (artificial intelligence act) and amending certain union legislative acts (2021)
10. European Parliament: Amendments adopted by the European parliament on 14 june 2023 on the proposal for a regulation of the european parliament and of the council on laying down harmonised rules on artificial intelligence (artificial intelligence act) and amending certain union legislative acts (2023)
11. Fawzi, A., Fawzi, O., Frossard, P.: Analysis of classifiers robustness to adversarial perturbations. Mach. Learn. **107**, 481–508 (2018)
12. Fowl, L., Goldblum, M., Chiang, P.-y., Geiping, J., Czaja, W., Goldstein, T.: Adversarial examples make strong poisons. In: Ranzato, M., Beygelzimer, A., Dauphin, Y., Liang, P., Vaughan, J.W. (eds.) Advances in Neural Information Processing Systems, vol. 34, pp. 30339–30351. Curran Associates, Inc. (2021)
13. Goodfellow, I.J., Shlens, J., Szegedy, C.: Explaining and harnessing adversarial examples. In: Bengio, Y., LeCun, Y. (eds.) 3rd International Conference on Learning Representations, San Diego, CA (2015)
14. Heusel, M., Ramsauer, H., Unterthiner, T., Nessler, B., Hochreiter, S.: GANs trained by a two time-scale update rule converge to a local Nash equilibrium. In: Guyon, I., von Luxburg, U., Bengio, S., Wallach, H.M., Fergus, R., Vishwanathan, S., Garnett, R. (eds.) Advances in Neural Information Processing Systems, Long Beach, CA, USA, pp. 6626–6637 (2017)
15. Higham, C.F., Higham, D.J., Grindrod, P.: Diffusion models for generative artificial intelligence: an introduction for applied mathematicians, SIAM Review (to appear)
16. Ho, J., Jain, A., Abbeel, P.: Denoising diffusion probabilistic models. In: Proceedings of the 34th International Conference on Neural Information Processing Systems, Red Hook, NY, USA. Curran Associates Inc. (2020)
17. Kim, H.: Torchattacks: a PyTorch repository for adversarial attacks. arXiv preprint arXiv:2010.01950 (2020)
18. Ktena, I., et al.: Generative models improve fairness of medical classifiers under distribution shifts. Nat. Med. **30**, 1166–1173 (2024)
19. LeCun, Y., et al.: Backpropagation applied to handwritten zip code recognition. Neural Comput. **1**, 541–551 (1989)
20. LeCun, Y., Bottou, L., Bengio, Y., Haffner, P.: Gradient-based learning applied to document recognition. Proc. IEEE **86**, 2278–2324 (1998)
21. LeCun, Y., Cortes, C., Burges, C.J.C.: The MNIST database of handwritten digits (2010)
22. Liu, S., Wei, Y., Lu, J., Zhou, J.: An improved evaluation framework for generative adversarial networks. arXiv preprint arXiv:1803.07474 (2018)
23. Luo, C.: Understanding diffusion models: a unified perspective. arXiv:2208.11970 (2022)
24. Madry, A., Makelov, A., Schmidt, L., Tsipras, D., Vladu, A.: Towards deep learning models resistant to adversarial attacks. In: 6th International Conference on Learning Representations, Vancouver, BC, OpenReview.net (2018)
25. Moosavi-Dezfooli, S., Fawzi, A., Frossard, P.: Deepfool: a simple and accurate method to fool deep neural networks. In: 2016 IEEE Conference on Computer Vision and Pattern Recognition, NV, USA, pp. 2574–2582. IEEE Computer Society (2016)

26. Rombach, R., Blattmann, A., Lorenz, D., Esser, P., Ommer, B.: High-resolution image synthesis with latent diffusion models. In: Proceedings of the IEEE/CVF Conference on Computer Vision and Pattern Recognition, pp. 10684–10695 (2022)
27. Ronneberger, O., Fischer, P., Brox, T.: U-Net: convolutional networks for biomedical image segmentation. In: Navab, N., Hornegger, J., Wells, W.M., Frangi, A.F. (eds.) MICCAI 2015. LNCS, vol. 9351, pp. 234–241. Springer, Cham (2015). https://doi.org/10.1007/978-3-319-24574-4_28
28. Shafahi, A., Huang, W., Studer, C., Feizi, S., Goldstein, T.: Are adversarial examples inevitable?. In: International Conference on Learning Representations, New Orleans, USA (2019)
29. Sutton, O.J., et al.: Stealth edits for provably fixing or attacking Large Language Models. In: Neural Information Processing Society (NeurIPS), Vancouver, Canada, December 2024
30. Szegedy, C., Vanhoucke, V., Ioffe, S., Shlens, J., Wojna, Z.: Rethinking the inception architecture for computer vision. In: Proceedings of the IEEE Conference on Computer Vision and Pattern Recognition, pp. 2818–2826 (2016)
31. Szegedy, C., et al.: Intriguing properties of neural networks. arXiv preprint arXiv:1312.6199 (2013)
32. Tyukin, I.Y., Higham, D.J., Bastounis, A., Woldegeorgis, E., Gorban, A.N.: The feasibility and inevitability of stealth attacks. IMA J. Appl. Math. **89**, 44–84 (2024)
33. Tyukin, I.Y., Higham, D.J., Gorban, A.N.: On adversarial examples and stealth attacks in artificial intelligence systems. In: 2020 International Joint Conference on Neural Networks (IJCNN), pp. 1–6. IEEE (2020)

TomoSelfDEQ: Self-supervised Deep Equilibrium Learning for Sparse-Angle CT Reconstruction

Tatiana A. Bubba[1], Matteo Santacesaria[2], and Andrea Sebastiani[3](✉)

[1] Department of Mathematics and Computer Science, University of Ferrara, Ferrara, Italy
tatiana.bubba@unife.it
[2] MaLGa Center, Department of Mathematics, University of Genoa, Genoa, Italy
matteo.santacesaria@unige.it
[3] Department of Physics, Computer Science and Mathematics, University of Modena and Reggio Emilia, Modena, Italy
andrea.sebastiani@unimore.it

Abstract. Deep learning has emerged as a powerful tool for solving inverse problems in imaging, including computed tomography (CT). However, most approaches require paired training data with ground truth images, which can be difficult to obtain, e.g., in medical applications. We present TomoSelfDEQ, a self-supervised Deep Equilibrium (DEQ) framework for sparse-angle CT reconstruction that trains directly on undersampled measurements. We establish theoretical guarantees showing that, under suitable assumptions, our self-supervised updates match those of fully-supervised training with a loss including the (possibly non-unitary) forward operator like the CT forward map. Numerical experiments on sparse-angle CT data confirm this finding, also demonstrating that TomoSelfDEQ outperforms existing self-supervised methods, achieving state-of-the-art results with as few as 16 projection angles.

Keywords: Inverse problems · Deep learning · Deep equilibrium models · Tomographic imaging

1 Introduction

In recent years, learning-based strategies for imaging inverse problems have become a popular alternative to more classical strategies, such as regularization theory or Bayesian approaches [2]. Initial deep learning methods, generally agnostic to the physical model describing an inverse problem, included convolutional neural networks trained either to directly map acquired measurements to the desired images or to post-process an initial "standard" reconstruction to remove noise and other artifacts [14,16]. More recent approaches strive to retain the information coming from the physics of the imaging methodology by incorporating it in the learned scheme. This is the case of deep unrolling or unfolding

T. A. Bubba et al. (Eds.): SSVM 2025, LNCS 15667, pp. 334–346, 2025.
https://doi.org/10.1007/978-3-031-92366-1_26

where a fixed number of iterations of an iterative scheme is regarded as layers of a deep architecture and trained end-to-end [1,10]. A particularly promising direction is the Deep Equilibrium (DEQ) paradigm [9], which allows unrolling an effectively infinite number of iterations by regarding minimizers of certain functionals as fixed points of suitable operators. Compared to unrolling, DEQ trains a deep implicit model, thus significantly reducing the associated memory cost.

While these approaches have shown impressive results, as supervised methods they typically require paired training data with ground truth images, which can be very limiting in, e.g., medical imaging applications. Self-supervised learning addresses this limitation while retaining the power of learning-based strategies. Well known methods include Plug-and-Play (PnP) [22], equivariant imaging [4] and Noise2Noise [15]. Very recently, building on the latter approach, a self-supervised deep equilibrium model, called SelfDEQ, has been introduced in [8] as a framework for training model-based implicit neural networks using Jacobian-Free Backpropagation (JFB). In [8] it is also proved that, for unitary operators (e.g., magnetic resonance imaging (MRI)), SelfDEQ updates match those of fully-supervised DEQ.

Among imaging inverse problems, computed tomography (CT) presents a particularly challenging case where learning without ground truth would be especially valuable. In sparse-angle CT, high-quality reconstruction from scarce angular measurements is crucial for reducing radiation dose in medical imaging and enabling rapid inspection in industrial applications. However, achieving such reconstructions without ground truth data remains challenging, and a framework like SelfDEQ is not directly applicable due to the non-unitary nature of the CT forward operator.

In this work, we introduce TomoSelfDEQ, a novel self-supervised deep equilibrium framework for sparse-angle CT reconstruction, which allows the following key contributions: (1) we extend the theoretical analysis of SelfDEQ to non-unitary forward operators, providing a more general relation between self- and fully-supervised losses (and corresponding JFB); (2) through extensive experiments, we demonstrate the effectiveness of TomoSelfDEQ on sparse-angle CT reconstruction, achieving state-of-the-art results even with severely undersampled data; (3) we provide theoretical guarantees establishing the equivalence between our self-supervised approach and fully-supervised training, ensuring optimal performance without the need for ground truth data.

This work fits into the line of research of training without the need for ground truth that is receiving growing interest in the imaging inverse problems community. In the context of tomographic imaging, several approaches have been proposed. Noise2Inverse [12], extends the ideas of Noise2Self [3] to image reconstruction in the case of fully sampled tomographic measurements. However, it does not explicitly address the challenges of sparse-angle reconstruction. More recently, Sparse2Inverse [11] was proposed to specifically handle undersampled projections by combining a neural network operating in the reconstruction domain with a loss computed in the projection domain. In particular, Sparse2Inverse aims to address limitations of previous methods by computing the loss in the projection domain while letting the neural network operate on images. Other approaches building on similar ideas are Proj2Proj [20], which focuses on low-dose data by applying self-supervision principles directly in the projection domain, and Self-Supervised Denoiser Framework (SDF) [21], which leverages pre-training on densely sampled data to enhance the image reconstruction quality from undersampled data. TomoSelfDEQ differs from all of these approaches because it leverages the Deep Equilibrium framework, and provides theoretical guarantees for the general case of non-unitary operators, i.e., it is not limited to the CT forward operator making it well-suited to other imaging modalities that are modeled by non-unitary transformations.

The paper is organized as follows. In Sect. 2, we introduce the mathematical formulation of undersampled linear inverse problems and provide a summary of relevant existing approaches. Section 3 presents DEQ and introduces our method, TomoSelfDEQ. In Sect. 4, we provide our theoretical results, establishing the equivalence between self-supervised and fully-supervised training for non-unitary operators. Section 5 demonstrates the effectiveness of our approach through numerical experiments on sparse-angle CT reconstruction, comparing it with conventional techniques and the recently proposed Sparse2Inverse. Finally, Sect. 6 summarizes our findings and discusses future research directions.

2 Background and Mathematical Formulation

We consider linear inverse problems of the form:

$$y = MAx + e, \tag{1}$$

where $y \in \mathbb{R}^m$ are the measurements, $e \in \mathbb{R}^m$ is additive white Gaussian noise (AWGN), $A \in \mathbb{R}^{p\times n}$ is a (non-unitary) measurement matrix, $x \in \mathbb{R}^n$ is the unknown target and $M \in \{0,1\}^{m\times p}$ is a matrix obtained from the identity matrix by keeping only those rows corresponding to the measured data. In CT, A represents the Radon transform [17], i.e., each row of A corresponds to a line integral (at a specific angle) modeling the X-ray attenuation through the object of interest. As such, A is a non-unitary operator. When the acquisition angles are sparse, M acts as a mask selecting (from the full measurements) only those projections corresponding to the actual acquisition angles. Due to the ill-posedness of the Radon transform, the sparser the acquisition protocol the more challenging the reconstruction problem becomes. Therefore, traditional methods based on directly inverting the forward map, such as the filtered back-projection (FBP), generally perform poorly. An alternative is provided by regularized approaches, which seek x by iteratively solving optimization problems of the form:

$$\hat{x} = \arg\min_{x\in\mathbb{R}^n_+} \{g(x) + h(x)\}, \tag{2}$$

where g is the data-fidelity term that quantifies the discrepancy between the measurements and the solution and h is a regularizer that imposes prior knowledge on the unknown image. In addition, in tomographic imaging it is natural to restrict the X-ray attenuation coefficients to be non-negative, that is, $\mathbb{R}^n_+$ denotes the non-negative orthant of $\mathbb{R}^n$. A common choice for the data fidelity is the least square term:

$$g(x) = \frac{1}{2}\|y - MAx\|_2^2, \tag{3}$$

and for the regularizer $h(x)$, Total Variation (TV) or other sparsity-promoting penalties are often employed to preserve edge-like features [2, Sect. 2].

While these classical approaches generally allow for accurate reconstructions from fewer tomographic measurements than usually required by FBP, they rely on hand-crafted regularizers (which can impact the mathematical tractability of the functional in (2)), and can be sensitive to parameter choices.

In recent years, the theoretical understanding that comes from regularized approaches has been combined with the practical advantages of learning-based methods showing that, in many instances, the hybrid approaches are able to significantly surpass both pure model- and more data-based methods retaining a sufficient degree of theoretical understanding and guarantees. Among many, we consider here Deep Equilibrium (DEQ), a paradigm initially introduced for supervised learning and very recently adapted in [8] to work as in the self-supervised framework using Jacobian-Free Backpropagation (JFB).

3 Deep Equilibrium Models

Deep Equilibrium (DEQ) enables training recursive networks with effectively infinite layers without storing intermediate variables [9]. The key idea is to find a fixed point $\bar{x}$ of a suitable operator T_θ rather than explicitly unrolling a finite number of iterations, that is:

$$\bar{x} = T_\theta(\bar{x}, y), \tag{4}$$

where y is the measurement vector and T_θ is defined as:

$$T_\theta(x) = \alpha f_\theta(s) + (1-\alpha)s \quad \text{with} \quad s = x - \gamma \nabla g(x), \tag{5}$$

being $\alpha, \gamma > 0$ hyperparameters, f_θ a (convolutional) neural network with learnable parameters θ, and $g(x)$ the data fidelity term from (3). DEQ is generally trained end-to-end in a supervised fashion, considering the empirical risk as minimization loss. In this setting, the neural network f_θ enforces a priori properties on the resulting fixed point $\bar{x}$, functioning as a denoiser similar to PnP [22].

In [8], the authors propose to combine DEQ with the self-supervised framework of Noise2Noise, which uses different noisy realizations of the same ground truth image to train a network without the need for the actual ground truth. The resulting method, named SelfDEQ, considers pairs of measurements $\{y_i, y'_i\}_{i=1}^N$ of the same underlying image x_i:

$$y_i = M_i A x_i + e_i \quad \text{and} \quad y'_i = M'_i A x_i + e'_i \tag{6}$$

where $N \geq 1$ denotes the number of training pairs and, in [8], A is the operator associated with MRI (i.e., the Fourier transform). Then, training the method consists of two steps in each training iteration.

Forward Pass. Starting from an initial guess x^0, the fixed point $\bar{x}$ is computed by running a fixed-point iteration of the form:

$$x^k = T_\theta(x^{k-1}, y) \tag{7}$$

until convergence or a maximum number of iterations is reached. Generally, an acceleration algorithm (e.g., Anderson acceleration) is used.

Backward Pass. Given a loss function $\mathcal{L}$, parameter updates are computed via Jacobian-Free Backpropagation (JFB), which has been theoretically shown to provide valid descent directions for training implicit networks [7], and avoids explicitly computing and storing the inverse Jacobian, leading to efficient training with constant memory complexity:

$$\text{JFB}\mathcal{L}(\theta) = \text{Real}\left((\nabla_\theta T_\theta(\bar{x}))^T \left[\frac{\partial \mathcal{L}}{\partial \bar{x}} \right]^T \right). \tag{8}$$

To enable self-supervised training, the authors of [8] consider the following weighted loss:

$$\mathcal{L}_{\text{self}}(\theta) = \mathbb{E}\left[\frac{1}{2} \| M' A \bar{x}(\theta, y) - y' \|_W^2 \right] \tag{9}$$

where $\| \cdot \|_W$ denotes the weighted norm given by $\|z\|_W^2 = z^T W z$ for $z \in \mathbb{R}^m$. In [8], $W = M' \overline{W} (M' \overline{W})^T \in \mathbb{R}^{m \times m}$ is a subsampling of the diagonal weighting matrix $\overline{W}$, whose entries are defined by:

$$\overline{w}_k = \begin{cases} \frac{1}{\sqrt{\mathbb{E}[M'^T M']_{k,k}}} & \text{if } \sqrt{\mathbb{E}[M'^T M']_{k,k}} \neq 0 \\ 0 & \text{otherwise} \end{cases}. \tag{10}$$

Note that, by minimizing $\mathcal{L}_{\text{self}}$, we implicitly learn the regularization through f_θ. The main theoretical result in [8] establishes that the JFB updates from the weighted self-supervised loss in (8) match those obtained using conventional supervised learning on DEQ, i.e., considering the supervised loss:

$$\mathcal{L}'_{\text{sup}} = \mathbb{E}\left[\|\bar{x} - x\|_2^2 \right]. \tag{11}$$

However, this result was only derived under the limiting assumption of A being a unitary operator.

3.1 TomoSelfDEQ

The framework of SelfDEQ can be easily extended to work with more general non-unitary forward operators A, as it is the case of tomographic imaging. To this end, we introduce TomoSelfDEQ, which is particularly well-suited for sparse-angle CT reconstruction, where both the physics of the imaging process and the self-supervised nature of the training need to be carefully considered. Starting from the original SelfDEQ framework, we modify (5) to include the projection onto the positive orthant Π_+, that is:

$$T_\theta(x) = \Pi_+(\alpha f_\theta(s) + (1-\alpha)s) \quad \text{with} \quad s = x - \gamma \nabla g(x). \tag{12}$$

The measurement pairs in (6) are obtained by splitting each measurement acquisition into two complementary subsets of angles. Each training iteration consists of a forward and a backward pass, as for SelfDEQ. The definition of the self-supervised loss $\mathcal{L}_{\text{self}}(\theta)$ and the diagonal weighting matrix $\overline{W}$, which is crucial for handling the possibly non-uniform sampling density across angles, are as in (9) and (10), respectively.

Our main theoretical result is presented in the next section. In particular, Theorem 1 establishes that the JFB updates from the weighted self-supervised loss (9) match those obtained using the following supervised loss:

$$\mathcal{L}_{\text{sup}} = \mathbb{E}\left[\frac{1}{2}\|A(\bar{x} - x)\|_2^2\right]. \tag{13}$$

Note that this loss differs from that of [8] (and more in general from standard supervised losses in deep learning), as it includes the forward operator A in its formulation. This choice is crucial for establishing the equivalence with the self-supervised loss and reflects the physics of the imaging process. The more common supervised loss (11) would not lead to the same theoretical guarantees due to the non-unitary nature of the Radon transform.

4 Theoretical Results

We now establish the relationship between self-supervised and supervised training of TomoSelfDEQ for non-unitary operators. Our analysis relies on two key assumptions.

Assumption 1. *The training samples correspond to the setting in* (6) *with* $x \sim \pi_x$, $M \sim \pi_M$, $M' \sim \pi_M$, $e \sim \mathcal{N}(0, \sigma^2 I)$, *and* $e' \sim \mathcal{N}(0, \sigma^2 I)$ *drawn i.i.d. from their respective distributions.*

Assumption 2. A *and* $\mathbb{E}_M[M^T M]$, *where the expectation is taken over* π_M, *have full rank.*

Assumption 1 is a mild statistical requirement stating that sampling matrices, images, and noise are sampled independently. Assumption 2 ensures that the sampling matrices provide complete coverage of the measurement domain, though each individual matrix can still represent undersampled measurements.

Our first result establishes a key property of the weighted measurements.

Proposition 1. *When Assumption 2 is satisfied,*

$$\mathbb{E}\left[(M'A)^T W M'A\right] = A^T A$$

where the expectation is with respect to π_M *and* $W = M'\overline{W}(M'\overline{W})^T$ *is defined through* (10).

Proof. Since Assumption 2 implies that $\mathbb{E}[M']_{k,k} \neq 0$,

$$\overline{w}_k = \frac{1}{\sqrt{\mathbb{E}[M'^T M']_{k,k}}}. \tag{14}$$

Given $M'^T M' \in \{0,1\}^{p \times p}$ and $\overline{W} \in \mathbb{R}^{p \times p}$ are both diagonal matrices, we have:

$$\mathbb{E}[M'^T W M'] = \mathbb{E}[M'^T M' \overline{W}\,\overline{W}^T M'^T M'] = \overline{W}\,\overline{W}^T \mathbb{E}[M'^T M' M'^T M'] = I. \tag{15}$$

Therefore,

$$\mathbb{E}[(M'A)^T W M'A] = A^T \mathbb{E}[M'^T W M'] A = A^T A, \tag{16}$$

where the last equation follows from (15). □

Next, our main result shows that TomoSelfDEQ updates match those of supervised training. This implies that TomoSelfDEQ can achieve the same performance as supervised training even for non-unitary operators like the Radon transform, provided the sampling pattern satisfies our assumptions.

Theorem 1. *Under Assumptions 1 and 2, the JFB update of the weighted self-supervised loss equals its supervised counterpart:*

$$JFB\mathcal{L}_{\text{self}}(\theta) = JFB\mathcal{L}_{\text{sup}}(\theta) \tag{17}$$

where $\mathcal{L}_{\text{sup}} = \mathbb{E}\left[\frac{1}{2}\|A(\bar{x} - x)\|_2^2\right]$ *and* $\mathcal{L}_{\text{self}}$ *is defined in* (9).

Proof. The supervised update $\text{JFB}\mathcal{L}_{\text{sup}}(\theta)$ is given by:

$$\text{JFB}\mathcal{L}_{\text{sup}}(\theta) = \mathbb{E}\left[(\nabla_\theta T_\theta(\bar{x}))^T (A^T A)(\bar{x} - x)\right].$$

For the self-supervised update, let $H' := \sqrt{W} M'A$. Then:

$$\begin{aligned}\text{JFB}\mathcal{L}_{\text{self}}(\theta) &= \mathbb{E}\left[(\nabla_\theta T_\theta(\bar{x}))^T H'^T (H'\bar{x} - \sqrt{W} y')\right] \\ &= \mathbb{E}\left[(\nabla_\theta T_\theta(\bar{x}))^T \mathbb{E}\left[H'^T (H'\bar{x} - \sqrt{W}(M'Ax + e')) \middle| x, M, e\right]\right]\end{aligned}$$

The conditional expectation can be decomposed as:

$$\begin{aligned}\mathbb{E}\left[H'^T (H'\bar{x} - \sqrt{W}(M'Ax + e')) \middle| x, M, e\right] &= \mathbb{E}[H'^T H'](\bar{x} - x) + \mathbb{E}[H'^T \sqrt{W}]\mathbb{E}[e'] \\ &= A^T A(\bar{x} - x)\end{aligned}$$

where we used Proposition 1 and $\mathbb{E}[e'] = 0$. Therefore,

$$\text{JFB}\mathcal{L}_{\text{self}}(\theta) = \mathbb{E}\left[(\nabla_\theta T_\theta(\bar{x}))^T (A^T A)(\bar{x} - x)\right] = \text{JFB}\mathcal{L}_{\text{sup}}(\theta)$$

which establishes the desired result. □

4.1 Properties of Random Sampling Patterns

For practical implementation in sparse-angle CT, we adopt uniform random sampling from $[0, \pi)$ to satisfy Assumption 2. The following proposition shows that this sampling strategy leads to a simple form for $\mathbb{E}[M^T M]$, which determines the weighting matrix W in our self-supervised loss.

Proposition 2. *When the sampling patterns M are drawn uniformly at random from a fixed subset of size s, the expectation $\mathbb{E}[M^T M]$ is a scalar multiple of the identity matrix. Specifically,*

$$\mathbb{E}[M^T M] = \frac{s}{n_\Sigma} I, \tag{18}$$

where n_Σ is the total number of possible measurements and s is the number of selected measurements.

Proof. For a masking matrix M with entries in $\{0, 1\}$, $M^T M$ is diagonal. Under uniform random sampling with size s, the probability of any index being selected is s/n_Σ, giving $\mathbb{E}[(M^T M)_{k,k}] = s/n_\Sigma$ for diagonal entries. Off-diagonal entries are always zero by construction. Therefore Eq. (18) follows. □

5 Numerical Experiments

In this section, we describe the implementation[1] of TomoSelfDEQ and we show some numerical results, comparing them with conventional techniques, i.e., FBP and TV regularization with non-negativity constraints (cf. Sect. 2) and the recently proposed self-supervised methods Sparse2Inverse [11], tailored to sparse CT problems. For a fair comparison with our method, we applied Sparse2Inverse on simulated measurements with AWGN, which differs from its original implementation. We conducted all the experiments on a workstation equipped with an Intel i9-12900 K processor and a NVIDIA RTX A5000 GPU.

The operator A is a discretization of the Radon transform, implemented with Tomosipo [13] (which relies on LION to specify the CT geometry[2]). We consider a discretization of the full angular range made of $n_\Sigma = 384$ equispaced angles. The matrix M is obtained sampling a fixed number s of angles from the total 384 angles and keeping only the rows corresponding to the selected angles. The forward pass is executed until the relative norm of the differences between iterations is less than 10^{-3}. In case the stopping criterion is not met, the maximum number of iterations is set to 100. Similarly to [8], we use a U-Net [18] for f_θ, considering the spectral normalization in all layers to improve the stability of the model [19]. We adaptively set the parameter γ in (5) depending on the number of selected angles s, to ensure the operator T_θ is non-expansive. Specifically, $\gamma = \frac{1}{\hat{\gamma}_s}$, where $\hat{\gamma}_s$ is the spectral norm of the operator $M_s A$, with

[1] https://github.com/sedaboni/TomoSelfDEQ/.
[2] https://github.com/CambridgeCIA/LION.

Table 1. Average PSNR and SSIM on the validation set with different sparsity levels s. The best results are highlighted in bold.

	$s = 16$		$s = 32$		$s = 64$	
	PSNR	SSIM	PSNR	SSIM	PSNR	SSIM
FBP	14.27	0.100	18.42	0.143	22.87	0.205
TV	23.83	0.722	26.14	0.828	27.39	0.863
Sparse2Inverse	19.21	0.314	21.14	0.675	26.80	0.760
TomoSelfDEQ	**26.25**	**0.884**	**28.26**	**0.911**	**29.41**	**0.927**

M_s being the sampling matrix obtained by selecting s equispaced angles. For the optimizer, we use a Schedule-Free version of Adam [5]. The learning rate is set to $2 * 10^{-4}$, the mini-batch size is 8 and we run 2000 training epochs.

The training dataset is generated on-the-fly, considering 32 2D slices from the CBCT Walnut dataset [6]. In detail, for each training batch the undersampled measurements are obtained randomly sampling s angles from $[0, \pi)$ with a uniform distribution. This data augmentation strategy ensures uniform coverage of the angular range during training as required by Assumption 2, enabling more robust learning. It is important to point out that this strategy is feasible since we have the full range data and it is not practical in a real medical settings. It has to be intended only for validation purposes. In contrast to the random sampling used during training, the validation set is fixed and uses equispaced angles, with the sampling mask M_s selecting s equispaced angles.

The numerical results are summarized in Table 1. These are evaluated in terms of peak signal-to-noise ratio (PSNR) and structural similarity index (SSIM) considering three different sampling settings, with different sparsity levels $s = 16, 32, 64$ angles, and noise set to 1% AWGN. It is evident that our method, TomoSelfDEQ, outperforms the competing approaches in terms of both PSNR and SSIM. In Fig. 1, we report the results obtained with the different methods considering 64 (top row), 32 (middle row) and 16 (bottom row) angles. We can see that TomoSelfDEQ was able to reconstruct sharp and neat boundaries, even though some finer details are lost due to over-smoothing in certain regions. In addition, TomoSelfDEQ accurately reconstructs constant patches, without adding any significant artifacts and it proves to be stable with respect to the reduction of angles. On the other hand, FBP reconstructions are noisy and show an increasing number of streaking artifacts as the number of angles decreases. In the TV regularized reconstructions, the staircasing behavior is evident, and the sharpness of the boundaries deteriorates as the number of angles decreases. Sparse2Inverse produces reconstructions with clear artifacts, particularly when only 16 angles are considered.

Similar consideration can be drawn for the reconstructions in Fig. 2, where we show reconstructions of different walnut slices in the most challenging setting, i.e., with 16 angles. From the close-ups it is clear that Sparse2Inverse introduces numerous noisy artifacts, compromising the reconstructions, while

TomoSelfDEQ better preserves objects shapes, significantly reducing noise presence. A note is, however, in order on the comparison with Sparse2Inverse. While both methods share key ideas such as operating in both image and projection domains, computing losses in the projection domain, and avoiding explicit nullspace assumptions, TomoSelfDEQ differs fundamentally in several important aspects. First, while Sparse2Inverse uses a standard U-Net architecture, our method leverages DEQ to implicitly represent an infinite-depth network. Second, TomoSelfDEQ provides theoretical guarantees about matching supervised performance, while Sparse2Inverse relies primarily on empirical validation. These theoretical and architectural advantages translate into practical improvements, as demonstrated in our numerical results where TomoSelfDEQ consistently outperforms Sparse2Inverse across different undersampling rates (cf. Table 1).

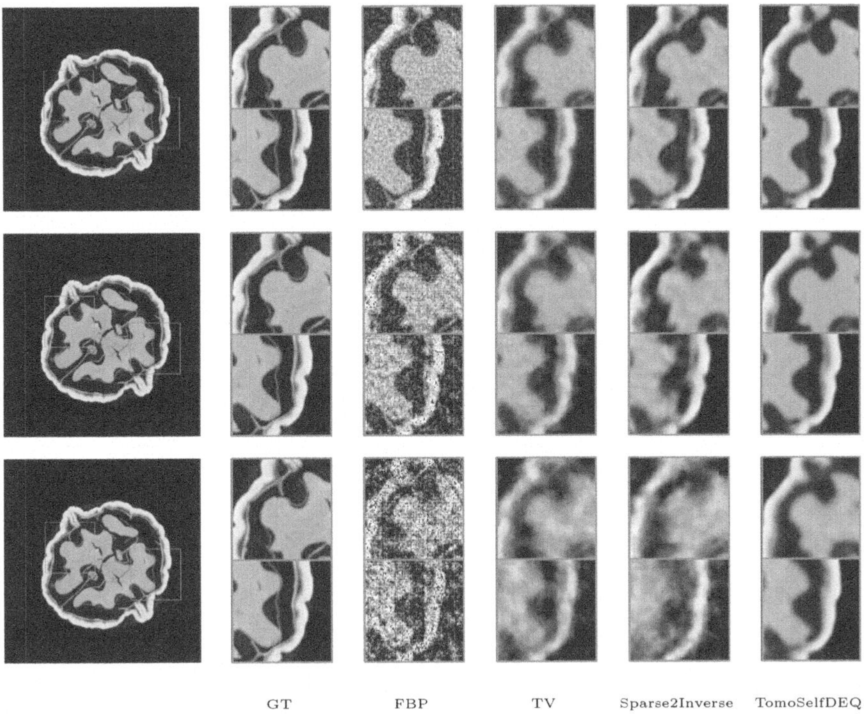

Fig. 1. Qualitative results on different sparsity levels s. Top row: $s = 64$ angles. Middle row: $s = 32$ angles. Bottom row: $s = 16$ angles.

Ablation Study. We trained the same infinite-depth network through DEQ, considering different losses to numerically validate the theoretical result in Theorem 1. In particular, we compare the self-supervised loss $\mathcal{L}_{\text{self}}$ (cf. (9)), the usual

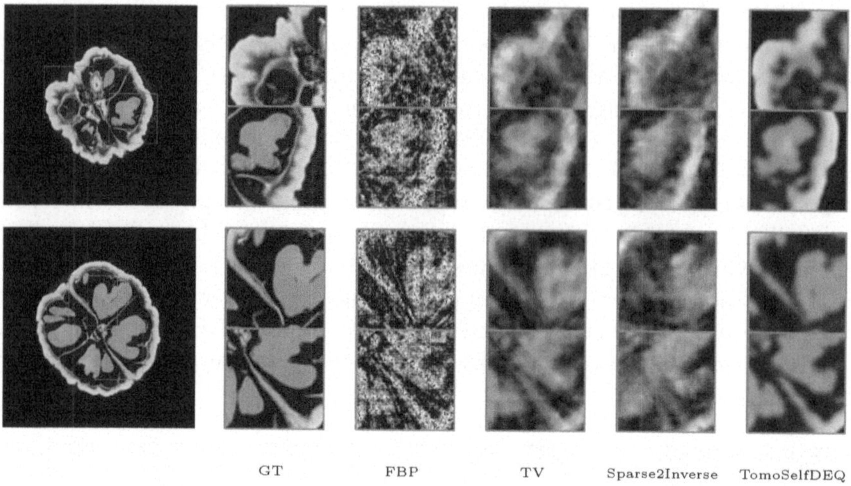

Fig. 2. Qualitative results on different walnut slices considering $s = 16$ angles.

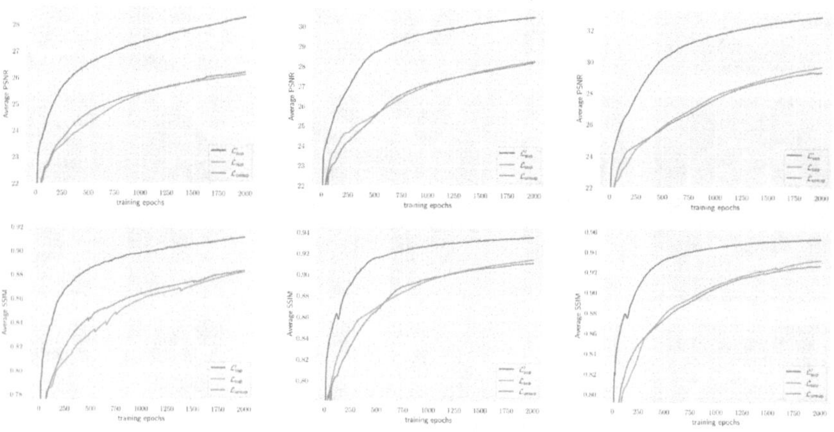

(a) $s = 16$ angles (b) $s = 32$ angles (c) $s = 64$ angles

Fig. 3. Quality metrics computed on the validation set and plotted with respect to the training epochs for different selection of loss. Top row: PSNR. Bottom row: SSIM.

supervised loss $\mathcal{L}'_{\text{sup}}$ (cf. (11)), and the supervised loss $\mathcal{L}_{\text{sup}}$ (cf. (13)). We report the behavior in terms of PSNR and SSIM during training with these different losses in Fig. 3. It can be seen that the unsupervised framework achieves comparable results with that of the supervised loss $\mathcal{L}_{\text{sup}}$, thus confirming the result of Theorem 1. It is evident that the results with the standard supervised loss $\mathcal{L}'_{\text{sup}}$ outperform the others, since in this loss the information concerning the undersampled data is not taken into account.

6 Conclusions

We introduced TomoSelfDEQ, a self-supervised DEQ framework for sparse-angle CT reconstruction. Our main theoretical contribution extends SelfDEQ training guarantees to non-unitary operators like the Radon transform, showing that under suitable assumptions, the self-supervised updates match those of fully-supervised training. Through extensive experiments on sparse-angle CT data, we demonstrated that TomoSelfDEQ achieves state-of-the-art reconstruction quality while requiring only undersampled measurements for training. The method consistently outperforms existing self-supervised approaches across different undersampling rates, achieving high-quality reconstructions even with as few as 16 projection angles. Future work could explore extensions to other non-unitary inverse problems and investigate the theoretical implications of different sampling strategies. Additionally, while our current validation strategy using equispaced angles demonstrates strong performance, further investigation, using different validation protocols and testing on out-of-distribution datasets, would provide valuable insights into the method's generalization capabilities.

Acknowledgments. The authors acknowledge (partial) support by the: Royal Society through the International Exchange scheme (grant n. IES\R3\223061 to TAB and MS); European Union - NextGeneration EU (FAIR "Future Partnership Artificial Intelligence Research", CUP J33C22002830006 to TAB, CUP J53C22003010006 to MS; PRIN P2022J9SNP "Advanced optimization METhods for automated central veIn Sign detection in multiple sclerosis from magneTic resonAnce imaging" (AMETISTA) CUP E53D23017980001 to AS, PRIN 2022B32J5C "Inverse problems in PDE: theoretical and numerical analysis", CUP D53D23005770006 to MS); Air Force Office of Scientific Research (award number FA8655-23-1-7083 to MS); INdAM-GNCS (CUP E53C24001950001 to TAB and AS, CUP E53C23001670001 to AS); MUR Excellence Department Project (awarded to Dipartimento di Matematica, Università di Genova, CUP D33C23001110001 to MS).

References

1. Adler, J., Öktem, O.: Solving ill-posed inverse problems using iterative deep neural networks. Inverse Prob. **33**(12), 124007 (2017)
2. Arridge, S., Maass, P., Öktem, O., Schönlieb, C.B.: Solving inverse problems using data-driven models. Acta Numer. **28**, 1–174 (2019)
3. Batson, J., Royer, L.: Noise2self: blind denoising by self-supervision. In: International Conference on Machine Learning, pp. 524–533. PMLR (2019)
4. Chen, D., Tachella, J., Davies, M.E.: Equivariant imaging: learning beyond the range space. In: Proceedings of the IEEE/CVF International Conference on Computer Vision, pp. 4379–4388 (2021)
5. Defazio, A., Yang, X.A., Mehta, H., Mishchenko, K., Khaled, A., Cutkosky, A.: The road less scheduled. arXiv:2405.15682 (2024)
6. Der Sarkissian, H., Lucka, F., van Eijnatten, M., Colacicco, G., Coban, S.B., Batenburg, K.J.: A cone-beam X-ray computed tomography data collection designed for machine learning. Sci. Data **6**(1), 215 (2019)

7. Fung, S.W., Heaton, H., Li, Q., McKenzie, D., Osher, S., Yin, W.: JFB: jacobian-free backpropagation for implicit networks. In: Proceedings of the AAAI Conference on Artificial Intelligence, vol. 36, pp. 6648–6656 (2022)
8. Gan, W., et al.: Self-supervised deep equilibrium models with theoretical guarantees and applications to MRI reconstruction. IEEE Trans. Comput. Imaging (2023)
9. Gilton, D., Ongie, G., Willett, R.: Deep equilibrium architectures for inverse problems in imaging. IEEE Trans. Comput. Imaging **7**, 1123–1133 (2021)
10. Gregor, K., LeCun, Y.: Learning fast approximations of sparse coding. In: Proceedings of the 27th International Conference on International Conference on Machine Learning, pp. 399–406 (2010)
11. Gruber, N., Schwab, J., Gizewski, E., Haltmeier, M.: Sparse2inverse: self-supervised inversion of sparse-view CT data. arXiv:2402.16921 (2024)
12. Hendriksen, A.A., Pelt, D.M., Batenburg, K.J.: Noise2inverse: self-supervised deep convolutional denoising for tomography. IEEE Trans. Comput. Imaging **6**, 1320–1335 (2020)
13. Hendriksen, A.A., et al.: Tomosipo: fast, flexible, and convenient 3D tomography for complex scanning geometries in python. Opt. Express **29**(24), 40494–40513 (2021)
14. Jin, K.H., McCann, M.T., Froustey, E., Unser, M.: Deep convolutional neural network for inverse problems in imaging. IEEE Trans. Comput. Imaging **26**(9), 4509–4522 (2017)
15. Lehtinen, J., et al.: Noise2noise: learning image restoration without clean data. arXiv:1803.04189 (2018)
16. Lucas, A., Iliadis, M., Molina, R., Katsaggelos, A.K.: Using deep neural networks for inverse problems in imaging: beyond analytical methods. IEEE Signal Process. Mag. **35**(1), 20–36 (2018)
17. Natterer, F.: The Mathematics of Computerized Tomography. SIAM (2001)
18. Ronneberger, O., Fischer, P., Brox, T.: U-Net: convolutional networks for biomedical image segmentation. In: Medical Image Computing and Computer-Assisted Intervention–MICCAI 2015: 18th International Conference, Munich, Germany, October 5–9, 2015, proceedings, part III 18, pp. 234–241. Springer (2015)
19. Ryu, E., Liu, J., Wang, S., Chen, X., Wang, Z., Yin, W.: Plug-and-play methods provably converge with properly trained denoisers. In: International Conference on Machine Learning, pp. 5546–5557. PMLR (2019)
20. Unal, M.O., Ertas, M., Yildirim, I.: Proj2proj: self-supervised low-dose CT reconstruction. PeerJ Comput. Sci. **10**, e1849 (2024)
21. Valat, E., Hauptmann, A., Öktem, O.: Self-supervised denoiser framework. arXiv:2411.19593 (2024)
22. Venkatakrishnan, S.V., Bouman, C.A., Wohlberg, B.: Plug-and-play priors for model based reconstruction. In: 2013 IEEE Global Conference on Signal and Information Processing, pp. 945–948. IEEE (2013)

Bilevel Learning with Inexact Stochastic Gradients

Mohammad Sadegh Salehi[1(✉)], Subhadip Mukherjee[2], Lindon Roberts[3], and Matthias J. Ehrhardt[1]

[1] Department of Mathematical Sciences, University of Bath, Bath BA2 7AY, UK
sadeghsalehi1997@gmail.com
[2] Department of Electronics and Electrical Communication Engineering, Indian Institute of Technology (IIT) Kharagpur, Kharagpur, India
[3] School of Mathematics and Statistics, University of Sydney, Camperdown, NSW 2006, Australia

Abstract. Bilevel learning has gained prominence in machine learning, inverse problems, and imaging applications, including hyperparameter optimization, learning data-adaptive regularizers, and optimizing forward operators. The large-scale nature of these problems has led to the development of inexact and computationally efficient methods. Existing adaptive methods predominantly rely on deterministic formulations, while stochastic approaches often adopt a doubly-stochastic framework with impractical variance assumptions, enforces a fixed number of lower-level iterations, and requires extensive tuning. In this work, we focus on bilevel learning with strongly convex lower-level problems and a nonconvex sum-of-functions in the upper-level. Stochasticity arises from data sampling in the upper-level which leads to inexact stochastic hypergradients. We establish their connection to state-of-the-art stochastic optimization theory for nonconvex objectives. Furthermore, we prove the convergence of inexact stochastic bilevel optimization under mild assumptions. Our empirical results highlight significant speed-ups and improved generalization in imaging tasks such as image denoising and deblurring in comparison with adaptive deterministic bilevel methods.

Keywords: Bilevel Learning · Stochastic Bilevel Optimization · Learning Regularizers · Machine Learning · Variational regularization

1 Introduction

Bilevel learning has emerged as a powerful framework in various machine learning domains such as hyperparameter optimization [15,25] and meta-learning [10]. In inverse problems and imaging applications, bilevel optimization plays a key role within variational regularization frameworks [16,19,27]. In this work, following the convention from [18], considering a sampling vector $v \in \mathbb{R}^m$ such that $v \sim \mathcal{D}$ and $\mathbb{E}_{\mathcal{D}}[v_i] = 1$ for all $i \in \{1, \ldots, m\}$, where $\mathcal{D}$ is a user-defined distribution

T. A. Bubba et al. (Eds.): SSVM 2025, LNCS 15667, pp. 347–359, 2025.
https://doi.org/10.1007/978-3-031-92366-1_27

(for example, the uniform distribution $\mathcal{D} = U(0, m)$), we focus on the bilevel optimization problem described as follows:

$$\min_{\theta} \mathbb{E}_{v \sim \mathcal{D}} \Big[f_v(\theta) := \frac{1}{m} \sum_{i=1}^{m} v_i f_i(\theta) = \frac{1}{m} \sum_{i=1}^{m} v_i g_i(\hat{x}_i(\theta)) \Big], \tag{1a}$$

$$s.t. \quad \hat{x}_i(\theta) := \arg\min_{x \in \mathbb{R}^n} h_i(x, \theta), \quad i = 1, 2, \ldots, m, \tag{1b}$$

Here, the upper-level problem is a stochastic optimization problem where the upper-level functions g_i measure the quality of the solution to the lower-level problem (1b), for example, $g_i(x) = \|x - x_i^*\|^2$, where x_i^* is the desired solution of the lower-level problem. In variational regularization, the lower-level objective is often formulated as $h_i(x, \theta) = D(Ax, y_i) + R_\theta(x)$, for all $i = 1, 2, \ldots, m$, where D denotes the data fidelity term, each y_i represents the noisy measurements, A is the forward operator (often linear) for the imaging problems, and R_θ is a regularizer parametrized by θ. Since the quality of reconstruction depends on the choice of θ, bilevel learning can be used to optimize the parameters in R_θ [4,8]. Additionally, data-adaptive regularizers have shown promising results in variational regularization frameworks [3,12,23]. Moreover, components of the data fidelity term, such as the forward operator, may also depend on θ. For instance, in magnetic resonance imaging (MRI), learning the sampling pattern [29] and in seismic imaging, learning design parameters [7] directly affect the reconstruction quality.

However, large-scale bilevel optimization is challenging, particularly in imaging. Gradient-based bilevel optimization algorithms [24] offer scalability, but the gradient of the upper-level depends on the solution of the lower-level problem, and computing the exact solution of the lower-level variational problem is impractical. Instead, approximate gradient methods, which rely on inexact solutions of the lower-level problem, have been investigated [9,25,28]. The existing imaging literature has primarily focused on methods for computing the approximate hypergradient and bounding its errors [1,9,22], with less attention given to the behavior and convergence of the upper-level solver in the presence of such errors. Recent works have addressed these gaps by proposing adaptive bilevel methods [28] or single-level reformulations [30], which reduce computational costs. However, these methods are predominantly deterministic.

In contrast, stochastic methods, particularly for the upper-level problem, can significantly accelerate computations and enable bilevel problems to be tackled on larger datasets, potentially enhancing generalization compared to methods that are limited to smaller datasets . Most existing stochastic bilevel optimization algorithms [11,17,26] adopt a doubly-stochastic formulation, where the lower-level problem focuses on a regularized training problem, and the upper-level problem handles hyperparameter optimization on a validation dataset. However, these methods often assume a fixed number of lower-level iterations, while paying less attention to the effects of inexact lower-level evaluations or the choice of the upper-level step size. Moreover, due to only considering machine learning applications such as multinomial logistic regression, stochastic methods are considered in their lower-level problems [15], whereas deterministic solvers are commonly used for lower-level problems arising in variational regularization [9,28].

Another class of bilevel algorithms employs the value-function approach [6], reformulating bilevel problems into a single-level constrained optimization problem, referred to as fully-first-order methods [20,21]. While these methods theoretically provide a flexible framework for the lower-level problem, they share similar assumptions regarding a fixed number of lower-level iterations, and convergence analysis as other stochastic bilevel methods.

Contributions: In this work, we consider inexact stochastic hypergradients within the bilevel optimization framework, as outlined in [9,25]. We establish the connection between these stochastic approximate hypergradients and the theoretical analysis of stochastic nonconvex optimization, particularly focusing on practical variance assumptions [5,18]. Specifically, we demonstrate that the inexact stochastic hypergradients satisfy the practical assumptions of biased stochastic gradient descent (SGD). Finally, we demonstrate the speed-up advantages of this framework compared to adaptive inexact deterministic bilevel methods in learning data-adaptive regularizers for image denoising. Additionally, we show that this approach leads to improved generalization when applied to learning data-adaptive regularizers for image deblurring problems.

2 The Algorithm

Considering the upper-level problem (1a), for the hypergradient we can write

$$\mathbb{E}_{v\sim\mathcal{D}}\Big[\nabla f_v(\theta) := \frac{1}{m}\sum_{i=1}^{m} v_i \nabla f_i(\theta) = \frac{1}{m}\sum_{i=1}^{m} v_i \partial \hat{x}_i(\theta)^T \nabla g_i(\hat{x}_i(\theta))\Big]. \qquad (2)$$

From the definition of the sampling vector, in case we have access to the exact lower-level solution $\hat{x}_i(\theta)$ for all $i \in \{1, 2, \ldots, m\}$, f_v and ∇f_v are the unbiased estimators of f and ∇f, respectively (i.e. $\mathbb{E}_{v\sim\mathcal{D}}[f_v(\theta)] = f(\theta)$ and $\mathbb{E}_{v\sim\mathcal{D}}\nabla f_v(\theta) = \nabla f(\theta)$). Each iteration $k \in \mathbb{N}$ of SGD takes the following form

$$\theta^{k+1} = \theta^k - \alpha_k \nabla f_{v^k}(\theta^k), \qquad (3)$$

where $v^k \sim \mathcal{D}$ i.i.d. and $\alpha_k > 0$ is the step size.

Since, in general, the lower-level solution $\hat{x}_i(\theta)$ can only be approximated, one typically has access only to the approximate lower-level solution $\tilde{x}_i(\theta)$, where $\|\hat{x}_i(\theta) - \tilde{x}_i(\theta)\| \le \epsilon$. This is ensured by the a posteriori bound $\|\nabla h_i(\tilde{x}_i(\theta), \theta)\| \le \mu_i \epsilon$, where μ_i is the strong convexity parameter of each h_i. Hence, corresponding to each sampling vector v_i for all $i \in \{1, 2, \ldots, m\}$, the stochastic hypergradient $v_i \partial \hat{x}_i(\theta)^T \nabla g_i(\hat{x}_i(\theta))$ can be approximated using methods described in [9,22] with an error of $\mathcal{O}(\epsilon)$. This results in an inexact stochastic hypergradient in the upper-level problem, which we denote by $z_v(\theta) = \nabla f_v(\theta) + e_v(\theta)$ for all $v \sim \mathcal{D}$. Note that ensuring the boundedness of this error can be done utilizing the error bounds provided in [9,14]. The steps of the inexact SGD algorithm are outlined in Algorithm 1.

Algorithm 1. Inexact Stochastic Gradient Descent (ISGD)

1: Input: Initial parameters $\theta^0 \in \mathbb{R}^d$, Sampling $\mathcal{D}$, step size $\alpha > 0$, accuracy $\epsilon \geq 0$.
2: **for** $k = 0, 1, \ldots$ **do**
3: Sample a mini-batch $v^k \sim \mathcal{D}$.
4: Compute the approximate stochastic gradient $z_{v^k}(\theta^k)$ with error $\mathcal{O}(\epsilon)$.
5: Update the iterate: $\theta^{k+1} = \theta^k - \alpha z_{v^k}(\theta^k)$

2.1 Convergence Analysis

To prove convergence of Algorithm 1, we consider the notations below and make the following regularity assumptions.

Notation. *For any $v \sim \mathcal{D}$ and for all $\theta \in \mathbb{R}^d$, the error of the hypergradient is shown as $\|e_v(\theta)\|$. Moreover, at each iteration $k \in \mathbb{N}$ of Algorithm 1, we denote $z_k := z_{v^k}(\theta^k)$, and $\epsilon_k^2 = \mathbb{E}_{v^k \sim \mathcal{D}}[\|e_{v^k}(\theta^k)\|^2]$.*

Assumption 1. *Each f_i is $L_{\nabla f_i}$-smooth, which means f and the f_i's are continuously differentiable with $L_{\nabla f}$ and $L_{\nabla f_i}$ Lipschitz gradients, respectively. Moreover, each f_i is bounded below.*

Theorem 1. ***Convergence:** Let assumption 1 hold and $\mathbb{E}[\epsilon_k^2]$ be bounded above for all $k \geq 0$. Let $\delta_0 \stackrel{def}{=} f(\theta^0) - f^*$. There exist constants $c_1, c_2, \ldots, c_5 > 0$ such that if the step size satisfies $0 < \alpha \leq \frac{c_1}{L_{\nabla f}}$, then the iterates $\{\theta^t\}_{t \geq 0}$ of Algorithm 1 (step 5) satisfy*

$$\min_{0 \leq t \leq T-1} \mathbb{E}\left[\|\nabla f(\theta^t)\|^2\right] \leq \frac{c_2(1 + c_3 L_{\nabla f} \alpha^2)^T}{\alpha T} \delta_0 + c_4 L_{\nabla f} \alpha + c_5.$$

Corollary 1. *For any fixed accuracy $\delta > 0$, there exists $\alpha_\delta > 0$ and $T_\delta \in \mathbb{N}$, such that for all $T \geq T_\delta$ we have*

$$\min_{0 \leq t \leq T-1} \mathbb{E}\left[\|\nabla f(\theta^t)\|^2\right] \leq \mathcal{O}(\delta). \tag{4}$$

Moreover, $T_\delta = \mathcal{O}(\delta^{-2})$ and $\alpha_\delta = \mathcal{O}(\delta)$.

Recent stochastic optimization analyses for nonconvex objectives have highlighted a more general, flexible, and practical bounded variance assumption, introducing the ABC assumption [5,13,18]. However, in the stochastic bilevel framework, the role of inexact stochastic hypergradients and their connection to this framework remains unexplored. We establish this connection by stating the following lemma.

Lemma 1. *[18, Proposition 2] Expected Smoothness (ES): Let Assumption 1 hold and $\mathbb{E}[v_i^2] < \infty$ for $1 \leq i \leq m$. There exist constants $A, B, C \geq 0$ such that for all $\theta \in \mathbb{R}^d$, the stochastic gradient satisfies*

$$\mathbb{E}_{v \sim \mathcal{D}}[\|\nabla f_v(\theta)\|^2] \leq 2A(f(\theta) - f^{inf}) + B\|\nabla f(\theta)\|^2 + C.$$

The following lemmas will be utilized in proving the convergence of ISGD. The strategy is to demonstrate the connection between the inexact stochastic gradient estimator and the biased ABC condition, enabling the use of convergence results under the biased ABC condition.

Lemma 2. *For all $\theta \in \mathbb{R}^d$, $v \sim \mathcal{D}$, and $\zeta > 0$ we have*

$$|\langle \nabla f(\theta), \mathbb{E}_{v\sim\mathcal{D}}[e_v(\theta)]\rangle| \leq \frac{\|\nabla f(\theta)\|^2}{2\zeta} + \frac{\zeta}{2}(\mathbb{E}_{v\sim D}[\|e_v(\theta)\|])^2 \tag{5}$$

Proof. For all $\theta \in \mathbb{R}^d$ and $v \sim \mathcal{D}$, utilizing Cauchy–Schwarz inequality we find

$$|\langle \nabla f(\theta), \mathbb{E}_{v\sim\mathcal{D}}[e_v(\theta)]\rangle| \leq \|\nabla f(\theta)\| \|\mathbb{E}_{v\sim\mathcal{D}}[e_v(\theta)]\|.$$

Now, since $\|\cdot\|$ is convex, using Jensen's inequality we have

$$\|\mathbb{E}_{v\sim\mathcal{D}}[e_v(\theta)]\| \leq \mathbb{E}_{v\sim\mathcal{D}}\left[\|e_v(\theta)\|\right].$$

Combining these inequalities we find

$$|\langle \nabla f(\theta), \mathbb{E}_{v\sim\mathcal{D}}[e_v(\theta)]\rangle| \leq \|\nabla f(\theta)\|\mathbb{E}_{v\sim D}[\|e_v(\theta)\|].$$

Applying Young's inequality on $\|\nabla f(\theta)\|\mathbb{E}_{v\sim D}[\|e_v(\theta)\|]$, with a parameter $\zeta > 0$ yields

$$\|\nabla f(\theta)\|\mathbb{E}_{v\sim D}[\|e_v(\theta)\|] \leq \frac{\|\nabla f(\theta)\|^2}{2\zeta} + \frac{\zeta}{2}(\mathbb{E}_{v\sim D}[\|e_v(\theta)\|])^2,$$

which implies (5). □

The following lemma relates the second moment of the inexact stochastic gradient with the stochastic gradient.

Lemma 3. *Let $\eta > 0$. The following relation between the second moment of the exact and inexact stochastic gradients hold*

$$\mathbb{E}_{v\sim\mathcal{D}}[\|z_v(\theta)\|^2] \leq \frac{1+\eta}{\eta}\mathbb{E}_{v\sim\mathcal{D}}[\|\nabla f_v(\theta)\|^2] + (1+\eta)\mathbb{E}_{v\sim\mathcal{D}}[\|e_v(\theta)\|^2]. \tag{6}$$

Proof. From the definition of $z_v(\theta)$, for all $\theta \in \mathbb{R}^d$, we have

$$\mathbb{E}_{v\sim\mathcal{D}}[\|z_v(\theta)\|^2] = \mathbb{E}_{v\sim\mathcal{D}}[\|\nabla f_v(\theta)\|^2] + \mathbb{E}_{v\sim\mathcal{D}}[\|e_v(\theta)\|^2] + 2\mathbb{E}_{v\sim\mathcal{D}}[\langle \nabla f_v(\theta), e_v(\theta)\rangle]. \tag{7}$$

Now, by applying Cauchy–Schwarz and then Young's inequality on the last term above, with a parameter $\eta > 0$, we get

$$\begin{aligned} 2\mathbb{E}_{v\sim\mathcal{D}}[\langle \nabla f_v(\theta), e_v(\theta)\rangle] &\leq 2\mathbb{E}_{v\sim\mathcal{D}}[\|\nabla f_v(\theta)\|\|e_v(\theta)\|] \\ &\leq \frac{1}{\eta}\mathbb{E}_{v\sim\mathcal{D}}[\|\nabla f_v(\theta)\|^2] + \eta\mathbb{E}_{v\sim\mathcal{D}}[\|e_v(\theta)\|^2]. \end{aligned}$$

Plugging it back in (7) yields (6) as desired. □

Proposition 1. *Suppose Assumption 1 hold and* $\mathbb{E}[v_i^2] < \infty$ *for* $1 \leq i \leq m$. *Let* $\eta > 0$, $\zeta > 0$, $\epsilon^2(\theta) := \mathbb{E}_{v\sim\mathcal{D}}[\|e_v(\theta)\|^2]$, *and* $b(\zeta) = 1 - \frac{1}{2\zeta}$, $c(\zeta, \epsilon^2(\theta)) = \frac{\zeta}{2}\epsilon^2(\theta)$, $\tilde{A}(\eta) = \frac{1+\eta}{\eta}A$, $\tilde{B}(\eta) = \frac{1+\eta}{\eta}B$, $\tilde{C}(\eta, \epsilon^2(\theta)) = \frac{1+\eta}{\eta}C + (1+\eta)\epsilon^2(\theta)$. *Then, for all* $\theta \in \mathbb{R}^d$, *the inexact stochastic gradient estimator satisfies*

$$\langle \nabla f(\theta), \mathbb{E}_{v\sim\mathcal{D}}[z_v(\theta)]\rangle \geq b(\zeta)\|\nabla f(\theta)\|^2 - c(\zeta, \epsilon^2(\theta)), \tag{8}$$

$$\mathbb{E}_{v\sim\mathcal{D}}[\|z_v(\theta)\|^2] \leq 2\tilde{A}(\eta)(f(\theta) - f^{inf}) + \tilde{B}(\eta)\|\nabla f(\theta)\|^2 + \tilde{C}(\eta, \epsilon^2(\theta)). \tag{9}$$

Proof. For the inexact stochastic gradient $z_v(\theta)$, we have

$$\langle \nabla f(\theta), \mathbb{E}_{v\sim\mathcal{D}}[z_v(\theta)]\rangle = \langle \nabla f(\theta), \mathbb{E}_{v\sim\mathcal{D}}[\nabla f_v(\theta)] + \mathbb{E}_{v\sim\mathcal{D}}[e_v(\theta)]\rangle.$$

Since the estimator of the exact gradient is unbiased i.e. $\mathbb{E}_{v\sim\mathcal{D}}[\nabla f_v(\theta)] = \nabla f(\theta)$, we have

$$\langle \nabla f(\theta), \mathbb{E}_{v\sim\mathcal{D}}[z_v(\theta)]\rangle = \|\nabla f(\theta)\|^2 + \langle \nabla f(\theta), \mathbb{E}_{v\sim\mathcal{D}}[e_v(\theta)]\rangle. \tag{10}$$

From Lemma 2 we know $\langle \nabla f(\theta), \mathbb{E}_{v\sim\mathcal{D}}[e_v(\theta)]\rangle \geq -\frac{\|\nabla f(\theta)\|^2}{2\zeta} - \frac{\zeta}{2}(\mathbb{E}_{v\sim\mathcal{D}}[\|e_v(\theta)\|])^2$. Combining it with (10) yields

$$\langle \nabla f(\theta), \mathbb{E}_{v\sim\mathcal{D}}[z_v(\theta)]\rangle \geq (1 - \frac{1}{2\zeta})\|\nabla f(\theta)\|^2 - \frac{\zeta}{2}(\mathbb{E}_{v\sim\mathcal{D}}[\|e_v(\theta)\|])^2.$$

Now utilizing Jensen's inequality $\mathbb{E}_{v\sim\mathcal{D}}[\|e_v(\theta)\|^2] \geq (\mathbb{E}_{v\sim\mathcal{D}}[\|e_v(\theta)\|])^2$ we find

$$\langle \nabla f(\theta), \mathbb{E}_{v\sim\mathcal{D}}[z_v(\theta)]\rangle \geq (1 - \frac{1}{2\zeta})\|\nabla f(\theta)\|^2 - \frac{\zeta}{2}\epsilon^2(\theta),$$

which gives us $b(\zeta)$ and $c(\zeta, \epsilon^2(\theta))$ as required. From Lemma 1, there exist $A, B, C \geq 0$ such that

$$\mathbb{E}_{v\sim\mathcal{D}}[\|\nabla f_v(\theta)\|^2] \leq 2A(f(\theta) - f^{\text{inf}}) + B\|\nabla f(\theta)\|^2 + C. \tag{11}$$

Utilizing Lemma 3, for all $\eta > 0$ we have

$$\mathbb{E}_{v\sim\mathcal{D}}[\|z_v(\theta)\|^2] \leq \frac{1+\eta}{\eta}\mathbb{E}_{v\sim\mathcal{D}}[\|\nabla f_v(\theta)\|^2] + (1+\eta)\mathbb{E}_{v\sim\mathcal{D}}[\|e_v(\theta)\|^2].$$

Using the upper bound of $\mathbb{E}_{v\sim\mathcal{D}}[\|\nabla f_v(\theta)\|^2]$ from (11) in the inequality above yields

$$\mathbb{E}_{v\sim\mathcal{D}}[\|z_v(\theta)\|^2] \leq 2\frac{1+\eta}{\eta}A(f(\theta) - f^{\text{inf}}) + \frac{1+\eta}{\eta}B\|\nabla f(\theta)\|^2 + \frac{1+\eta}{\eta}C + (1+\eta)\epsilon^2(\theta). \tag{12}$$

Comparing (11) and (12), we find the required $\tilde{A}(\eta)$, $\tilde{B}(\eta)$, and $\tilde{C}(\eta, \epsilon^2(\theta))$. □

Remark 1. Since [18, Proposition 2 and 3] which states there exist certain A, B, $C \geq 0$ for each sampling strategy, holds under the same assumptions as Proposition 1, using the connection between $\tilde{A}$, $\tilde{B}$, and $\tilde{C}$, and A, B, and C, [18, Proposition 2 and 3] holds for the inexact stochastic gradient.

Now, utilizing the proposition above, we discuss the proofs of Theorem 1 and Corollary 1.

***Proof of Theorem* 1.** Since Proposition 1 demonstrates, under assumption 1, that the inexact stochastic gradient estimator in our bilevel setting satisfies the biased ABC condition [5, Assumption 9], the convergence theorem for biased SGD [5, Theorem 3] directly applies to ISGD. Hence, utilising Proposition 1 by setting $c_1 = \frac{b(\zeta)}{\tilde{B}(\eta)}$, $c_2 = \frac{2}{b(\zeta)}$, $c_3 = \tilde{A}(\eta)$, $c_4 = \frac{(1+\eta)C}{\eta b(\zeta)} + \frac{1+\eta}{b(\zeta)}\tau$, $c_5 = \frac{\zeta}{b(\zeta)}\tau$, where $\tau \geq 0$ such that $\sum_{k=0}^{T} \mathbb{E}[\epsilon_k^2] \leq \tau$, $\zeta > \frac{1}{2}$ and $\eta > 0$, Theorem 1 holds. □

***Proof of Corollary* 1.** By setting

$$T(\mathbb{E}[\epsilon_T^2]) = \frac{12\delta_0 L_{\nabla f}}{\zeta \mathbb{E}[\epsilon_T^2]} \max\left\{\frac{\tilde{B}(\eta)}{b(\zeta)}, \frac{12\delta_0 \tilde{A}(\eta)}{\zeta\mathbb{E}[\epsilon_T^2]}, \frac{2(1+\eta)(C+\eta\mathbb{E}[\epsilon_T^2])}{\eta\zeta\mathbb{E}[\epsilon_T^2]}\right\}, \text{ and}$$

$$\alpha(\mathbb{E}[\epsilon_T^2]) = \min\left\{\frac{1}{\sqrt{L_{\nabla f}\tilde{A}(\eta)T}}, \frac{b(\zeta)}{L_{\nabla f}\tilde{B}(\eta)}, \frac{\eta\zeta\mathbb{E}[\epsilon_T^2]}{2L_{\nabla f}(1+\eta)(C+\eta\mathbb{E}[\epsilon_T^2])}\right\},$$

by using Theorem 1 with constants $c_1 = \frac{b(\zeta)}{\tilde{B}(\eta)}$, $c_2 = \frac{2}{b(\zeta)}$, $c_3 = \tilde{A}(\eta)$, $c_4 = \frac{(1+\eta)C}{\eta b(\zeta)} + \frac{1+\eta}{b(\zeta)}\tau$, $c_5 = \frac{\zeta}{b(\zeta)}\tau$, where $\tau \geq 0$ such that $\sum_{k=0}^{T} \mathbb{E}[\epsilon_k^2] \leq \tau$, $\zeta > \frac{1}{2}$ and $\eta > 0$, for $T \geq T(\mathbb{E}[\epsilon_T^2])$ we have

$$\min_{0\leq t\leq T-1} \mathbb{E}\left[\|\nabla f(\theta^t)\|^2\right] \leq \frac{3\zeta\mathbb{E}[\epsilon_T^2]}{2b(\zeta)}.$$

Now for a fixed accuracy $\delta > 0$, substituting $\delta = \mathbb{E}[\epsilon_T^2]$, setting $T_\delta := T(\delta)$, and $\alpha_\delta := \alpha(\delta)$, we find the desired result. □

Remark 2. Note that if $\{\epsilon_t\}_{t=0}^{\infty}$ is chosen as a deterministic non-negative sequence such that $\sum_{t=0}^{\infty} \epsilon_t^2 < \infty$, and $\zeta \geq \frac{1}{2}$, then following similar steps as the proof of Corollary 1, we have $\lim_{T\to\infty}\left(\min_{0\leq t\leq T-1} \mathbb{E}\left[\|\nabla f(\theta^t)\|^2\right]\right) = 0$.

3 Numerical Experiments

The implementation codes are available on the GitHub repository[1].

3.1 Learning Field of Experts Regularizer for Denoising

In this section, we consider Field of Experts [3] (FoE) priors as a data-adaptive approach. In all experiments, accelerated gradient descent [2, Algorithm 3] is used as the lower-level solver, while the conjugate gradient method is used to approximate the inverse of the Hessian of the lower-level objective involved in computing the hypergradient. Given m pairs $\{(y_i, x_i^*)\}_{i=1}^m$ of noisy images with

[1] https://doi.org/10.5281/zenodo.14987053.

Gaussian noise and clean images, the bilevel problem correspond to learning the FoE regularizer for image denoising takes the following form

$$\min_{\theta \in \mathbb{R}^d} \frac{1}{m} \sum_{i=1}^{m} \|\hat{x}_i(\theta) - x_i^*\|^2 \quad s.t. \quad \hat{x}_i(\theta) = \arg\min_{x \in \mathbb{R}^n} \frac{1}{2}\|x - y_i\|^2 + R_\theta(x), \qquad (13)$$

where $R_\theta(x) = e^{\theta_0} \sum_{j=1}^{J} e^{\theta_j} \|c_j * x\|_{\theta_{J+j}}$, each c_j represents a $2D$ convolutional kernel with parameters dependent on θ, and $\|x\|_\nu = \sum_{i=1}^{N} \sqrt{x_i^2 + \nu^2} - \nu$ is the smoothed 1-norm. We take $J = 10$ kernels of size 7×7 with 3 channels, which yields to $d = 1482$ parameters. Moreover we chose $m = 1024$ colored training images of size 96×96 from STL10 dataset[2] in mini-batches of size $S = 32$, and to obtain each y_i, we add Gaussian noise drawn from $\mathcal{N}(0, \sigma)$ with noise level 25 (i.e. $\sigma = \frac{25}{255}$). In Fig. 1, the left subfigure shows the loss per iteration for ISGD with fixed step sizes plotted against the total lower-level computations for solving (13). The right subfigure presents the same results for ISGD with decreasing step sizes $\alpha_k = \frac{\alpha_0}{\sqrt{k}}$. For fixed $\alpha_0 = 10^{-3}$, oscillations were high without sustained loss reduction, while the decreasing step size version reduced oscillations, achieving the lowest loss temporarily before increasing at the end. Fixed $\alpha_0 = 10^{-4}$ initially fluctuated but eventually achieved the lowest loss among fixed settings. Its decreasing variant was more stable early on, eventually matching the fixed version's success without diverging. Both $\alpha_0 = 10^{-5}$ and 10^{-6} showed strong initial loss reduction, plateauing after 10^5 computations, with decreasing variants demonstrating similar trends but with less fluctuation. Overall, smaller step sizes provide stable initial decreases but limited long-term progress, whereas larger step sizes, especially with decreasing schedules, improve long-term performance, underscoring the importance of their theoretical analysis. In Fig. 2, the left subfigure compares the upper-level loss captured per batch for ISGD variants and per epoch for Method of Adaptive Inexact Descent (MAID) [28], plotted against

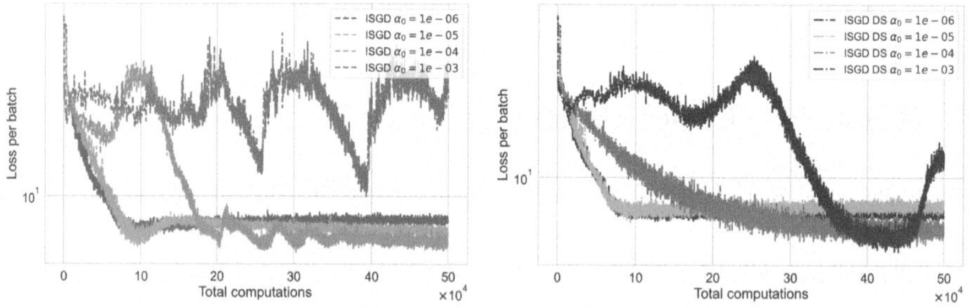

Fig. 1. Loss per mini-batch versus total lower-level computational cost for ISGD with a constant step size, $\alpha_k = \alpha_0$, (left) , and a decreasing step size (DS) , $\alpha_k = \frac{\alpha_0}{\sqrt{k}}$, (right) . While the decreasing step size variant exhibits more stability, a fixed small step size achieves comparable performance.

[2] https://cs.stanford.edu/~acoates/stl10/.

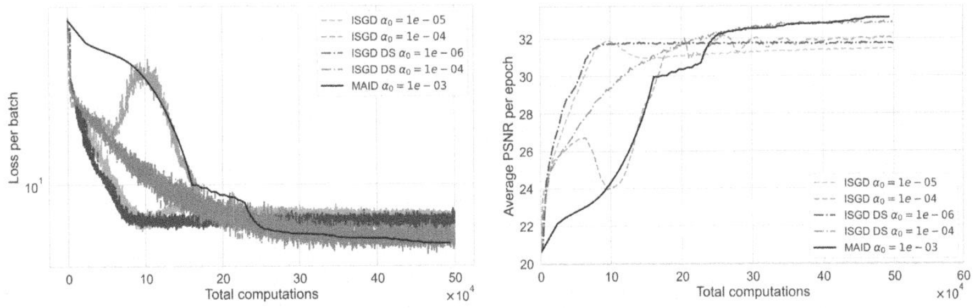

Fig. 2. Comparison of ISGD (with and without decreasing step sizes) and MAID in terms of loss per mini-batch for ISGD variants and upper-level loss for MAID, plotted against computational cost, as well as average training PSNR per epoch vs total computations.

total lower-level computations. The hyperparameters of MAID are selected based on empirical tuning. ISGD variants, except fixed step size $\alpha_0 = 10^{-4}$, reached a neighborhood of the local minima faster than MAID within 2×10^5 computations. However, MAID's adaptivity allowed it to perform comparably to the best ISGD variant ($\alpha_0 = 10^{-4}$ with decreasing step size) by the end. The right subfigure shows a similar trend for average PSNR per epoch, demonstrating ISGD's faster performance on the training dataset. Up to 2×10^5 computations, ISGD achieved significantly higher PSNR than MAID, highlighting its speed advantage during early training stages.

To qualitatively demonstrate this difference, Fig. 3 compares the performance of the best ISGD variants (fixed step size $\alpha_0 = 10^{-5}$, and $\alpha_0 = 10^{-4}$ with decreasing step size) to MAID on a training image. Before 2×10^5 total computations, ISGD achieves a higher PSNR.

3.2 Learning Field of Experts Regularizer for Deblurring

In this section, using the FoE model as in the previous section, we consider an image deblurring problem as the lower-level problem. For training, we use $m = 1536$ color images with three channels, each of size 256×256, from the Oxford-IIIT Pet Dataset[3]. In this case, while the upper-level problem remains the same as in (13), the lower-level problem is replaced with $\hat{x}_i(\theta) = \arg\min_{x \in \mathbb{R}^n} \frac{1}{2}\|Ax - y_i\|^2 + R_\theta(x)$, where A is the convolution operator corresponding to a Gaussian blur kernel of size 7×7 with $\sigma = 2$. Each y_i is obtained by applying A to x_i^* and adding Gaussian noise with a noise level of 5. In this experiment, for ISGD, we use all $m = 1536$ images in mini-batches of size $S = 32$. By contrast, MAID requires access to the full training dataset, which in this case exceeds the available GPU memory (8 GB). To allow MAID's entire dataset to be stored in memory, we only use the first m = 256 images from the full dataset for MAID.

[3] https://www.robots.ox.ac.uk/~vgg/data/pets/.

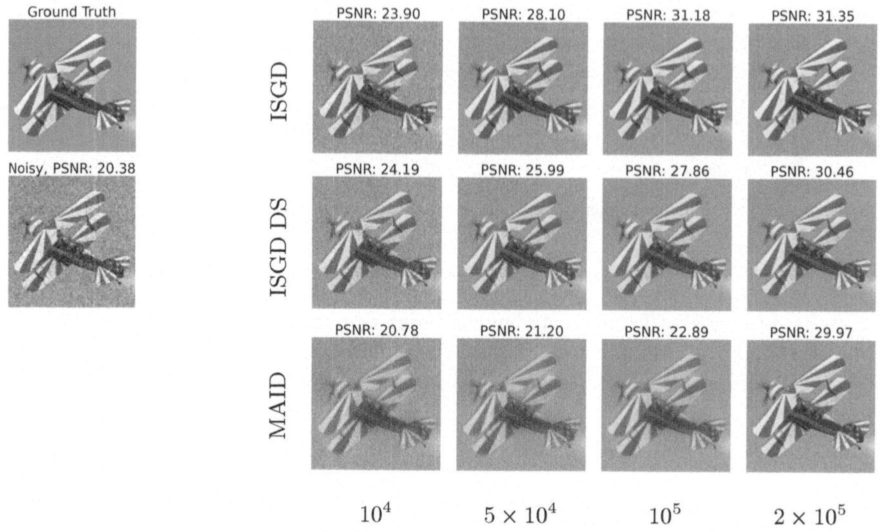

Fig. 3. First row: A denoised training image after 10^4, 5×10^4, 10^5, and 2×10^5 computations (lower-level + linear solver iterations) using ISGD with fixed step size $\alpha = 10^{-5}$. Second row: Same image with ISGD using decreasing step size (DS) , $\alpha_k = \frac{\alpha_0}{\sqrt{k}}$, $\alpha_0 = 10^{-4}$. Third row: Same image with MAID, $\alpha_0 = 10^{-3}$.

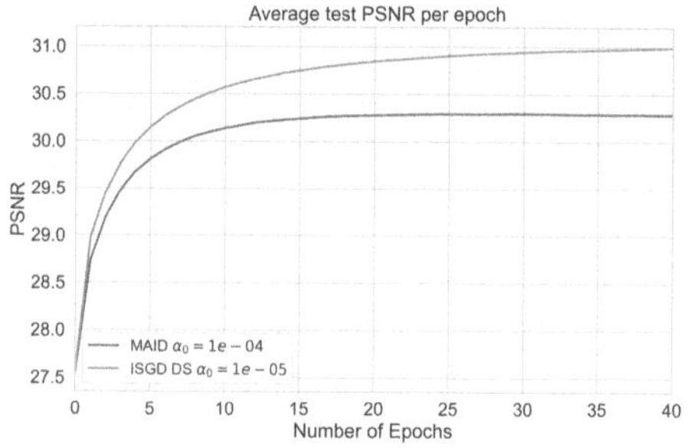

Fig. 4. Comparison on test dataset between the average PSNR per MAID iteration (full-batch) and average PSNR per epoch of ISGD with decreasing step size $\alpha_k = \frac{\alpha_0}{\sqrt{k}}$.

Hence in this setting the data sub-sampling in ISGD allows it to use a much larger training dataset than MAID.

As shown in Fig. 4, ISGD with $\alpha_0 = 10^{-4}$ and decreasing step size achieved a higher average PSNR on 32 test images after each epoch of training compared to MAID with $\alpha_0 = 10^{-3}$. This demonstrates ISGD's ability to handle larger datasets, leading to better generalization. A quantitative comparison of the deblurred test image using the learned regularizer from MAID and ISGD is provided in Fig. 5, where ISGD achieved a higher PSNR.

(a) Ground truth (b) Blurry: 24.78 dB (c) MAID: 27.06 dB (d) ISGD: 27.86 db

Fig. 5. Comparison of deblurred images using the learned FoE regularizer, learned by MAID and ISGD, on test images.

4 Conclusions and Future Work

In this work, we exploited the connection between inexact stochastic hypergradients in bilevel problems with upper-level stochasticity on data and the practical assumptions of biased SGD. We proposed the first convergence proof for this framework under practical assumptions. Our numerical results showcase the superiority of this approach in terms of convergence speed compared to state-of-the-art deterministic bilevel learning algorithms. Moreover, it leads to a better generalization, driven by its scalability to larger datasets.

In the future, as seen in Sect. 3, it would be worthy to extend our analysis to include decaying and adaptive step sizes, as well as exploring adaptive line search methods to determine suitable step sizes automatically, potentially enhancing both performance and robustness to the initial choice of the step size.

Acknowledgments. The work of Mohammad Sadegh Salehi was supported by a scholarship from the EPSRC Centre for Doctoral Training in Statistical Applied Mathematics at Bath (SAMBa), under the project EP/S022945/1. Matthias J. Ehrhardt acknowledges support from the EPSRC (EP/S026045/1, EP/T026693/1, EP/V026259/1). Lindon Roberts is supported by the Australian Research Council Discovery Early Career Award DE240100006.

References

1. Bogensperger, L., Chambolle, A., Pock, T.: Convergence of a piggyback-style method for the differentiation of solutions of standard saddle-point problems. SIAM J. Math. Data Sci. **4**(3), 1003–1030 (2022)
2. Chambolle, A., Pock, T.: An introduction to continuous optimization for imaging. Acta Numer. **25**, 161–319 (2016)
3. Chen, Y., Ranftl, R., Pock, T.: Insights into analysis operator learning: from patch-based sparse models to higher order MRFs. IEEE Trans. Image Process. **23**(3), 1060–1072 (2014)
4. Crockett, C., Fessler, J.A.: Bilevel methods for image reconstruction. Found. Trends Signal Process. **15**(2–3), 121–289 (2022)

5. Demidovich, Y., Malinovsky, G., Sokolov, I., Richtárik, P.: A guide through the zoo of biased SGD. In: Thirty-seventh Conference on Neural Information Processing Systems (2023)
6. Dempe, S., Zemkoho, A. (eds.): Bilevel Optimization: Advances and Next Challenges, vol. 161. Springer International Publishing, Cham (2020)
7. Downing, S., Gazzola, S., Graham, I.G., Spence, E.A.: Optimising seismic imaging design parameters via bilevel learning. Inverse Probl. **40**(11), 115008 (2024)
8. Ehrhardt, M.J., Gazzola, S., Scott, S.J.: On optimal regularization parameters via bilevel learning. In: Data-driven Models in Inverse Problems. Radon Series on Computational and Applied Mathematics - RICAM, De Gruyter (2024)
9. Ehrhardt, M.J., Roberts, L.: Analyzing inexact hypergradients for bilevel learning. IMA J. Appl. Math. **89**(1), 254–278 (2023)
10. Franceschi, L., Frasconi, P., Salzo, S., Grazzi, R., Pontil, M.: Bilevel programming for hyperparameter optimization and meta-learning. In: Proceedings of the 35th ICML, vol. 80, pp. 1568–1577 (2018)
11. Ghadimi, S., Wang, M.: Approximation methods for bilevel programming (2018)
12. Goujon, A., Neumayer, S., Bohra, P., Ducotterd, S., Unser, M.A.: A neural-network-based convex regularizer for inverse problems. IEEE Trans. Comput. Imaging **9**, 781–795 (2022)
13. Gower, R.M., Loizou, N., Qian, X., Sailanbayev, A., Shulgin, E., Richtárik, P.: SGD: general analysis and improved rates. In: Proceedings of the 36th ICML, vol. 97, pp. 5200–5209 (2019)
14. Grazzi, R., Franceschi, L., Pontil, M., Salzo, S.: On the iteration complexity of hypergradient computation. In: Proceedings of the 37th ICML, vol. 119, pp. 3748–3758 (2020)
15. Grazzi, R., Pontil, M., Salzo, S.: Convergence properties of stochastic hypergradients. In: Proceedings of the 24th AISTATS, vol. 130, pp. 3826–3834 (2021)
16. Holler, G., Kunisch, K., Barnard, R.C.: A bilevel approach for parameter learning in inverse problems. Inverse Prob. **34**(11), 115012 (2018)
17. Ji, K., Yang, J., Liang, Y.: Bilevel optimization: convergence analysis and enhanced design. In: Proceedings of the 38th ICML, vol. 139, pp. 4882–4892 (2021)
18. Khaled, A., Richtárik, P.: Better theory for SGD in the nonconvex world. Trans. Mach. Learn. Res. (2023)
19. Kunisch, K., Pock, T.: A bilevel optimization approach for parameter learning in variational models. SIAM J. Imaging Sci. **6**, 938–983 (2013)
20. Kwon, J., Kwon, D., Wright, S., Nowak, R.D.: A fully first-order method for stochastic bilevel optimization. In: Proceedings of the 40th ICML, vol. 202, pp. 18083–18113 (2023)
21. Liu, B., Ye, M., Wright, S., Stone, P., Liu, Q.: BOME! bilevel optimization made easy: a simple first-order approach. In: Oh, A.H., Agarwal, A., Belgrave, D., Cho, K. (eds.) Advances in Neural Information Processing Systems (2022)
22. Mehmood, S., Ochs, P.: Automatic differentiation of some first-order methods in parametric optimization. In: Proceedings of the 23rd AISTATS, vol. 108, pp. 1584–1594 (2020)
23. Mukherjee, S., Dittmer, S., Shumaylov, Z., Lunz, S., Öktem, O., Schönlieb, C.B.: Data-driven convex regularizers for inverse problems. In: ICASSP 2024 - 2024 IEEE International Conference on Acoustics, Speech and Signal Processing (ICASSP), pp. 13386–13390 (2024)
24. Ochs, P., Ranftl, R., Brox, T., Pock, T.: Techniques for gradient-based bilevel optimization with non-smooth lower level problems. J. Math. Imaging Vis. **56**(2), 175–194 (2016)

25. Pedregosa, F.: Hyperparameter optimization with approximate gradient (2016)
26. Ramzi, Z., Mannel, F., Bai, S., Starck, J.L., Ciuciu, P., Moreau, T.: SHINE: sharing the inverse estimate from the forward pass for bi-level optimization and implicit models. In: International Conference on Learning Representations (2022)
27. Reyes, J.C.D.l., Villacís, D.: Bilevel Optimization Methods in Imaging, pp. 909–941. Springer International Publishing, Cham (2023)
28. Salehi, M.S., Mukherjee, S., Roberts, L., Ehrhardt, M.J.: An adaptively inexact first-order method for bilevel optimization with application to hyperparameter learning (2024)
29. Sherry, F., et al.: Learning the sampling pattern for MRI. IEEE Trans. Med. Imaging **39**(12), 4310–4321 (2020)
30. Suonperä, E., Valkonen, T.: General single-loop methods for bilevel parameter learning (2024)

Deep Generative Clustering with VAEs and Expectation-Maximization

Michael Adipoetra(✉) and Ségolène Martin

Technische Universität Berlin, Institute of Mathematics, Straße des 17. Juni 136, 10623 Berlin, Germany
michael.adipoetra@campus.tu-berlin.de, martin@math.tu-berlin.de

Abstract. We propose a novel deep clustering method that integrates Variational Autoencoders (VAEs) into the Expectation-Maximization (EM) framework. Our approach models the probability distribution of each cluster with a VAE and alternates between updating model parameters—by maximizing the Evidence Lower Bound (ELBO) of the log-likelihood—and refining cluster assignments based on the learned distributions. This enables effective clustering and generation of new samples from each cluster. Unlike existing VAE-based methods, our approach eliminates the need for a Gaussian Mixture Model (GMM) prior or additional regularization techniques. Experiments on MNIST and FashionMNIST demonstrate superior clustering performance compared to state-of-the-art methods.

Keywords: Variational Autoencoders · Clustering

1 Introduction

Clustering of images is a fundamental challenge in machine learning research [7,27]. While classical non-neural methods like K-means [14] or the Expectation-Maximization (EM) algorithm with a Gaussian Mixture Model [20] are still widely used in many applications, these traditional approaches struggle to effectively capture the complex structures of data, such as image distributions. This limitation arises from the underlying assumption that the distribution of each cluster is Gaussian, centered around a parameter learned from the data. However, this assumption is often too simplistic.

To address these limitations, dimensionality reduction techniques can be used to project data into a lower-dimensional space where separation becomes easier. These transformations can be linear, such as Principal Component Analysis (PCA) [21], or nonlinear, using neural networks to learn complex data representations [6,26]. When the network, often referred to as an *encoder*, is learned independently of the clustering task, classical methods like K-means can still be applied on the latent representations. However, the results may be suboptimal since the encoding is not tailored for clustering on the given data. Designing an encoder specifically for clustering is challenging in the absence of labeled data.

T. A. Bubba et al. (Eds.): SSVM 2025, LNCS 15667, pp. 360–373, 2025.
https://doi.org/10.1007/978-3-031-92366-1_28

In recent years, generative models have gained popularity for clustering tasks. These models enable the learning of the (multi-modal) data distribution using only unlabeled training data [4,19]. Interestingly, the use of a generative model for clustering goes beyond the clustering tasks and also opens up the possibility of generating new data points from the learned distribution, thus enriching classification datasets without the need of human labeling.

In particular, Variational Autoencoders (VAE) [10,11], which consist of an encoder that maps data to a low-dimensional latent space and a decoder that reconstructs the data, have been used as a means to simultaneously learn meaningful latent representations and cluster data. Several VAE-based clustering methods have been proposed in recent years. Variational Deep Embedding (VaDE) [8] employs a Gaussian Mixture Model (GMM) prior in the latent space with trainable parameters, initialized using a pretrained model. The Gaussian Mixture VAE (GMVAE) method [3] offers a similar approach but eliminates the need for pretraining. However, GMVAE often suffers from cluster aggregation in the latent space, necessitating a cut-off trick during training to stabilize performance. To address such issues, some studies have introduced regularization terms in the loss function. For example, [13] proposed a similarity-based loss as a graph embedding extension to ensure a more compact and well-separated latent space. Similarly to VaDE, [22] introduced a model (referred to here as VAEIC) that avoids pretraining and incorporates an augmentation loss to maintain consistent cluster assignments between original and augmented data.

In this paper, we propose a novel VAE-based clustering method. Unlike existing methods such as VaDE, GMVAE, and VAEIC, our method eliminates the need for a GMM prior in the latent space and avoids any regularization strategies. Our method integrates VAE with an EM clustering framework, inspired by the few-shot classification method proposed in [16]. We formalize our method by deriving a lower bound for the expected log-likelihood, expressed as the Evidence Lower Bound (ELBO). The EM algorithm is then used to maximize the bound, where it alternately updates the clustering assignments and the VAE parameters for each cluster.

The remainder of this paper is organized as follows. In Sect. 2, we provide a brief overview of the EM algorithm and VAE. In Sect. 3, we describe how our approach integrates the EM algorithm with the VAE generative framework for clustering. Finally, in Sect. 4, we present experimental results, comparing our method's clustering performance with other VAE-based clustering methods and showcasing the generated samples.

2 Background

Our method leverages the EM algorithm to learn cluster assignments and VAE parameters, enabling sample generation from each cluster. Section 2.1 reviews the EM algorithm, and Sect. 2.2 covers Variational Autoencoders (VAEs), a key component of our approach.

2.1 Expectation-Maximization Algorithm

The Expectation-Maximization (EM) algorithm provides an iterative framework to maximize log-likelihood in the presence of latent variables. Let $\mathcal{X}$ denote the data space. Here, we assume $\mathcal{X} = \mathbb{R}^d$ and model the data using a random variable $X : \Omega \to \mathcal{X} = \mathbb{R}^d$ with unknown probability distribution P_X. We aim to find a distribution P_α parameterized by model parameters $\boldsymbol{\alpha} \in \mathcal{A}$, where $\mathcal{A}$ represents the parameter space, such that $P_\alpha \approx P_X$. The corresponding density function (assumed to exist) is denoted as p_α. A common approach to determine the parameter $\boldsymbol{\alpha}$ is *maximum likelihood estimation* (MLE) which involves maximizing the expected log-likelihood function, defined as

$$\ell(\boldsymbol{\alpha}) := \mathbb{E}_{\boldsymbol{x} \sim P_X}[\log p_\alpha(\boldsymbol{x})].$$

MLE works effectively when the dataset is fully observed. However, the dataset might include hidden variables that are not directly observed, referred to as latent variables. These variables, denoted as $\boldsymbol{z}$, are realizations of a random variable $Z : \Omega \to \mathcal{Z}$ defined on some latent space $\mathcal{Z}$, which may differ from the data space $\mathcal{X}$. Although $\boldsymbol{z}$ is unobserved, it influences the modeling of $\boldsymbol{x}$. For example, in a mixture model, the latent variable $\boldsymbol{z}$ may represent an indicator that assigns data points to their respective clusters. Meanwhile, in VAE framework that we present next, the latent variable $\boldsymbol{z}$ is a low-dimensional latent representation that captures the underlying structure of the data $\boldsymbol{x}$. To account for its contribution, we marginalize over all possible values of $\boldsymbol{z}$, leading to the marginal log-likelihood

$$\ell(\boldsymbol{\alpha}) = \log p_\alpha(\boldsymbol{x}) = \log \int_z p_\alpha(\boldsymbol{x}, \boldsymbol{z}) \mathrm{d}\boldsymbol{z}.$$

Maximizing this function over $\boldsymbol{\alpha}$ may be very difficult or impossible to do analytically due to the integral. To address this issue, the Expectation-Maximization (EM) algorithm was introduced in [2]. Instead of directly maximizing the log-likelihood, it maximizes a lower bound on the log-likelihood, known as the Evidence Lower Bound (ELBO), which we re-derive for clarity. Using Bayes' formula, the joint density of observed data $\boldsymbol{x}$ and latent variable $\boldsymbol{z}$ is factorized as $p_\alpha(\boldsymbol{x}, \boldsymbol{z}) = p_\alpha(\boldsymbol{z}|\boldsymbol{x}) p_\alpha(\boldsymbol{x})$, where $p_\alpha(\boldsymbol{z}|\boldsymbol{x})$ is the posterior of $\boldsymbol{z}$ given $\boldsymbol{x}$. For any probability density function

$$q_x \in \mathcal{Q} := \left\{ q : \mathcal{Z} \to [0, \infty), \int_{z \in \mathcal{Z}} q(z) dz = 1 \right\},$$

the log-likelihood function can be decomposed as

$$\log(p_\alpha(\boldsymbol{x})) = \mathbb{E}_{\boldsymbol{z} \sim q_x} \left[\log \left(\frac{p_\alpha(\boldsymbol{x}, \boldsymbol{z})}{q_x(\boldsymbol{z})} \right) \right] + \mathrm{KL}\left(q_x(\cdot) \| p_\alpha(\cdot|\boldsymbol{x}) \right), \quad (1)$$

where $\mathrm{KL}(\cdot||\cdot)$ denotes the Kullback-Leibler divergence. Here, both the model parameter $\boldsymbol{\alpha}$ and the density q_x are unknown. This decomposition introduces the term Evidence Lower Bound (ELBO), a function of $\boldsymbol{\alpha}$ and q_x, as

$$\mathrm{ELBO}(\boldsymbol{\alpha}, q_x; \boldsymbol{x}) := \mathbb{E}_{z\sim q_x}\left[\log\left(\frac{p_{\boldsymbol{\alpha}}(\boldsymbol{x},\boldsymbol{z})}{q_x(\boldsymbol{z})}\right)\right]. \tag{2}$$

Since KL is non-negative, ELBO provides a lower bound for $\log(p_{\boldsymbol{\alpha}}(\boldsymbol{x}))$ and the bound is tight if $q_x = p_{\boldsymbol{\alpha}}(\cdot|\boldsymbol{x})$ for all $\boldsymbol{x}$.

To use the EM algorithm, it is important to choose a model where the posterior distribution $p_{\boldsymbol{\alpha}}(\cdot|\boldsymbol{x})$ can be computed in a closed form (e.g., a Gaussian Mixture Model [20]). In order to optimize ELBO, the EM algorithm alternates between updates on the density function $q_x \in \mathcal{Q}$ and the model parameter $\boldsymbol{\alpha} \in \mathcal{A}$. For each iteration j, the following steps are performed.

E-Step. Given the current parameter $\boldsymbol{\alpha}^{(j)}$, the density q_x is updated by maximizing $\mathrm{ELBO}(\boldsymbol{\alpha}^{(j)}, q_x; \boldsymbol{x})$ with respect to q_x. It is equivalent to minimizing the KL divergence $\mathrm{KL}\left(q_x \| p_{\boldsymbol{\alpha}^{(j)}}(\cdot|\boldsymbol{x})\right)$, which achieves its minimum value of zero for $q_x = p_{\boldsymbol{\alpha}^{(j)}}(\cdot|\boldsymbol{x})$.

M-Step. In the maximization step, we aim to update $\boldsymbol{\alpha}$ while keeping q fixed from the E-step. Using q from the E-step (depending on $\boldsymbol{\alpha}^{(j)}$), ELBO can be decomposed as

$$\mathbb{E}_{z\sim q_x}\left[\log\left(\frac{p_{\boldsymbol{\alpha}}(\boldsymbol{x},\boldsymbol{z})}{q_x(\boldsymbol{z})}\right)\right] = \underbrace{\mathbb{E}_{z\sim q_x}\left[\log\left(p_{\boldsymbol{\alpha}}(\boldsymbol{x},\boldsymbol{z})\right)\right]}_{=:U(\boldsymbol{\alpha};\boldsymbol{\alpha}^{(j)},\boldsymbol{x})} - \mathbb{E}_{z\sim q_x}\left[\log\left(q_x(\boldsymbol{z})\right)\right].$$

Since the second term only depends on $\boldsymbol{\alpha}^{(j)}$ and not on $\boldsymbol{\alpha}$, the M-step reduces to maximizing $\mathbb{E}_{x\sim P_X}\left[U(\boldsymbol{\alpha};\boldsymbol{\alpha}^{(j)},\boldsymbol{x})\right]$, which can nevertheless be tedious. Dempster, Laird and Rubin [2] introduced the Generalized Expectation-Maximization (GEM) algorithm, which relaxes this requirement by only requiring that the updated parameter $\boldsymbol{\alpha}^{(j+1)}$ satisfies the increasing condition

$$\mathbb{E}_{x\sim P_X}\left[U(\boldsymbol{\alpha}^{(j+1)};\boldsymbol{\alpha}^{(j)},\boldsymbol{x})\right] \geq \mathbb{E}_{x\sim P_X}\left[U(\boldsymbol{\alpha}^{(j)};\boldsymbol{\alpha}^{(j)},\boldsymbol{x})\right].$$

Similarly to the EM algorithm, in the VAE framework, one optimizes a likelihood-based objective to learn the data distribution. However, while the EM algorithm assumes that the posterior distribution $p_{\boldsymbol{\alpha}}(\cdot|\boldsymbol{x})$ can be computed, VAE use neural networks to approximate it.

2.2 VAE

Variational Autoencoders (VAEs), introduced by Kingma and Welling [11], are a generative modeling framework that incorporates a latent variable $Z : \Omega \to \mathcal{Z}$, where $\mathcal{Z} = \mathbb{R}^n$ represents a low-dimensional latent space ($n \ll d$). The latent variable Z follows a simple prior distribution P_Z, typically a standard normal distribution, making sampling tractable even in higher dimensions.

A VAE approximates the data distribution P_X using P_Z, employing a pair of stochastic mappings: the encoder and decoder, denoted by (E, D). These mappings are used in the inference process (to estimate latent variable $\boldsymbol{z}$ from data $\boldsymbol{x}$) and the generation process (to reconstruct or generate $\boldsymbol{x}$ from $\boldsymbol{z}$), akin to dimensionality reduction techniques. The latent variable $\boldsymbol{z}$ captures a low-dimensional representation of the data, similar to the role of unobserved variables in the EM algorithm.

Inference Process. Given an observed data $\boldsymbol{x}$, this process aims to estimate the latent variable $\boldsymbol{z}$. Specifically, for $\boldsymbol{x} \in \mathbb{R}^d$ drawn from the data distribution P_X, the neural network for the encoder $E = E_{\boldsymbol{\phi}} : \mathbb{R}^d \to \mathbb{R}^n \times \mathbb{R}^{n \times n}$ outputs the parameters $(\mu_{\boldsymbol{\phi}}(\boldsymbol{x}), \Sigma_{\boldsymbol{\phi}}(\boldsymbol{x}))$ of a multivariate Gaussian distribution $\mathcal{N}(\mu_{\boldsymbol{\phi}}(\boldsymbol{x}), \Sigma_{\boldsymbol{\phi}}(\boldsymbol{x}))$ with $E_{\boldsymbol{\phi}}(\boldsymbol{x}) := (\mu_{\boldsymbol{\phi}}(\boldsymbol{x}), \Sigma_{\boldsymbol{\phi}}(\boldsymbol{x}))$. The density of the stochastic encoder $E_{\boldsymbol{\phi}}$, often called inference model is given by

$$q_{\boldsymbol{\phi}}(\boldsymbol{z}|\boldsymbol{x}) = \mathcal{N}(\boldsymbol{z}; \mu_{\boldsymbol{\phi}}(\boldsymbol{x}), \Sigma_{\boldsymbol{\phi}}(\boldsymbol{x})).$$

In practice, the covariance $\Sigma_{\boldsymbol{\phi}}(\boldsymbol{x})$ is often assumed diagonal, with the encoder outputting the mean and log variance, i.e., $E_{\boldsymbol{\phi}}(\boldsymbol{x}) = (\mu_{\boldsymbol{\phi}}(\boldsymbol{x}), \log(\sigma^2_{\boldsymbol{\phi}}(\boldsymbol{x})))$. This results in the density:

$$q_{\boldsymbol{\phi}}(\boldsymbol{z}|\boldsymbol{x}) = \mathcal{N}\left(\boldsymbol{z}; \mu_{\boldsymbol{\phi}}(\boldsymbol{x}), \mathrm{diag}(\sigma^2_{\boldsymbol{\phi}}(\boldsymbol{x}))\right).$$

Generation Process. After obtaining the latent variable $\boldsymbol{z}$ from data $\boldsymbol{x}$ in the inference process, we can reconstruct $\boldsymbol{x}$ using $\boldsymbol{z}$ or generate new data by sampling $\boldsymbol{z}$. For reconstruction, $\boldsymbol{z} \sim q_{\boldsymbol{\phi}}(\cdot|\boldsymbol{x})$, and for random generation, $\boldsymbol{z} \sim \mathcal{N}(\mathbf{0}, I)$. The decoder neural network $D_{\boldsymbol{\theta}} : \mathbb{R}^n \to \mathbb{R}^d$ maps $\boldsymbol{z}$ to the mean $\mu_{\boldsymbol{\theta}}(\boldsymbol{z})$ of a multivariate Gaussian distribution with identity covariance, where $D_{\boldsymbol{\theta}}(\boldsymbol{z}) := \mu_{\boldsymbol{\theta}}(\boldsymbol{z})$, and the density of the decoder, known as generation model, is

$$p_{\boldsymbol{\theta}}(\boldsymbol{x}|\boldsymbol{z}) = \mathcal{N}(\boldsymbol{x}; \mu_{\boldsymbol{\theta}}(\boldsymbol{z}), I). \tag{3}$$

The reconstructed or generated data $\boldsymbol{x}$ is obtained as $\boldsymbol{x} = \mu_{\boldsymbol{\theta}}(\boldsymbol{z})$. Some approaches learn a diagonal covariance similar as the encoder to mitigate mode collapse [23], however we fix it to the identity matrix in this paper.

Training. In the following, we denote $\boldsymbol{\alpha} := (\boldsymbol{\phi}, \boldsymbol{\theta})$, where $\boldsymbol{\phi}$ and $\boldsymbol{\theta}$ are the neural network parameters for encoder and decoder of the VAE respectively. In order to learn these parameters $\boldsymbol{\alpha}$, Kingma [11] seeks to maximizing the log-likelihood $\mathbb{E}_{\boldsymbol{x} \sim P_X}[\log p_{\boldsymbol{\theta}}(\boldsymbol{x})]$ by optimizing ELBO, which can further be derived as

$$\mathrm{ELBO}(\boldsymbol{\alpha}; \boldsymbol{x}) = \mathbb{E}_{\boldsymbol{z} \sim q_{\boldsymbol{\phi}}(\cdot|\boldsymbol{x})}[\log p_{\boldsymbol{\theta}}(\boldsymbol{x}|\boldsymbol{z})] - \mathrm{KL}(q_{\boldsymbol{\phi}}(\boldsymbol{z}|\boldsymbol{x}) || p_Z(\boldsymbol{z})). \tag{4}$$

For a detailed derivation of the ELBO and an overview of VAE, see [5]. The first term, known as reconstruction loss, encourages the decoder to reconstruct $\boldsymbol{x}$ accurately for a given $\boldsymbol{z}$. This is expressed as

$$\mathbb{E}_{\boldsymbol{z} \sim q_{\boldsymbol{\phi}}(\cdot|\boldsymbol{x})}[\log p_{\boldsymbol{\theta}}(\boldsymbol{x}|\boldsymbol{z})] = -\frac{1}{2}\mathbb{E}_{\boldsymbol{z} \sim \mathcal{N}([E_{\boldsymbol{\phi}}(\boldsymbol{x})]_1, \mathrm{diag}(\exp[E_{\boldsymbol{\phi}}(\boldsymbol{x})]_2))}\left[||\boldsymbol{x} - D_{\boldsymbol{\theta}}(\boldsymbol{z})||^2\right].$$

The second term regularizes the inference model to align with the prior. The KL divergence between two multivariate Gaussians has the closed-form expression

$$\mathrm{KL}(q_\phi(\boldsymbol{z}|\boldsymbol{x})||p_Z(\boldsymbol{z})) = \frac{1}{2}\sum_{i=1}^{n}\left(-1 - [[E_\phi(\boldsymbol{x})]_2]_i + [\exp([E_\phi(\boldsymbol{x})]_2)]_i + [[E_\phi(\boldsymbol{x})]_1]_i^2\right).$$

However, the gradient of the ELBO with respect to the parameter ϕ is intractable, so we use the *reparameterization trick* introduced in [11], enabling computation of the reconstruction loss as:

$$\begin{aligned}&\mathbb{E}_{\boldsymbol{z}\sim q_\phi(\cdot|\boldsymbol{x})}\left[\log p_{\boldsymbol{\theta}}(\boldsymbol{x}|\boldsymbol{z})\right]\\ &= -\frac{1}{2}\mathbb{E}_{\boldsymbol{\varepsilon}\sim\mathcal{N}(\mathbf{0},I)}\left[\|x - D_\theta\left([E_\phi(\boldsymbol{x})]_1 + \exp([E_\phi(\boldsymbol{x})]_2/2)\odot\boldsymbol{\varepsilon}\right)\|^2\right].\end{aligned}$$

3 Proposed Approach

We assume access to N data points $\boldsymbol{x}_1, \ldots, \boldsymbol{x}_N$ from $K \in \mathbb{N}$ distinct clusters, with K assumed to be known. The objective is twofold: (1) to partition the data into K clusters without access to labels, and (2) to learn the distribution of each cluster. At inference, the learned distributions can be leveraged to (i) assign test samples to a cluster and (ii) generate new samples from each cluster.

Note that training a generic generative model (e.g., VAE, GAN, flow-based model) on a multi-class dataset allows for generating new samples from the overall dataset but not from specific classes. A common approach to address this is to condition the model on class labels [18]. In contrast, our method enables generation of samples from specific clusters without requiring labels.

Our method builds on the transductive few-shot clustering algorithm introduced by Martin et al. [16] and extended in [17]. While their method addresses a different problem—few-shot classification, where the goal is to classify with very limited training samples—it shares key methodological similarities with ours. In these works, a pretrained encoder (trained on a generic dataset) is used to obtain low-dimensional representations of images. The distributions of the clusters in this latent space are then modeled using a parametric distribution, such as Dirichlet distributions. Their algorithm alternates between optimizing cluster assignments and estimating the parameters of these cluster distributions. In contrast, while we adopt a similar alternating optimization strategy, our approach learns the distribution of each cluster directly in the data space using a generative model. This enables us to avoid dependence on a fixed latent representation.

Assume that each data point $\boldsymbol{x}$ is generated from a mixture model [20] with fixed mixing coefficients $\boldsymbol{\pi} = \left(\frac{1}{K}, \ldots, \frac{1}{K}\right)$. The probability density function of the mixture model is given by:

$$p_{\boldsymbol{\alpha}}(\boldsymbol{x}) = \sum_{k=1}^{K}\pi_k p_{\boldsymbol{\alpha}_k}(\boldsymbol{x}), \tag{5}$$

where $\pi_k = \frac{1}{K}$ denotes the mixing coefficient for cluster k, $p_{\boldsymbol{\alpha}_k}(\boldsymbol{x})$ is the probability density function of cluster k, and $\boldsymbol{\alpha}_k$ represents the parameters specific to that cluster. We fix the mixing coefficients to $\frac{1}{K}$ for simplicity. In principle, if cluster sizes are significantly unbalanced or unknown, π_k can be treated as an additional trainable parameters.

The expected log-likelihood corresponding to the mixture model is:

$$\mathbb{E}_{x \sim P_X}[\log(p_{\boldsymbol{\alpha}}(\boldsymbol{x}))] = \sum_{i=1}^{N} \log(p_{\boldsymbol{\alpha}}(\boldsymbol{x}_i)) = \sum_{i=1}^{N} \log\left(\sum_{k=1}^{K} \frac{1}{K} p_{\boldsymbol{\alpha}_k}(\boldsymbol{x}_i)\right). \tag{6}$$

Furthermore, we introduce for each data point $\boldsymbol{x}_i$ a soft assignment variable $\boldsymbol{u}_i \in \Delta_K$, where Δ_K denotes the probability simplex over K classes:

$$\Delta_K = \left\{ \boldsymbol{u} = (u_k)_{1 \le k \le K} \in \mathbb{R}^K \;\middle|\; u_k \ge 0,\ \forall k \in \{1, \dots, K\}, \text{ and } \sum_{k=1}^{K} u_k = 1 \right\}.$$

Each element $u_{i,k}$ in the vector $\boldsymbol{u}_i$ represents the probability that sample $\boldsymbol{x}_i$ belongs to cluster k. Leveraging the concavity of the log function and starting from Eq. (5), we use the property that $\boldsymbol{u}_i \in \Delta_K$ to derive a lower bound for the expected log-likelihood:

$$\begin{aligned}
\mathbb{E}_{x \sim P_X}[\log(p_{\boldsymbol{\alpha}}(\boldsymbol{x}))] &\simeq \sum_{i=1}^{N} \log\left(\sum_{k=1}^{K} u_{i,k} \frac{p_{\boldsymbol{\alpha}_k}(\boldsymbol{x}_i)}{K u_{i,k}}\right) \\
&\ge \sum_{i=1}^{N} \sum_{k=1}^{K} u_{i,k} \log\left(\frac{p_{\boldsymbol{\alpha}_k}(\boldsymbol{x}_i)}{K u_{i,k}}\right) \\
&= \sum_{i=1}^{N} \sum_{k=1}^{K} u_{i,k} \log(p_{\boldsymbol{\alpha}_k}(\boldsymbol{x}_i)) - \sum_{i=1}^{N} \sum_{k=1}^{K} u_{i,k} \log(u_{i,k}) - N \log(K).
\end{aligned}$$

This lower bound of the log-likelihood leads to the optimization problem:

$$\begin{aligned}
\underset{(\boldsymbol{u}_i)_{1 \le i \le N}, (\boldsymbol{\alpha}_k)_{1 \le k \le K}}{\text{minimize}} \quad & -\sum_{i=1}^{N} \sum_{k=1}^{K} u_{i,k} \log p_{\boldsymbol{\alpha}_k}(\boldsymbol{x}_i) + \sum_{i=1}^{N} \sum_{k=1}^{K} u_{i,k} \log u_{i,k}, \\
\text{subject to} \quad & \boldsymbol{u}_i \in \Delta_K \quad \forall i \in \{1, \dots, N\}.
\end{aligned} \tag{7}$$

Remark 1. The minimization problem (7) aligns with the EM framework described in Sect. 2.1. In this context, the latent space is the discrete space of clusters, i.e., $\mathcal{Z} = \{1, \dots, K\}$. The random variable Z is the uniform cluster assignment, and for all $i \in \{1, \dots, N\}$, the approximate posterior $q_{\boldsymbol{x}_i}$ is defined as $q_{\boldsymbol{x}_i} = \sum_{k=1}^{K} u_{i,k} \delta_k$, where δ_k is the Dirac delta function centered at k. The Expectation (E) step involves minimizing the following expression:

$$-\sum_{i=1}^{N}\mathbb{E}_{z\sim q_{x_i}}\left[\log\left(\frac{p_{\boldsymbol{\alpha}}(z|\boldsymbol{x}_i)}{q_{\boldsymbol{x}_i}(z)}\right)\right] = -\sum_{i=1}^{N}\sum_{k=1}^{K} u_{i,k}\log\left(\frac{p_{\boldsymbol{\alpha}}(z=k|\boldsymbol{x}_i)}{u_{i,k}}\right),$$
$$= -\sum_{i=1}^{N}\sum_{k=1}^{K} u_{i,k}\log\left(\frac{p_{\boldsymbol{\alpha}}(\boldsymbol{x}_i|z=k)}{p_{\boldsymbol{\alpha}}(\boldsymbol{x}_i)u_{i,k}}\frac{1}{K}\right),$$
$$= -\sum_{i=1}^{N}\sum_{k=1}^{K} u_{i,k}\left(\log\left(\frac{p_{\boldsymbol{\alpha}_k}(\boldsymbol{x}_i)}{K u_{i,k}}\right)\right) + c,$$

where c is a constant independent on $(\boldsymbol{u}_i)_{1\leq i\leq N}$. We thus retrieve (7).

We now propose modeling the log-probability function $\log(p_{\boldsymbol{\alpha}_k}(\cdot))$ for each cluster $k \in \{1,\ldots,K\}$ using K distinct VAEs (one per cluster) with encoder and decoder parameters $\boldsymbol{\alpha}_k = (\boldsymbol{\phi}_k, \boldsymbol{\theta}_k)$, as defined in Sect. 2.2. Leveraging the ELBO from (4), the minimization problem (7) simplifies to:

$$\begin{aligned}&\underset{(\boldsymbol{u}_i)_{1\leq i\leq N},(\boldsymbol{\alpha}_k)_{1\leq k\leq K}}{\text{minimize}} && -\sum_{i=1}^{N}\sum_{k=1}^{K} u_{i,k}\text{ELBO}(\boldsymbol{\alpha}_k;\boldsymbol{x}_i) + \sum_{i=1}^{N}\sum_{k=1}^{K} u_{i,k}\log u_{i,k}, \\ &\text{subject to} && \boldsymbol{u}_i \in \Delta_K \quad \forall i \in \{1,\ldots,N\}.\end{aligned} \tag{8}$$

This optimization is performed iteratively via the EM algorithm.

Expectation Step. In the expectation step, we assume that the data $\boldsymbol{x}_i$ and the parameters $\boldsymbol{\alpha}_k$ are fixed. It is straightforward to observe that the minimization of (8) with respect to $\boldsymbol{u}_i$ is equivalent to minimizing a Kullback-Leibler divergence over the simplex. The solution to this problem has a known closed-form expression [1], given by:

$$\boldsymbol{u}_i = \text{softmax}\Big(\left\{\text{ELBO}(\boldsymbol{\alpha}_k^{(j)};\boldsymbol{x}_i)\right\}_{1\leq k\leq K}\Big)$$

Maximization Step. In the maximization step, the variables $\boldsymbol{u}_i$ are fixed, and the focus shifts to optimizing the parameters $\boldsymbol{\alpha}_k$ for each class $k \in \{1,\ldots,K\}$. Since the second term in the objective function does not depend on the neural network parameters, the optimization for each class simplifies to:

$$\underset{\boldsymbol{\alpha}_k}{\arg\min} -\sum_{i=1}^{N} u_{i,k}\text{ELBO}(\boldsymbol{\alpha}_k;\boldsymbol{x}_i).$$

The parameters $\boldsymbol{\alpha}_k$ are updated using the ADAM optimizer [9]. As exact minimization of the ELBO is not guaranteed with gradient-based approaches, this procedure aligns with the Generalized Expectation-Maximization framework, which requires only that the ELBO decreases at each iteration. The detailed process for solving the minimization problem in (8) is presented in Algorithm 1.

Algorithm 1: EM-Based Clustering with VAE

Input: $\{\boldsymbol{x}_1, \dots, \boldsymbol{x}_N\}$ train data, VAE models $(E_{\boldsymbol{\phi}_k}, D_{\boldsymbol{\theta}_k})$, for $k \in \{1, \dots, K\}$.
Output: Assignments $(u_{i,k})_{\substack{1 \le i \le N \\ 1 \le i \le K}}$ and model weights $\boldsymbol{\alpha} = \{(\boldsymbol{\phi}_k, \boldsymbol{\theta}_k)\}_{1 \le k \le K}$.
Init: Initialize soft assignments $\boldsymbol{u}_i \in \Delta_K$ randomly, initialize model weights $\boldsymbol{\alpha}_k^{(0)} = (\boldsymbol{\phi}_k^{(0)}, \boldsymbol{\theta}_k^{(0)})$, $k \in \{1, \dots, K\}$, randomly.
for $j = 1, 2, \dots,$ **do**

M-Step: Update VAE parameters for each cluster $k \in \{1, \dots, K\}$;

$$\boldsymbol{\alpha}_k^{(j)} = \arg\min_{\boldsymbol{\alpha}_k} - \sum_{i=1}^{N} u_{i,k}^{(j-1)} \text{ELBO}(\boldsymbol{\alpha}_k; \boldsymbol{x}_i)$$

E-Step: Update soft assignments for $i \in \{1, \dots, N\}$;

$$\boldsymbol{u}_i^{(j)} = \text{softmax}\Big(\Big\{ \text{ELBO}(\boldsymbol{\alpha}_k^{(j)}; \boldsymbol{x}_i) \Big\}_{1 \le k \le K} \Big)$$

4 Numerical Experiments

In this section, we present a comparative analysis of our proposed clustering method against several VAE-based clustering models. The code of our model is available at https://github.com/madipoetra20/VAE_EM_clustering.

4.1 Experimental Setup

Datasets. We created a synthetic 2D dataset with 5000 points derived from arcs of 5 half-moon shapes. Additionally, we used the MNIST [12] and FashionMNIST [25] datasets, each comprising 70,000 grayscale images of 28×28 pixels. MNIST features handwritten digits, while FashionMNIST consists of fashion items.

Baselines. We compare our approach with clustering VAE-based methods, namely VaDE [8], GMVAE [3], and VAEIC [22]. Other clustering approaches, such as hierarchical VAE (e.g., LadderVAE [24], TreeVAE [15]), GAN-based methods like ClusterGAN [19], also exist and have demonstrated strong performance in related tasks. However, in this experiment, we focus solely on comparisons with VAE-based methods. For VaDE, which relies on pretraining, we also report the accuracy of the pretraining step. The pretraining follows the original approach, using an autoencoder to reduce the data dimensionality to 10, followed by clustering with a Gaussian Mixture Model (GMM). This pretraining method is referred to as AE+GMM. GMVAE is implemented using 10 Monte Carlo (MC) samples and we set the number of clusters to be 10. All baseline methods are implemented using the architectures and hyperparameters proposed in their respective papers.

Hardware. All tests were run on a single NVIDIA GeForce RTX 2080 Ti with 16 GB of memory.

Training of Our Method. Our implementation uses a fully connected neural network architecture for both the encoder and decoder. The implementation details for our method are summarized in Table 1.

Table 1. Implementation Details for MNIST, FashionMNIST, and Synthetic Dataset. We put a '–' if we do not use the component.

Component	MNIST and FashionMNIST	Synthetic Dataset
Network Architecture	Fully connected	Fully connected
Encoder Dimensions	784 – 500 – 20	2 – 100 – 2
Decoder Dimensions	20 – 500 – 784	2 – 100 – 2
Activation Function	Leaky ReLU (slope 0.2)	Leaky ReLU (slope 0.2)
Dropout Rate	0.2	–
Learning Rate	0.001	0.0001
Decay	0.9 every 20 EM iterations	–
Optimizer	ADAM [9]	ADAM [9]
Batch Size	256	256
Number MC samples (E-step)	10	10
Number MC samples (M-step)	1	1
Reconstruction loss	1	5
Training Epochs (M-step)	20 per iteration	5 per iteration
Total EM Iterations	300	1000

4.2 Clustering Results

Table 2 shows the average and best clustering accuracy over 10 random runs for the MNIST and FashionMNIST datasets. AE+GMM performs worse, highlighting the effectiveness of VAE-based models. For MNIST, our method achieves the highest accuracy on both average and best runs. On FashionMNIST, all methods perform similarly, with GMVAE achieving the best single-run accuracy, while our approach leads in average accuracy, demonstrating greater consistency.

Table 2. Clustering accuracy (%). Best is shown in boldface and second best is underlined.

	***MNIST*-train**		***MNIST*-test**		***Fashion*-train**		***Fashion*-test**	
Method	Avg	Best	Avg	Best	Avg	Best	Avg	Best
GMVAE	85.4±4.1	93.1	85.4±4.1	93.5	<u>58.3</u>±5.8	**70.1**	<u>58.1</u>±5.6	**69.2**
VAEIC	86.8±4.9	93.2	87.1±4.8	93.7	57.6±4.0	<u>63.6</u>	57.3±3.9	63.5
AE+GMM	78.2±1.7	81.4	78.6±2.5	82.5	55.5±1.4	56.8	56.4±2.2	59.8
VaDE	<u>88.1</u>±2.9	<u>93.4</u>	<u>88.3</u>±3.0	<u>94.2</u>	57.2±1.6	59.4	57.0±1.7	59.8
Ours	**88.4**±5.2	**94.3**	**88.5**±5.1	**94.6**	**59.1**±2.9	63.3	**58.9**±3.2	<u>63.6</u>

4.3 Generated Samples

One key advantage of our proposed model is its ability to independently generate samples from each cluster. To generate new samples, we first draw random samples from the standard Gaussian distribution and pass them through the decoder of the corresponding cluster-specific VAE. In Figs. 1 and 2, we visualize the generated images from each cluster across the three datasets. We observe that: 1) our method effectively samples from each cluster independently with few class mismatch; 2) the generated samples exhibit high variability, and as shown in Fig. 1, our method covers the entire distribution of each cluster.

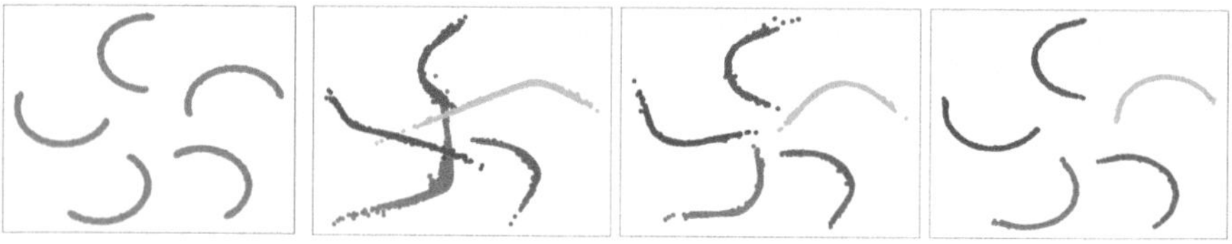

(a) Training samples (b) Iter 100 (c) Iter 200 (d) Iter 1000

Fig. 1. Training samples (unlabeled) and generated samples after 100, 200, and 1000 EM iterations. Each color represents samples from the same cluster-specific VAE. Our method effectively clusters and generates samples.

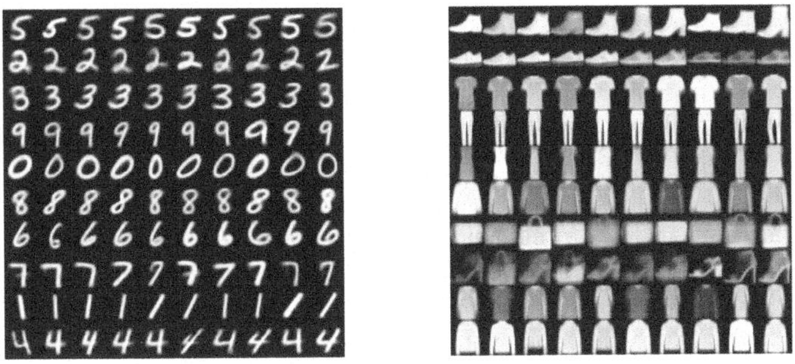

Fig. 2. Generated samples with our method on MNIST (left) and Fashion-MNIST (right). Images in the same row come from the same cluster.

5 Conclusion

In this paper, we introduced a novel clustering framework that integrates VAEs with the EM algorithm for unsupervised clustering. Our approach alternately optimizes soft cluster assignments and VAE parameters for each cluster, enabling both effective clustering and the generation of new samples from each cluster using the respective VAE models. Empirical evaluations on synthetic and real datasets demonstrate that our model outperforms existing VAE-based clustering methods in terms of average clustering accuracy. Future work could extend this generative EM framework to other models, such as normalizing flows, which allow for easy computation of log-likelihoods without relying on lower bounds.

Acknowledgments. Ségolène Martin's work is carried out in the framework of the DFG funded Cluster of Excellence EXC 2046 MATH+ (project ID: AA5-8).

References

1. Benamou, J.D., Carlier, G., Cuturi, M., Nenna, L., Peyré, G.: Iterative Bregman projections for regularized transportation problems. SIAM J. Sci. Comput. **37**(2), A1111–A1138 (2015)
2. Dempster, A.P., Laird, N.M., Rubin, D.B.: Maximum likelihood from incomplete data via the EM algorithm. J. Roy. Stat. Soc. B **39**(1), 1–22 (1977)
3. Dilokthanakul, N., et al.: Deep unsupervised clustering with gaussian mixture variational autoencoders. arXiv preprint arXiv:1611.02648 (2016)
4. Ghasedi, K., Wang, X., Deng, C., Huang, H.: Balanced self-paced learning for generative adversarial clustering network. In: IEEE/CVF Conference on Computer Vision and Pattern Recognition, pp. 4391–4400 (2019)
5. Hagemann, P.L., Hertrich, J., Steidl, G.: Generalized normalizing flows via markov chains. In: Elements in Non-local Data Interactions: Foundations and Applications. Cambridge University Press (2023)

6. Hinton, G.E., Salakhutdinov, R.R.: Reducing the dimensionality of data with neural networks. Science **313**(5786), 504–507 (2006)
7. Jain, A.K., Murty, M.N., Flynn, P.J.: Data clustering: a review. ACM Comput. Surv. **31**(3), 264–323 (1999)
8. Jiang, Z., Zheng, Y., Tan, H., Tang, B., Zhou, H.: Variational deep embedding: an unsupervised and generative approach to clustering. In: International Joint Conference on Artificial Intelligence, (IJCAI), pp. 1965–1972 (2017)
9. Kingma, D.P.: ADAM: a method for stochastic optimization. arXiv preprint arXiv:1412.6980 (2014)
10. Kingma, D.P.: Auto-encoding variational bayes. In: International Conference on Learning Representations (2014)
11. Kingma, D.P., Welling, M.: An introduction to variational autoencoders. Found. Trends Mach. Learn. **12**(4), 307–392 (2019)
12. LeCun, Y., Bottou, L., Bengio, Y., Haffner, P.: Gradient-based learning applied to document recognition. Proc. IEEE **86**(11), 2278–2324 (1998)
13. Ma, H.: Achieving deep clustering through the use of variational autoencoders and similarity-based loss. Math. Biosci. Eng. **19**(10), 10344–10360 (2022)
14. MacQueen, J.: Some methods for classification and analysis of multivariate observations. In: Berkeley Symposium on Mathematical Statistics and Probability (1967)
15. Manduchi, L., Vandenhirtz, M., Ryser, A., Vogt, J.: Tree variational autoencoders. In: Advances in Neural Information Processing Systems, vol. 36, pp. 54952–54986 (2023)
16. Martin, S., Boudiaf, M., Chouzenoux, E., Pesquet, J.C., Ayed, I.: Towards practical few-shot query sets: transductive minimum description length inference. In: Advances in Neural Information Processing Systems, vol. 35, pp. 34677–34688 (2022)
17. Martin, S., Huang, Y., Shakeri, F., Pesquet, J.C., Ben Ayed, I.: Transductive zero-shot and few-shot CLIP. In: IEEE/CVF Conference on Computer Vision and Pattern Recognition, pp. 28816–28826 (2024)
18. Mirza, M.: Conditional generative adversarial nets. arXiv preprint arXiv:1411.1784 (2014)
19. Mukherjee, S., Asnani, H., Lin, E., Kannan, S.: ClusterGAN: latent space clustering in generative adversarial networks. In: Proceedings of the AAAI Conference on Artificial Intelligence, vol. 33, pp. 4610–4617 (2019)
20. Pearson, K.: Contributions to the mathematical theory of evolution. Philos. Trans. R. Soc. A **185**, 71–110 (1894)
21. Pearson, K.: On lines and planes of closest fit to systems of points in space. London Edinburgh Dublin Philos. Mag. J. Sci. **2**(11), 559–572 (1901)
22. Prasad, V., Das, D., Bhowmick, B.: Variational clustering: leveraging variational autoencoders for image clustering. In: International Joint Conference on Neural Networks, pp. 1–10. IEEE (2020)
23. Rybkin, O., Daniilidis, K., Levine, S.: Simple and effective VAE training with calibrated decoders. In: International Conference on Machine Learning, pp. 9179–9189. PMLR (2021)
24. Sønderby, C.K., Raiko, T., Maaløe, L., Sønderby, S.K., Winther, O.: Ladder variational autoencoders. In: Advances in Neural Information Processing Systems, vol. 29 (2016)
25. Xiao, H., Rasul, K., Vollgraf, R.: Fashion-MNIST: a novel image dataset for benchmarking machine learning algorithms. arXiv preprint arXiv:1708.07747 (2017)

26. Xie, J., Girshick, R., Farhadi, A.: Unsupervised deep embedding for clustering analysis. In: International Conference on Machine Learning, pp. 478–487. PMLR (2016)
27. Xu, R., Wunsch, D.: Survey of clustering algorithms. IEEE Trans. Neural Networks **16**(3), 645–678 (2005)

Sampling from Boltzmann Densities with Physics Informed Low-Rank Formats

Paul Hagemann[1], Janina Schütte[2(✉)], David Sommer[2], Martin Eigel[2], and Gabriele Steidl[1]

[1] Technische Universität Berlin, Berlin, Germany
[2] Weierstraß Institute of Applied Analysis and Stochastics, Berlin, Germany
schuette@wias-berlin.de

Abstract. Our method enables the efficient generation of samples from an unnormalized Boltzmann density by solving the underlying continuity equation in the low-rank tensor train (TT) format. It is based on the annealing path commonly used in MCMC literature, which is given by the linear interpolation in the space of energies. Inspired by Sequential Monte Carlo, we alternate between deterministic time steps from the TT representation of the flow field and stochastic steps, which include Langevin and resampling steps. These adjust the relative weights of the different modes of the target distribution and anneal to the correct path distribution. We showcase the efficiency of our method on multiple numerical examples.

1 Introduction

Sampling from Boltzmann densities with unknown normalization constant is an important and timely research problem [1,9,25]. The main challenge is that no samples are given. It hence is difficult to explore the probability space where the mass is concentrated. In the context of MCMC, this leads to the so-called energy barriers [4], which means that it takes chains a long time to jump from one mode of the distribution to another with uninformed proposals. In the context of (conditional) normalizing flows, the training of sampling problems is usually carried out using the reverse Kullback-Leibler (KL) divergence [25], which leads to unwanted mode-collapse. This in principle can be fixed by using MCMC steps to either favor exploration via path learning ideas [3,15,38] or by considering a mode-covering objective [21].

Recently, the path-based learning objective literature has not only been explored via variational objectives, which essentially rely on reformulations of the KL divergence, but also by learning the distributions via the corresponding PDEs. This can for instance be the continuity equation [19] or the Hamilton-Jacobi-Bellman equation [5]. Conceptionally, it is often done in a way similar to physics-informed neural networks (PINN) [29], which is known to suffer from stability and optimization issues [7] as well as slow training.

In this work, we address these shortcomings. Concretely, we learn the velocity field of the continuity equation inspired by [19,34]. To overcome stability issues

T. A. Bubba et al. (Eds.): SSVM 2025, LNCS 15667, pp. 374–386, 2025.
https://doi.org/10.1007/978-3-031-92366-1_29

and speed up training, we explore the tensor train (TT) framework [18,26,27], which has shown great success in solving PDEs [11,13,30,37], and which has recently been used for sampling with diffusion paths [33]. Here, we follow a classical approach often used in Sequential Monte Carlo methods [10,12,23] and reformulate the (discrete) continuity equation by expansion into a product basis. This reduces training to solving linear systems of equations. To facilitate expressivity, we enrich the flow with stochastic steps combining ideas from [3,12,38], in particular resampling and Langevin steps. In contrast to [3], we replace their normalizing flow layers with TT constructions with associated functional basis.

Our contributions can be summarized as follows:

- We introduce a functional tensor train (FTT) format for solving the continuity equation for the annealing path. In particular, we derive formulas for the action of the continuity equation on vector fields represented by FTTs with arbitrary basis functions.
- Using the action of the continuity equation on FTT, we propose a numerical optimization scheme involving empirical risk minimization with the alternating linear scheme (ALS) [17]. Our approximation is naturally rank-adaptive due to an exponential moving average of successive TT approximations.
- We apply an H^2 (Sobolev) orthonormal Fourier basis opposed to the commonly used polynomials within the TT format [11,14,30]. Using a Sobolev regularization has been concurrently explored in [36] in the context of stochastic control.
- We add stochastic steps, namely Langevin postprocessing and resampling, to improve expressivity.
- We illustrate the efficiency of the format on two-dimensional Gaussian mixture problems and the "many well problem" introduced in [21].

2 Logarithmic Continuity Equation

We want to sample from a Boltzmann density p depending on functions $f : [0,1] \times \mathbb{R}^d \to \mathbb{R}$ satisfying mild conditions such that for $f_t(\cdot) := f(t, \cdot)$

$$p_t := \frac{e^{-f_t}}{Z_t}, \quad Z_t := \int_{\mathbb{R}^d} e^{-f_t}\, \mathrm{d}x.$$

Here, p_t describes a family of (Boltzmann) densities, indexed by a time t, f_t describes its unnormalized negative log likelihood and Z_t the corresponding normalization constant.

We follow the considerations in [6,19] and refer to [6] for a mathematically rigorous treatment. Starting with an easy-to-sample latent distribution p_0, the goal is to sample from a target distribution p_1 knowing only f_1. In this paper, we follow the annealing path

$$f_t = tf_1 + (1-t)f_0, \quad t \in [0,1], \tag{1}$$

which is commonly used in the MCMC literature [3,12,23,38]. Now the task is to find a measurable function $v : [0,1] \times \mathbb{R}^d \to \mathbb{R}^d$ such that (p_t, v_t) satisfies the continuity equation

$$\partial_t p_t + \nabla \cdot (p_t v_t) = 0, \tag{2}$$

where $\nabla\cdot$ denotes the divergence. For the Boltzmann density, the continuity equation becomes [19]

$$(\partial_t f_t + \langle \nabla f_t, v_t \rangle - \nabla \cdot v_t + C_t) p_t = 0 \quad \text{with} \quad C_t := \partial_t(\log Z_t).$$

With the linear interpolation (1) we obtain

$$f_1 - f_0 + \langle \nabla f_t, v_t \rangle - \nabla \cdot v_t + C_t = 0 \quad \text{a.e. } t \in [0,1]. \tag{3}$$

Indeed, it was shown under mild assumptions on p_0 and p_1 that appropriate Borel measurable vector fields v_t exist, see [6] and the references therein. Fortunately, absolutely continuous curves p_t with corresponding vector fields v_t in (2) can be described by the solution $\phi : [0,1] \times \mathbb{R}^d \to \mathbb{R}^d$ of an ODE

$$\partial_t \phi(t,x) = v_t(\phi(t,x)), \quad \phi(0,x) = x \tag{4}$$

and $\phi(t,\cdot)_{\#} p_0 = p_t$, see [2, Theorem 8.1.8]. Therefore, once the velocity field satisfying (3) is known, one can sample from p_1 by solving the above ODE with classical numerical solvers. In this paper, we propose to approximate v_t by a low-rank TT representation v_t^θ that minimizes by (3) the least squares problem

$$\mathcal{J}_t(\theta) := |f_1 - f_0 + \langle \nabla f_t, v_t^\theta \rangle - \nabla \cdot v_t^\theta + C_t|^2.$$

3 Low-Rank Tensor Train Decomposition

We propose to approximate the vector field v_t with a tensor product ansatz. Choosing a basis $\{\varphi_i : \mathbb{R} \to \mathbb{R} : i \in [n]\}$, $[n] := \{1, \dots, n\}$, of a space of univariate functions, $v_t^\theta : \mathbb{R}^d \to \mathbb{R}^d$ is then represented in a product basis expansion. Collecting the basis coefficients in an array yields a high-dimensional tensor $T \in \mathbb{R}^{d \times n \times \cdots \times n}$ with dn^d entries. Hence, the storage complexity quickly becomes prohibitive in higher dimensions. To alleviate this, an approximation is directly learned in a lower dimensional tensor manifold. There are different low-rank formats used in the literature, such as the canonic-polyadic (CP), the hierarchical (HT) or TT format. For a comprehensive overview, we refer to [24]. The TT format is particularly useful for approximation tasks: contrary to the CP format, the set of tensors with so-called "TT rank" smaller than a fixed upper bound is closed. Moreover, there exists a gradient-free alternating optimization procedure known as *alternating linear scheme* (ALS), which is used in this work.

3.1 Tensor Train Architecture

For a formal derivation of the TT format, we note that each entry of a tensor $T \in \mathbb{R}^{d \times n \times \cdots \times n}$ can be represented by a multiplication of matrices. For a multi-index $(i_1, \ldots, i_d) \in [n] \times \cdots \times [n]$, we write

$$T_{:,i_1,\ldots,i_d} = K_{i_1}^{(1)} \cdot \ldots \cdot K_{i_d}^{(d)} \in \mathbb{R}^d, \tag{5}$$

where $K_{i_1}^{(1)} \in \mathbb{R}^{d \times r_1}$, $K_{i_d}^{(d)} \in \mathbb{R}^{r_{d-1} \times 1}$ and $K_{i_k}^{(k)} \in \mathbb{R}^{r_{k-1} \times r_k}$ for $k = 2, \ldots, d-1$ are matrices. The smallest values $r_1, \ldots, r_{d-1} \in \mathbb{N}$ such that a representation (5) exists are called the *TT-ranks*. The three dimensional tensors $K^{(k)} \in \mathbb{R}^{r_{k-1} \times n \times r_k}$ defined by $K_{:,i_k,:}^{(k)} = K_{i_k}^{(k)}$ for $k \in [d]$ and $i_k \in [n]$ are called the *cores* (or *component tensors*) of the TT. Using a graphical notation, we can write the tensor as a contraction of the TT cores along the rank dimensions

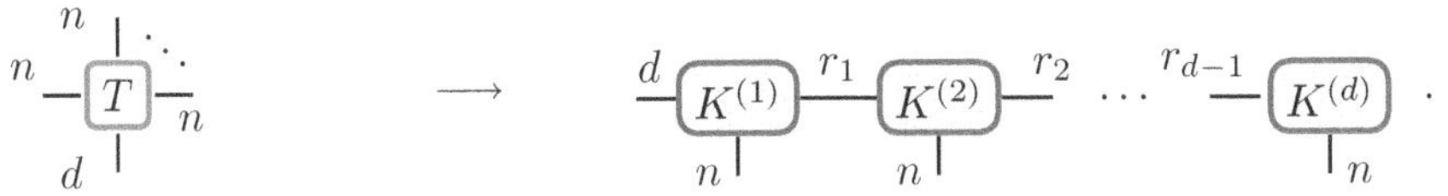

The basis expansion with coefficient tensor T can be rewritten based on the TT representation for any $x = (x_1, \ldots, x_d)^\mathsf{T} \in \mathbb{R}^d$ as follows:

$$v(x) = \sum_{i_1=1}^{n_1} \cdots \sum_{i_d=1}^{n_d} T_{:,i_1,\ldots,i_d} \prod_{k=1}^{d} \varphi_{i_k}(x_k) \quad \longrightarrow \quad v(x) = \sum_{i_1=1}^{n_1} \cdots \sum_{i_d=1}^{n_d} \prod_{k=1}^{d} K_{i_k}^{(k)} \varphi_{i_k}(x_k).$$

Imposing a low-rank structure on the coefficient tensor T results in a manifold of functions with low-rank tensors of rank at most $\mathbf{r} = (r_1, \ldots, r_{N-1})$. In practice this can be understood as neglecting spurious features by an implicit feature selection. Analytical bounds, which quantify this behavior rigorously are given in [26]. Explicit constructions of some notable functions in TT format can be found in [27]. The combination of a TT with an associated set of basis functions defines an FTT.

3.2 Vector Field Learning with the Alternating Linear Scheme

We approximate the vector field v_t at fixed times t as an FTT v_t^θ of order d and with d-dimensional output. The parameter θ is short-hand notation for the collection of all TT components $K_t^{(1)}, \ldots, K_t^{(d)}$ defining v_t^θ. The FTT is the contraction of the TT components with the basis functions $\Phi(x_k) := (\varphi_1(x_k), \ldots, \varphi_n(x_k))^T$ for each $k \in [d]$, yielding a d-dimensional output,

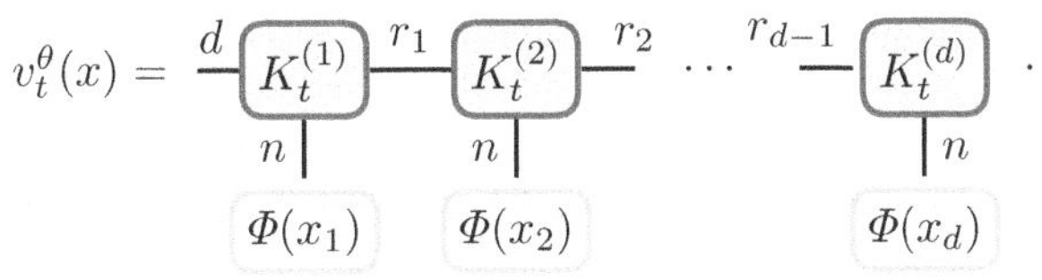

For the training process, the empirical L^2 error in (3) on points $\{x^{(\ell)}\}_{\ell=1}^N$ with

$$J_{t,N}(\theta) = \sum_{\ell=1}^{N} \left| \partial_t f_t(x^{(\ell)}) + \left\langle \nabla f_t(x^{(\ell)}), v_t^\theta(x^{(\ell)}) \right\rangle - \nabla \cdot v_t^\theta(x^{(\ell)}) + C_t \right|^2 \tag{6}$$

is minimized. In the alternating linear scheme [17], the mappings $K_t^{(i)} \mapsto J_{t,N}(\theta)$ are iteratively minimized over $K_t^{(i)}$, while keeping the other components fixed. Once $K_t^{(i)}$ is optimized, the next component $K_t^{(i+1)}$ is optimized and so on, sweeping forward and backward through the chain of TT components until a (local) minimum is reached.

By (6), there exist linear operators $\mathcal{L}_i \colon \mathbb{R}^{r_{i-1} \times n \times r_i} \to \mathbb{R}^N$ such that the minimization over one core can be written as

$$\min_{K_t^{(i)}} J_{t,N}(v_t^\theta) = \min_{K_t^{(i)}} \left\| \mathcal{L}_i(K_t^{(i)}) - y \right\|^2, \tag{7}$$

where $y \in \mathbb{R}^N$ with $y_\ell = -\partial_t f_t(x^{(\ell)}) - C_t$ and the entries of $\mathcal{L}_i(K_t^{(i)})$ are defined for $\ell = 1, \ldots, N$ in graphical notation by

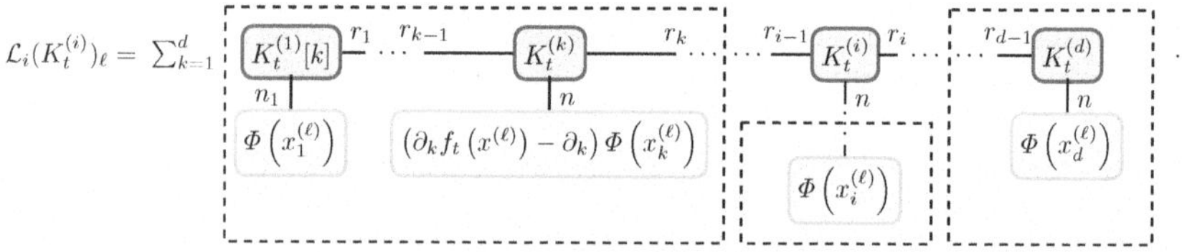

Here, $K_t^{(1)}[k]$ denotes the $\mathbb{R}^{1\times r_1}$-vector defined by $(K_t^{(1)}[k])_{i,j} := K_t^{(1)}{}_{k,i,j}$. Due to the linearity of $\mathcal{L}_i$, (7) is a simple quadratic program, which can be solved by standard techniques. The ALS solves this quadratic program iteratively for all components until convergence. The convergence of ALS to local minima is investigated in [31].

4 Methodology

The vector field $v_t = v_t^\theta$ is learned on a fixed time grid $t_0 = 0, \ldots, t_m = m \cdot \frac{1}{M}, \ldots, t_M = 1$ by the procedure described in Sect. 3.2. In the following, we discuss the critical question of how to choose the samples defining the empirical loss (6) at any given time.

4.1 Stochastic Steps

There are two main obstacles when solving the PDE (6). First, it is unclear how to choose the samples on which to solve the PDE. We could use uniform samples in space. However, this scales unfavorably. Alternatively, we can follow [19], which always draws from the current trained flow. This is similar to a *mode-seeking* reverse KL in the context of normalizing flows [28] and our method of choice. Pushing samples along the trajectories of the learned vector field is later denoted by `ODEsolve`. Second, the path of f_t is well-known to exhibit large jumps in the velocity field close to the final time, where relative mode weights get reassigned. This is called "teleportation issue" in [6], where it is investigated analytically. See also [19] for visualizations of this issue.

Both obstacles can be mitigated by adding stochastic steps. This has been crucial in many sampling related works, such as [3,16,38]. After the TT steps, two kinds of stochastic steps are added as outlined in Algorithm 1:

- *Langevin steps*: Given an energy f of some density p, we can sample asymptotically from p by following the dynamics

$$X_{\ell+1} = X_\ell - h\,\nabla f(X_\ell) + \sqrt{2h}\mathcal{N}(0, I),$$

  where $(X_\ell)_\ell$ denotes a Markov chain that converges asymptotically to samples from $p \sim e^{-f}$ [8,22,39]. Required for this are growth and regularity conditions on f such as strong convexity and Lipschitz gradient of the energy. Only a few steps of this update rule are applied and denoted by `LangevinStep` in Algorithm 1.
- *Resampling steps*: Since Langevin sampling struggles with mode mixing [38], we also perform so-called "resampling steps" as commonly used in Sequential Monte Carlo methods [3,10,12]. Here, we are given some points $\{x^{(\ell)}\}_{\ell=1}^N$ drawn from some density q, which can be evaluated up to the normalizing constant. Assume we target density p. Then, for each of the points $x^{(\ell)}$ the importance weights [23] $w^{(\ell)} = \frac{p(x^{(\ell)})}{q(x^{(\ell)})}$ are calculated. Then, samples are drawn from the categorical distribution proportional to $\sum_\ell w^{(\ell)} x^{(\ell)}$. This however yields the "same" sample multiple times, which is why we also add Langevin steps after resampling steps. This procedure is denoted by `ResamplingStep`.

While the resampling steps help to overcome teleportation issues, the Langevin dynamics deter a collapse of the resampling steps to point masses. For the resampling step, we still have to discuss how q should be chosen. Advancing step $k \to k+1$ with a known transport $\mathcal{T}_k = \phi(t_{k+1}, \cdot) \circ \phi(t_k, \cdot)^{-1}$ with $(\mathcal{T}_k)_{\#} p_k = q \approx p_{k+1}$ as in (4), the density q can be written as $q(x) = p_k(\mathcal{T}_k^{-1}(x))|\det \nabla \mathcal{T}_k^{-1}(x))|$ by leveraging the change of variables formula. The map $\mathcal{T}_k$ given a velocity field v can be defined by $\mathcal{T}_k(x) = x + \int_{t_k}^{t_{k+1}} v_s(x_s)\,\mathrm{d}s$, which we approximate using our discrete formulation (and similarly for $\mathcal{T}_k^{-1}$). In practice we find it beneficial to use Langevin steps before and after resampling steps. However, assessing the effect of the Langevin layers on the density q is left for future work and we only use the flow layers.

4.2 Adaptive TT Ranks

The training on one set of prior samples, which is transported by the vector field to serve as training samples for the TT in the next time step, is repeated to train the TTs on more samples. In each iteration, the distribution of the samples in later iteration stages should be closer to the interpolated distribution, since the TTs approximate the solution of the PDE increasingly better in each iteration. They therefore serve as better training samples. When training on a new set of samples, the ALS updates each core without consideration of the information from the last iteration. To not deviate too far from the previously learned TT in the last iteration, the old and newly learned TTs are added in a weighted manner similar to an *exponential moving average (EMA)*. The TT representing the addition can exhibit doubled ranks, which can be recompressed by applying a singular value decomposition on each core and truncating small singular values (TT-SVD, [26]). This procedure is denoted by `UpdateTT` in Algorithm 1 and bears similarities with the CRAFT framework [20]. By this, the ranks of the TT do not have to be set a priori but are adjusted dynamically during training. The resulting algorithm is described in Algorithm 1.

Algorithm 1: Training the vector field

Input: latent density p_0, target energy f_1
while *not converged* **do**
  sample prior $x \leftarrow \{x_i\}_{i\in[N]} \sim p_0$ ▷ initialize training data set
  for $k = 1, \dots, M$ **do**
    calculate $\partial_t f_{t_k}(x), \nabla f_t(x)$
    train $v_{t_k}^{\text{new}}$ with ALS to minimize (7) ▷ train vector field
    $v_{t_k} \leftarrow$ `UpdateTT`$(v_{t_k}, v_{t_k}^{\text{new}})$ ▷ update and adjust TT ranks
    $x \leftarrow$ `ODEsolve`$(x, t_k, t_{k+1}, v_{t_k})$ ▷ update samples for next time step
    $x \leftarrow$ `ResamplingStep`(x)
    $x \leftarrow$ `LangevinStep`(x)
  end
end
return: $v_{t_1}, \dots, v_{t_M}$

5 Numerical Experiments

The proposed architecture is tested with different target distributions. For comparison with other works, the energy distance as in [32,35] is used. The energy distance (also called MMD of the negative distance kernel) is a two-sample test and therefore a good way to compare distributions. Further, our toy problems are chosen in such a way that we are able to sample from the Boltzmann density, although our algorithm does not use this information.

Target Distributions. As targets, we choose the following three distributions:

1. *Gaussian mixture with two modes*: The two-dimensional Gaussian mixture is considered with two modes. Both have same weights and variance 0.01. One Gaussian is with mean $(2, 2)$ and other with mean $(-2, -2)$.

2. *Gaussian mixture with* 40 *modes*: The two-dimensional Gaussian mixture as in [6,21] is considered, where 40 different normal distributions with means uniformly distributed in $[-40, 40]^2$ are mixed with equal weights.
3. *Many well problem (MWP)*: The target density in the d-dimensional MWP as in [6,21] consists of the multiplication of m copies of the two-dimensional double well potential and $d - 2m$ Gaussians given by

$$f_1^{m,d} = \sum_{i=0}^{m} \left(x_{2i}^4 - 6x_{2i}^2 - \tfrac{1}{2}x_{2i} + \tfrac{1}{2}x_{2i+1}^2\right) + \sum_{i=2m+1}^{d} \tfrac{1}{2}x_i^2.$$

Methods. To assess the importance of the different algorithmic steps, we test the following methods:

- **flow**: a pure TT normalizing flow is trained without stochastic steps.
- **flow+**: after the full TT flow, resampling and Langevin steps are carried out.
- **flow+stochastic**: a flow is trained with stochastic steps in the training process and the evaluation process after every time step as in Algorithm 1.
- **stochastic**: the results are compared with samples generated by the same amount of stochastic steps without steps of a trained flow, where we take the ratio for the resampling step to be $\frac{p_{k+1}}{p_k}$. Hence, this reduces to an alternation of Langevin and resampling steps. We found that including naive resampling steps with respect to the density path increases the performance as opposed to simple Langevin sampling.

In all examples, the ground truth is known. The models are tuned by comparison with a validation set of samples. The tuning procedures without ground truth samples is left for future work.

In all experiments, the basis is chosen to be an H^2 orthonormalized Fourier basis as it performed best in this setting compared to Legendre polynomials, B-splines and a H^1 orthonormalized Fourier basis. A comparison of the architectures is not included in this work. The number of time steps is $M \leq 35$ in each example. The number of Fourier basis coefficients is $n \leq 10$ in all experiments except for Problem 2, in which we use $n = 41$. In Problems 1 and 3, the basis is orthonormalized on $[-5, 5]^d$ and the latent distribution is chosen to be a normal distribution. In Problem 2, the parameters are $[-50, 50]^2$ and we start in a zero mean Gaussian with standard deviation 500.

Table 1. Average energy distance ± standard deviation over 100 independent draws of 50.000 samples for the methods flow, flow+, flow+stochastic and stochastic. Best method in bold.

	flow	flow+	flow+stochastic	stochastic
GM 2 modes	3.7e−3 ± 2e−4	**1.2e−3 ± 5e−4**	**1.1e−3 ± 1e−3**	2.7e−3 ± 3e−3
GM 40 modes	4.5e−1 ± 3e−2	**7.6e−2 ± 2e−2**	**7.6e−2 ± 4e−2**	1.4e−1 ± 9e−2
MWP $f_1^{4,8}$	1.6e−1 ± 1e−4	1.2e−2 ± 2e−4	**1.9e−3 ± 2e−4**	8.1e−2 ± 9e−3
MWP $f_1^{4,16}$	2.1e−1 ± 2e−3	1.3e−1 ± 6e−3	**6.6e−3 ± 2e−3**	7.4e−2 ± 1e−2

Table 1 contains the results for the four methods and four different target distributions. It can be seen that the pure flow already achieves a good fit. Especially in higher dimensions, the stochastic steps are needed to achieve very good performance. Usually either *flow+* or *flow+stochastic* take the lead, where *flow+stochastic* seems to perform especially well in the high-dimensional MWP examples. We also see that the flow improves upon the stochastic steps since *flow+stochastic* usually outperforms *stochastic*. We show the marginal distributions of the 4 methods in Fig. 2 for some marginals of the MWP. Here, continuous TT flow (*flow*) struggles to separate the mass perfectly, which is resolved by the stochastic steps. Figure 1 depicts the samples generated for the 40-mode GM problem. The 40-mode GM and MWP $f_1^{4,8}$ were chosen as target distributions to be able to compare to results for the linear interpolation in [6]. In both cases the presented method outperforms the model presented there for the fixed interpolation. In high dimensional MWP, it can be seen that better results can be achieved for a learned interpolation. This motivates future research extending the method presented here to an algorithm also learning f_t.

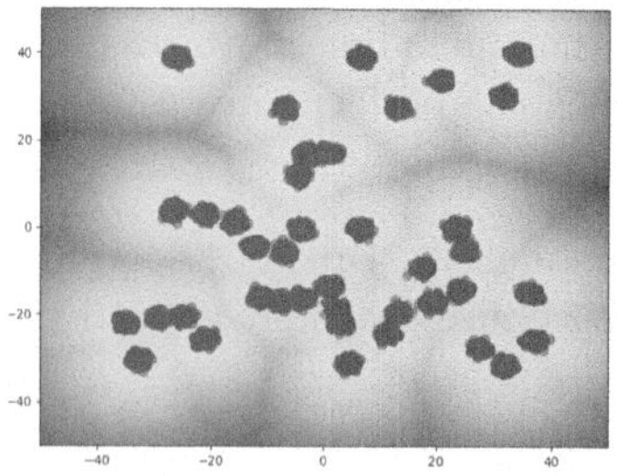

(a) ground truth

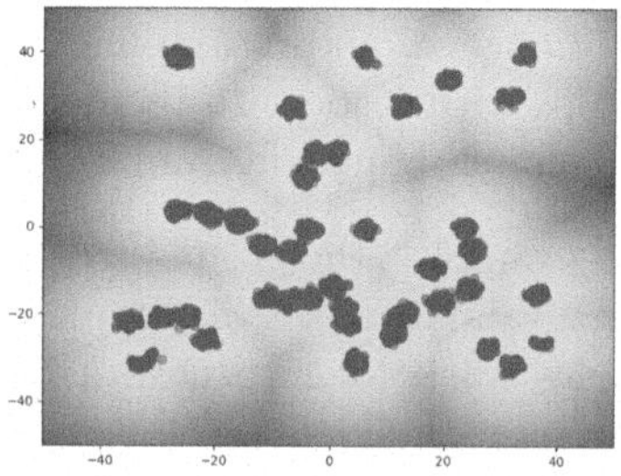

(b) flow+

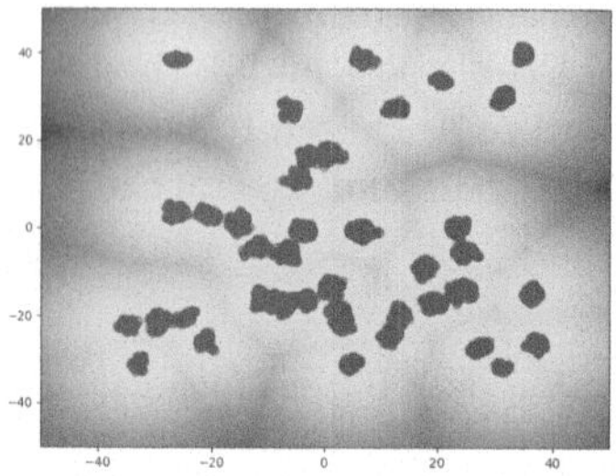

(c) flow+stochastics

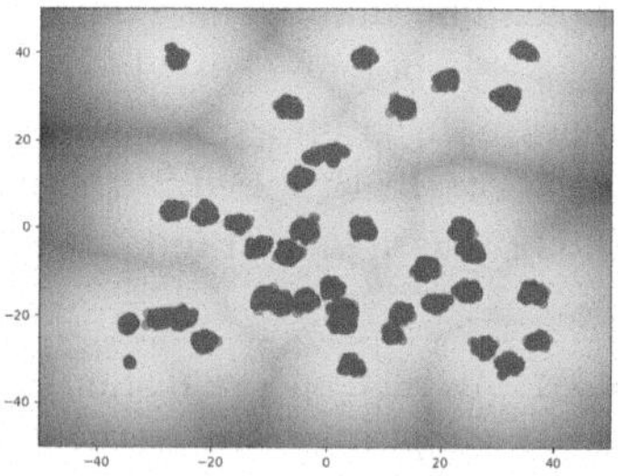

(d) stochastics

Fig. 1. GM density sampled with the different methods. While the methods perform similarly on this problem, the TT flow in (b) and (c) helps to distribute the mass correctly compared to a purely stochastic approach (as can e.g. be seen for the mode on the lower left part of the domain).

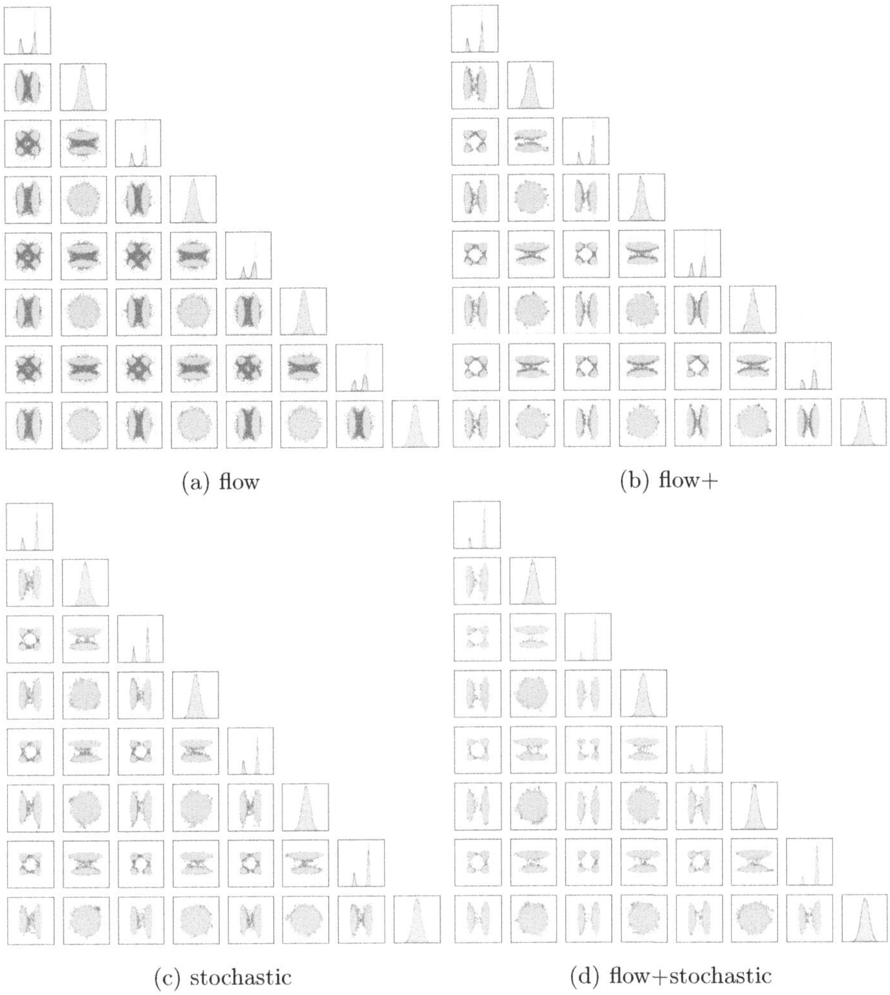

Fig. 2. Histogram corner plots showing the 1-d marginals on the diagonal and the 2-d joint distributions on the off-diagonal for the MWP $f_1^{4,16}$ for different methods (blue: competing method, orange ground truth). A strong match of the orange and blue plots indicates closeness of distributions.

6 Conclusions

In this work, the continuity equation of the annealing path is solved in a functional low-rank TT format to draw samples from a given target density. In contrast to other works, a Fourier basis orthogonalized with respect to the H^2 scalar product is used for the FTT architecture. This choice works well to solve the PDE, while common basis functions such as polynomials perform poorly. This observation could be attributed to the lack of Lipschitz continuity of higher-order polynomials and will be examined in future work. The TT architecture is adapted during training similar to an EMA, leading to a more stable training procedure and the advantage that the architecture does not have to be set a priori. In future work, non-sample-based techniques could be developed to choose the best architecture during training, e.g., based on the residual of the PDE.

Acknowledgment. P.H., M.E., G.S. & J.S. gratefully acknowledge the funding from DFG SPP 2298 "Theoretical Foundations of Deep Learning". M.E. & D.S. acknowledge support by the ANR-DFG project *COFNET: Compositional functions networks - adaptive learning for high-dimensional approximation and uncertainty quantification.*

Disclosure of Interests. The authors have no competing interests to declare that are relevant to the content of this article.

References

1. Albergo, M., Kanwar, G., Shanahan, P.: Flow-based generative models for Markov chain Monte Carlo in lattice field theory. Phys. Rev. D **100**(3), 034515 (2019)
2. Ambrosio, L., Gigli, N., Savaré, G.: Gradient flows. In: Metric Spaces and in the Space of Probability Measures. Springer, Basel (2008). https://doi.org/10.1007/978-3-7643-8722-8
3. Arbel, M., Matthews, A., Doucet, A.: Annealed flow transport Monte Carlo. In: International Conference on Machine Learning, pp. 318–330. PMLR (2021)
4. Bandeira, A., Maillard, A., Nickl, R., Wang, S.: On free energy barriers in gaussian priors and failure of cold start MCMC for high-dimensional unimodal distributions. Philos. Trans. Ser. A Math. Phys. Eng. Sci. **381**, 20220150 (2023)
5. Berner, J., Richter, L., Ullrich, K.: An optimal control perspective on diffusion-based generative modeling. Trans. Mach. Learn. Res. (2024)
6. Chemseddine, J., Wald, C., Duong, R., Steidl, G.: Neural sampling from Boltzmann densities: Fisher-Rao curves in the Wasserstein geometry. arXiv preprint arXiv:2410.03282 (2024)
7. Chuang, P.Y., Barba, L.A.: Experience report of physics-informed neural networks in fluid simulations: pitfalls and frustration. arXiv preprint arXiv:2205.14249 (2022)
8. Dalalyan, A.S., Karagulyan, A.G.: User-friendly guarantees for the Langevin Monte Carlo with inaccurate gradient. ArXiv **abs/1710.00095** (2017)
9. Del Debbio, L., Marsh Rossney, J., Wilson, M.: Efficient modeling of trivializing maps for lattice ϕ^4 theory using normalizing flows: a first look at scalability. Phys. Rev. D **104**(9), 094507 (2021)
10. Del Moral, P., Doucet, A., Jasra, A.: Sequential Monte Carlo samplers. J. Royal Stat. Soc. Ser. B Statist. Method. **68**(3), 411–436 (2006)
11. Dolgov, S., Kalise, D., Kunisch, K.K.: Tensor decomposition methods for high-dimensional Hamilton-Jacobi-Bellman equations. SIAM J. Sci. Comput. **43**(3), A1625–A1650 (2021)
12. Doucet, A., Godsill, S., Andrieu, C.: On sequential monte Carlo sampling methods for Bayesian filtering. Stat. Comput. **10**, 197–208 (2000)
13. Eigel, M., Neumann, J., Schneider, R., Wolf, S.: Non-intrusive tensor reconstruction for high-dimensional random PDEs. Comput. Methods Appl. Math. **19**(1), 39–53 (2019). https://doi.org/10.1515/cmam-2018-0028

14. Eigel, M., Schneider, R., Sommer, D.: Dynamical low-rank approximations of solutions to the Hamilton-Jacobi-Bellman equation. Numer. Linear Algebra Appl. **30**(3), e2463 (2023)
15. Hagemann, P., Hertrich, J., Steidl, G.: Stochastic normalizing flows for inverse problems: a Markov chains viewpoint. SIAM/ASA J. Uncertain. Quant. **10**(3), 1162–1190 (2022)
16. Hertrich, J., Gruhlke, R.: Importance corrected neural JKO sampling. arXiv preprint arXiv:2407.20444 (2024)
17. Holtz, S., Rohwedder, T., Schneider, R.: The alternating linear scheme for tensor optimization in the tensor train format. SIAM J. Sci. Comput. **34**(2), A683–A713 (2012)
18. Holtz, S., Rohwedder, T., Schneider, R.: On manifolds of tensors of fixed TT-rank. Numer. Math. **120**(4), 701–731 (2012)
19. Máté, B., Fleuret, F.: Learning interpolations between Boltzmann densities. Trans. Mach. Learn. Res. (2023)
20. Matthews, A.G.D.G., Arbel, M., Rezende, D.J., Doucet, A.: Continual repeated annealed flow transport Monte Carlo. In: Proceedings of the 39th International Conference on Machine Learning. Proceedings of Machine Learning Research (2022)
21. Midgley, L.I., Stimper, V., Simm, G.N.C., Schölkopf, B., Hernández-Lobato, J.M.: Flow annealed importance sampling bootstrap. arXiv preprint arXiv:2208.01893 (2023)
22. Miranda, C., Schütte, J., Sommer, D., Eigel, M.: Approximating Langevin Monte Carlo with ResNet-like neural network architectures. arXiv preprint arXiv:2311.03242 (2024)
23. Neal, R.M.: Annealed importance sampling. Statist. Comp. **11**, 125–139 (2001)
24. Nouy, A.: Chapter 4: Low-rank methods for high-dimensional approximation and model order reduction, pp. 171–226. Society for Industrial and Applied Mathematics (2017). https://doi.org/10.1137/1.9781611974829.ch4
25. Noé, F., Olsson, S., Köhler, J., Wu, H.: Boltzmann generators: sampling equilibrium states of many-body systems with deep learning. Science **365**(6457), 982 (2019)
26. Oseledets, I.V.: Tensor-train decomposition. SIAM J. Sci. Comput. **33**(5), 2295–2317 (2011)
27. Oseledets, I.V.: Constructive representation of functions in low-rank tensor formats. Constr. Approx. **37**, 1–18 (2013)
28. Papamakarios, G., Nalisnick, E., Rezende, D.J., Mohamed, S., Lakshminarayanan, B.: Normalizing flows for probabilistic modeling and inference. J. Mach. Learn. Res. **22**(57), 1–64 (2021)
29. Raissi, M., Perdikaris, P., Karniadakis, G.: Physics-informed neural networks: a deep learning framework for solving forward and inverse problems involving nonlinear partial differential equations. J. Comp. Phys. **378**, 686–707 (2019)
30. Richter, L., Sallandt, L., Nüsken, N.: Solving high-dimensional parabolic PDEs using the tensor train format. In: International Conference on Machine Learning, pp. 8998–9009. PMLR (2021)
31. Rohwedder, T., Uschmajew, A.: On local convergence of alternating schemes for optimization of convex problems in the tensor train format. SIAM J. Numer. Anal. **51**(2), 1134–1162 (2013)
32. Sejdinovic, D., Sriperumbudur, B., Gretton, A., Fukumizu, K.: Equivalence of distance-based and RKHS-based statistics in hypothesis testing. Ann. Stat. **41**(5), 2263–2291 (2013)

33. Sommer, D., Gruhlke, R., Kirstein, M., Eigel, M., Schillings, C.: Generative modelling with tensor train approximations of Hamilton–Jacobi–Bellman equations. arXiv preprint arXiv:2402.15285 (2024)
34. Sun, J., et al.: Dynamical measure transport and neural PDE solvers for sampling. arXiv preprint arXiv:2407.07873 (2024)
35. Székely, G.J., Rizzo, M.L.: Energy statistics: a class of statistics based on distances. J. Stat. Plan. Inference **143**(8), 1249–1272 (2013)
36. Tang, X., Sheng, N., Ying, L.: Solving high-dimensional Hamilton-Jacobi-Bellman equation with functional hierarchical tensor. arXiv preprint arXiv:2408.04209 (2024)
37. Vemuri, S.K., Büchner, T., Niebling, J., Denzler, J.: Functional tensor decompositions for physics-informed neural networks. arXiv preprint arXiv:2408.13101 (2024)
38. Wu, H., Köhler, J., Noé, F.: Stochastic normalizing flows. Adv. Neural. Inf. Process. Syst. **33**, 5933–5944 (2020)
39. Zhang, Y., Akyildiz, Ö.D., Damoulas, T., Sabanis, S.: Nonasymptotic estimates for stochastic gradient Langevin dynamics under local conditions in nonconvex optimization. Appl. Math. Optim. **87**(2), 25 (2023)

Parameter-Free Structure-Texture Image Decomposition by Unrolling

Laura Girometti[1(✉)], Jean-François Aujol[2], Antoine Guennec[2], and Yann Traonmilin[2]

[1] Department of Mathematics, University of Bologna, Bologna, Italy
laura.girometti2@unibo.it
[2] University of Bordeaux, Bordeaux INP, CNRS, IMB, UMR 5251, F-33400 Talence, France
{jean-francois.aujol,antoine.guennec,yann.traonmilin}@math.u-bordeaux.fr

Abstract. In this work, we propose a parameter-free and efficient method to tackle the structure-texture image decomposition problem. In particular, we present a neural network LPR-NET based on the unrolling of the Low Patch Rank model. On the one hand, this allows us to automatically learn parameters from data, and on the other hand to be computationally faster while obtaining qualitatively similar results compared to traditional iterative model-based methods. Moreover, despite being trained on synthetic images, numerical experiments show the ability of our network to generalize well when applied to natural images.

Keywords: image decomposition · unrolling

1 Introduction

The problem of structure-texture image decomposition has been widely investigated in the last decades and successfully applied to tasks such as detail enhancement, image abstraction, texture transfer, HDR tone mapping [13,25]. In addition, it has also been exploited to solve inverse problems in imaging such as image fusion [24], inpainting [11] and super-resolution [5]. It aims to find an additive decomposition of an observed image f of the form

$$f = u + v, \tag{1}$$

where u represents the structure component of f containing the geometrical elements identified in edges and homogeneous regions, whereas the texture component v contains repeated patterns of small scale details. This problem is usually formulated as a constrained minimization problem as follows:

$$\min_{u,v \in \mathbb{R}^N \ s.t. f=u+v} \mu \mathcal{J}_1(u) + \mathcal{J}_2(v), \tag{2}$$

where $\mathcal{J}_1$ and $\mathcal{J}_2$ are regularization functions that enforce the desired properties on the components u and v, and μ is a positive weighting parameter that plays

T. A. Bubba et al. (Eds.): SSVM 2025, LNCS 15667, pp. 387–399, 2025.
https://doi.org/10.1007/978-3-031-92366-1_30

a crucial role in separating u from v, adapting to the image content. While for the structure component the natural choice for $\mathcal{J}_1$ falls on functions promoting gradient-sparsity - Total Variation [19], Non convex Total Variation [12] - various proposals for $\mathcal{J}_2$ have been deeply investigated: high-oscillatory pattern promoting functions - G-norm [2,16] - low patch rank penalties that promote similarity between texture patches - [8,18,20], a generative texture prior inspired by generative convolutional neural networks [11], convolutional sparse coding with dictionary learning-based sparsity prior [5,23]. Another line of research focuses on optimal parameter selection based on separability between components- minimize the cross-correlation between pairs of them [3,6] - or on adjusting the tuning parameters depending on the right relationship between the gradient sparsity level of u and the rank of the texture patches [8]. Recently, motivated by progress in machine learning techniques, various data-driven approaches have been presented. The problem of lack of available datasets containing pairs of structure-texture components has been addressed by applying an unsupervised strategy [25]. The Plug-and-Play framework, where a denoiser is trained to learn a more significant regularizer for the structure u on natural image datasets [13] enables the learning of a joint regularizer $R_x(u, v)$ for structure and texture on a generated synthetic dataset [9]. Among the hybrid techniques that combine model-based methods and deep-learning approaches, algorithm unrolling is a recent promising paradigm to build efficient and interpretable neural networks. These techniques date back to Gregor et al.'s paper [7], where they proposed to improve the computational efficiency of sparse coding algorithms through end-to-end training. Following their idea, the guiding principle in building unrolled neural networks is to represent each iteration of an iterative algorithm as one layer of the network. Consequently, mimicking model-based methods results in high interpretability of network layers, fewer parameters compared with popular neural networks and, therefore, less training data, as well as better generalization power than generic networks [17]. Recently, it has been applied to different signal and image processing tasks such as image denoising [14], blind image deblurring [15] and CT [1].

This paper proposes a novel strategy to achieve an efficient and automatic structure-texture image decomposition by leveraging the unrolling technique: model and algorithm parameters are learned in a supervised fashion making use of a synthetic generated dataset. In particular, in Sect. 2, we propose to unroll an extension of the Low Patch Rank model [20] and we present a numerical solution based on the Alternating Direction Method of Multipliers (ADMM). In Sect. 3, we propose our neural network LPR-NET and we investigate various design choices. Finally, in Sect. 4, numerical results validate that our proposal outperforms optimally hand-tuned classical variational models and gives a competitive end-to-end alternative to the Plug-and-Play framework where optimization of the learned regularization must be performed.

2 Proposed Structure-Texture Decomposition Model

In this section, we first describe an extended version of the Low Patch Rank model for structure-texture decomposition and we present an ADMM-based algorithm to solve it. Then, we propose our LPR-NET network based on the unrolling of the proposed ADMM iterative scheme.

2.1 Low Patch Rank Model for Structure-Texture Decomposition

Given an image f possibly degraded by a linear operator $\mathcal{M}$ (e.g. deblurring, mask operator), the Low Patch Rank model [20] recovers the structure component u and the texture component v by enforcing u to be gradient sparse and v to be of low patch rank. Specifically, they consider the classical Total Variation as regularization function $\mathcal{J}_1$ and the nuclear norm of texture patches as $\mathcal{J}_2$. We propose to extend this model by considering, instead of the ℓ_1-norm, a non-convex non-smooth sparsity promoting function $\phi(\cdot;a)$ parametrized by $a \geq 0$. The resulting variational problem reads as:

$$\min_{u\in\mathbb{R}^N, v\in\mathbb{R}^N \, s.t. \, f=\mathcal{M}(u+v)} \mu \sum_{i=1}^{N} \phi(||(\mathrm{D}u)_i||_2; a) + ||\mathcal{P}v||_*, \tag{3}$$

where μ is a positive balancing parameter, $(\mathrm{D}u)_i$ represents the discrete gradient of u at pixel i and the nuclear norm $||\cdot||_*$ is chosen as a convex surrogate of the rank function. The operator $\mathcal{P}$ represents the patch operator characterized by the size p of the $p \times p$ patches and by the overlap o between two adjacent patches as shown in Fig. 1:

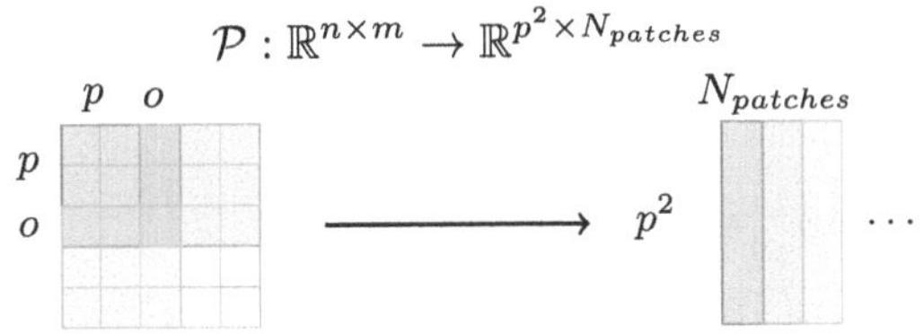

Fig. 1. Graphical action of the patch operator $\mathcal{P}$.

The considered non-convex penalty function $\phi(\cdot;a)$ is the minimax concave penalty (MCP) [22] defined as

$$\phi(t;a) = \begin{cases} |t| - \frac{a}{2}t^2 & \text{if } |t| \leq \frac{1}{a} \\ \frac{1}{2a} & \text{if } |t| \geq \frac{1}{a} \end{cases}. \tag{4}$$

Note that as $a \rightarrow 0$, we recover the absolute value function $|t|$ as a special case of the penalty function $\phi(\cdot;a)$ thus the classical Total Variation regularizer in (3). The MCP function was first introduced in [22] in the context of unbiased variable selection to overstep the bias introduced by the ℓ_1 penalty on large coefficients. Therefore, we aim to less penalize high gradients in the structure component u.

2.2 ADMM-Based Algorithm for LPR

To numerically solve the constrained minimization problem in (3), we equivalently reformulate the problem introducing two auxiliary variables $t := Du \in \mathbb{R}^{2N}$ and $s := v \in \mathbb{R}^N$, which represent the discrete gradient of u and the texture v respectively. Problem (3) is rewritten as

$$\min_{\substack{t \in \mathbb{R}^{2N}, u, v, s \in \mathbb{R}^N \\ \text{s.t. } t = Du,\, s = v}} \mu \sum_{i=1}^{N} \phi(||t_i||_2; a) + ||\mathcal{P}s||_* + i_{\mathcal{C}}(u, v) \tag{5}$$

where the constraint $f = \mathcal{M}(u+v)$ is imposed by adding $i_{\mathcal{C}}$, the indicator function of the convex set $\mathcal{C} := \{x := (u, v) \in \mathbb{R}^{2N} \text{ s.t. } f = \mathcal{M}(u+v)\}$, to the cost function in (3). We propose the following Alternating Direction Method of Multipliers iterative scheme, where we denote by ρ_t and ρ_s the ADMM penalty parameters, by $x := \begin{pmatrix} u \\ v \end{pmatrix}$ and $y := \begin{pmatrix} y_t \\ y_s \end{pmatrix}$ the vector of Lagrange multipliers where $y_t \in \mathbb{R}^{2N}$ and $y_s \in \mathbb{R}^N$. Given the initial guesses $x^0 := (u^0, v^0), y^0 := (y_t^0, y_s^0)$, the k-th iteration reads as

$$t^k \in \arg\min_{t \in \mathbb{R}^{2N}} \sum_{i=1}^{N} \mu\phi(||t_i||_2; a) + \frac{\rho_t}{2} \left\| t - \left(\mathrm{D}u^{k-1} + \frac{y_t^{k-1}}{\rho_t} \right) \right\|^2 \tag{6}$$

$$s^k \in \arg\min_{s \in \mathbb{R}^N} ||\mathcal{P}s||_* + \frac{\rho_s}{2} \left\| s - \left(v^{k-1} + \frac{y_s^{k-1}}{\rho_s} \right) \right\|^2 \tag{7}$$

$$x^k \in \arg\min_{x \in C} \frac{\rho_t}{2} \left\| t^k - \frac{y_t^{k-1}}{\rho_t} - \left[\mathrm{D}\; 0_N \right] x \right\|^2 + \frac{\rho_s}{2} \left\| s^k - \frac{y_s^{k-1}}{\rho_s} - \left[0_N \; \mathrm{I}_N \right] x \right\|^2 \tag{8}$$

$$y^k = y^{k-1} + \begin{bmatrix} \rho_t \mathrm{D} \\ \rho_s \mathrm{I}_N \end{bmatrix} x^k - \begin{pmatrix} \rho_t t^k \\ \rho_s s^k \end{pmatrix} \tag{9}$$

In the following, we describe the numerical solution of problems (6)–(8).

t-Subproblem, Estimation of Gradients: The minimization problem (6) is separable with respect to $t_i = (\mathrm{D}u)_i \in \mathbb{R}^2$ hence is equivalent to N optimization problems of the form:

$$\min_{t_i \in \mathbb{R}^2} \mu\phi(||t_i||_2; a) + \frac{\rho_t}{2} \left\| t_i - \left((\mathrm{D}u^{k-1})_i + \frac{(y_t^{k-1})_i}{\rho_t} \right) \right\|^2 . \tag{10}$$

Sufficient conditions to ensure that problem (10) has a unique solution are presented in Proposition 1. For a detailed proof, see [12].

Proposition 1. *Given $v \in \mathbb{R}^2$, $a \geq 0$, $\mu, \rho_t > 0$, the function $\frac{\mu}{\rho_t}\phi(\cdot; a) + \frac{1}{2}||t_i - v||^2$ is strongly convex if $a \leq \frac{\rho_t}{\mu}$. Morever, it has a unique global minimizer that* *has the following closed-form expression:*

$$t_i = \begin{cases} \min\left(1, \frac{1}{1-a\frac{\mu}{\rho_t}} \max\left(1 - \frac{\mu}{\rho_t ||v||_2}, 0\right)\right) v & \text{if } ||v||_2 > 0 \\ 0 & \text{if } ||v||_2 = 0 \end{cases} \tag{11}$$

Equation (11) is a generalization of the proximal operator of the ℓ_2-norm (recovered for $a \to 0$) where a second shrinkage value, that depends on a, is introduced.

s-Subproblem, Estimation of LR Texture: The solution of the minimization problem in (7) is the evaluation of the proximal operator of the nuclear norm in $v^{k-1} + \frac{y_s^{k-1}}{\rho_s}$. This proximal operator has the following closed-form solution:

$$\text{prox}_{||\cdot||_*, \beta}(x) = U\max(\text{S} - \beta\text{I})V^T \tag{12}$$

where $x = \text{USV}^\text{T}$ is the Singular Value Decomposition (SVD) of x and the maximum is taken elementwise.

x-Subproblem, Enforcing the Constraint: The minimization problem in (8) is a linearly constrained quadratic programming that does not admit a general closed-form solution but which can be efficiently solved by means of the Projected Gradient Descent. The cost function is differentiable with respect to x and the projection onto the set $\mathcal{C}$ can be computed as in [9]. Starting from x_0, the algorithm reads as

$$x_{i+1} = \mathcal{P}_\mathcal{C}\left(x_i - \tau\left(\text{A}x_i - \text{B}z\right)\right), \; i = 0, \ldots, n-1,$$

where τ is the gradient-step parameter, $\mathcal{P}_\mathcal{C}$ denotes the projection onto the set $\mathcal{C}$ and

$$\text{A} = \begin{bmatrix} \rho_t \text{D}^\text{T}\text{D} & \text{O} \\ \text{O} & \rho_s \text{I} \end{bmatrix}, \; \text{B} = \begin{bmatrix} \rho_t \text{D} & \text{O} \\ \text{O} & \rho_s \text{I} \end{bmatrix} \text{ and } z = \begin{pmatrix} t^k - \frac{y_t^{k-1}}{\rho_t} \\ s^k - \frac{y_s^{k-1}}{\rho_s} \end{pmatrix}. \tag{13}$$

We remark that the proposed ADMM algorithm has no guarantee of convergence, except for $a = 0$, when the cost function is convex. For the general case, as detailed in [21], a Lipschitz-gradient term is required. This would imply the replacing of the strong constraint by some L^2 penalization, that could also handle noisy cases. Nevertheless, we prefer to hardly force u and v to satisfy the constraint.

3 LPR-NET Architecture

Starting from the ADMM iterative scheme proposed in (6)–(9), we propose an unrolled neural network by stacking K blocks, each one resembling one iteration of the iterative algorithm considered. As described in Sect. 2, one iteration of the proposed ADMM consists of four subproblems to be solved, corresponding to the

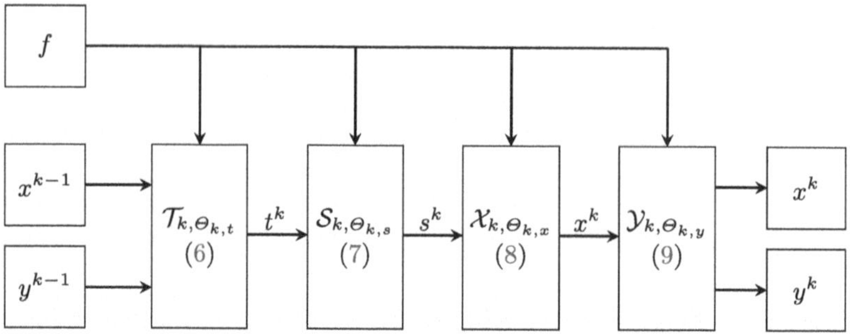

Fig. 2. Flow of the k-th block of the network. Each operation performed in a block corresponds to a step of the ADMM algorithm described in (6)–(9).

t, s, x and y updates. Hence, the generic k-th block of our network will be divided into four consecutive subblocks reproducing respectively the four subproblems. In Fig. 2, the flow of a generic block of the network is shown.

We define the output of the k-th block as

$$(x^k, y^k) = \mathcal{L}^k_{\Theta_k}(x^{k-1}, y^{k-1}) := \mathcal{Y}_{\Theta_{k,y}} \circ \mathcal{X}_{\Theta_{k,x}} \circ \mathcal{S}_{\Theta_{k,s}} \circ \mathcal{T}_{\Theta_{k,t}}(x^{k-1}, y^{k-1}) \quad (14)$$

where $\Theta_{k,t}$, $\Theta_{k,s}$, $\Theta_{k,x}$ and $\Theta_{k,y}$ denote the set of learnable parameters in $\mathcal{T}_{k,\Theta_{k,t}}$, $\mathcal{S}_{k,\Theta_{k,s}}$, $\mathcal{X}_{k,\Theta_{k,x}}$, $\mathcal{Y}_{k,\Theta_{k,y}}$ subblocks respectively. Finally, composing K blocks, we have the final relation between the input and the output of the network:

$$(x^K, y^K) = \mathcal{F}_\Theta(x^0, y^0) = \mathcal{L}^K_{\Theta_K} \circ \cdots \circ \mathcal{L}^1_{\Theta_1}(x^0, y^0), \quad (15)$$

where $\Theta_k = \{\Theta_{k,t}, \Theta_{k,s}, \Theta_{k,x}, \Theta_{k,y}\}$ and $\Theta := \cup_{k=1}^{K} \Theta_k$ denotes the set of the learnable parameters of the network $\mathcal{F}_\Theta$. Various design options can be explored when constructing the subblocks $\mathcal{T}_{k,\Theta_{k,t}}$, $\mathcal{S}_{k,\Theta_{k,s}}$, $\mathcal{X}_{k,\Theta_{k,x}}$, $\mathcal{Y}_{k,\Theta_{k,y}}$ depending on the desired level of flexibility and complexity of $\mathcal{F}_\Theta$.

$\mathcal{T}_{\mathbf{k},\Theta_{\mathbf{k},\mathbf{t}}}$-Subblock, Estimation of Gradients: The closed-form expression of (11) can be recast as a linear layer followed by a non linear function as follows:

$$t^k = \eta_1 \left(\mathrm{D}^{(k)} u^{k-1} + \frac{y_t^{k-1}}{\rho_t}; \mu^{(k)}, \rho_t^{(k)} \right) \quad (16)$$

where η_1 is the non linear function defined by the proximal operator of the function $\phi(\cdot; a)$ defined in (11). We replace the discrete operator D with a convolutional operator $\mathrm{D}_{\mathcal{T}}^{(k)}$ with kernel size 2×2 and 2 output channels. We also learn the model parameter $\mu^{(k)}$ that is free to vary in different blocks while the penalty parameter ρ_t is shared across them. To ensure the condition in Proposition 1 is satisfied, we parametrize $a^{(k)}$ as $a^{(k)} = \frac{\rho_t}{\mu^{(k)}} \left(1 - \frac{1}{b^{(k)}}\right)$ where $b^{(k)}$ is a learnable parameter constrained to be larger than 1.

$\mathcal{S}_{\mathbf{k},\Theta_{\mathbf{k},\mathbf{s}}}$-Subblock, Estimation of LR Texture: As described in Eq. (12), updating s involves the computation of the SVD decomposition of the input since we need to apply a threshold on its singular values. In this work, we do not explore possible linear approximations of the SVD decomposition and we compute it exactly. Therefore, we preserve the same structure of (12) adding flexibility by learning a shared ρ_s. Finally, the patch operator requires the size of the patch p and the overlap o between adjacent patches and a good choice is to fix $p = 4$ and $o = 2$.

$\mathcal{X}_{\mathbf{k},\mathbf{\Theta}_{\mathbf{k},\mathbf{x}}}$-Subblock, Enforcing the Constraint: The algorithm described in (13) can be viewed as a composition of n layers, each one made of a linear layer followed by a non linear function, as illustrated in detail in Fig. 3. To preserve the specific form of the algorithm while gaining flexibility, we replace the discrete gradient operator D and its transpose D^{T} with two convolutional operators with kernel size 2×2 and 2 and 1 output channels respectively. We denote them with $\mathrm{D}_{\mathcal{X}}^{(k)}$ and $\tilde{\mathrm{D}}_{\mathcal{X}}^{(k)}$ allowing them to change in each block $\mathcal{X}_{k,\Theta_{k,x}}$ and relaxing the matching between $\tilde{\mathrm{D}}_{\mathcal{X}}^{(k)}$ and $(\mathrm{D}_{\mathcal{X}}^{(k)})^{\mathrm{T}}$. Finally, the parameter τ is learned and shared across different blocks.

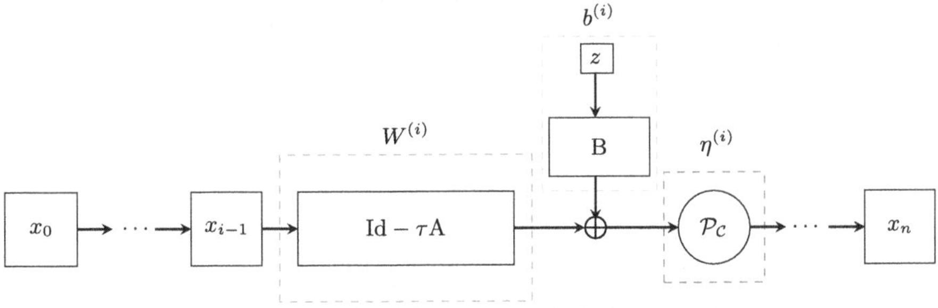

Fig. 3. Unrolling of the Projected Gradient Descent algorithm.

$\mathcal{Y}_{\mathbf{k},\mathbf{\Theta}_{\mathbf{k},\mathbf{y}}}$-Subblock: in the updating of the Lagrange multipliers, we only replace the discrete gradient operator D with a convolutional kernel of size 2×2 with two channels denoted as $\mathrm{D}_{\mathcal{Y}}^{(k)}$.

Specifically, we evaluate and compare four distinct configurations of parameters with the twofold aim of studying the improvement gained with:

- a non-convex penalty ϕ ($\mathrm{a} > 0$) with respect to the Total Variation ($\mathrm{a} = 0$);
- a data-dependent $\mu^{(k)}$, estimated on x^k after each block $\mathcal{L}_{\Theta_k}^k$ with respect to a fixed $\mu^{(k)}$. In the former, the estimation is performed by adding a small CNN composed of three convolutional layers with kernel size 3×3 and 4,8,16 channels respectively, each one followed by a ReLU, and in the end, we have a fully connected layer and a Softplus to enforce the positivity of the estimated parameter $\mu^{(k)}$.

Table 1. Four different parameter configurations with an estimate of the number of parameters.

	$\Theta_{k,t}$	$\Theta_{k,s}$	$\Theta_{k,x}$	$\Theta_{k,y}$	$\|\Theta\|$
Θ_1	$a^{(k)} = 0, \mu^{(k)}, \rho_t, \mathrm{D}_{\mathcal{T}}^{(k)}$	ρ_s	$\mathrm{D}_{\mathcal{X}}^{(k)}, \tilde{\mathrm{D}}_{\mathcal{X}}^{(k)}, \tau$	$\mathrm{D}_{\mathcal{Y}}^{(k)}$	$\approx 50\mathrm{K}$
Θ_2	$b^{(k)}, \mu^{(k)}, \rho_t, \mathrm{D}_{\mathcal{T}}^{(k)}$	ρ_s	$\mathrm{D}_{\mathcal{X}}^{(k)}, \tilde{\mathrm{D}}_{\mathcal{X}}^{(k)}, \tau$	$\mathrm{D}_{\mathcal{Y}}^{(k)}$	$\approx 50\mathrm{K}$
Θ_3	$a^{(k)} = 0, \mathrm{CNN}^{(k)}, \rho_t, \mathrm{D}_{\mathcal{T}}^{(k)}$	ρ_s	$\mathrm{D}_{\mathcal{X}}^{(k)}, \tilde{\mathrm{D}}_{\mathcal{X}}^{(k)}, \tau$	$\mathrm{D}_{\mathcal{Y}}^{(k)}$	$\approx 2 \cdot 10^3\mathrm{K}$
Θ_4	$b^{(k)}, \mathrm{CNN}^{(k)}, \rho_t, \mathrm{D}_{\mathcal{T}}^{(k)}$	ρ_s	$\mathrm{D}_{\mathcal{X}}^{(k)}, \tilde{\mathrm{D}}_{\mathcal{X}}^{(k)}, \tau$	$\mathrm{D}_{\mathcal{Y}}^{(k)}$	$\approx 2 \cdot 10^3\mathrm{K}$

By combining a convex ($a = 0$) or non-convex Total Variation ($a > 0$) and a data-dependent or fixed $\mu^{(k)}$, we obtain four configurations. For each configuration, the learnable parameters are summarized as follows in Table 1: Configurations Θ_1 and Θ_2 contain less parameters and a lower representation power than Θ_3 and Θ_4, whereas configurations Θ_2 and Θ_4 should take advantage of the non-convex setting to better recover piecewise constant images. A detailed analysis will be carried out in Sect. 4.

4 Numerical Experiments

In this section, the training procedure of our LPR-NET is explained in detail and some numerical results are presented. In the first example, we will investigate the role that K, n and the four different configurations play in the final output results while in the second example, we present decomposition of synthetic and natural images compared with state-of-the-art methods.

Training Strategy. We generate a synthetic image dataset composed of 400 images of dimension 64×64 that are the sum of a piecewise constant image u^{GT} and a texture image v^{GT} eventually degraded by a mask operator $\mathcal{M}$. The structure component is obtained by randomly positioning connected supports generated via the Lane-Riesenfeld algorithm while sparse Fourier texture generation has been used for the texture component. For further details, see [9]. Given our synthetic dataset and our architecture $\mathcal{F}_\Theta$, the estimate of the set of parameters Θ is carried out via the minimization of the Mean Square Error over the training dataset:

$$\widehat{\Theta} \in \arg\min_{\Theta} \frac{1}{N_{train}} \sum_{j=1}^{N_{train}} \left\|\mathcal{F}_\Theta(x_j^0, y_j^0) - x_j^{\mathrm{GT}}\right\|^2, \tag{17}$$

where $x_j^0 = (u_j^0, v_j^0)$ and y_j^0 are the initial guesses of the ADMM set as $u_j^0 = f_j$, $v_j^0 = 0$ and $y_j^0 = 0$. At the k-th block of the network, the initial guess for x_j in the Projected Gradient Descent is chosen as x_j^{k-1}. The network has been implemented on Pytorch and trained using Adam optimizer with learning rate $= 10^{-3}$ halved after forty epochs, training batch size $= 16$ and for a total of 250 epochs. We initialized the parameters as: $(\mu^{(k)})^0 = 0.05$, $\rho_t^0 = 1$, $\rho_s^0 = 1$, $\tau^0 = 0.1$,

$(b^{(k)})^0 = 1.1$, $\mathrm{D}_{\mathcal{T}}^{(k)}, \mathrm{D}_{\mathcal{X}}^{(k)}, \mathrm{D}_{\mathcal{Y}}^{(k)}$ initialized as the discrete gradient, $\tilde{\mathrm{D}}_{\mathcal{X}}^{(k)}$ initialized as the transpose of the discrete gradient and the parameters of the small CNN are initialized such that the output of the network is equal to $(\mu^{(k)})^0 = 0.05$ (all zeros except the bias of the last fully connected layer).

Example 1: Ablation Study of the LPR-NET. In this example, we trained our LPR-NET for different values of the number of unrolled iterations of the ADMM K and of the Projected Gradient Descent n over the training dataset of 400 images considering no degradation $\mathcal{M}$. In particular, we chose $\mathrm{K} = 5, 10, 15, 20$ and $n = 5, 10, 15, 20$ and quantitatively compared the different networks on a test dataset of 1000 images generated from the synthetic generator of structure and texture images available at [10]. In Table 2, the mean PSNR values of the dataset are reported for the four configurations of parameters.

Table 2. Mean PSNR over the test dataset for different values of K and n and for the four different configurations.

	Θ_1						Θ_2			
n	5	10	15	20		n	5	10	15	20
K 5	38.91	39.03	39.11	39.23		K 5	39.14	39.13	38.93	39.09
K 10	40.44	40.38	40.48	40.49		K 10	40.78	40.61	40.60	40.61
K 15	41.08	41.11	41.01	40.94		K 15	41.44	41.34	41.28	41.24
K 20	41.60	41.44	41.35	41.38		K 20	42.00	41.75	41.68	41.74
	Θ_3						Θ_4			
n	5	10	15	20		n	5	10	15	20
K 5	39.23	39.34	39.38	39.40		K 5	39.52	39.51	39.55	39.54
K 10	40.64	40.49	40.41	40.39		K 10	41.02	40.93	40.82	40.93
K 15	41.33	41.19	41.13	41.00		K 15	41.69	41.69	41.60	41.46
K 20	41.81	41.51	41.43	41.41		K 20	42.13	41.96	41.76	41.65

As expected, increasing the number of blocks of the network K leads to better quality decomposition results while higher values of n show a small deterioration. Among the four configurations, adding a non-convex penalty ϕ improved the decomposition with respect to $a = 0$ while adding a small CNN to make $\mu^{(k)}$ data-dependent brings a slight improvement when compared to a variable $\mu^{(k)}$ shared for all images. Note that we limit our ablation study to K and n but it could be extended to the size of filters, on matching and untied adjoints and that more flexibility could be achieved by allowing larger filters and more channels.

Example 2: Comparison with State-of-the-Art Methods. In this example, we compare the best configuration of our LPR-NET with three variational models - TV-L_2, TV-G, LPR - and with the joint structure-texture decomposition model $R_x(u, v)$ proposed in [9], reporting the mean PSNR over the same test dataset in Table 3. The best configuration of our LPR-NET outperforms classical

Table 3. Comparison of mean PSNR values with classical variational models and [9]. Notice that the regularization parameter μ was set by trial and error for the 3 variational models to achieve the best possible PSNR.

	Θ_4	TV-L_2 [4]	TV-G [2]	LPR [20]	$R_x(u,v)$ [13]
PSNR	42.13	38.84	38.86	41.61	42.69

variational models whose parameters are selected in order to get the best PSNR values for each distinct image. This means that, despite, on average, our results are worse than the $R_x(u, v)$ model, our network is able to automatically set the optimal parameters adapting to the image content avoiding having to set-up the parameters of the Plug-and-Play optimization algorithm. We also remark that, with respect to [9], our network has a lighter architecture (around 4×10^4 parameters) that requires a smaller training dataset and less training hours (in the worst case, around 4 h). In Fig. 4 and Fig. 5, we illustrate decomposition results of synthetic and natural images compared with $R_x(u, v)$.

While for the synthetic image the results are visually comparable, we note how on natural images the behaviour is slightly different: the LPR-NET removes most of the texture but also some structural edges (e.g. the face of the pirate) while the $R_x(u, v)$ model better preserves the geometric of the image but removes few tiles in the mosaic.

Extension to Inverse Problems. Finally, we also carry out some inpainting experiments. With this aim, we trained our LPR-NET ((K, n) = (20, 20)) fol-

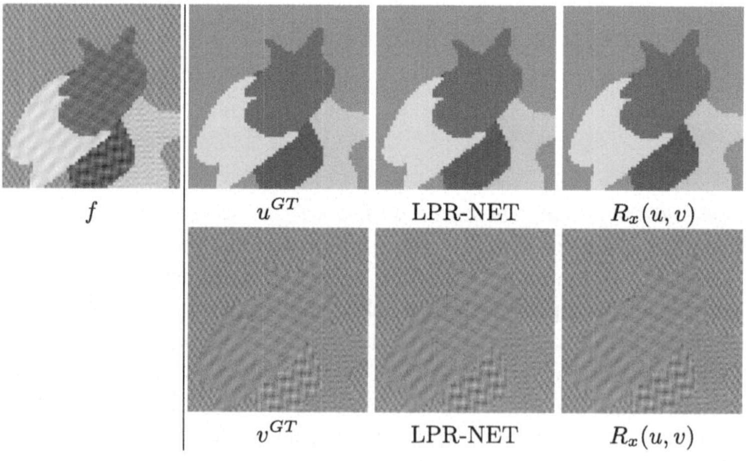

Fig. 4. Decomposition results of a synthetic image obtained with our u - LPR-NET with (K, n) = (20, 5) and with the joint structure-texture model proposed in [9] (PSNR($u_{\text{LPR-NET}}, u^{GT}$) = 41.93, PSNR($v_{R_x(u,v)}, u^{GT}$) = 40.53, PSNR($v_{\text{LPR-NET}}, v^{GT}$) = 41.80, PSNR($v_{R_x(u,v)}, v^{GT}$) = 40.25).

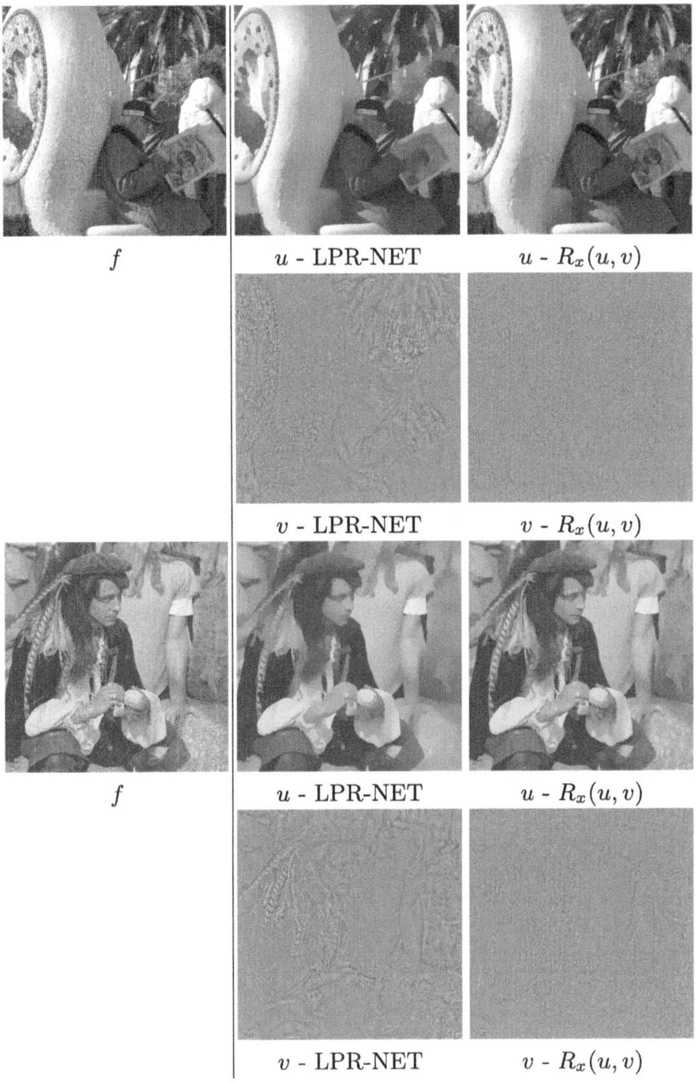

Fig. 5. Decomposition results on natural images obtained with our LPR-NET with $(\mathrm{K}, n) = (20, 5)$ (left) and with the model proposed in [9] (right).

lowing the training strategy described in Sect. 4 except for the training dataset where each image is masked with some random shapes. In Fig. 6, we compare our model with [9] on Barbara where we randomly remove 65% of pixels. The results are promising when the measurement operator is fixed and well described by $\mathcal{M}$ as our network achieves a more "natural" decomposition with respect to the Plug-and-Play method $R_x(u, v)$ where some artifacts are visible in Barbara's arm (recall that $R_x(u, v)$ is trained independently of the measurement method).

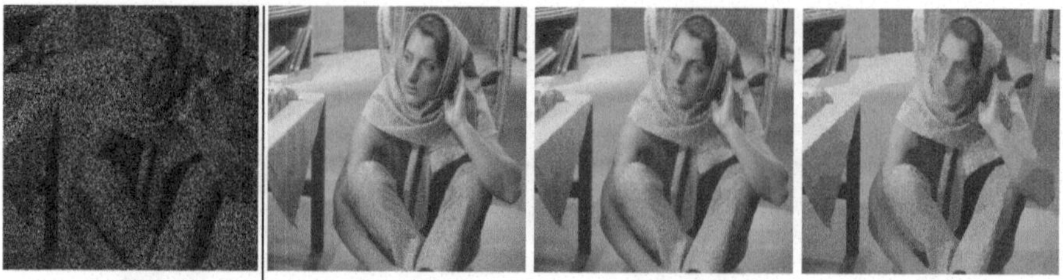

Fig. 6. Inpainting results on Barbara with 65% of pixels randomly removed, and comparison with the model proposed in [9].

5 Conclusion

In this paper, we presented a fully automatic way of separating images into structure and texture components based on the unrolling of a generalized low patch rank model. We proposed a novel unrolled network LPR-NET and investigated various architectural choices. Numerical experiments validate the potentiality of our network to achieve high quality decomposition results without tuning parameters.

References

1. Adler, J., Öktem, O.: Learned primal-dual reconstruction. IEEE Trans. Med. Imaging **37**(6), 1322–1332 (2018)
2. Aujol, J.-F., Aubert, G., Blanc-Feraud, L., Chambolle, A.: Image decomposition into a bounded variation component and an oscillating component. J. Math. Imaging Vis. **22**(571), 71–88 (2005)
3. Aujol, J.-F., Giboa, G., Chan, T., Osher, S.: Structure-texture image decomposition - modeling, algorithms, and parameter selection. Int. J. Comput. Vision **67**, 11–136 (2006)
4. Chambolle, A.: An algorithm for total variation minimization and applications. J. Math. 586 Imaging Vis. **20**, 89–97 (2004)
5. Ducotterd, S., Neumayer, S., Unser, M.: Learning of Patch-Based Smooth-Plus-Sparse Models for Image Reconstruction, arXiv preprint arXiv:2412.13070 (2024)
6. Girometti, L., Huska, M., Lanza, A., Morigi, S.: Quaternary image decomposition with crosscorrelation-based multi-parameter selection. In: Scale Space and Variational Methods in Computer Vision: 9th International Conference, SSVM 2023, pp. 120–133 (2023)
7. Gregor, K., LeCun, Y.: Learning fast approximations of sparse coding. In: Proceedings of International Conference on Machine Learning (2010)
8. Guennec, A., Aujol, J.-F., Traonmilin, Y.: Adaptive parameter selection for gradient-sparse plus low patch-rank recovery: application to image decomposition. In: 32nd European Signal Processing Conference (EUSIPCO), Lyon, France, pp. 2672–2676 (2024)

9. Guennec, A., Aujol, J.-F., Traonmilin, Y.: Joint structure-texture low dimensional modeling for image decomposition with a plug and play framework, ffhal-04648963 (2024)
10. Guennec, A., Aujol, J.-F., Traonmilin, Y. (2024). https://github.com/aguennecjacq/joint_decomposition
11. Habring, A., Holler, M.: A generative variational model for inverse problems in imaging. SIAM J. Math. Data Sci. **4**(1), 306–335 (2022)
12. Huska, M., Lanza, A., Morigi, S., Selesnick, I.: A convex-nonconvex variational method for the additive decomposition of functions on surfaces. Inverse Prob. **35**, 124008 (2019)
13. Kim, Y., Ham, B., Do, M.N., Sohn, K.: Structure-texture image decomposition using deep variational priors. IEEE Trans. Image Process. **28**(6), 2692–2704 (2019)
14. Le, H.T.V., Pustelnik, N., Foare, M.: The faster proximal algorithm, the better unfolded deep learning architecture? The study case of image denoising. In: 30th European Signal Processing Conference (EUSIPCO), Belgrade, Serbia, pp. 947–951 (2022)
15. Li, Y., Tofighi, M., Geng, J., Monga, V., Eldar, Y.C.: Efficient and interpretable deep blind image deblurring via algorithm unrolling. IEEE Trans. Comput. Imaging **6**, 666–681 (2020)
16. Meyer, Y., Lewis, D.: Oscillating Patterns in Image Processing and Nonlinear Evolution Equations: The Fifteenth Dean Jacqueline B. Lewis Memorial Lectures. Memoirs of the American Mathematical Society, American Mathematical Society (2001)
17. Monga, V., Li, Y., Eldar, Y.C.: Algorithm unrolling: interpretable, efficient deep learning for signal and image processing. IEEE Signal Process. Mag. **38**(2), 18–44 (2021)
18. Ono, S., Miyata, T., Yamada, I.: Cartoon-texture image decomposition using blockwise low-rank texture characterization. IEEE Trans. Image Process. **23**(3), 1128–1142 (2014)
19. Rudin, L., Osher, S., Fatemi, E.: Nonlinear total variation based noise removal algorithms. Physica D **60**, 259–268 (1992)
20. Schaeffer, H., Osher, S.: A low patch-rank interpretation of texture. SIAM J. Imag. Sci. **6**(1), 226–262 (2013)
21. Themelis, A., Patrinos, P.: Douglas-Rachford splitting and ADMM for nonconvex optimization: tight convergence results. SIAM J. Optim. **30**(1), 149–181 (2020)
22. Zhang, C.H.: Nearly unbiased variable selection under minimax concave penalty. Ann. Stat. **38**(2), 894–942 (2010)
23. Zhang, H., Patel, V.M.: Convolutional sparse and low-rank coding-based image decomposition. IEEE Trans. Image Process. **27**(5), 2121–2133 (2018)
24. Zhang, Y., Yang, M., Li, N., Yu, Z.: Analysis-synthesis dictionary pair learning and patch saliency measure for image fusion. Signal Process. **167**, 107327 (2020)
25. Zhou, F., Chen, Q., Liu, B., Qiu, G.: Structure and texture-aware image decomposition via training a neural network. IEEE Trans. Image Process. **29**, 3458–3473 (2019)

Joint Manifold Learning and Optimal Transport for Dynamic Imaging

Sven Dummer(✉), Puru Vaish, and Christoph Brune

University of Twente, 7522 NB Enschede, Overijssel, The Netherlands
{s.c.dummer,p.vaish,c.brune}@utwente.nl

Abstract. Dynamic imaging is critical for understanding and visualizing dynamic biological processes in medicine and cell biology. These applications often encounter the challenge of a limited amount of time series data and time points, which hinders learning meaningful patterns. Regularization methods provide valuable prior knowledge to address this challenge, enabling the extraction of relevant information despite the scarcity of time-series data and time points. In particular, low-dimensionality assumptions on the image manifold address sample scarcity, while time progression models, such as optimal transport (OT), provide priors on image development to mitigate the lack of time points. Existing approaches using low-dimensionality assumptions disregard a temporal prior but leverage information from multiple time series. OT-prior methods, however, incorporate the temporal prior but regularize only individual time series, ignoring information from other time series of the same image modality. In this work, we investigate the effect of integrating a low-dimensionality assumption of the underlying image manifold with an OT regularizer for time-evolving images. In particular, we propose a latent model representation of the underlying image manifold and promote consistency between this representation, the time series data, and the OT prior on the time-evolving images. We discuss the advantages of enriching OT interpolations with latent models and integrating OT priors into latent models.

Keywords: Dynamic imaging · Neural network · Optimal transport · Autoencoder

1 Introduction

Dynamic imaging is crucial in many fields, particularly medical imaging and cell biology. It enables visualization and spatiotemporal modeling of time-dependent biological processes, offering a deeper understanding of biological processes and dynamics. These insights help experts to better understand and potentially influence how cells and diseases develop over time.

Dynamic imaging models for visualization and understanding require data, which is often scarce. The number of image time series is generally small due to a limited number of patients or cells. Moreover, these time series typically contain

T. A. Bubba et al. (Eds.): SSVM 2025, LNCS 15667, pp. 400–411, 2025.
https://doi.org/10.1007/978-3-031-92366-1_31

relatively few data points over time. This scarce data makes it challenging to develop accurate dynamic imaging models that generalize well across cases and capture the underlying trends of disease progression or cell development.

The small data scenario requires prior knowledge about the underlying image manifold and the plausible development of the images. One well-known prior assumption on the image manifold is its relatively low dimensionality. State-of-the-art models in fields such as generative modeling [25] and inverse problems [28] utilize this prior and learn representations of plausible images via low-dimensional latent vectors.

Another important prior is Optimal Transport (OT). OT can be used as prior knowledge on the underlying image manifold via a patch prior [18] or via a template-based prior [13]. For the time-related data scarcity, OT provides priors for image development via the dynamical Benamou-Brenier formulation [4–6,8,20].

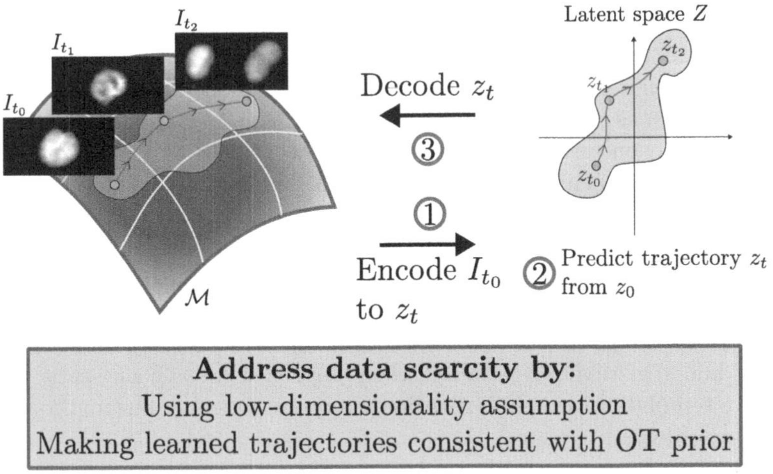

Fig. 1. Model overview. An initial image is encoded into latent space, where a trajectory is predicted and decoded back into image space. To address the scarcity of time series data and time points, existing methods often rely on priors such as a low-dimensional manifold assumption or a dynamic OT prior. Our model combines these approaches by constructing a low-dimensional latent space and regularizing the predicted image trajectories to align with the OT prior.

While existing approaches based on the low-dimensionality assumption analyze the data without accounting for such temporal priors, methods leveraging OT priors focus solely on regularizing individual time series, neglecting information from other time series lying on the same data manifold. As depicted in Fig. 1, this work explores combining the low-dimensionality assumption with a temporal OT prior. In particular, we introduce a latent model that uses the dynamic OT prior and is applicable to dynamic imaging. Using an autoencoder as a (learned) low-dimensional latent representation of plausible images, we map predicted latent-space trajectories to image-space trajectories, which are regularized via the dynamical OT prior. We use this model and cell biology-inspired experiments to examine how combining latent space models and dynamical optimal transport priors affects manifold learning of dynamic and static images and how it affects image interpolation and extrapolation.

2 Related Works

2.1 Dynamic Imaging via Optimal Transport, LDDMM, and Metamorphosis

Optimal transport has been used as template-based regularization [13] for static imaging and as a regularizer of temporal dynamics in dynamic imaging [3,5,6,8,20]. Similar models, such as Large Deformation Diffeomorphic Metric Mapping (LDDMM) and metamorphosis, have also been applied as template-based regularizer [10,14,19,22,23] and time dynamic regularizer [7,9,17].

Dynamic imaging methods that employ OT, LDDMM, or metamorphosis typically focus on a single dynamic image and use an additional predefined static image prior, such as total variation (TV) or a segmentation mask that constrains the location of non-zero pixel intensities [3]. These priors constrain the OT, LDDMM, or metamorphosis geodesics, which are used to regularize temporal dynamics.

In contrast, our approach introduces a static image manifold prior, where the manifold of plausible (static) images is not predefined but learned from multiple dynamic images. Our model shares similarities with a recent model [11] that combines latent modeling and LDDMM to regularize models for joint shape latent modeling and registration. However, their approach focuses on static objects, adopting a template-based regularization viewpoint, whereas our method applies to dynamic objects.

2.2 Dimensionality Reduction in Wasserstein Space

Combining OT and dimensionality reduction has recently gained considerable interest. For instance, one approach learns a dictionary within Wasserstein space [26], while another [15] extends Isomap to this setting. The differential geometric structure of (learned) manifolds in Wasserstein space has also been analyzed mathematically [16]. In the context of latent space models for images, OT-based methods have been used to enhance latent interpolation. For instance, a recent approach [27] applies an OT-based scheme after the model has been trained on the data, while another [12] adds a regularization term that promotes linear latent interpolations to follow OT geodesics. While these works explore combining autoencoders and OT, they do not address the integration of time series data and OT geodesics.

3 Dynamical Optimal Transport and Barycenters

Optimal transport is defined as optimally moving a probability distribution to another probability distribution.

Definition 1 (Wasserstein space $\mathcal{W}_2(\mathbb{R}^d)$ [30]). *Assume we have $\mu, \nu \in P(\mathbb{R}^d)$ with $P(\mathbb{R}^d)$ the space of Borel probability measures. For an arbitrary $x_0 \in \mathbb{R}^d$, $(\mathcal{W}_p(\mathbb{R}^d), W_p)$ is the p-Wasserstein space with:*

$$W_p(\mu,\nu) := \left(\inf_{\pi \in \Pi(\mu,\nu)} \int_{\mathbb{R}^d} \|x-y\|_p^p d\pi(x,y)\right)^{\frac{1}{p}}$$

$$\mathcal{W}_p(\mathbb{R}^d) := \{\mu \in P(\mathbb{R}^d); \quad \int_{\mathbb{R}^d} \|x_0 - x\|_p^p d\mu(x) < \infty\}$$

where $\Pi(\mu,\nu) := \{\pi \in P(\mathbb{R}^d \times \mathbb{R}^d); \int_{\mathbb{R}^d} \pi(dx, y) = \nu, \int_{\mathbb{R}^d} \pi(x, dy) = \mu\}$.

For $p = 2$, the Wasserstein distance has a time-dependent reformulation.

Theorem 1 ([1,2]). *The 2-Wasserstein distance can be rewritten as:*

$$W_2(\mu_0,\mu_1) = \inf_{\mu_t} \int_0^1 |\dot{\mu}_t| dt$$

where the infimum is over absolutely continuous curves with starting point μ_0, endpoint μ_1, and $|\dot{\mu}_t|$ the metric derivative of the curve μ_t:

$$|\dot{\mu}| := \lim_{h \to 0} \frac{W_2(\mu_{t+h}, \mu_t)}{|h|}.$$

The reformulation resembles a Riemannian geodesic and yields a weak Riemannian structure on $\mathcal{W}_2$ via definitions of tangent vectors and Riemannian metrics [1,2,24]. Instead of this time-dependent formulation of Riemannian OT geodesics, we can alternatively define OT geodesics via the concept of Wasserstein barycenters.

Definition 2 (Wasserstein barycenters and interpolations). *Let $\mu_i \in P(\mathbb{R}^d)$ for $i \in \{1, \ldots, n\}$. Then for some sequence α_i with $\sum_{i=1}^n \alpha_i = 1$, a Wasserstein barycenter is defined as:*

$$\mu := \underset{\mu \in P(\mathbb{R}^d)}{\arg\min} \left(\sum_{i=1}^n \alpha_i W_2(\mu_i, \mu)\right).$$

In particular, for $t \in [0,1]$, we define the barycentric interpolation as:

$$B(\mu_1, \mu_2, t) := \underset{\mu \in P(\mathbb{R}^d)}{\arg\min} \left((1-t) W_2(\mu_1,\mu) + t W_2(\mu_2,\mu)\right) = \mu_t$$

with μ_t the geodesic defined in Theorem 1.

4 Autoencoder Model for Dynamic Imaging

As shown in Fig. 1, our approach uses an autoencoder as a low-dimensional manifold prior and combines this prior with dynamical OT regularization. Given N time series $\{I_j^i\}_{j=1}^{N_t}$ of length N_t, indexed by $i \in \{1, \ldots, N\}$, our model first encodes each time point j in time series i into a latent code $z_{ij} = E(I_j^i)$, where the encoder $E\colon \mathbb{R}^m \to Z$ maps images to a latent space $Z = \mathbb{R}^d$.

To model temporal evolution, we then evolve the initial latent codes z_{i1} via a neural ordinary differential equation in latent space. In particular, given a neural network $v_\varphi\colon Z \to Z$ parameterizing the right-hand side of the neural ODE, we evolve the initial latent codes z_{i1} via:

$$\frac{\mathrm{d}z^i(t)}{\mathrm{d}t} = v_\varphi(z^i(t)), \quad z^i(0) = z_1^i.$$

Since v_φ depends only on the current state, supplying an initial condition determines the future trajectory, enabling image prediction.

Subsequently, a decoder $D\colon Z \to \mathbb{R}^m$ generates images $D(z_{ij})$ and $D(z^i(t_j))$ using the static and dynamic encodings, respectively. Both reconstructions should match the ground truth data I_j^i. At the same time, the latent codes z_{ij} and $z^i(t_j)$ should be consistent since they represent the same image I_j^i. Our model ensures accurate reconstructions and latent consistency via the loss function:

$$\|D(z^i(t_j)) - I_j^i\|_2^2 + \gamma_1 \|D(z_j^i) - I_j^i\|_2^2 + \gamma_2 \|z^i(t_j) - z_j^i\|_2^2,$$

where $\gamma_1, \gamma_2 \in \mathbb{R}$.

Finally, we introduce a Riemannian dynamical OT regularizer, which imposes an OT Riemannian structure on the learned manifold. This regularizer encourages the reconstructed time series, once normalized, to approximate OT geodesics. In particular, using $n(I) := I / \int_\Omega I(x)\mathrm{d}x$ as normalization function, the normalized reconstruction $\mu_t^i := n(D(z^i(t)))$ for $t_j < t < t_{j+1}$ should be close to the Wasserstein barycentric interpolation $B\left(\mu_{t_j}^i, \mu_{t_{j+1}}^i, \frac{t-t_j}{t_{j+1}-t_j}\right)$, as defined in Definition 2. The normalization is required since OT assumes unit mass distributions. However, as the decoder does not enforce the unit mass constraint, this leaves a degree of freedom in the mass of the image. To address this, we encourage a linear interpolation between the masses of the boundary images using:

$$s_j^i(t) = \left(1 - \frac{t - t_j}{t_{j+1} - t_j}\right) \|D(z^i(t_j))\|_1 + \left(\frac{t - t_j}{t_{j+1} - t_j}\right) \|D(z^i(t_{j+1}))\|_1,$$

In particular, the Riemannian dynamical OT regularizer enforces proximity to the scaled barycentric interpolation:

$$\left\| D(z^i(t)) - s_j^i(t) B\left(\mu_{t_j}^i, \mu_{t_{j+1}}^i, \frac{t - t_j}{t_{j+1} - t_j}\right) \right\|_2^2$$

Bringing everything together, our model corresponding to Fig. 1 solves the following optimization problem:

$$\begin{aligned}\min_{E,D,v_\varphi} \quad & \frac{1}{N \cdot N_t} \sum_{i=1}^{N} \sum_{j=1}^{N_t} \left(\|D(z^i(t_j)) - I_j^i\|_2^2 + \gamma_1 \|D(z_j^i) - I_j^i\|_2^2 + \gamma_2 \|z^i(t_j) - z_j^i\|_2^2 \right) \\ & + \lambda \sum_{i=1}^{N} \sum_{j=1}^{N_t - 1} \int_{t_j}^{t_{j+1}} \left\| D(z^i(t)) - s_j^i(t) B\left(\mu_{t_j}^i, \mu_{t_{j+1}}^i, \frac{t - t_j}{t_{j+1} - t_j} \right) \right\|_2^2 \mathrm{d}t \\ \text{s. t.} \quad & z_j^i = E(I_j^i), \\ & \frac{\mathrm{d}z^i(t)}{\mathrm{d}t} = v_\varphi(z^i(t)), \quad z^i(0) = z_1^i, \\ & \mu_t^i = n(D(z^i(t))), \\ & s_j^i(t) = \left(1 - \frac{t - t_j}{t_{j+1} - t_j}\right) \|D(z^i(t_j))\|_1 + \left(\frac{t - t_j}{t_{j+1} - t_j}\right) \|D(z^i(t_{j+1}))\|_1,\end{aligned}$$

5 Numerical Experiments

This section presents numerical examples[1] to illustrate the impact of joining manifold learning and dynamic OT. In particular, we test the OT regularizer against an l_2 regularizer $\|\frac{\mathrm{d}}{\mathrm{d}t} D(z(t))\|_2$ on the change in the image and an l_2 regularizer$\|\frac{\mathrm{d}}{\mathrm{d}t} z(t)\|_2$ on the change in the latent code. We focus on these regularizers because more complex ones, such as brightness consistency in optical flow, require solving an additional variational problem, making efficient backpropagation challenging. While Wasserstein regularization also imposes a constraint in the form of mass conservation, it benefits from efficient entropy-regularized Sinkhorn solvers and their differentiation properties, enabling efficient integration into our model. Besides investigating the benefit of OT regularization, we also investigate the benefit of the learned manifold by comparing our neural ODE prediction model to simple l_2 and Wasserstein interpolations of the images.

We use a synthetic Gaussian dataset and the HeLa cells dataset from the Cell Tracking Challenge [21,29] for the experiments. The synthetic Gaussian dataset consists of image time series depicting Gaussians that move from the left to the right while growing in size. The movement and growth accelerate in the middle, simulating a discontinuous transition from one state to another when the time series has insufficient temporal resolution. The HeLa dataset contains time series that transition from a single cell to two cells, presenting a challenge similar to, but slightly more complex than, the synthetic Gaussians dataset.

5.1 Effect of OT Regularization on Manifold-Based Interpolation

To investigate the impact of the OT regularizer, we first explore its effect on image interpolations on the learned manifold. We train the models on time series

[1] Our code is available at: https://github.com/SCdummer/joint-manifold-learning-and-ot.

whose data is subsampled every 5 time steps. This subsampling ensures missing images and can be used to assess the quality of the interpolations.

Figures 2 and 3 show images at locations where a Gaussian performs a quick jump and a cell divides, respectively. The interpolations with the latent space model result from the neural ODE, which uses the latent code of the initial image in the time series as the initial condition.

All methods reconstruct the presented initial and final frames reasonably well, but the intermediate frames differ. In contrast to the other regularizations, OT regularization preserves consistent Gaussian shapes. For the HeLa cells in Fig. 3, OT regularization smoothly separates the single cell into two. The other regularizers create a smeared transition, reducing intensity in the middle rather than distinctly splitting the cell. These results highlight that the OT prior can help manifold learning preserve the data's structure.

5.2 Static and Dynamic Image Reconstruction

To quantitatively assess the OT regularizer's impact, we explore its effect on static and dynamic reconstructions. Static reconstruction involves direct encoding and decoding, while dynamic reconstruction uses latent codes from the neural ODE model.

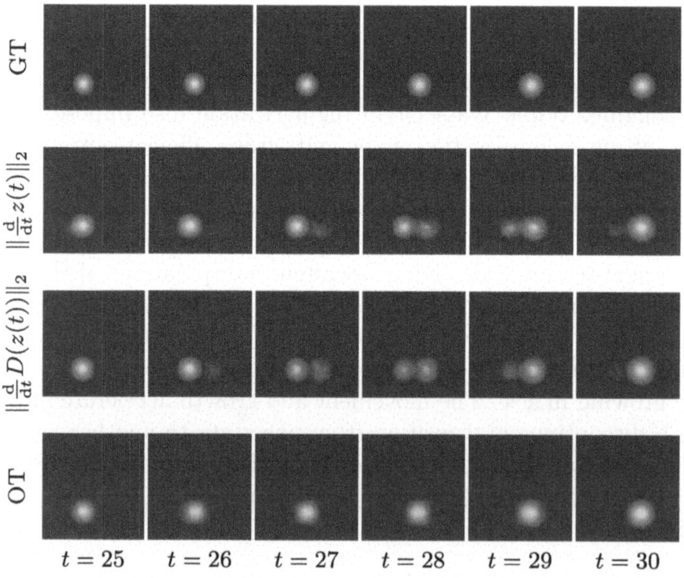

Fig. 2. Comparison of manifold-based interpolations of Gaussians. The ground truth images and the estimated images by the neural ODE model when trained with different regularizers: a penalty on the time derivative of latent vectors, a penalty on the time derivative of decoded images, or our OT regularization. During training, the models only see images subsampled at an interval of 5 time points. OT regularization is the only regularizer maintaining consistent Gaussian shapes across time.

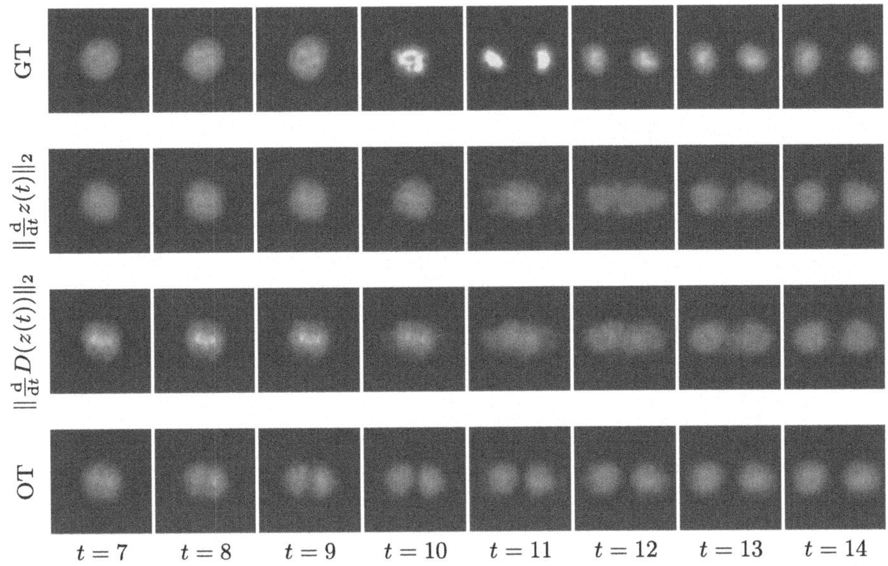

Fig. 3. Comparison of manifold-based interpolations of HeLa cells. The ground truth images and the estimated images by the neural ODE model when trained with different regularizers: a penalty on the time derivative of latent vectors, a penalty on the time derivative of decoded images, or our OT regularization. During training, the models only see images subsampled at an interval of 5 time points. Only OT regularization obtains a clear and smooth cell division.

Tables 1 and 2 summarize the performance of static and dynamic reconstruction, respectively. For the static reconstruction of the Gaussian dataset, the OT prior preserves intensity and enforces shape awareness, leading to slightly improved reconstruction metrics compared to the other regularizers. In the more challenging HeLa datasets, all regularizers show similar performance in the static reconstruction.

Table 1. Static reconstruction evaluation. We evaluate the models trained with various regularization types by encoding-decoding images and calculating the average value of the mean squared error (MSE) and average and standard deviation (right column) of the structural similarity index (SSIM). The best values are in bold.

Regularizer	Dataset (metric)			
	Gaussian (MSE)	Gaussian (SSIM)	HeLa (MSE)	HeLa (SSIM)
$\|\frac{\mathrm{d}}{\mathrm{d}t}z(t))\|_2$	2.24×10^{-3}	0.956 (0.040)	1.41×10^{-3}	**0.907** (0.065)
$\|\frac{\mathrm{d}}{\mathrm{d}t}D(z(t))\|_2$	1.96×10^{-3}	0.957 (0.038)	1.38×10^{-3}	**0.905** (0.067)
OT	$\mathbf{1.66} \times 10^{-3}$	**0.960** (0.029)	$\mathbf{1.26} \times 10^{-3}$	**0.905** (0.061)

Table 2. Dynamic reconstruction evaluation. We evaluate the models trained with various regularization types by using our neural ODE model for obtaining reconstructions and calculating the average value of the mean squared error (MSE) and average and standard deviation (right column) of the structural similarity index (SSIM). The best values are in bold.

Regularizer	Dataset (metric)			
	Gaussian (MSE)	Gaussian (SSIM)	HeLa (MSE)	HeLa (SSIM)
$\Vert \frac{\mathrm{d}}{\mathrm{d}t} z(t)) \Vert_2$	1.67×10^{-3}	0.961 (0.030)	9.14×10^{-3}	**0.757** (0.191)
$\Vert \frac{\mathrm{d}}{\mathrm{d}t} D(z(t)) \Vert_2$	1.57×10^{-3}	0.959 (0.033)	8.74×10^{-3}	**0.762** (0.176)
OT	$\mathbf{0.98} \times 10^{-3}$	**0.966** (0.019)	$\mathbf{8.31} \times 10^{-3}$	**0.773** (0.189)

Comparing Tables 1 and 2 shows that for each regularizer, dynamic reconstruction outperforms static reconstruction on the Gaussian dataset but underperforms on the HeLa dataset. This difference mainly stems from errors in the latent prediction process rather than the decoding step. In particular, dynamic reconstruction performs well on the Gaussian dataset due to the dataset's relatively simple temporal behavior. Combined with the fact that we emphasize dynamic reconstruction more than static reconstruction during training, this results in improved reconstruction metrics compared to the static case. In contrast, predicting future states from an initial image of the HeLa dataset is significantly more challenging, especially due to its heterogeneous intracellular motion. As a result, inaccuracies in latent prediction primarily drive reconstruction errors, resulting in worse reconstruction metric values.

5.3 Interpolations Consistent with Manifold Structure

This section investigates the benefits of manifold learning compared to optimal transport interpolations. Specifically, we perform a Euclidean and Wasserstein interpolation between two time points and compare it with the trajectory predicted by our neural ODE model. This comparison sheds light on the learned structure of the data manifold. The idea is that an Euclidean or Wasserstein interpolation does not consider this structure and, hence, does not follow the structure of the image manifold. In contrast, the neural ODE model offers an alternative path that more faithfully aligns with the dynamic data and its underlying manifold structure. In other words, while Euclidean and Wasserstein interpolations are the shortest paths not consistent with the image manifold and the temporal dynamics of the data, our model learns the shortest paths consistent with the geometry of the data manifold and the temporal behavior of the images.

The results on the HeLa dataset, shown in Fig. 4, demonstrate that the neural ODE model effectively captures the dynamics of the data while respecting the image manifold. It accurately follows the structure of cell division, and the generated images resemble the ground truth divided cells. On the other hand, the Euclidean interpolation fails to account for the manifold structure and the temporal evolution of the data, resulting in images that neither properly resemble

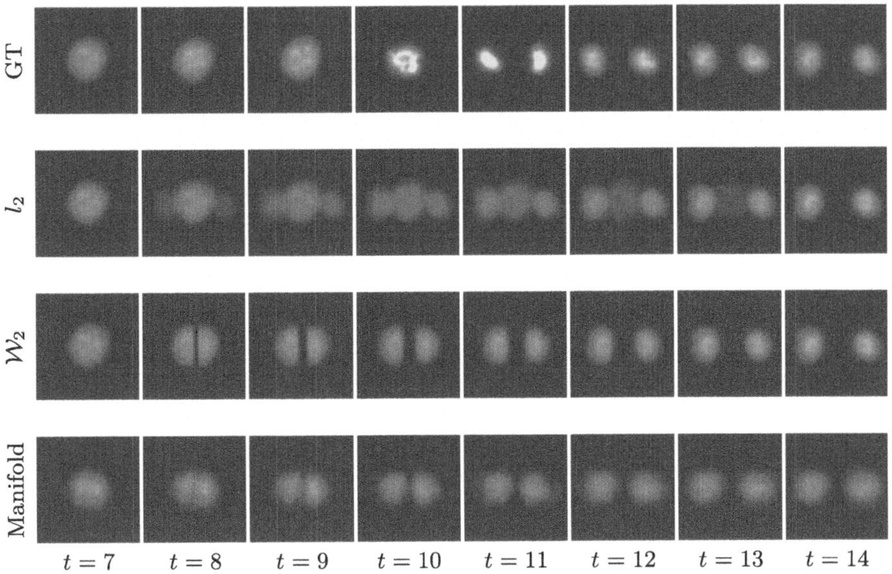

Fig. 4. Barycentric vs. Manifold interpolation. The ground truth images, the estimated images by l_2 or Wasserstein $\mathcal{W}_2$ interpolation, and the estimated images by the neural ODE manifold model. l_2 interpolation does not resemble a (divided) cell, $\mathcal{W}_2$ splits the cells prematurely, and manifold interpolation avoids both issues.

a divided cell nor represent the underlying dynamics. The Wasserstein interpolations better align with the image manifold. However, at the early time points, the cells appear as if cut in half, which deviates from the smoother shapes seen in realistic cells. Furthermore, the cell division occurs too early compared to the ground truth.

These observations suggest that manifold learning can help align the non-learned methods discussed in Sect. 2.1 with the geometry of the data manifold and the temporal behavior of the data.

6 Conclusion and Future Work

This work explores the effects of combining a low-dimensional manifold assumption with a temporal OT prior, specifically in the context of dynamic imaging. We designed an autoencoder model where movement in latent space amounts to movement in image space. We regularize the resulting manifold of static images by combining a neural ordinary differential equation, dynamic image data, and the dynamic interpretation of OT. Our results demonstrate that the OT prior enhances manifold learning approaches, while the manifold improves OT by incorporating data structure.

While this work focuses on balanced OT, the images we deal with have varying masses. The varying mass highlights the relevance of unbalanced OT, which deals with objects with different masses, making it an interesting direction for future research. Additionally, our experiments do not consider scenarios involving a forward operator. Future work could explore these problems and extend the investigation of the interplay between OT and manifold learning.

Disclosure of Interests. The authors have no competing interests to declare that are relevant to the content of this article.

References

1. Ambrosio, L., Gigli, N.: A user's guide to optimal transport, pp. 1–155. Springer, Heidelberg (2013). https://doi.org/10.1007/978-3-642-32160-3_1
2. Ambrosio, L., Gigli, N., Savaré, G.: Gradient Flows: In Metric Spaces and in the Space of Probability Measures. Springer (2005)
3. Boink, Y.E.: Combined modelling of optimal transport and segmentation revealing vascular properties. Master's thesis, University of Twente (2016)
4. Bon, D., Pai, G., Bellaard, G., Mula, O., Duits, R.: Optimal transport on the lie group of roto-translations. arXiv preprint arXiv:2402.15322 (2024)
5. Bredies, K., Carioni, M., Fanzon, S., Romero, F.: A generalized conditional gradient method for dynamic inverse problems with optimal transport regularization. Found. Comput. Math. **23**(3), 833–898 (2023)
6. Brune, C.: 4D imaging in tomography and optical nanoscopy. Ph.D. thesis, Münster (Westfalen), Univ., Diss., 2010 (2010)
7. Burger, M., Dirks, H., Schonlieb, C.B.: A variational model for joint motion estimation and image reconstruction. SIAM J. Imag. Sci. **11**(1), 94–128 (2018)
8. Chen, C.: Spatiotemporal imaging with diffeomorphic optimal transportation. Inverse Prob. **37**(11), 115004 (2021)
9. Chen, C., Gris, B., Oktem, O.: A new variational model for joint image reconstruction and motion estimation in spatiotemporal imaging. SIAM J. Imag. Sci. **12**(4), 1686–1719 (2019)
10. Chen, C., Oktem, O.: Indirect image registration with large diffeomorphic deformations. SIAM J. Imag. Sci. **11**(1), 575–617 (2018)
11. Dummer, S., Strisciuglio, N., Brune, C.: RDA-INR: Riemannian diffeomorphic autoencoding via implicit neural representations. SIAM J. Imag. Sci. **17**(4), 2302–2330 (2024). https://doi.org/10.1137/24M1644730
12. Feng, X., Strohmer, T.: Improving autoencoder image interpolation via dynamic optimal transport. arXiv preprint arXiv:2404.08900 (2024)
13. Gao, Y., Jin, Z., Li, X.: Template-based CT reconstruction with optimal transport and total generalized variation. Inverse Prob. **39**(9), 095007 (2023)
14. Gris, B., Chen, C., Öktem, O.: Image reconstruction through metamorphosis. Inverse Prob. **36**(2), 025001 (2020)
15. Hamm, K., Henscheid, N., Kang, S.: Wassmap: Wasserstein isometric mapping for image manifold learning. SIAM J. Math. Data Sci. **5**(2), 475–501 (2023)
16. Hamm, K., Moosmüller, C., Schmitzer, B., Thorpe, M.: Manifold learning in wasserstein space. arXiv preprint arXiv:2311.08549 (2023)
17. Hauptmann, A., Öktem, O., Schönlieb, C.: Image reconstruction in dynamic inverse problems with temporal models. In: Handbook of Mathematical Models and Algorithms in Computer Vision and Imaging: Mathematical Imaging and Vision, pp. 1–31 (2021)

18. Hertrich, J., Houdard, A., Redenbach, C.: Wasserstein patch prior for image super-resolution. IEEE Trans. Comput. Imaging **8**, 693–704 (2022). https://doi.org/10.1109/TCI.2022.3199600
19. Lang, L.F., Neumayer, S., Öktem, O., Schönlieb, C.B.: Template-based image reconstruction from sparse tomographic data. Appl. Math. Optim. **82**, 1081–1109 (2020)
20. Maas, J., Rumpf, M., Schönlieb, C., Simon, S.: A generalized model for optimal transport of images including dissipation and density modulation. ESAIM Math. Model. Numer. Anal. **49**(6), 1745–1769 (2015)
21. Maška, M., et al.: The cell tracking challenge: 10 years of objective benchmarking. Nat. Methods **20**(7), 1010–1020 (2023)
22. Neumayer, S., Persch, J., Steidl, G.: Regularization of inverse problems via time discrete geodesics in image spaces. Inverse Prob. **35**(5), 055005 (2019)
23. Neumayer, S., Topalovic, A.: Template-based image reconstruction facing different topologies. J. Sci. Comput. **98**(3), 57 (2024)
24. Otto, F.: The geometry of dissipative evolution equations: the porous medium equation (2001)
25. Rombach, R., Blattmann, A., Lorenz, D., Esser, P., Ommer, B.: High-resolution image synthesis with latent diffusion models. In: Proceedings of the IEEE/CVF Conference on Computer Vision and Pattern Recognition, pp. 10684–10695 (2022)
26. Schmitz, M.A., et al.: Wasserstein dictionary learning: optimal transport-based unsupervised nonlinear dictionary learning. SIAM J. Imag. Sci. **11**(1), 643–678 (2018)
27. Simon, D., Aberdam, A.: Barycenters of natural images constrained wasserstein barycenters for image morphing. In: Proceedings of the IEEE/CVF Conference on Computer Vision and Pattern Recognition, pp. 7910–7919 (2020)
28. Song, B., Kwon, S.M., Zhang, Z., Hu, X., Qu, Q., Shen, L.: Solving inverse problems with latent diffusion models via hard data consistency. In: The Twelfth International Conference on Learning Representations (2024). https://openreview.net/forum?id=j8hdRqOUhN
29. Ulman, V., et al.: An objective comparison of cell-tracking algorithms. Nat. Methods **14**(12), 1141–1152 (2017)
30. Villani, C., et al.: Optimal Transport: Old and New, vol. 338. Springer (2009)

Author Index

T. A. Bubba et al. (Eds.): SSVM 2025, LNCS 15667, pp. 413–415, 2025.
https://doi.org/10.1007/978-3-031-92366-1

The manufacturer's authorised representative in the EU is Springer Nature Customer Service Centre GmbH, Europaplatz 3, 69115 Heidelberg, Germany. If you have any concerns regarding our products, please contact ProductSafety@springernature.com

Printed and bound by CPI Group (UK) Ltd, Croydon, CR0 4YY
30/09/2026
02225597-0018